Molecular Biology of Hematopoiesis 6

Molecular Biology of Hematopoiesis 6

Edited by

Nader G. Abraham

New York Medical College
Valhalla, New York

Antonio Tabilio
Massimo Martelli

University of Perugia
Perugia, Italy

Shigetaka Asano

University of Tokyo
Tokyo, Japan

and

Alberto Donfrancesco

Bambino Gesu Hospital
Rome, Italy

Kluwer Academic / Plenum Publishers
New York, Boston, Dordrecht, London, Moscow

Library of Congress Cataloging-in-Publication Data

Molecular biology of hematopoiesis 6 / edited by Nader G. Abraham ...
[et al.].
p. cm.
"Proceedings of the Eleventh Symposium on the Molecular Biology of Hematopoiesis, held June 25-29, 1998, in Bormio, Italy"--T.p. verso.
Includes bibliographical references and index.
ISBN 978-0-306-46136-1
1. Myeloproliferative disorders Congresses. 2. None marrow--Transplantation Congresses. 3. Hematopoiesis--Regulation Congresses. 4. Hematopoietic stem cells Congresses. I. Abraham, Nader G. II. Symposium on the Molecular Biology of Hematopoiesis (11th 1998 : Bormio, Italy) III. Title: Molecular biology of hematopoiesis VI. IV. Title: Molucular biology of hematopoiesis six.
[DNLM: 1. Hematopoiesis Congresses. 2. MolecularBiology Congresses. WH 140 M7177 1999]
RC645.75.M652 1999
616.4'1--dc21
DNLM/DLC
for Library of Congress 99-31967
CIP

Proceedings of the Eleventh Symposium on the Molecular Biology of Hematopoiesis,
held June 25–29, 1998, in Bormio, Italy

ISBN 978-0-306-46136-1

233 Spring Street, New York, N.Y. 10013

10 9 8 7 6 5 4 3 2 1

A C.I.P. record for this book is available from the Library of Congress

SPECIAL GUESTS

Prof. Dr. Fumimaro Takaku
President, Jichi Medical School
Tochigi-ken, Japan

Prof. Dr. E. Donnall Thomas
Nobel Laureate in Medicine
Fred Hutchinson Cancer Research Center
Seattle, WA, USA

PREFACE

This volume of *Molecular Biology of Hematopoiesis* is dedicated to many international scientists and clinicians for their contribution to the field of Hematology/Oncology presented at the 11th International Symposium on Molecular Biology of Hematopoiesis, which was held in Bormio, Italy, June 25–29, 1998. The continuous support of the Presidents of the meeting, Professor F. Takaku, President of Jichi University, and E. D. Thomas, Nobel Laureate, was greatly acknowledged, especially Professor Takaku, for his vision and support for development of gene therapy in Japan.

New information on BMT for autoimmune disease and organ transplantation was presented at the symposium and is published in this volume. Several new findings on gene therapy/transfer into HSC were presented by E. F. Vanin and A. Nienhuis, K. Humphries, J. A. Nolta, H. E. Heslop, and M. K. Brenner. Professors S. Asano and K. Tani presented new studies on gene transfer into primates. Among the highlights were the new papers on gene transfer presented by G. Wagemaker, N. G. Abraham, and M. Onoderea from R. M. Blaese's group. The use of BMT for organ transplant and autoimmune disease was updated and a representative paper is presented in this volume. Exciting new scientific and clinical results on iron overload and hematopoiesis were presented, and selected papers are included in this volume. Among the highlights of presented papers were presentations by N. C. Andrews and C. Camaschella on iron and erythropoiesis, and on the search for candidate genes. Updated information on molecular diagnosis and chemotherapy is included in Part IV of this book. The most recent information on the use of cord blood stem cells for transplantation was presented in a session chaired by Professor J. Adamson. New information on various transcriptional factors regulating hematopoiesis were presented and selected manuscripts are included in this book. Papers submitted for publication after the deadline of the publication of the proceedings have not been included. Our sincerest apologies for not being able to hold up the printing of this volume until a later date.

Many thanks to the Scientific Committee and the abstract reviewers for their support and organization for what promises to be another outstanding year for the understanding of novel cancer therapy and immunotherapy for molecular biology of hematopoiesis.

We thank the Holy Father John Paul II for his message and welcome address, the Mayor of the city of Bormio for his support, and Mr. M. Gandolfi for making Bormio the ideal environment for holding the MBH '98 meeting.

N. G. Abraham
A. Tabilio
M. Martelli
S. Asano
A. Donfrancesco

ACKNOWLEDGMENTS

The Organizing Committee wishes to acknowledge the following companies whose support has made this meeting possible:

MAJOR SPONSORS
Chugai Rhone-Poulenc

EXHIBITORS AND CONTRIBUTORS
Associazione Ciavattini
Baxter
Beckman Analytical
Bristol-Myers Squibb
Cariplo
Chugai-Rhone Poulenc
Chugai Pharmaceutical Co., Ltd.
CellPro Deutschland GmbH
Dompe'Biotec
Eli Lilly Italia
Fujisawa Pharmaceutical Co., Ltd.
Genetics Institute
Glaxo Wellcome
Hoechst Marion Roussel
Italfarmaco
S. Karger Publishers
Kirin Co.
Miltenyi Biotec
Monsanto Searle
NeXstar Pharmaceuticals Italia
Novartis
Pharmacia & Upjohn
Rhone Poulenc Italia
Roche
Sankyo Co., Ltd.
Schering
Schering-Plough
Stemcell Technologies, Inc.
Tsumura & Co.

The Organizing Committee gratefully acknowledges the patience and support, including editorial assistance, of Joyce Eshet, Symposium Secretary, and the patience of Dr. J-L. da Silva, Symposium Coordinator.

For further information, please write to:

N. G. Abraham
Director of Gene Therapy
Professor of Pharmacology
New York Medical College
Valhalla, NY 10595 USA
Telephone: 914-594-4132/Fax: 914-594-4119

Editors'/Publisher's Disclaimer

CONTENTS

Update on Solid Tumor and Bone Marrow Transplantation

Regulation of Hematopoiesis

Novel Approaches to Gene Transfer and Gene Therapy

Iron Metabolism and Hematopoiesis

1

THE DEVELOPMENT OF HEMATOPOIETIC CELL TRANSPLANTATION AND ITS APPLICATION TO ORGAN TRANSPLANTATION AND AUTOIMMUNE DISEASES

E. Donnall Thomas

Jacobson,[1] in the course of studies of radiation biology, found that shielding the spleen of a mouse during otherwise lethal irradiation permitted survival. Subsequently, Lorenz et al.[2] found that irradiated mice could be protected by an infusion of spleen or marrow cells. Confirmation that the "irradiation protection" phenomenon was due to reconstitution by donor cells came from the observation of Main and Prehn[3] that irradiated mice protected by an infusion of allogeneic marrow subsequently displayed tolerance of a donor skin graft. Their observation was followed shortly by the report of Ford et al.[4] that the marrow of such animals had the cytogenetic characteristics of the marrow donor.

Canine Hematopoietic Cell Transplantation (HCT). In the 1950s and 1960s, many of the problems of transplantation biology were explored in the canine model. Using lethally irradiated recipients, the problems of marrow engraftment, graft rejection, and graft-versus-host disease (GVHD) were defined.[5] Despite the problems, some canine recipients of allogeneic marrow became long-term survivors. Mannick et al. showed that dogs given successful marrow grafts would then accept a kidney graft from the marrow donor without further immunosuppressive treatment.[6] Following the development of typing methods for identifying the dog leukocyte antigen (DLA) system, it was shown that grafts between histocompatible littermate pairs were often successful, especially if followed by a few months of methotrexate (MTX) to suppress GVHD.[7,8]

HCT in Human Patients. At the end of the 1960s developments in supportive care and, particularly, in the knowledge of human histocompatibility typing led to renewed attempts at allogeneic marrow grafting in human patients. The demonstration that some patients with immunodeficiency diseases, aplastic anemia or advanced leukemia could be cured by marrow transplantation from a human leukocyte antigen (HLA) matched sibling donor[9–11] rapidly led to the application of marrow grafting to patients with a variety of malignant and non-malignant diseases having in common a high probability of failure of other forms of therapy.

Molecular Biology of Hematopoiesis 6, edited by Abraham *et al.*
Kluwer Academic/Plenum Publishers, New York, 1999.

HCT for Malignant Disease. HCT after intensive chemo-irradiation therapy is now the treatment of choice for many types of leukemia, lymphoma, and other malignant diseases of the marrow and is being applied to the treatment of solid tumors. If transplantation is carried out early in the disease, for example, chronic myeloid leukemia (CML) within the first year of diagnosis, the long-term survival is 70–80% with HLA matched family member donors.[12]

HCT for Nonmalignant Diseases. HCT is being applied to aplastic anemia, Thalassemia, sickle cell disease, and a variety of genetically determined diseases, of the marrow. Patients treated before the development of complications show approximately a 90% long-term disease-free survival.[13,14] Even with advanced disease, many patients can be cured.

Source of Cells of HCT. In the past decade two types of donor cells have been used increasingly. The first are HLA matched unrelated marrow donors.[15] The results of grafts from unrelated donors have improved so that, for example, patients with CML who are transplanted within a year of diagnosis and who have had less than 6 months of treatment with interferon-alpha show a 5-year disease-free survival of 74%.[16]

The second is the use of the patient's own marrow, an autologous graft. The success of autologous grafts had long been demonstrated in animal models.[17] Autologous marrow grafts are now being used following high-dose chemo-irradiation therapy of both hematological malignancies and solid tumors, especially breast cancer.[18,19]

The Hematopoietic Cell Donor. In the recent past two major developments in regard to the cell donor have occasioned the shift in terminology from bone marrow transplantation (BMT) to hematopoietic cell transplantation (HCT). The first is the use of peripheral blood stem cells rather than marrow.[20] The second is the use of cord blood stem cells.[21]

Graft-versus-host Disease. Billingham and Brent defined the immunological basis of the graft-versus-host reaction in murine studies.[22] The magnitude of the problem in human patients was not appreciated until the early 1970s when consistent engraftment of donor marrow was finally achieved. Even with an HLA matched donor, GVHD occurred in one-half of the patients. Prevention or treatment with MTX and/or glucocorticoids was only partially effective.[23] The addition of cyclosporine to a short course of MTX resulted in improved prevention of GVHD,[24] but the disease has continued to be a major source of morbidity and mortality. Removal of T-cells from the graft can prevent GVHD but at the cost of graft failure, delayed immunological recovery and loss of the graft-versus-leukemia reaction.[25] Advances in immunology are providing new approaches to prevention of GVHD without necessarily impairing immunological recovery or inhibiting the graft-versus-leukemia reaction (review in).[26]

Newer Applications of HCT. Long-term disease-free survival of 80 to 90% for patients treated early for malignant diseases and of 80 to 95% for patients treated for nonmalignant diseases indicates that HCT is now safe enough to consider it for application to other problems.

Hematopoietic Cell Transplantation for Autoimmune Diseases. Studies in animal systems have shown that autoimmune diseases can be reversed by HCT (reviewed in).[27,28] Cautious clinical studies are already underway, initially with autologous hematopoietic stem cells. It remains to be determined whether or not autologous HCT will provide prolonged relief of the disease. Allogeneic HCT is more likely to provide a permanent cure but at the risk of GVHD and its complications.[29]

HCT for Solid Organ Grafting. In the 1950s, marrow grafting was being considered as a means of inducing tolerance for a kidney graft. However, this approach was not pursued because of the development of effective immunosuppressive drugs and regimens. The increasing safety of HCT suggests that it should again be considered for transplantation of solid organs including islet cells. Tolerance induced by HCT does not involve the long-term use of immunosuppressive agents which now complicate and limit the survival of solid organ transplant recipients. The possibility of inducing tolerance to solid organ grafts by concurrent administration of donor marrow without myeloablation of host marrow is being actively investigated.[30]

HCT Without the Use of a Myeloablative Preparative Regimen. Achievement of an HCT without the risks of regimen-related toxicity associated with myeloablative preparative regimens would obviously be desirable, especially for genetic or autoimmune diseases. Even for leukemia, low-dose antileukemic regimens accompanied by reconstitution with cells reactive against the leukemia might provide therapeutic benefit. Stable mixed chimerism has been achieved in the dog using sub-lethal irradiation combined with mycophenolate-mofetil.[31] Clinical studies have been initiated in elderly human patients with leukemia who would be expected to have unacceptable regimen-related toxicity with the usual preparative regimens.[32] The achievement of stable mixed chimerism with non-myeloablative regimens should revive the possibility of a kidney graft from the donor without subsequent long-term immunosuppressive therapy.[33]

SUMMARY

Looking ahead to the near future, chimerism will be achieved in stages so that patients will not be exposed to regimen related morbidity. The abnormal cells will be eliminated by techniques that specifically target these cells such as monoclonal antibodies in conjunction with radioisotopes or toxins. Reconstitution will be with stem cells and cloned immune cells that react immunologically with the abnormal cells while at the same time inducing donor-specific tolerance and restoring normal immune reactivity. Reconstitution may also involve cells modified by gene therapy techniques. Thus, HCT will be applied not only to genetic or malignant diseases of the marrow but also to other cancers, to solid organ transplants including islet cells, and to autoimmune diseases.

REFERENCES

1. Jacobson LO, Simmons EL, Marks EK, Eldredge JH: Revovery from radiation injury. Science 113:510, 1951.
2. Lorenz E, Uphoff D, Reid TR, Shelton E: Modification of irradiation injury in mice and guinea pigs by bone marrow injections. J Natl Cancer Inst 12:197, 1951.
3. Main JM, Prehn RT: Successful skin homografts after the administration of high dosage X radiation and homologous bone marrow. J Natl Cancer Inst 15:1023, 1955.
4. Ford CE, Hamerton JL, Barnes DWH, Loutit JF: Cytological identification of radiation-chimaeras. Nature 177:452, 1956.
5. Thomas ED, Collins JA, Herman EC, Jr., Ferrebee JW: Marrow transplants in lethally irradiated dogs given methotrexate. Blood 19:217, 1962.
6. Mannick JA, Lochte HL, Jr., Ashley CA, Thomas ED, Ferrebee JW: A functioning kidney homotransplant in the dog. Surgery 46:821, 1959.
7. Epstein RB, Storb R, Ragde H, Thomas ED: Cytotoxic typing antisera for marrow grafting in littermate dogs. Transplantation 6:45, 1968.

8. Storb R, Epstein RB, Ragde H, Thomas ED: Marrow grafts by combined marrow and leukocyte infusions in unrelated dogs selected by histocompatibility typing. Transplantation 6:587, 1968.
9. Gatti RA, Meuwissen JH, Allen HD, Hong R, Good RA: Immunological reconstitution of sex-linked lymphopenic immunological deficiency. Lancet ii:1366, 1968.
10. Thomas ED, Storb R, Clift RA, Fefer A, Johnson FL, Neiman PE, Lerner KG, Glucksberg H, Buckner CD: Bone-marrow transplantation. N Engl J Med 292:832–843, 895, 1975.
11. Thomas ED, Flournoy N, Buckner CD, Clift RA, Fefer A, Neiman PE, Storb R: Cure of leukemia by marrow transplantation. Leuk Res 1:67, 1977.
12. Clift RA, Buckner CD, Thomas ED, Bensinger WI, Bowden R, Bryant E, Deeg HJ, Doney KC, Fisher LD, Hansen JA, Martin P, McDonald GB, Sanders JE, Schoch G, Singer J, Storb R, Sullivan KM, Witherspoon RP, Appelbaum FR: Marrow transplantation for chronic myeloid leukemia: A randomized study comparing cyclophosphamide and total body irradiation with busulfan and cyclophosphamide. Blood 84:2036, 1994.
13. Storb R, Leisenring W, Anasetti C, Appelbaum FR, Buckner CD, Bensinger WI, Chauncey T, Clift RA, Deeg HJ, Doney KC, Flowers MED, Hansen JA, Martin PJ, Sanders JE, Sullivan KM, Witherspoon RP: Long-term follow-up of allogeneic marrow transplants in patients with aplastic anemia conditioned by cyclophosphamide combined with antithymocyte globulin (Letter to the Editor). Blood 89:3890, 1997.
14. Lucarelli G, Giardini C, Baronciani D: Bone Marrow transplantation in thalassemia (Review). Semin Hematol 32:297, 1995.
15. Bone Marrow Donors Worldwide (BMDW): http://bmdw.leidenuniv.nl. 1998 (unppub).
16. Hansen JA, Gooley TA, Martin PJ, Appelbaum F, Chauncey TR, Clift RA, Petersdorf EW, Radich J, Sanders JE, Storb RF, Sullivan KM, Anasetti C: Bone marrow transplants from unrelated donors for patients with chronic myeloid leukemia. N Engl J Med 338:962, 1998.
17. Mannick JA, Lochte HL, Jr., Ashley CA, Thomas ED, Ferrebee JW: Autografts of bone marrow in dogs after lethal total-body radiation. Blood 15:255, 1960.
18. Peters WP, Ross M, Vredenburgh JJ, Meisenberg B, Marks LB, Winer E, Kurtzberg J, Bast RC, Jr., Jones R, Shpall E, Wu K, Rosner G, Gilbert C, Mathias B, Coniglio D, Petros W, Henderson IC, Norton L, Weiss RB, Budman D, Hurd D: High-dose chemotherapy and autologous bone marrow support as consolidation after standard-dose adjuvant therapy for high-risk primary breast cancer. J Clin Oncol 11:1132, 1993.
19. Shpall EJ, Jones RB, Bearman SI, Franklin WA, Archer PG, Curiel T, Bitter M, Claman HN, Stemmer SM, Purdy M, Myers SE, Hami L, Taffs S, Heimfeld S, Hallogan J, Berenson RJ: Transplantation of enriched CD34-positive autologous marrow into breast cancer patients following high-dose chemotherapy: Influence of CD34-positive peripheral-blood progenitors and growth factors on engraftment. J Clin Oncol 12:28, 1994.
20. Gianni AM, Siena S, Bregni M, Tarella C, Stern AC, Pileri A, Bonadonna G: Granulocyte-macrophage colony-stimulating factor to harvest circulating haemopoietic stem cells for autotransplantation. Lancet ii:580, 1989.
21. Gluckman E, Broxmeyer HE, Auerbach AD, Friedman HS, Douglas GW, Devergie A, Esperou H, Thierry D, Socie G, Lehn P, Cooper S, English D, Kurtzberg J, Bard J, Boyse EA: Hematopoietic reconstitution in a patient with Fanconi's anemia by means of umbilical-cord blood from an HLA-identical sibling. N Engl J Med 321:1174, 1989.
22. Billingham RE, Brent L: Quantitative studies on tissue transplantation immunity. IV. Induction of tolerance in newborn mice and studies on the phenomenon of runt disease. Philos Trans R Soc Lond B Biol Sci 242:477, 1959.
23. Thomas ED, Buckner CD, Clift RA, Fefer A, Johnson FL, Neiman PE, Sale GE, Sanders JE, Singer JW, Shulman H, Storb R, Weiden PL: Marrow transplantation for acute nonlymphoblastic leukemia in first remission. N Engl J Med 301:597, 1979.
24. Storb R, Deeg HJ, Whitehead J, Farewell V, Appelbaum FR, Beatty P, Bensinger W, Buckner CD, Clift R, Doney K, Hansen J, Hill R, Lum LG, Martin P, McGuffin R, Sanders JE, Singer J, Stewart P, Sullivan KM, Witherspoon RP, Thomas ED: Marrow transplantation for leukemia and aplastic anemia: Two controlled trials of a combination of methotrexate and cyclosporine versus cyclosporine alone or methotrexate alone for prophylaxis of acute graft-verus-host disease. Transplant Proc 19:2608, 1987.
25. Martin PJ, Kernan NA: T-cell depletion for the prevention of graft-vs.-host disease, in Burakoff SJ, Deeg HJ, Ferrara J, Atkinson K (eds): Graft-vs.-Host Disease: Immunology, Pathophysiology, and Treatment, New York, Marcel Dekker, 1990, p 371.
26. Blazar BR, Korngold R, Vallera D: Recent advances in graft-versus-host disease (GVHD) prevention. Immunol Rev 157:109, 1997.

27. van Bekkum DW: BMT in experimental autoimmune diseases. Bone Marrow Transplant 11:183, 1993.
28. Marmont AM: Immune ablation followed by allogeneic or autologous bone marrow transplantation: A new treatment for severe autoimmune diseases. Stem Cells 12:125, 1994.
29. Ikehara S: Autoimmune diseases as stem cell disorders: normal stem cell transplant for their treatment (Review). Int J Mol Med 1:5, 1998.
30. Kawai T, Cosimi AB, Colvin RB, Powelson J, Eason J, Kozlowski T, Sykes M, Monroy R, Tanaka M, Sachs DH: Mixed allogeneic chimerism and renal allograft tolerance in Cynomolgus monkeys. Transplantation 59:256, 1995.
31. Storb R, Yu C, Wagner JL, Deeg HJ, Nash RA, Kiem H-P, Leisenring W, Shulman H: Stable mixed hematopoietic chimerism in DLA-identical littermate dogs given sublethal total body irradiation before and pharmacological immunosuppression after marrow transplantation. Blood 89:3048, 1997.
32. Giralt S, Estey E, Albitar M, van Besien K, Rondon G, Anderlini P, O'Brien S, Khouri I, Gajewski J, Mehra R, Claxton D, Andersson B, Beran M, Przepiorka D, Koller C, Kornblau S, Korbling M, Keating M, Kantarjian H, Champlin R: Engraftment of allogeneic hematopoietic progenitor cells with pruine analog-containing chemotherapy: harnessing graft-versus-leukemia without myeloablative therapy. Blood 89:4531, 1997.
33. Colson YL, Wren SM, Schuchert MJ, Patrene KD, Johnson PC, Boggs SS, Ildstad ST: A nonlethal conditioning approach to achieve durable mulilineage mixed chimerism and tolerance across major, minor, and hematopoietic histocompatibility barriers. J Immunol 155:4179, 1995.

2

HEMOPOIETIC STEM CELLS—NORMAL AND ABNORMAL

Susumu Ikehara*

First Department of Pathology
Kansai Medical University
Osaka, Japan

ABSTRACT

We have succeeded in purifying pluripotent hemopoietic stem cells (P-HSCs) from the bone marrow of 5-fluorouracil (5-FU)-treated normal mice. The P-HSCs were found to be lineage-negative (Lin^-)/$CD71^-$/major histocompatibility complex (MHC) class I^{high}/c-$kit^{<low}$. The cells formed colony-forming units in spleen (CFU-S) on day 16, although c-kit^{low} or c-kit^+cells formed CFU-S on day 14 or day 10, respectively.

Serial BMT revealed that c-$kit^{<low}$ cells have long-term-reconstituting activity (LTRA) for more than 2 years, whereas c-kit^{low} cells have LTRA (<1.5 years).

Since we have previously found that autoimmune diseases are stem cell disorders, we compared the qualitative differences between normal and abnormal P-HSCs. Although normal P-HSCs showed a poor proliferative response under MHC-incompatible microenvironments (stromal cells), abnormal P-HSCs showed a good proliferative response under such conditions. In addition, the abnormal P-HSCs were found to proliferate much faster than normal P-HSCs. These findings indicate that abnormal P-HSCs are more resilient than normal P-HSCs.

**Correspondence to*: Susumu Ikehara, M.D., Ph.D., First Department of Pathology, Kansai Medical University, 10-15 Fumizono-cho, Moriguchi City, Osaka 570-8506, Japan. Telephone 81-6-993-9429; Fax: 81-6-994-8283; E-mail: ikehara@takii.kum.ac.jp

Molecular Biology of Hematopoiesis 6, edited by Abraham *et al.*
Kluwer Academic/Plenum Publishers, New York, 1999.

PLURIPOTENT HEMOPOIETIC STEM CELLS (P-HSCs) ARE C-KIT$^{<LOW}$

Mouse P-HSCs in the Bone Marrow

P-HSCs are defined as cells with the capacity to self-renew eternally and to differentiate into cells in all lineages, including lymphoid cells. We have previously demonstrated that HSCs in the Go phase can be purified by both in vivo and in vitro 5-FU treatments, followed by sorting wheat germ agglutinin-binding (WGA^{+}) cells.[1] Bone marrow cells (BMCs) from 5-FU-treated mice were first fractionated by Percoll density gradient. The cells with densities between 1.063 and 1.075 were further separated into WGA-positive and -negative cells using fluorescent-activated cell sorter (FACS) after staining with fluorescein isothiocyanate-conjugated WGA (FITC-WGA). An assay for CFU-S revealed that the WGA-positive cells (1×10^6) produced 1380 CFU-S (about 150 times the number in the original BMCs) on day 12 (but no CFU-S on day 8), whereas the WGA-negative cells produced no CFU-S. Thus, HSCs in the Go phase were found to be enriched 150 times in 5-FU-treated WGA-positive cells. However, the 5-FU treatments (both in vivo and in vitro) have cytotoxic effects even on HSCs in the Go phase. We have therefore modified the method to include only in vivo 5-FU treatment followed by sorting CD71^{-}/class I^{high} cells from Lin^{-} cells. The sorted cells (only 4 cells) have LTRA (~1 year) in the assay of (male → female) chimeras.[2]

The HSCs (Lin^{-}/CD71^{-}/class I^{high} cells) were further fractionated on the basis of the expression of c-kit. Fig. 1a shows the flow cytometric profile of Lin^{-}/CD71^{-} cells stained with anti-H-2K^{b} and anti-c-kit mAbs. H-2K$^{b\ high}$ cells were gated, and two cell populations—c-kit^{low} and c-kit$^{<low}$ (which are comparable to the negative control level)—were fractionated by sorting on the basis of c-kit expression. Sorting gates and histogram are shown in Fig. 1a and 1b; the sorting gate of c-kit$^{<low}$ cells was set not to overlap that of c-kit^{low} cells. The FACS profile is very similar to that reported by Katayama et al.;[3] the c-kit^{low} and c-kit$^{<low}$ populations were identified as shown in Fig. 1a and 1b. Purified Lin^{-}/CD71^{-}/class I^{high}/c-kit^{low} cells or Lin^{-}/CD71^{-}/class I^{high}/c-kit$^{<low}$ cells were $\approx$0.006% or 0.008% of the 5-FU-treated whole BMCs, respectively.

The next step was to examine the expression of c-kit transcripts in these cells. The c-kit transcripts in Lin^{-}/CD71^{-} cells, Lin^{-}/CD71^{-}/ class I^{high}/c-kit^{low} cells, and Lin^{-}/CD71^{-}/class I^{high}/c-kit$^{<low}$ cells were therefore compared by RT-PCR analyses. As shown in Fig. 1c, substantial expression of c-kit mRNA was detected even in the class I^{high}/c-kit$^{<low}$ cells, although the level in these cells was lower than that in class I^{high}/c-kit^{low} cells. Based on these findings, we defined the cells as "c-kit$^{<low}$" instead of "c-kit^{-}", although the cells appeared to be "c-kit^{-}" in FACS analyses, as shown in Fig. 1a and 1b.

Five hundred cells of each fraction from 5-FU-treated B6 Ly5.1 mice were intravenously (iv) injected into irradiated B6 Ly5.2 recipients along with B6 Ly5.2 compromized cells. Six months after the transplantation, cells from several organs were double-stained with a panel of mAbs against mature myeloid/lymphoid cells and anti-Ly5.1 mAb to check for donor-derived cells. Ly5.1^{+} donor-derived mature cells—B220^{+} or surface Ig^{+} B cells, CD4^{+} or CD8^{+} T cells, Gr-1^{+} granulocytes, and Mac-1^{+} macrophages—were detected in both recipients that received either Lin^{-}/CD71^{-}/class

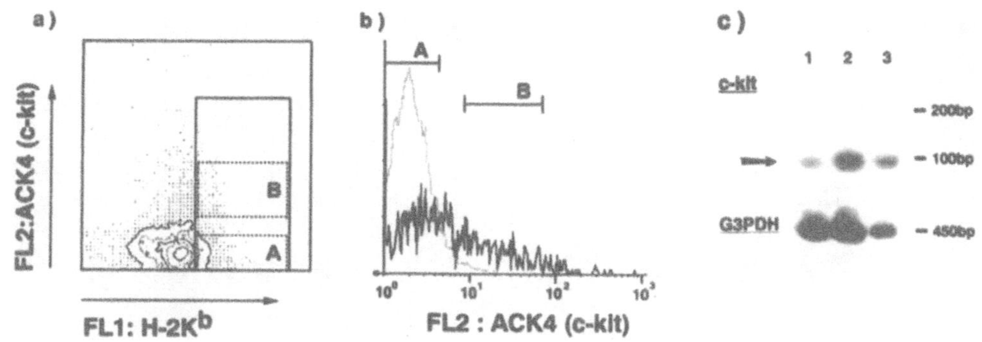

Figure 1. Staining profiles (a,b) and RT-PCR analyses (c) of Lin$^-$/CD71$^-$/class I^{high}/c-kit$^{<low}$ cells. Lin$^-$/CD71$^-$ cells were stained with anti-H-2K^b and anti-c-kit mAbs, and cells with a high expression of MHC class I were collected (represented in a). These cells were further divided into two populations based on the expression of c-kit [a and b (histogram)], and c-kit$^{<low}$ (A) and c-kit^{low} (B) cells were sorted by a FACStar. The dotted line in b represents a negative control stained by isotype-matched rat mAb. (c) RT-PCR-amplified products from a c-kit gene (upper) and from a glyceraldehyde 3-phosphate dehydrogenase (G3PDH) gene as an internal standard (lower) were electrophoresed, transferred to a nylon membrane, and probed with internal oligonucleotides. Lanes: 1, Lin$^-$/CD71$^-$/class I^{high}/c-kit$^{<low}$ cells; 2, Lin$^-$/CD71$^-$/class I^{high}/c-kit^{low} cells; and 3, Lin$^-$/CD71$^-$ unsorted BMCs.

I^{high}/c-kit^{low} cells or Lin$^-$/CD71$^-$/class I^{high}/c-kit$^{<low}$ cells, in all organs tested such as the bone marrow, thymus, spleen, lymph nodes, peripheral blood and peritoneal cavity (data not shown).

To examine the LTRA of these two cell populations, 1×10^6 BMCs either from recipients of Lin$^-$/CD71$^-$/class I^{high}/c-kit^{low} cells or Lin$^-$/CD71$^-$/class I^{high}/c-kit$^{<low}$ cells were re-transplanted into irradiated B6 Ly5.2 mice. Six months later, several organs were again examined to check for original donor-derived Ly5.1$^+$cells. In contrast to the results observed in the primary recipients, original donor-derived Ly5.1$^+$ cells could be detected only in secondary recipients originally reconstituted with Lin$^-$/CD71$^-$/class I^{high}/c-kit$^{<low}$ cells; Ly5.1$^+$ cells were observed in the myeloid and lymphoid cells in all organs tested (Fig. 2a). However, in the secondary recipients of BMCs from the mice that originally received Lin$^-$/CD71$^-$/class I^{high}/c-kit^{low} cells, no Ly5.1$^+$ donor-derived cells were detected at all (Fig. 2b). These findings clearly show that P-HSCs with LTRA are Lin$^-$/CD71$^-$/class I^{high}/c-kit$^{<low}$ cells, but not Lin$^-$/CD71$^-$/class I^{high}/c-kit^{low} cells.

Tertiary transplantation was carried out to confirm the LTRA of the original Lin$^-$/CD71$^-$/class I^{high}/c-kit$^{<low}$ cells. Six months after the secondary transplantation, BMCs (1×10^6) from the secondary recipients were again transplanted into irradiated B6 Ly5.2 tertiary recipients. All the tertiary recipients that had received BMCs from secondary recipients originally reconstituted with Lin$^-$/CD71$^-$/class I^{high}/c-kit^{low} cells died within 10 days. However, all the tertiary recipients that had received BMCs from secondary recipients originally reconstituted with Lin$^-$/CD71$^-$/class I^{high}/c-kit$^{<low}$ cells showed donor (Ly5.1$^+$)-derived cells in their peripheral blood; cells expressing Ly5.1 phenotype were detected in the bone marrow, spleen, lymph nodes, thymus, and peritoneal cavity 6 months after the tertiary BMT (2 years after the first BMT).[4]

We next carried out CFU-S assays using c-kit^{low} or c-kit$^{<low}$ cells. c-kit^{low} or c-kit$^{<low}$ cells (1×10^3) from male B6 Ly5.1 mice were transplanted into allotype-disparate

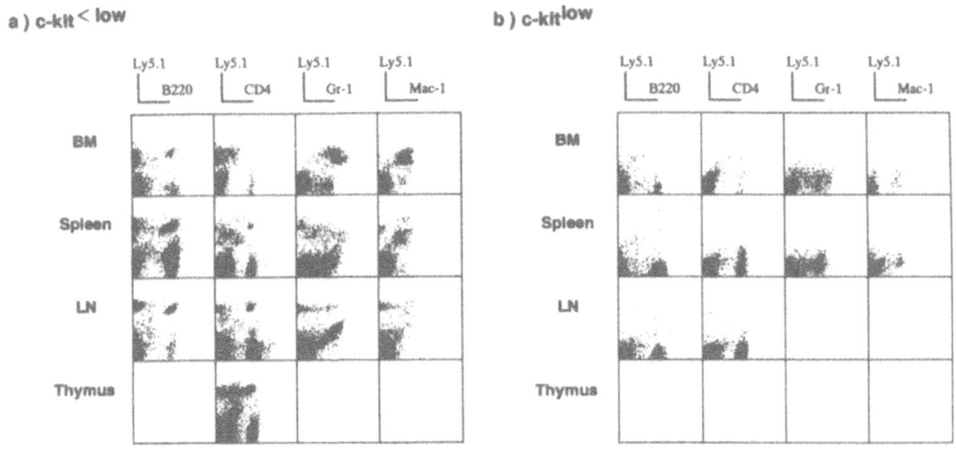

Figure 2. Detection of donor-derived cells in the secondary recipients. Six months after the primary transplantation, BMCs from the primary recipients were transferred to the irradiated secondary B6 Ly5.2 recipients. Six months later, cells from various organs were stained with a panel of mAbs and donor-specific anti-Ly5.1 mAb to determine whether the original donor-derived cells could be detected among the multilineage cells.

female B6 Ly5.2 mice, and CFU-S counts were measured 8 to 20 days later. As shown in Table 1, the recipients of c-kit^{low} cells formed visible CFU-S on day 12. CFU-S counts (and also spleen weight) gradually increased, and became uncountable as a result of fusion on days 16 and 20. In contrast, no CFU-S were detected in the recipients of c-kit$^{<low}$ cells on days 12 and 14, but first appeared on day 16. The numbers had increased on day 20, though each colony was small. Thus, c-kit$^{<low}$ cells can form CFU-S, although colony formation is delayed. It has also been reported that c-kit^{+} cells are P-HSCs.[5-9] Therefore, we next carried out CFU-S assays using c-kit^{+} cells (Lin^{-}/Thy-1^{low}/Sca-1^{+} cells). As shown in Table 1, the recipients of c-kit^{+} cells formed CFU-S on days 10 and 12. In contrast, all the irradiated mice died within 14 days without the transplantation of HSCs (c-kit^{+}, c-kit^{low}, or c-kit$^{<low}$ cells), and no CFU-S were detected after the fractionated irradiation (5.5 Gy × 2 = 11 Gy). This indicates that CFU-S in the recipients are actually donor-derived, and are not derived from the radioresistant progenitors of the recipients.[10]

Table 1. CFU-S counts after injection of c-kit^{+}, c-kit^{low}, or c-kit$^{<low}$ cells

Cells Injected	Days after Injection 10	12	14	16	20
c-kit^{+}	11.67 ± 1.15 (93.2 ± 3.7)	12.33 ± 11.15 (144.4 ± 22.9)	Uncountable (210.9 ± 39.1)		
c-kit^{low}	0 (20.2 ± 2.9)	4.33 ± 1.21 (31.7 ± 14.4)	7.00 ± 1.00 (65.5 ± 15.6)	Uncountable (166.8 ± 15.6)	
c-kit$^{<low}$	0 (16.6 ± 1.5)	0 (18.6 ± 1.6)	0 (16.4 ± 2.2)	8.40 ± 1.52 (27 ± 6.4)	22.25 ± 10.55

Numbers in parentheses represent spleen weights measured just before the fixation.

Human P-HSCs in the Cord Blood

The next step was to examine whether human P-HSCs are c-kit$^{<low}$, c-kit^{low}, or c-kit^{+}. We purified CD34^{+}/c-kit$^{<low}$ cells (phenotypically c-kit-negative but only detectable at the message level) from human cord blood and examined their maturational steps in relation to the expression of c-kit molecules. When the CD34^{+}/c-kit$^{<low}$ cells were cultured with cytokines (flt3-ligand, IL-6, and IL-7) plus immobilized anti-CD34 monoclonal antibody (mAb) (to cross-link CD34 molecules), c-kit molecules were clearly induced within 24 hrs. The c-kit expression gradually increased until day 8. When CD34^{+}/c-kit^{low} or CD34^{+}/c-kit^{+} cells that had been induced from CD34^{+}/c-kit$^{<low}$ cells were re-sorted and re-cultured using a methylcellulose culture system, they showed the same colony-forming ability as the freshly isolated CD34^{+}/c-kit^{low} or CD34^{+}/c-kit^{+} cells, respectively. Furthermore, CD34^{+}/c-kit$^{<low}$ cells have a similar hemopoietic potential to CD34^{+}/c-kit^{low} cells in assays for LTC-IC and CFU-C generated from long-term cultures. These findings suggest that CD34^{+}/c-kit$^{<low}$ cells differentiate into CD34^{+}/c-kit^{low} and CD34^{+}/c-kit^{+} cells, and acquire the reactivity to various humoral hematopoietic stimuli. Moreover, CD34^{+}/c-kit$^{<low}$ cells showed a low level of rhodamine 123 (Rh123) retention, suggesting that CD34^{+}/c-kit$^{<low}$ cells have multidrug resistance. Therefore, we conclude that human cord blood contains CD34^{+}/c-kit$^{<low}$ cells, which correspond to mouse P-HSCs.[11]

Peripheral Blood Stem Cells (PBSCs) in Mice

PBSCs were mobilized in mice by treatment with cytosine-arabinoside on day 0, followed by the administration of granulocyte colony-stimulating factor for 4 days. There were remarkable increases in the numbers of cells with Lin^{-}/c-kit^{+} markers, cells with colony-forming unit-cell (CFU-C) and CFU-S activities, and cells with marrow-repopulating ability (MRA) in the extramedullary sites (the spleen, peripheral blood, and liver) on day 5, whereas the number of these immature hematopoietic cells decreased in the bone marrow (BM) on day 5. This finding suggests the mobilization of immature hematopoietic cells from the BM to the extramedullary sites. Three-color flow cytometric analyses showed that CD4 antigen was not expressed on the Lin^{-}/Sca-1^{+} cells in the mobilized PB cells (PBCs), although CD4^{low} cells were found in those of normal BM cells. Lin^{-}/c-kit^{+} cells in the mobilized PBCs contained more cells with immature phenotypes (Sca-1^{+}/Thy1.2^{low}/CD71^{-}/Rh123^{dull}) than in normal BMCs, indicating an alteration of the hierarchical composition of the Lin^{-}/c-kit^{+} cells. We could find neither Lin^{-}/Sca-1^{+}/c-kit^{low} cells nor Lin^{-}/Sca^{+}/c-kit$^{<low}$ cells in the PBCs. Electron microscopic studies of these cells in the mobilized PBCs showed that only 10% to 20% of these cells had a thin rim of cytoplasm with poorly developed organelles (Fig. 3a). In contrast, P-HSCs present in the bone marrow had a large nucleus with narrow cytoplasm. Their chromatin pattern was dispersed, but small aggregates appeared at nuclear margins. There were few cytoplasmic organellas but abundant free ribosomes (Fig. 3b). It should be noted that P-HSCs possess microvilli; they show active movement like neutrophils when observed on video tape. Thus, PBSCs seem to be more mature than P-HSCs in the bone marrow.[12]

Allogeneic PBSCT [B6 → C3H] showed long-term reconstituting activity across the MHC barrier within 1 year of transplantation. However, all the [B6 → C3H] mice with PBSCT died 60 weeks after PBSCT, although the [B6 → C3H] mice with allogeneic BMT showed a 70% survival rate (Fig. 4). These findings suggest that PBSCs are not true P-HSCs (manuscript in preparation).

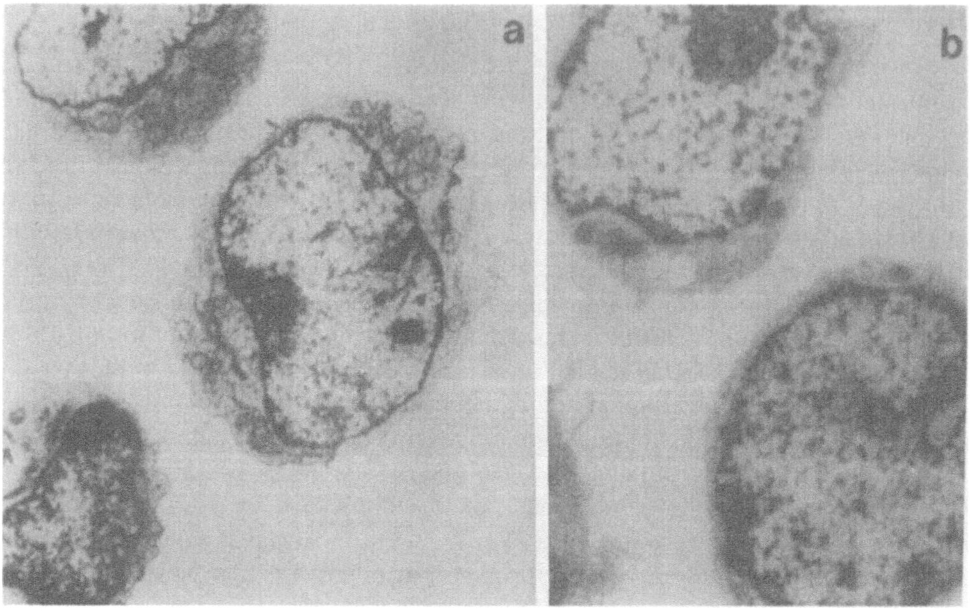

Figure 3. (a) Electron microscopic findings of mobilized PBSCs ($Lin^-/Sca\text{-}1^+/c\text{-}kit^+$ cells). (b) Electron microscopic findings of P-HSCs in the bone marrow. P-HSCs show large nuclei with narrow cytoplasm and microvilli.

MHC RESTRICTION EXISTS BETWEEN NORMAL P-HSCs AND STROMAL CELLS

In vitro studies revealed that P-HSCs cannot proliferate in the presence of putative cytokines such as GM-CSF, stem cell factor (SCF) and IL-3, whereas they can do so by direct interaction with stromal cells without adding any cytokines (Fig. 5). This

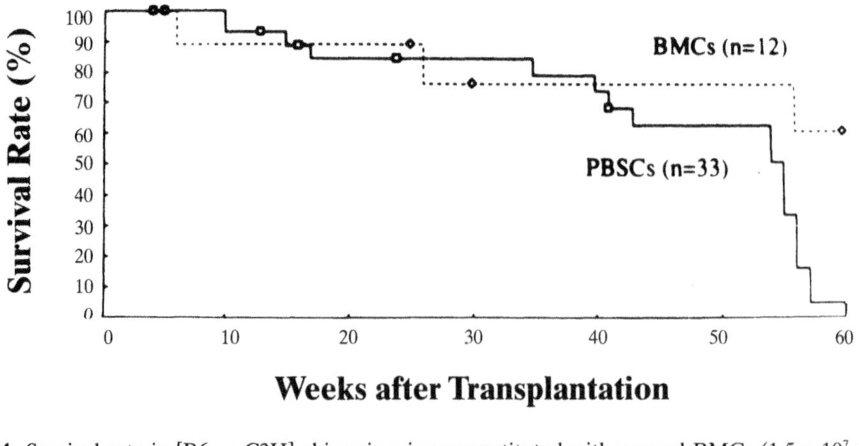

Figure 4. Survival rate in [B6 → C3H] chimeric mice reconstituted with normal BMCs (1.5×10^7 cells) or PBSCs (7×10^6 cells).

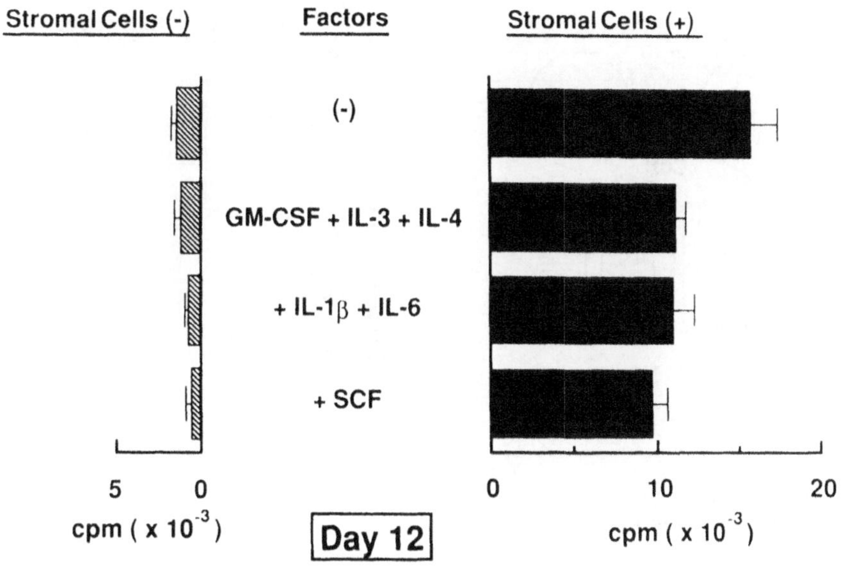

Figure 5. Necessity of stromal cells for P-HSCs to proliferate. When P-HSCs were cocultured with stromal cells, the P-HSCs proliferated without the addition of cytokines, whereas they did not without stromal cells.

finding indicates that the direct interaction of P-HSCs with stromal cells is essential for P-HSCs to proliferate.

We have previously found that donor-derived stromal cells play a crucial role in successful BMT across MHC barriers.[13,14] This finding prompted us to examine whether there is MHC restriction between P-HSCs and stromal cells. As shown in Fig. 6, hemopoiesis was observed only in the bone marrow engrafted with the BALB/c bone when BALB/c BMCs (T cell-depleted and adherent cell-depleted) were i.v. injected into irradiated C3H/HeN mice which had been engrafted with bones of C3H/HeN, B6, and BALB/c mice or with a teflon tube as a control. This finding strongly suggests that an MHC restriction exists between P-HSCs and stromal cells *in vivo*. This was confirmed in *in vitro* experiments; when B10 (H-2^b) P-HSCs were cocultured with B10 stromal cells, the P-HSCs proliferated, whereas, when B10 P-HSCs were cocultured with B10D2 (H-2^d) stromal cells, the P-HSCs showed poor proliferative responses (Fig. 7).[15]

QUALITATIVE DIFFERENCES EXIST BETWEEN NORMAL AND ABNORMAL P-HSCs

The next question was whether there are any qualitative differences between normal and abnormal P-HSCs. To answer this question, we first carried out BMT between normal and autoimmune-prone mice using partially purified P-HSCs. The transplantation of bones plus abnormal P-HSCs obtained from autoimmune-prone mice induced autoimmune diseases in normal mice, as did transplantation of T cell-depleted bone marrow cells (Fig. 8). However, the transplantation of bones plus normal P-HSCs could not reconstruct hemopoiesis in autoimmune-prone mice due to graft rejection (Fig. 9),[16] although the transplantation of T cell-depleted bone marrow cells from normal mice can be used to prevent and treat autoimmune diseases in autoimmune-prone mice.[17-19] This finding suggests that abnormal P-HSCs are more

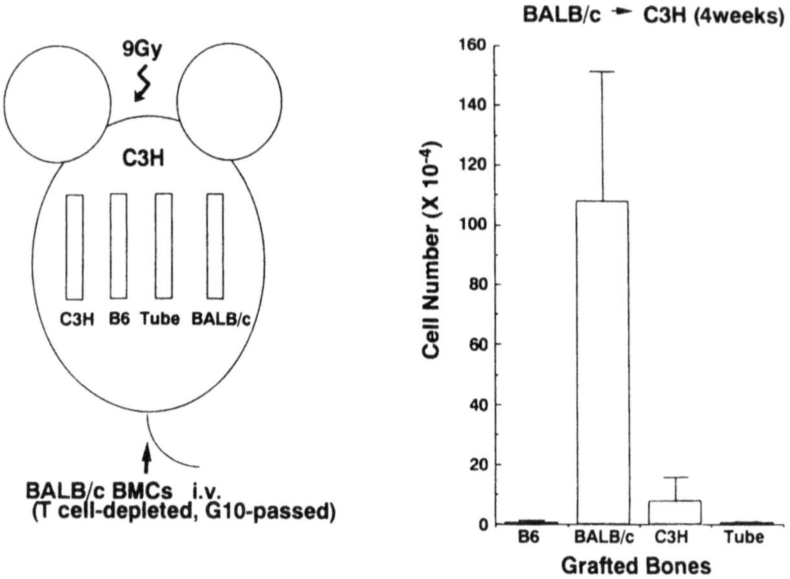

Figure 6. *In vivo* MHC restriction between P-HSCS and stromal cells. Hemopoiesis is observed only in the bone marrow of the BALB/c bone which has been engrafted subcutaneously, when BALB/c BMCs (T cell-depleted and adherent cell-depleted) are i.v. injected into the irradiated (9 Gy) C3H/HeN mouse.

resilient than normal P-HSCs; the former can proliferate in MHC-mismatched microenvironments, while the latter cannot.

To confirm that abnormal HSCs can survive even in allogeneic microenvironments, we carried out CFU-S assays on day 12 using P-HSCs (Lin$^-$/CD71$^-$/class I^{high}). As shown in Fig. 10, both (B6 → B6) and (C3H → C3H) chimeric mice formed high numbers of colonies in the spleen on day 12 (B6 → 3.55 ± 0.47/spleen, C3H → 5.6 ±

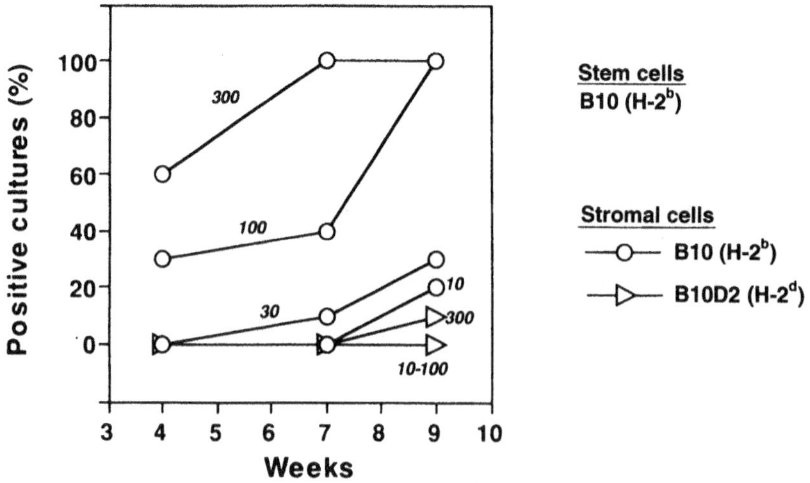

Figure 7. *In vitro* MHC restriction between P-HSCs and stromal cells. When B10 (H-2^b) P-HSCs (300 or 100 cells) are cocultured with B10 stromal cells, the P-HSCs proliferate. In contrast, when B10 P-HSCs are cocultured with B10D2 (H-2^d) stromal cells, the P-HSCs (10–300 cells) show poor proliferative responses.

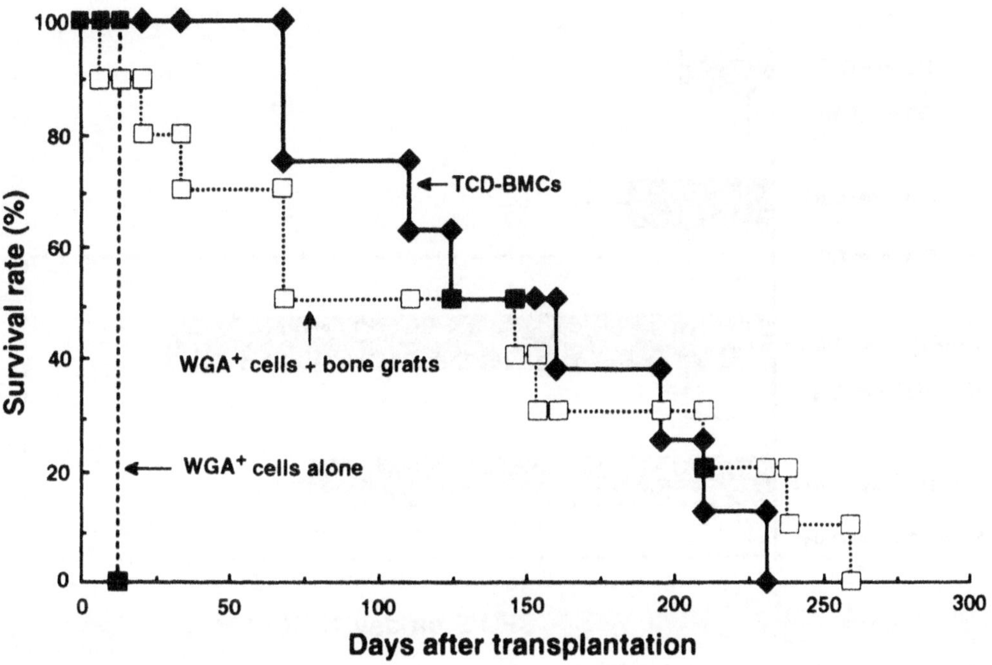

Figure 8. Survival rate in [W/BF1 → C3H] chimeric mice. C3H/HeN mice are irradiated (9.5 Gy) and then reconstituted with either 1–2 × 10^7 T cell-depleted (TCD)-BMCs (◆---◆) or 1–2 × 10^5 WGA$^+$ cells plus 40 Gy-irradiated bone grafts (□-□) or WGA$^+$ cells alone from male W/BF1 mice (<2 months) (■---■).

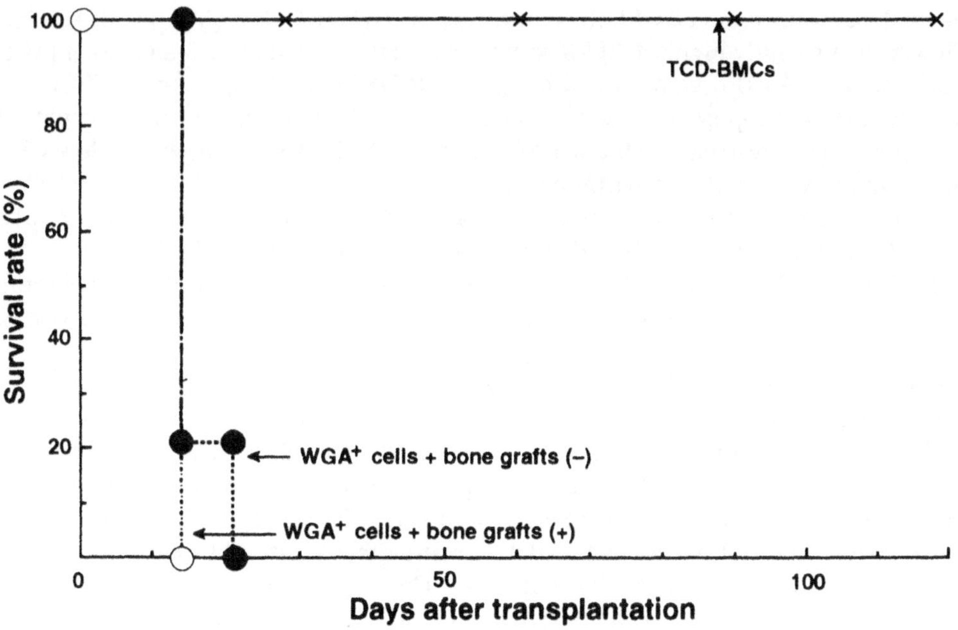

Figure 9. Survival rate in [C3H→W/BF1] chimeric mice. Male W/F1 mice (<2 months) are irradiated (9.0 Gy) and reconstituted with 1–2 × 10^7 T cell-depleted (TCD)-BMCs (×---×), 1–2 × 10^5 WGA$^+$ cells with (○-○) or without (•---•) bone grafts from C3H mice.

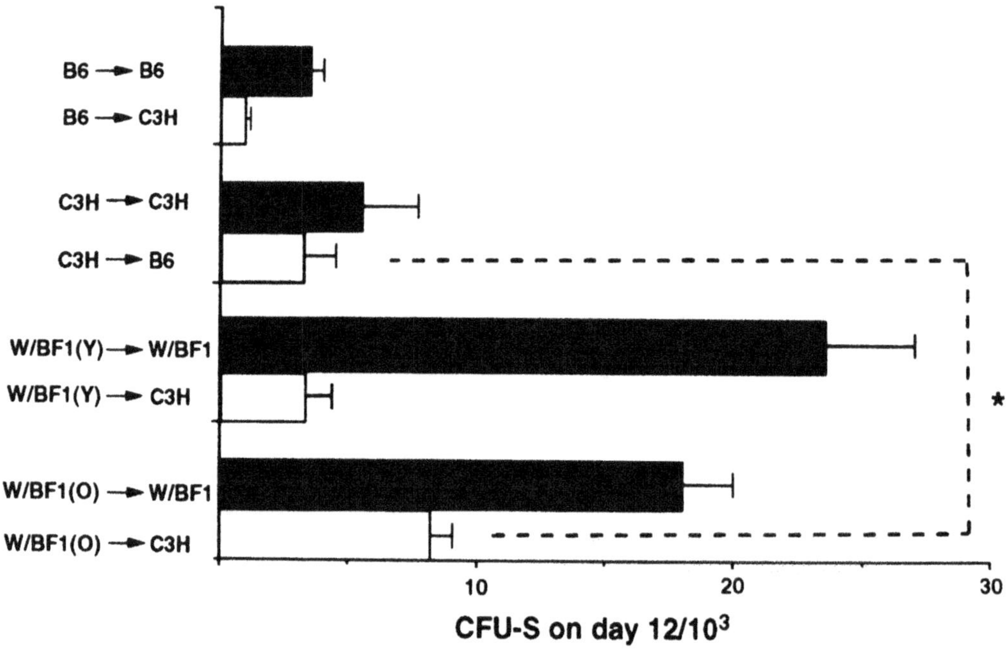

Figure 10. Comparison of CFU-S counts on day 12 between allogeneic (□) and syngeneic (■) combinations using P-HSCs from autoimmune-prone young (Y: <2 mo) or old (O: >4 mo) W/BF1 and normal mice. (W/BF1(O) → C3H) chimeric mice show significantly higher CFU-S counts on day 12 than (B6 → C3H) or (C3H → B6) chimeric mice. Error bars indicate standard error. $*p < 0.05$.

2.14), although (B6 → C3H) (0.95 ± 0.24) and (C3H → B6) (3.33 ± 1.25) chimeras showed markedly reduced CFU-S counts. However, (W/BF1 (O: old) → C3H) chimeras showed signigicantly higher CFU-S counts (8.29 ± 0.87) on day 12 than those in (B6 → C3H) or (C3H → B6) chimeras, although the P-HSCs of young (<2 mo) W/BF1 mice (before the development of autoimmune diseases) show major histocompatibility complex (MHC) restriction (the P-HSCs of young W/BF1 mice show very low CFU-S counts in the C3H mouse environment).[16]

To further confirm the existence of qualitative differences between abnormal and normal HSCs, we have established an in vitro assay system. When P-HSCs of B6($H\text{-}2^b$) mice were cocultured with B6 BM adherent cells, they showed a good proliferative response (Table 2). In contrast, when the P-HSCs of B6 mice were cocultured with C3H ($H\text{-}2^k$) mouse stromal cells, they showed a poor proliferative response. However, the P-HSCs of old (>4 mo) W/BF1 mice showed no MHC restriction: they could proliferate in collaboration with either W/BF1 or C3H stromal cells. This was the case when P-HSCs were cocultured with a BM stromal cell line, the MS-5 ($H\text{-}2^k$). As shown in Fig. 11, the P-HSCs of B6 mice showed a poor proliferative response in collaboration with MS-5. However, abnormal P-HSCs of W/BF1 mice showed a much greater proliferative response in collaboration with MHC-incompatible MS-5 cells than when B6 HSCs were cocultured with B6 BM adherent cells.[16] This was the case when P-HSCs of other autoimmune-prone mice such as BXSB and MRL/lpr mice were cocultured with BM adherent cells (data not shown). We therefore conclude that abnormal P-HSCs are more resilient than normal P-HSCs both in vivo and in vitro.

Table 2. No MHC restriction between abnormal P-HSCs and stromal cells

P-HSCs/Stromal Cells	Per Flask: No. of Non-Adherent Cells ($\times 10^4$)		Per Flask: No. of Cobblestone Colonies	
	Mean ± SE			
B6 (O)/B6 (Y)	12.5 ± 1.4	*	28.0 ± 1.1	*
B6 (O)/C3H (Y)	4.2 ± 0.2		7.5 ± 3.8	
W/BFl (O)/W/BFl (Y)	17.7 ± 6.4	**	99.7 ± 8.4	**
W/BFl (O)/C3H (Y)	12.0 ± 0.1		69.5 ± 5.5	

P-HSCs were purified from old (O: >4 mo) W/BFl or B6 mice. Stromal cells were obtained from young (<2 mo) W/BFl and normal (B6 or C3H) mice. P-HSCs (1×10^3) per flask were cultured on syngeneic or allogeneic bone marrow adherent cells (20 Gy-irradiated) of young (Y) mice. Two weeks later, the numbers of non-adherent cells and cobblestone colonies per flask were counted. The data are expressed as mean ± SE of triplicates.
*<0.05; **not significant.

ABNORMAL P-HSCs ARE MORE RESILIENT THAN NORMAL P-HSCs

To further confirm that abnormal P-HSCs of autoimmune-prone mice are more resilient that nomal P-HSCs in vivo, we transplanted 1×10^7 BALB/c bone marrow cells mixed with 1×10^6 NOD bone marrow cells into NOD mice. The [(NOD + BALB/c) → NOD] chimeras developed insulitis and diabetes, indicating that BALB/c bone marrow cells do not interfere with the development of the disease in NOD mice. Surprisingly, these mice have been reconstituted with only NOD hematolymphoid cells:

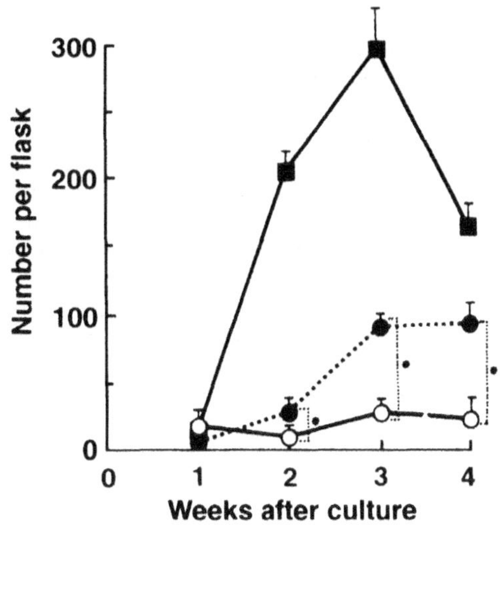

Figure 11. No MHC restriction between abnormal P-HSCs and stromal cells. Abnormal P-HSCs obtained from W/BF1 (H-2^z/H-2^b) mice show a significantly higher responsiveness in coculture with MHC-incompatible stromal cells [MS-5 (H-2^k):□-□] than do normal B6 P-HSCs cocultured with B6 stromal cells (●---●). B6 P-HSCs (H-2^b) show a poor proliferative response when cocultured with MS-5 (H-2^k) (○-○). Error bars indicate standard deviation. *$p < 0.005$.

BALB/c hematolymphoid cells were completely rejected by NOD hematolymphoid cells. When the pancreatic tissues from newborn NOD and BALB/c mice were grafted into [(NOD + BALB/c) → NOD] chimeras, the BALB/c pancreatic tissues were rejected, whereas the NOD grafts showed insulitis. Furthermore, the spleen cells of the chimeras showed responsivenes to BALB/c spleen cells in MLR and generated the CTLs for the $H\text{-}2^d$ and third party targets. These findings indicate that the hematolymphoid cells (including hemopoietic stem cells) of NOD mice are more resilient than those of normal BALB/c mice, and that insulin-dependent diabetes mellitus will recur after bone marrow transplantation unless the hematolymphoid cells of the NOD mice are completely destroyed by irradiation.[20]

Based on these findings, we propose that allogeneic BMT, instead of mixed allogeneic BMT or autologous BMT, should be carried out for the treatment of autoimmune diseases.

ACKNOWLEDGMENTS

These experiments were carried out in collaboration with researchers who appear in the references of this paper. I would like to express my deep appreciation to them.

I would also like to thank Ms. Yoko Tokuyama and Ms. Mari Shinkawa for their skillful technical assistance, and Mr. Hilary Eastwick-Field and Ms. Keiko Ando for preparing this manuscript.

This work was supported by a grant from the Japanese Ministry of Health and Welfare; the Ministry of Education, Science, and Culture, Japan; a grant from "Haiteku Research Center" of the Ministry of Education, Science, and Culture, Japan; the "Traditional Oriental Medical Science Program" of the Public Health Bureau of the Tokyo Metropolitan Government; the Mochida Memorial Foundation for Medical and Pharmaceutical Research (1986); the Naito Foundation (1986); the Mitsubishi Foundation (1986); the Suzuken Memorial Foundation (1987); the Takeda Science Foundation (1989); the Foundation for Rheumatic Diseases (1990), and the Science Research Promotion Fund of the Japan Private School Promotion Foundation, Japan.

REFERENCES

1. Miyama-Inaba M, Ogata H, Toki J, Kuma S, Sugiura K, Yasumizu R, Ikehara S: Isolation of murine pluripotent hemopoietic stem cells in the Go phase. Biochem Biophys Res Commun 75: 1809, 1987.
2. Ogata H, Bradley WG, Inaba M, Ogata N, Ikehara S, Good RA: Long-term repopulation of hematolymphoid cells with only a few hemopoietic stem cells in mice. Proc Natl Acad Sci USA 92: 5945, 1995.
3. Katayama N, Shih JP, Nishikawa S, Kinaa T, Clark SC, Ogawa M: Stage-specific expression of c-kit protein by murine hematopoietic progenitors. Blood 82: 2353, 1993.
4. Doi H, Inaba M, Yamamoto Y, Taketani S, Mori S-I, Sugihara A, Ogata H, Toki J, Hisha H, Inaba K, Sogo S, Adachi M, Matsuda T, Good RA, Ikehara S: Pluripotent hemopoietic stem cells are c-kit$^{<low}$. Proc Natl Acad Sci USA 94: 2513, 1997.
5. Orlic D, Fischer R, Nishikawa S-I, Nienhuis AW, Bodine DM: Purification and characterization of heterogeneous pluripotent hematopoietic stem cell populations expressing high levels of c-kit receptor. Blood 82: 762, 1993.
6. Ikuta K, Weissman IL: Evidence that hematopoietic stem cells express c-kit, but do not depend on steel factor for their generation. Proc Natl Acad Sci USA 89: 1502, 1992.
7. Ogawa M, Matsuzaki Y, Nishikawa S, Hayashi S-I, Kunisada T, Sudo T, Kina T, Nakauchi H, Nishikawa S-I: Expression and function of c-kit in hemopoietic progenitor cells. J Exp Med 174: 63, 1991.
8. Li CL, Johnson GR: Murine hematopoietic stem and progenitor cells: I. Enrichment and biologic characterization. Blood 85: 1472, 1995.

9. Taniguchi H, Toyoshima T, Fukao K, Nakauchi H: Presence of hematopoietic stem cells in the adult liver. Nature Med 2: 198, 1996.
10. Lian Z, Sugiura K, Inaba M, Yu C, Jin T, Feng B, Fan T, Yasumizu R, Toki J, Adachi Y, Hisha H, Ikehara S: c-kit$^{<low}$ pluripotent hemopoietic stem cells form CFU-S on day 16. (Submitted to "Stem Cells")
11. Sogo S, Inaba M, Ogata H, Hisha H, Adachi Y, Mori S-I, Toki J, Yamanishi K, Kanzaki H, Adachi M, Ikehara S: Induction of c-kit molecules on human CD34$^+$/c-kit$^{<low}$ cells: Evidence for CD34$^+$/c-kit$^{<low}$ cells as primitive hematopoietic stem cells. Stem Cells 15: 420, 1997.
12. Yamamoto Y, Yasumizu R, Amou Y, Watanabe N, Nishio N, Toki J, Fukuhara S, Ikehara S: Characterization of peripheral blood cells in mice. Blood 88: 445, 1996.
13. Ishida T, Inaba, Hisha H, Sugiura K, Adachi Y, Nagata N, Ogawa R, Good RA, Ikehara S: Requirement of donor-derived stromal cells in the bone marrow for successful allogeneic bone marrow transplantation: complete prevention of recurrence of autoimmune diseases in MRL/MP-lpr/lpr mice by transplantation of bone marrow plus bone (stromal cells) from the same donor. J Immunol 152: 3119, 1994.
14. Hisha H, Nishino T, Kawamura M, Adachi S, Ikehara S: Successful bone marrow transplantation by bone grafts in chimeric-resistant combination. Exp Hematol 23: 347, 1995.
15. Hashimoto F, Sugiura K, Inoue K, Ikehara S: Major histocompatibility complex restriction between hematopoietic stem cells and stromal cells in vivo. Blood 89: 49, 1997.
16. Kawamura M, Hisha, Li Y, Fukuhara S, Ikehara S: Distinct qualitative differences between normal and abnormal hemopoietic stem cells in vivo and in vitro. Stem Cells 15: 56, 1997.
17. Ikehara S, Good RA, Nakamura T, Sekita K, Inoue S, Maung Maung Oo, Muso E, Ogawa K, Hamashima Y: Rationale for bone marrow transplantation in the treatment of autoimmune diseases. Proc Natl Acad Sci USA 1985; 82: 2483, 1985.
18. Ikehara S, Ohtsuki H, Good RA, Asamoto H, Nakamura T, Sekita K, Muso E, Tochino Y, Ida T, Kuzuya H, Imura H, Hamashima Y: Prevention of type I diabetes in nonobese diabetic mice by allogeneic bone marrow transplantation. Proc Natl Acad Sci USA 82: 7743, 1985.
19. Ikehara S, Yasumizu R, Inaba M, Izui S, Hayakawa K, Sekita K, Toki J, Sugiura K, Iwai H, Nakamura T, Muso E, Hamashima Y, Good RA: Long-term observations of autoimmune-prone mice treated for autoimmune disease by allogeneic bone marrow transplantation. Proc Natl Acad Sci USA 86: 3306, 1989.
20. Takao F, Yasumizu R, Ohnishi-Inoue Y, Sugiura K, Inaba M, Ikehara S: Development of insulin-dependent diabetes mellitus in [(NOD + BALB/c) → NOD] mixes allogeneic bone marrow chimeras. Immunobiol 194: 376, 1995.

3

IMMUNITY AND TOLERANCE ARE RELATED, AND GOVERNED BY ANTIGEN MIGRATION AND LOCALIZATION

Thomas E. Starzl,[1] Noriko Murase,[1] Angus W. Thomson,[1]
Massimo Trucco,[2] and Abdul Rao[1]

From the Thomas E. Starzl Transplantation Institute
and the Departments of Surgery[1]
and Pediatrics[2]
University of Pittsburgh Medical Center
Pittsburgh, Pennsylvania, 15213

INTRODUCTION

The purpose of this communication is to describe a recently delineated principle of immunology[1] that can be summarized in one line: *immunity, or alternatively tolerance, to any antigen is governed by the migration and localization of that antigen.* This concept, which has been developed with transplantation models[2-4] and from observations after clinical and experimental infections[5-7] has been described elsewhere in more detail.[1] Here, we will be concerned mainly with the relation between donor leukocyte chimerism and transplantation tolerance, focusing at first on clinicopathologic observations in transplant recipients made in 1991–1992 that prompted reconsideration of a number of long held convictions.

Reprint Address: Thomas E. Starzl, M.D., Ph.D., Department of Surgery, 3601 Fifth Avenue, 5C Falk Clinic, University of Pittsburgh, Pittsburgh, Pennsylvania, 15213.
Presented at the 11th International Symposium of Molecular Biology of Hematopoiesis, June 28, 1998, and at the XVII World Congress of The Transplantation Society, Montreal, July 15, 1998.

Molecular Biology of Hematopoiesis 6, edited by Abraham *et al.*
Kluwer Academic/Plenum Publishers, New York, 1999.

HISTORICAL CLUES IN TRANSPLANTATION

The most important new finding reported in 1992 was the presence of multilineage donor leukocyte microchimerism in the blood, skin, lymph nodes, and other tissues of organ allograft recipients up to 30 years after successful transplantation.[2-4] The persistence of the disseminated donor leukocytes for this long implied (as was subsequently proved)[8,9] that precursor or stem cells are included in the burst of donor leukocytes that briefly constitute 1–20% of the recipient circulating mononuclear cells after organ transplantation (Fig. 1, upper). Although the number of these donor cells is greatest with transplantation of the intestine or liver,[10,11] the same events on a smaller scale occur after transplantation of organs like the kidney and heart.[12]

Another change during the early postoperative period is disappearance of the resident donor mononuclear leukocytes from the graft and their replacement by recipient cells of the same lineages (Fig. 1, lower). Although this was observed in hepatic allografts nearly 30 years ago,[13] it was assumed at first to be a unique feature of liver transplants. After the same changes were found in intestinal allografts,[11,14] evidence quickly accrued that this was a generic phenomenon; i.e. it occurred in all successfully transplanted organs.[3,15]

These discoveries, showing that both the "accepted" allograft and recipient became genetic composites, suggested an explanation for 2 enigmatic observations reported for the first time in 1963[16] that had allowed the development of organ transplantation as a practical and reproducible clinical service. First, kidney rejection developing in patients immunosuppressed with azathioprine could be consistently reversed with large doses of prednisone. The second, and more important, observation was that renal allografts self-induced variable degrees of donor-specific tolerance, manifested by the subsequent ability to reduce immunosuppression to below doses that had been unable to prevent or control rejection at the outset.[16] It was obvious that a host-versus-graft immune reaction gathered strength at first despite anti-rejection therapy, but then

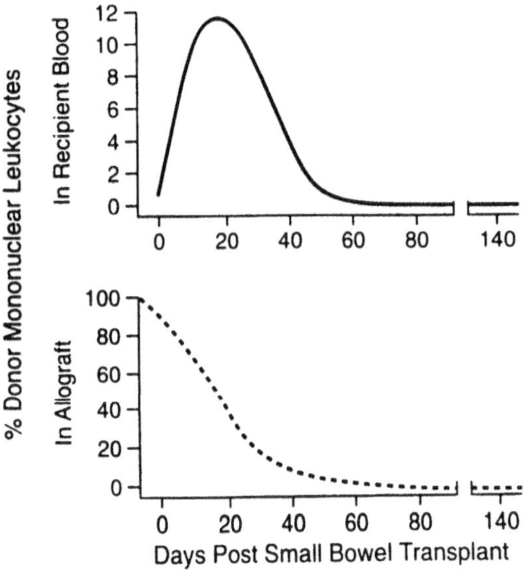

Figure 1. The dynamic events after transplantation of all organs, exemplified by intestinal engraftment. **Upper**—The surge of migratory donor mononuclear leukocytes in the recipient blood during the first 2 or 3 weeks post-transplantation. **Lower**—The contemporaneous disappearance of the passenger leukocytes from the allograft; these are replaced by recipient cells of the same lineages (schematically redrawn from data in).[11]

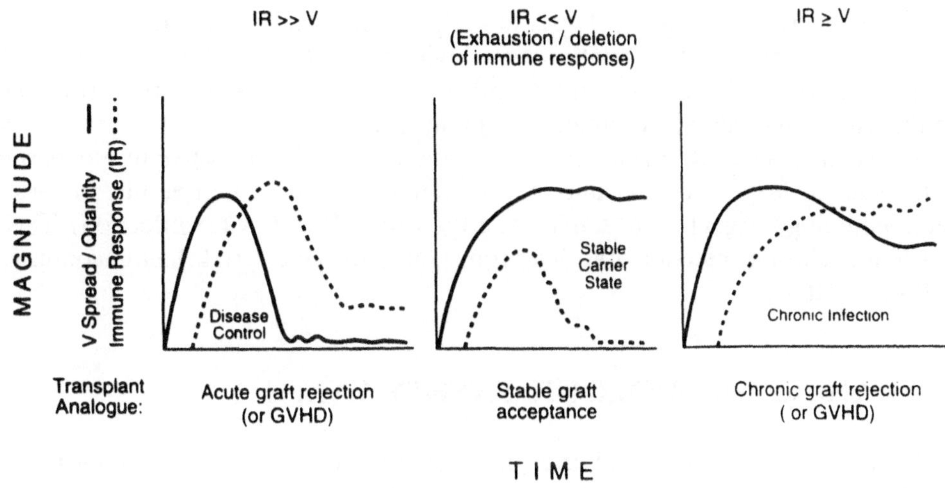

Figure 2. Variable outcomes after infection with widely disseminated non-cytopathic viruses (or other microorganisms) and analogies (in the subscripts) to organ and bone marrow transplantation. The horizontal axis denotes time, and the vertical axis shows the viral load (v, solid line), and the host immune response (IR, dashed line).

collapsed with temporarily intensified immunosuppression. The same events were soon documented after liver transplantation[17,18] and eventually with all other kinds of organ allografts. They can be observed in many animal models *without* the need for immunosuppression (summarized in Reference 4), most commonly when the liver is the transplanted organ.[19]

THE HISTORICAL BRIDGE TO INFECTION

Early workers in transplantation recognized the resemblance of allograft rejection to the response against infections associated with delayed hypersensitivity, as exemplified by tuberculosis.[20,21] With the demonstration of the MHC-restricted mechanisms of adaptive infectious immunity in 1973,[22-25] it seemed obvious that allograft rejection must be the physiologic equivalent of the response to this kind of infection. Microorganisms that generate such an adaptive immune response are generally intracellular, and have no or low cytopathic qualities. Although MHC-restricted host cytolytic T lymphocytes recognize only infected cells, elimination of all the infected cells could disable or even kill the host. Consequently, mechanisms have evolved that can temper or terminate the immune response, allowing both host and pathogen to survive.[5-7]

The highly variable clinical manifestations of these mechanisms can be illustrated by the different outcomes that may follow an infection with disseminated non-cytopathic microorganisms (e.g. the common hepatitis viruses). In one scenario, the pathogen (antigen) load may rapidly increase during the so-called latent period, but then be dramatically and efficiently reduced by antigen-specific effector T-cells. Following control of the infection, the cytotoxic T lymphocytes (CTL) subside (Fig. 2, left panel). The events are similar to those of irreversible organ rejection in the unmodified or ineffectively treated recipient (Fig. 2, subscript to left panel). Alternatively, however, such infections may lead to a continuously high antigen load and an antigen specific immunologic collapse (Fig. 2, second panel). The resulting asymptomatic carrier state is equivalent to unqualified acceptance of an allograft.

Between these 2 extremes of protective immunity and a carrier state, a persistent infectious agent may induce an unrelenting immune response that results in serious immunopathology (e.g. chronic active hepatitis with a B or C virus infection). This is equivalent to chronic rejection after liver transplantation (Fig. 2, right), or uncommonly GVHD (see later).

TOLERANCE: THE BIOLOGIC SAFETY VALVE

Under both infectious and transplant circumstances, only 2 mechanisms are needed to explain the compromise outcomes shown in the second and third panels of Fig. 2 which allow co-existence of live antigen and the host. One is clonal exhaustion leading to clonal deletion. The other is immune indifference. Although these 2 mechanisms will be considered separately, they tend to be mutually reinforcing. Both are regulated by antigen migration and localization.

Clonal Exhaustion/Deletion

The exhaustion and deletion of an antigen specific clone was postulated as early as 1959–60 to explain the acquisition of tolerance in animal models to heterologous protein (with the aid of 6-mercaptopurine),[26] and to allogeneic splenocytes (without the need for immunosuppression).[27] Clonal exhaustion also was invoked at an early time to explain organ allograft acceptance,[28,29] as shown in the illustration and caption reproduced in Fig. 3. In this figure, which was published in 1969, induction of the activated clone by alloantigen was depicted via host macrophages rather than by dendritic cells which would not be described until 4 years later.[30]

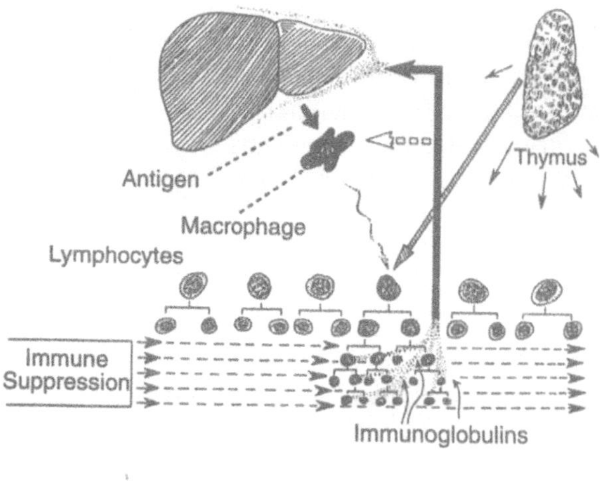

Figure 3. Hypothesis published in 1969 of allograft acceptance by clonal exhaustion. Antigen presentation was depicted via the macrophages rather than by the dendritic cells (which had not yet been described). A gap in this hypothesis was the failure to stipulate the location of the immune activation (by permission of W.B. Saunders Co.).[28]

The Eclipse of Clonal Exhaustion. Despite circumstantial evidence of its existence as a tolerogenic mechanism (summarized in),[31,32] clonal exhaustion disappeared from the literature between 1970 and 1990, ostensibly because it was only a theory. Since 1990, however, clonal exhaustion/deletion has been formally demonstrated in many infectious, transplantation, and other models.[33–38] A subpopulation of T cells is induced by the antigen within a few days, end-differentiates to effector cells, and disappears.

Death of the activated cells by apoptosis has been demonstrated in a mouse transplantation model by Qian, Lu, and Thomson,[37] possibly due to interleukin deprivation, and associated with telemere shortening.[39] Although clonal exhaustion is the most efficient way to eliminate maturing self-reactive T cells in ontogeny and throughout the life of many higher vertebrates, purging of T cells (and also apparently B cells) also occurs in the peripheral lymphoid organs. The peripheral sites may be the principal site of clonal deletion after successful transplantation in humans.[4]

Alternative Tolerogenic Mechanisms. Although lack of proof contributed to the dismissive reception of clonal exhaustion, a more pervasive factor in the context of transplantation immunology was a lack of understanding about the role of the organ's "passenger leukocytes". These donor cells of bone marrow origin have been known for more than 3 decades to be the principal immunogenic component of allografts.[40,41] Because they disappear from successfully transplanted organs,[11,14,15,40–43] it was assumed until recently that they had been selectively destroyed by the recipient immune system, with selective sparing of the specialized parenchymal cells.

As a corollary, organ allograft acceptance was assumed to be by fundamentally different mechanisms than the chimerism-dependent acquired neonatal tolerance of Billingham, Brent, and Medawar.[44] Proposed chimerism-exclusionary mechanisms have included suppressor, veto, and other immune regulatory cells; cytokine profile changes; various antibodies; and failure of delivery of a second (co-stimulatory) signal following primary antigen presentation. Contrary to these hypotheses, the discovery of microchimerism in organ recipients made it possible in 1992 to explain allograft acceptance by "... [widespread] responses of co-existing donor and recipient immune cells, each to the other, causing reciprocal clonal expansion, followed by peripheral clonal deletion".[2]

The Double Immune Reaction. This feature distinguishes the allograft response from the single immune reaction induced by an infection.[1] If some degree of reciprocal clonal exhaustion is not induced and maintained (requiring an umbrella of immune suppression in humans), one cell population usually will destroy the other, or both may be destroyed together. Following organ transplantation, the dominant host system usually rejects the graft (Fig. 4). However, serious or lethal GVHD is not rare after transplantation of leukocyte-rich organs like the liver.[3]

In contrast, the recipient cytoablation used in preparation for bone marrow transplantation transfers dominance to the donor system (Fig. 5). Consequently, GVHD is the most common complication in bone marrow recipients, but the graft may be rejected instead, or simultaneously.

The Governance of Clonal Exhaustion. Migration of spreading noncytopathic microorganisms to host lymphoid organs, and localization there, are well known to be critical in initiating and sustaining protective immune activation.[5–7,45] The similar lymphoid-oriented traffic of passenger leukocytes[46–48] is equally acknowledged to be the essential basis of host allo-sensitization. Carried one step farther to clonal exhaustion, it also is the means by which specific immunologic tolerance is induced.

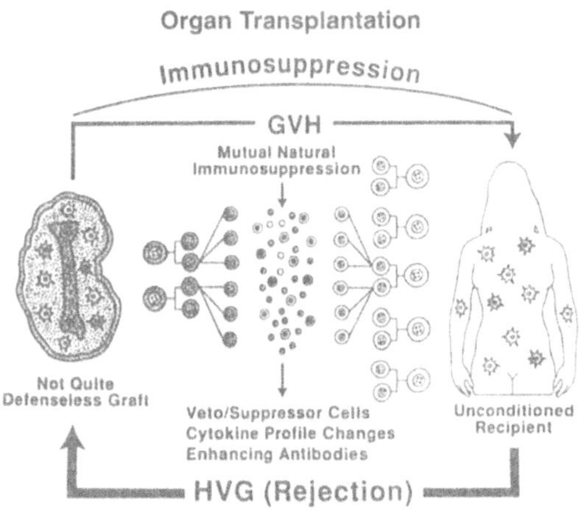

Figure 4. Organ transplantation, seen as a bidirectional and mutually cancelling immune reaction that is predominantly host vs. graft in most cases.

Immune Indifference

Although clonal exhaustion/deletion is the seminal mechanism of tolerance, survival of either allo- or infectious antigen may be promoted by a second non-tolerogenic mechanism, called "immune indifference".[5,7] Like clonal exhaustion, immune indifference is controlled by antigen migration and localization.

Primary. Pure examples of *de novo* immune indifference are provided by the rabies and wart viruses which elicit little or no immune response (Fig. 6), simply by avoiding migration through, or to, host lymphoid organs. This has been mimicked in numerous transplant models by depletion of donor leukocytes from allografts. Graft survival is thereby prolonged (subscript of Fig. 6). However, tolerance is not induced, as shown by the fact that rejection can be readily precipitated with an injection of donor leukocytes.[41,49,50]

Secondary. Pristine examples of immune indifference are not seen in the usual setting of clinical organ transplantation. Nevertheless, immune indifference can evolve secondarily. As also occurs with microorganisms following a widespread non-cytopathic

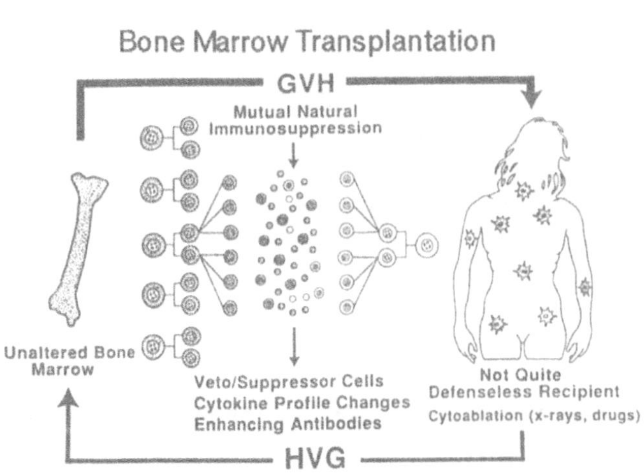

Figure 5. Bone marrow transplantation. Because the recipient is cytoablated, the immune reaction is predominantly graft vs. host, resulting in a mirror image of that shown in Fig. 4.

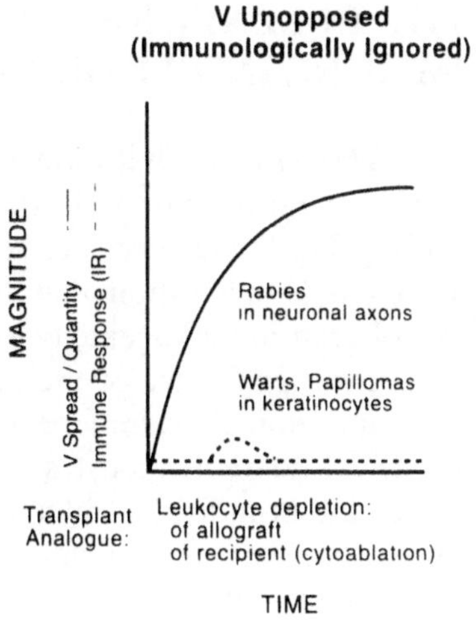

Figure 6. Immune indifference when an infectious agent does not migrate through or to an organized lymphoid collection. Transplantation analogues are shown in the subscript. The horizontal and vertical axes are as in Fig. 2. Note the minimal or absent host immune response (IR, dashed line).

infection,[7,51] migratory donor leukocytes that have not been eliminated by passage through lymphoid organs may leave the lymphoid compartment, having induced various stages of incomplete and/or reversible antigen specific exhaustion. In experimental organ transplant models, this begins after about 2 weeks, and by 100 days, the most prominent donor leukocyte population has shifted from lymphoid sites to non-lymphoid sites (e.g. skin and heart).[52] Maintenance clonal exhaustion apparently occurs subsequently by leakage of donor leukocytes from the non-lymphoid to the lymphoid compartment (Fig. 7).

The balance that develops between destructive and non-destructive immunity as the result of lymphoid/non-lymphoid leakage has been difficult to quantitate in transplantation models. However, a stable equilibrium has been demonstrated by Ohashi and Zinkernagel in transgenic mouse preparations.[53,54] In these models, pancreatic islets expressing viral antigens are not destroyed by low level CTL activity, but are rejected with resulting diabetes by the induction of high virus-specific reactivity.

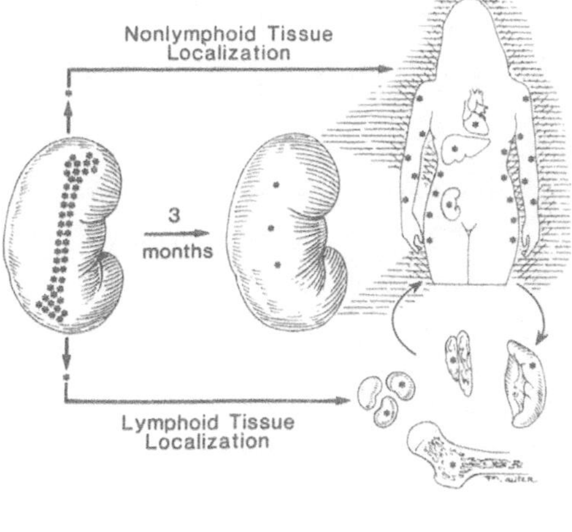

Figure 7. Reconstruction of events after organ transplantation, epitomized by the heart with its "passenger leukocyte" component depicted as a bone silhouette. Most, but not all, of the donor leukocytes leave the graft and are largely replaced by recipient cells of the same lineages. The early migration of the donor leukocytes is oriented to the host lymphoid organs, but by 100 days, the surviving donor cells have largely relocated to the non-lymphoid tissues (here shown as skin and native heart). Subsequent leakage between the lymphoid and non-lymphoid compartments is thought to maintain clonal exhaustion (see text).

THE COLLABORATION OF CLONAL EXHAUSTION AND IMMUNE INDIFFERENCE

With clonal exhaustion and immune indifference *in combination*, both regulated by the migration and localization of antigen, the 4 inter-related events shown schematically in Fig. 8 must occur close together if organ transplantation is to succeed: double acute clonal exhaustion, maintenance clonal exhaustion which waxes and wanes, and loss of graft immunogenicity as the organ is depleted of its passenger leukocytes.

The significance of the microchimerism observed at the end of this process has been questioned (summarized in Reference 55) because, as we also have emphasized,[2–4,56] donor leukocytes may be detectable during rejection, and are often not detectable in individual blood or tissue samples in patients bearing stable allografts. These observations are readily fitted into the concept that "... Donor leukocyte chimerism is a prerequisite for but *not* synonymous with and not a consequence of, organ allograft acceptance".[1] This conclusion applies equally to macro and microchimerism.

BONE MARROW VERSUS ORGAN TRANSPLANTATION

Conventional bone marrow transplantation (Fig. 5) is only a mirror image of the events after organ transplantation (Fig. 4), with the same governance of the immune events by antigen migration and localization. Although pretransplant cytoablation renders the recipient subject to GVHD, the host leukocytes are not all eliminated.[57] The weak host versus graft reaction mounted by the remaining recipient cells, and the parallel GVH reaction of the donor cells can eventually result in reciprocal tolerance.

SELF/NON-SELF DISCRIMINATION

Because the fetus possesses very early T cell immune function,[31,58,59] the ontogeny of self/non-self discrimination during fetal development can be explained by the same mechanisms as acquired tolerance in later life. Autoimmune diseases then reflect unac-

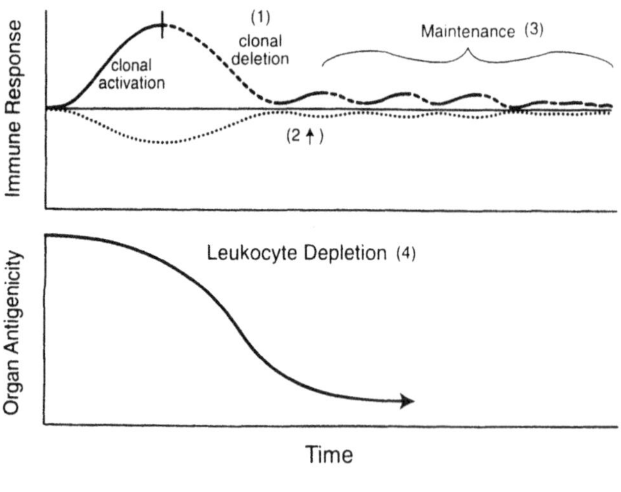

Figure 8. The four events that occur in close temporal approximation when there is successful organ engraftment: above, double acute clonal exhaustion[1,2] and subsequent maintenance clonal exhaustion[3] plus, below, loss of organ immunogenicity due to depletion of the graft's passenger leukocytes[4].

Table 1. Effectors involved in response to cytopathic parasites and discordant xenografts

The First Line of Defense
Interferons
Macrophages
γ/δ T cells
Natural killer (NK) cells
B cells
Non-Specific or Less Specific Effectors
Complement
Early interleukins
Phagocytes

ceptable post-natal perturbations of the prenatally established localization of self antigens in non-lymphoid versus lymphoid compartments.

XENOTRANSPLANTATION

There is no MHC-restricted safety valve for cytopathic microorganisms which are typically extracellular and generate the full resources of the innate as well as the adaptive immune system.[1,5] An uncontrollable innate immune response involving the effectors shown in Table 1 is provoked by discordant xenografts expressing the Gal α Gal epitope, an epitope which also is found on numerous cytopathic bacteria, protozoa, and viruses. The clinical use of such discordant animal donors will require changing the xenogeneic epitope to one that mimics a non-cytopathic profile, or else elimination of the xenogeneic epitope.[1]

ACKNOWLEDGMENT

Aided by Project Grant No. DK 29961 from the National Institutes of Health, Bethesda, Maryland.

REFERENCES

1. Starzl TE, Zinkernagel RM. Antigen localization and migrationn in immunity and tolerance. New Eng J Med 1999;339:1905–1913.
2. Starzl TE, Demetris AJ, Murase N, Ildstad S, Ricordi C, Trucco M. Cell migration, chimerism, and graft acceptance. Lancet 1992;339:1579–1582.
3. Starzl TE, Demetris AJ, Trucco M, et al. Cell migration and chimerism after whole-organ transplantation: The basis of graft acceptance. Hepatology 1993;17:1127–1152.
4. Starzl TE, Demetris AJ, Murase N, Trucco M, Thomson AW, Rao AS. The lost chord: Microchimerism. Immunol Today 1996;17:577–584;588.
5. Zinkernagel RM. Immunology taught by viruses. Science 1996;271:173–178.
6. Zinkernagel RM, Bachmann MF, Kundig TM, Oehen S, Pircher H, Hengartner H. On immunologic memory. Annu Rev Immunol 1996;14:333–367.
7. Zinkernagel RM, Ehl S, Aichele P, Oehen S, Kundig T, Hengartner H. Antigen localization regulates

immune responses in a dose- and time-dependent fashion: a geographical view of immune reactivity. Immunol Reviews 1997;156:199–209.
8. Murase N, Starzl TE, Ye Q, et al. Multilineage hematopoietic reconstitution of supralethally irradiated rats by syngeneic whole organ transplantation: with particular reference to the liver. Transplantation 1996;61:1–4.
9. Taniguchi H, Toyoshima T, Fukao K, Nakauchi H. Presence of hematopoietic stem cells in the adult liver. Nature Medicine 1996;2:198–203.
10. Grant D, Wall W, Mimeault R, Zhong R, Ghent C, Garcia B, Stiller C, Duff J. Successful small bowel/liver transplantation. Lancet 1990;335:181–184.
11. Iwaki Y, Starzl TE, Yagihashi A, Taniwake S, Abu-Elmagd K, Tzakis A, Fung J, Todo S. Replacement of donor lymphoid tissue in human small bowel transplants under FK 506 immunosuppression. Lancet 1991;337:818–819.
12. Schlitt HJ, Hundrieser J, Hisanaga M, Uthoff K, Karck M, Wahlers T, Wonigeit K, Pichlmayr R. Patterms of donor-type microchimerism after heart transplantation. Lancet 1994;343:1469–1471.
13. Kashiwagi N, Porter KA, Penn I, Brettschneider L, Starzl TE. Studies of homograft sex and of gamma globulin phenotypes after orthotopic homotransplantation of the human liver. Surg Forum 1969;20:374–376.
14. Murase N, Demetris AJ, Matsuzaki T, Yagihasi A, Todo S, Fung J, Starzl TE. Long survival in rats after multivisceral *versus* isolated small bowel allotransplantation under FK 506. Surgery 1991;110:87–98.
15. Starzl TE, Demetris AJ, Trucco M, Zeevi A, Ramos H, Terasaki P, Rudert WA, Kocova M, Ricordi C, Ildstad S, Murase N. Chimerism and donor-specific nonreactivity 27 to 29 years after kidney allotransplantation. Transplantation 1993;55:1272–1277.
16. Starzl TE, Marchioro TL, Waddell WR. The reversal of rejection in human renal homografts with subsequent development of homograft tolerance. Surg Gynecol Obstet 1963;117:385–395.
17. Starzl TE, Marchioro TL, Von Kaulla KN, Hermann G, Brittain RS, Waddell WR. Homotransplantation of the liver in humans. Surg Gynecol Obstet 1963;117:659–676.
18. Starzl TE, Marchioro TL, Porter KA, Taylor PD, FAris TD, Herrmann TJ, Hlad CJ, Waddell WR. Factors determining short- and long-term survival after orthotopic liver homotransplantation in the dog. Surgery 1965;58:131–155.
19. Calne RY, Sells RA, Pena Jr., Davis DR, Millard PR, Herbertson BM, Binns RM, Davies DAL. Induction of immunological tolerance by porcine liver allografts. Nature 1969;223:472–474.
20. Lawrence HS. Homograft sensitivity. An expression of the immunologic origins and consequences of individuality. Physiological Reviews 1959;39:811–859.
21. Medawar PB. In: The Harvey Lectures, Series 52, New York, Acad. Press, 1956–57. pp: 144
22. Doherty PC, Zinkernagel RM. Enhanced immunological surveillance in mice heterozygous at the H-Z gene complex. Nature 1975;256:50–52.
23. Zinkgernagel RM. Restriction by H-2 gene complex of transfer of cell-mediated immunity to Listeria monocytogenes. Nature 1974;251:230–233.
24. Doherty PC, Zinkernagel RM. A biological role for the major histocompatibility antigens. Lancet 1975;i:1406–1409.
25. Zinkernagel RM, Doherty PC. The discovery of MHC restriction. Immunology Today 1997;18:14–17.
26. Schwartz R, Damashek W. Drug induced immunologic tolerance. Nature (London) 1959;183:1682.
27. Simonsen M. On the acquistion of tolerance by adult cells. Ann NY Acad Sci 1960;87:382–390.
28. Starzl TE. Host Graft Adaptation. *Experience In Renal Transplantation* WB Saunders Company, Philadelphia, PA, 1964. pp: 168–170.
29. Starzl TE. Efforts to Migigate or Prevent Rejection. *In: Experience in Hepatic Transplantion*. W.B. Saunders Company, Philadelphia, PA. 1969. pp: 228–233.
30. Steinman RM, Cohn ZA. Identification of a novel cell type in peripheral lymphoid organs of mice. I. Morphology, quantitation, tissue distribution. J Exp Med 1973;137:1142–1162.
31. Sterzl J, Silverstein AM. Development aspects of immunity. Adv Immunol 1967;6:337–459.
32. Dresser DW, Mitchison NA. The mechanism of immunological paralysis. Adv Immunol 1968;8:129–181.
33. Webb S, Morris C, Sprent J. Extrathymic tolerance of mature T cells: clonal elimination as a consequence of immunity. Cell 1990;63:1249–1256.
34. Moskophidis D, Lechner F, Pircher H, Zinkernagel RM. Virus persistence in acutely infected immunocompetent mice by exhaustion of antiviral cytotoxic effector T cells. Nature 1993;362:758–761.
35. Critchfield J, Racke M, Zuniga-Pflucker J, et al. T cell deletion in high antigen dose therapy of autoimmune encephalomyelitis. Science 1994;263:1139–1143.

36. Aichele P, Brduscha-Riem K, Zinkernagel RM, Hengartner H, Pircher H. T cell priming versus T cell tolerance induced by synthetic peptides. J Exp Med 1995;182:261–266.
37. Qian S, Lu L, Fu F, et al. Apoptosis within spontaneously accepted mouse liver allografts: evidence for deletion of cytotoxic T cells and implications for tolerance induction. J Immunol 1997;158:4654–4661.
38. Bishop AG, Sun J, Sheil AGR, McCaughan GW. High-dose/activation-associated tolerance. Transplantation 1997;64:1377–1382.
39. Effros RB, Pawelec G. Replicative senescence of T cells: does the Hayflick limit lead to immune exhaustion? Immunology Today 1997;18:450–454.
40. Steinmuller D. Immunization with skin isografts taken from tolerant mice. Science 1967;158:127–129.
41. Elkins WL, Guttmann RD. Pathogenesis of a local graft versus host reaction: Immunogenicity of circulating host leukocytes. Science 1968;159:1250,1251.
42. Hart DNJ, Winearls CG, Fabre JW. Graft adaptation: studies on possible mechanisms in long-term surviving rat renal allografts. Transplantation 1980;30:73–80.
43. Lechler RI, Batchelor JR. Restoration of immunogenicity to passenger cell-depleted kidney allografts by the addition of donor-strain dendritic cells. J Exp Med 1982;155:31.
44. Billingham RE, Brent L, Medawar PB. "Actively acquired tolerance" of foreign cells. Nature 1953;172:603–606.
45. Kundig TM, Bachmann MF, Oehen S, et al. On the role of antigen in maintaining cytotoxic T cell memory. Proc Natl Acad Sci, USA 1996;93:9716–9723.
46. Nemlander A, Soots A, Willebrand EV, Husberg B, Hayry P. Redistribution of renal allograft responding leukocytes during rejection. II. Kinetics and specificity. J Exp Med 1982;156:1087–1100.
47. Larsen CP, Morris PJ, Austyn JM. Migration of dendritic leukocytes from cardiac allografts into host spleens: a novel route for initiation of rejection. J Exp Med 1990;171:307–314.
48. Demetris AJ, Qian S, Sun H, et al. Early events in liver allograft rejection: Delineation of sites simultaneous intragraft and recipient lymphoid tissue sensitization. Am J Pathol 1991;138:609–618.
49. Talmage DW, Dart G, Radovich J, Lafferty KJ. Activation of transplant immunity: effect of donor leukocytes on thyroid allograft rejection. Science 1976;191:385–387.
50. Lafferty KJ, Prowse SJ, Simeonovic CJ. Immunobiology of tissue transplantation: a return to the passenger leukocyte concept. Ann Rev Immunol 1983;1:143–173.
51. Zinkgernagel RM, Hengartner H. Antiviral immunity. Immunology Today 1997;18:258–260.
52. Terakura M, Murase N, Demetris AJ, Ye Q, Thomson A, Starzl TE. The lymphoid/non-lymphoid compartmentalization of donor leukocyte chimerism in rat recipients of heart allografts, with or without adjunct bone marrow: correlations with chronic rejection. Transplantation, in press.
53. Ohashi PS, Oehen S, Buerki K, et al. Ablation of "tolerance" and induction of diabetes by viruses infection in viral antigen transgenic mice. Cell 1991;65:305–317.
54. Ohashi PS, Oehen S, Aichele P, et al. Induction of diabetes is influenced by the infectious virus and local expression of MHC class I and tumor necrosis factor-α. J Immunol 1993;150:5185–5194.
55. Wood K, Sachs DH. Chimerism and transpantation tolerance: cause and effect. Immunol Today 1996;17:584–587;588.
56. Starzl TE, Murase N, Demetris AJ, Qian S. Microchimerism: inaccurate literature citations. Transplantation 1996;62:703.
57. Przepiorka D, Thomas ED, Durham DM, Fisher L. Use of a probe to repeat sequence of the Y chromosome for detection of host cells in peripheral blood of bone marrow transplant recipients. Am J Clin Pathol 1991;95:201–206.
58. Matzinger P. Tolerance, danger, and the extended family. Ann Rev Immunol 1994;12:991–1045.
59. Ridge JP, Fuchs EJ, Matzinger P. Neonatal tolerance revisited: turning on newborn T cells with dendritic cells. Science 1996;271:1723–1726.

4

GENETIC APPROACHES TO THE INVESTIGATION OF THE MECHANISMS OF HEMATOPOIETIC PROGENITOR CELL MOBILIZATION INTO THE BLOOD

Andrew W. Roberts, Simon Foote[#], and Donald Metcalf

Division of Cancer and Haematology, and [#]Genetics & Bioinformatics Group
The Walter and Eliza Hall Institute of Medical Research
Post Office, Royal Melbourne Hospital
3050 Victoria, Australia

1. ABSTRACT

Despite an enormous body of knowledge regarding hematopoietic cell growth and differentiation and the regulation of these processes by hematopoietic growth factors, the phenomenon of progenitor cell mobilization remains an enigma. With the expectation that clinical development of blood stem cell transplantation will be greatly benefited by advances in knowledge about the specific processes which underpin mobilization, we have sought to define the limiting steps which regulate blood progenitor cell numbers following *in vivo* stimulation with G-CSF. This paper reviews the rationale behind the genetic approach we have chosen to take and summarises our recent promising, but preliminary, results.

2. RATIONALE

Several key problems face investigators addressing the questions of how and why progenitor cells circulate. Firstly, mobilization is an *in vivo* phenomenum which has proved difficult to model *in vitro*. Indeed what is observed as "mobilization" is almost

---

Correspondence to: Dr Andrew Roberts, The Walter and Eliza Hall Institute of Medical Research, Post Office, Royal Melbourne Hospital, 3050 Victoria, Australia. Telephone: 61-3-9345-2555, Fax: 61-3-9347-0852, Email: roberts@wehi.edu.au

Molecular Biology of Hematopoiesis 6, edited by Abraham *et al.*
Kluwer Academic/Plenum Publishers, New York, 1999.

certainly the nett result of several interrelated phenomena: egress from the bone marrow, circulation in the peripheral blood, and clearance into various tissues. Investigation of each of these *in vivo* processes is challenging in isolation, and more so when dynamic balances between them need to be considered. Secondly, mobilization is induced by agents with diverse properties. This suggests that there are multiple effector pathways leading to an increased number of progenitor cells in the blood, and that the search for a final common pathway may be futile. Related to this is the third major hurdle for investigators: mobilization does not necessarily occur immediately following administration of an agent, and consequently both direct and indirect mechanisms of action of inducing agents on target cells need consideration.

Much information regarding peripheral blood progenitor cell (PBPC) mobilization has been derived from the careful investigation of "candidate" processes such as the interaction of progenitor cells with the extracellular matrix and the actions of growth factors and chemokines on bone marrow progenitor cells.[1–7] This traditional approach is anchored by current dogma regarding the regulation of hematopoiesis, and progress occurs by the gradual extension of present knowledge. Recent developments in mammalian genetics provide an alternative and hopefully complementary strategy for identifying the mechanisms of progenitor cell mobilization. Just as the key step in defining the biological basis of genetic disease is now identification of the gene itself by genetic mapping rather than by initial identification of the gene product,[8,9] identification of genes influencing progenitor cell mobilization is theoretically possible without assumptions being made as to the nature or function of the gene products. To test the practicality of this approach, we have focussed attention initially on progenitor cell mobilization induced by G-CSF as a single agent.

3. PROGENITOR CELL MOBILIZATION INDUCED BY G-CSF IS A COMPLEX GENETIC TRAIT

There is now a large experience with the induction of progenitor cell mobilization by G-CSF in normal adult volunteers. A consistent finding by multiple groups has been broad inter-individual variation in the concentration of progenitor cells in the blood following injection of standardised regimens of G-CSF.[10–13] In the absence of obvious environmental factors, the most likely explanation of the diversity observed (over a 10–15-fold range) is genetic heterogeneity in determinants of *in vivo* responses to G-CSF. Previous studies have largely excluded other alternative explanations such as assay imprecision,[14] altered *in vitro* progenitor cell responsiveness to G-CSF[6] and variable pharmacokinetics.[10]

Formal proof of a genetic basis in humans for high, intermediate, and low mobilization responses to G-CSF requires careful family and twin studies. As yet, such studies have not been reported. However, direct evidence of a genetic basis for the determination of progenitor cell mobilization has been obtained by the study of mice. Inbred strains of mice are characterised by (i) genetic identity among mice of a given strain and (ii) homozygosity of alleles at each locus throughout the genome. Consequently, they provide extremely useful tools for the study of genetic traits. Among six inbred strains injected for 5 days with a weight-standardised course of G-CSF, we observed broad variability in three measures of *in vivo* response: PBPC numbers, leukocytosis, and splenomegaly.[6] A 10-fold range in circulating progenitor cell numbers was observed, remarkably similar to that observed for normal human subjects receiving G-CSF.

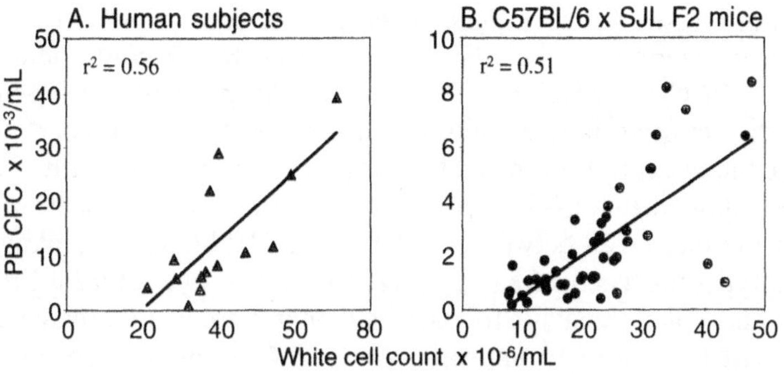

Figure 1. Circulating progenitor cell numbers and leukocyte counts correlate in both humans and mice injected with G-CSF for five days. In panel A, individual data are shown for 14 normal volunteers receiving filgrastim 10 ug/kg/d for five days.[10] In panel B, individual data are shown for 48 F2 mice injected with G-CSF 5 ug/d for five days.[6]

In order to determine the genetic basis of the strain differences, intercrosses of low-responding C57BL/6 and SJL (an intermediate-responding strain) mice and C57BL/6 and DBA/2 (the highest-responding strain identified) mice were performed. Analysis of the patterns of distribution of responses among F1 and F2 progeny provides information as to whether the traits under study are Mendelian (ie single gene) or complex traits with multiple loci involved. Phenotypic data from F1 mice of both intercrosses indicated that the low mobilization response was dominant.[6,15] The distributions of PBPC and leukocytosis phenotypes among F2 mice of both intercrosses was heavily weighted towards that of low-response C57BL/6 mice, but did extend over a wide range. Again this was similar to that observed in normal human subjects. These patterns were indicative of complex genetic traits. Although the exact number of genes involved in regulation of mobilization could not be determined, the data were suggestive of the involvement of three or more genetic loci. It is probable that manifestation of the full high-response mobilization phenotype requires co-inheritance of high-response alleles at each of these loci.

In both normal humans[10] and F2 mice from either intercross,[6,15] we observed moderately strong correlations between white cell count (WCC) and PBPC (Fig. 1). As well as providing further support for the contention that strain differences in mice are valid models for inter-individual variations in PBPC mobilization in humans, these data suggest that at least one genetic locus co-modulates both progenitor cell and mature leukocyte mobilization.

4. GENETIC DISSECTION OF *IN VIVO* RESPONSES TO G-CSF: A USEFUL APPROACH TO THE INVESTIGATION OF THE PHENOMENON OF PROGENITOR CELL MOBILIZATION?

Genetic mapping followed by positional cloning is now routine for diseases with simple Mendelian (ie single gene) inheritance and >60 human genes have been cloned in this fashion.[9] A major challenge over the last few years has been to adapt this successful approach to traits where genetic inheritance is complex and where non-genetic factors may be influential eg most human diseases such as hypertension, diabetes, schizophrenia, and some cancers. One approach is quantitative trait locus (QTL) mapping of experimental crosses between genetically distinct organisms.[8,9] The term quantitative trait refers to complex traits which can be measured as continuous variables (eg blood pressure, WCC, body weight). The power of this particular mapping approach is its ability to dissect quantitative traits which represent the combined action of individual genetic factors, into their discrete genetic factors or quantitative trait loci.[8]

We have applied this QTL approach to the analysis of data from 192 F2 mice derived from the intercross between C57BL/6 and DBA/2 mice. Linkage was sought between genotype and each of three *in vivo* response phenotypes: PBPC numbers, leukocytosis, and splenomegaly. To date the only significant linkage discovered is between a 22 cM locus on Chromosome 1 and G-CSF-induced splenomegaly.[15] Surprisingly, the high-response allele was not associated with DBA genotype as expected, but with C57BL/6 genotype. Nevertheless, the relevance of this locus was able to be confirmed by phenotypic study of mice congenic for this region. Intriguingly, the locus identified for spleen size did appear to modulate the other two *in vivo* responses as well, with significant differences observed between congenic mice in the number of circulating progenitor cells and leukocytes.[15] The apparent biological effect of this locus is minor however, and does not explain the 10-fold differences in progenitor cell mobilization observed between parental DBA/2 and C57BL/6 mice.

The identification of the locus on Chromosome 1 provides proof of principle, but also illustrates the problems associated with the genetic mapping approach. Large numbers of mice are required for the analyses and determination of phenotypes for each genetically unique F2 mouse is restricted to a single measurement. Study power is the critical determinant of success for the utility of QTL analysis.[9] In the current setting, data from 192 F2 mice has provided insufficient power to resolve the multiple genetic loci involved. Consequently we are extending the study to include data from 600 mice.

Recombinant inbred (RI) strains of mice potentially provide an alternative source of data for this study, and have been used with some recent success in experimental hematology.[16,17] These strains are fixed hybrids of two parental strains and for the genetic mix of C57BL/6 and DBA/2 (BXD), 26 RI strains are available. Although the number of strains is restricted (which reduces power), high-density mapping has been previously established for each strain and the *in vivo* phenotype of interest can be assayed in multiple mice of the same genotype. These features were used to advantage by two groups to identify genetic loci associated with the frequency of hematopoietic stem cells in the bone marrow. Like G-CSF-induced PBPC mobilization, the frequency of putative stem cells in bone marrow is a complex trait with DBA/2 mice maintaining a frequency of 3–10 times that of C57BL/6 mice.[16–18] Müller-Sieburg and Riblet[16] identified two loci on Chromosome 1 (which are distinct from the locus we identified for mobilization) while deHaan and Van Zant[17] identified a single locus on Chromosome 18 as being statistically associated with stem cell frequency. Although both groups used the same BXD RI mice, similar *in vitro* surrogate assays for stem cell frequency and the same analytical approach, different loci were identified. Further experience with this technology is required to interpret the significance of obtaining disparate results, but it seems likely that each locus contributes to the complex trait of stem cell frequency.

Could strain differences in stem cell frequency in the bone marrow simply explain the differences we have observed in progenitor cell mobilization? Based on previous data from humans and mice, we currently believe that stem cell frequency is likely to influence the mobilization phenotype, but that other regulatory processes are as, or

Table 1. Comparison of the patterns of interstrain differences reported for progenitor cell mobilization and stem cell phenotypes

	G-CSF *in vivo* Response	Stem Cell Phenotypes			
		Müller-Sieburg	Van Zant & Colleagues		
Strain	PBPC Mobilisation	LTC-IC Frequency	Day 35 CAFC Frequency	Response to Flt 3 Ligand	Proportion Progenitor Cells in S Phase
DBA/2	+++	+++	+++	++	+++
C57BL/6	+	+	+/++	–	+
B6D2F1	+	++			+++*
BALB/c	++	++	+	–	++
C3H/He	+		+	–	+++
129Sv	+++		+	–	+

Data is summarized from references 6,15,16–19*.

more, important. Table 1 compares strain characteristics for PBPC mobilization phenotype with those for stem cell phenotypes (frequency, responsiveness to an early acting growth factor, and proportion of progenitor cells in S phase). As is readily apparent, there is not a tight correlation between mobilization response and any stem cell phenotype.

5. FUTURE PROSPECTS

It is premature to judge whether genetic dissection will succeed in defining the mechanisms of progenitor cell mobilization induced by G-CSF and other agents. Currently, we and others are still learning how to use this potentially powerful approach in sufficiently sophisticated manners to address complicated *in vivo* phenomena. Validation of our preliminary results of a QTL on Chromosome 1 will require independent supporting evidence and ultimately the positional cloning and characterization of the implicated gene product.

ACKNOWLEDGMENT

This work was supported by the National Health and Medical Research Council of Australia, the Carden Fellowship Fund of the Anti-Cancer Council of Victoria, the Wellcome Trust, and Grant No. CA22556 from the National Institutes of Health, Bethesda.

REFERENCES

1. Aiuti A, Webb IJ, Bleul C, Springer T, Gutierrez-Ramos JC: The chemokine SDF-1 is a chemoattractant for human CD34+ hematopoietic progenitor cells and provides a new mechanism to explain the mobilization of CD34+ progenitors to peripheral blood. J Exp Med 185:111, 1997.
2. Laterveer L, Lindley IJD, Hamilton MS, Willemze R, Fibbe WE: Interleukin-8 induces rapid mobilization of hematopoietic stem cells with radioprotective capacity and long-term myelolymphoid repopulating ability. Blood 85:2269, 1995.

3. Liu F, Poursine-Laurent J, Link DC: The granulocyte colony-stimulating factor receptor is required for the mobilization of murine hematopoietic progenitors into peripheral blood by cyclophosphamide or interleukin-8 but not flt-3 ligand. Blood 90:2522, 1997.
4. Papayannopoulou T and Nakamoto B: Peripheralization of hemopoietic progenitors in primates treated with anti-VLA4 integrin. Proc Natl Acad Sci USA 90:9374, 1993.
5. Papayannopoulou T, Priestley G, Nakamoto B: Anti-VLA4/VCAM-1-induced mobilization requires cooperative signaling through kit/mkit ligand pathway. Blood 91:2231, 1998.
6. Roberts AW, Foote S, Alexander WS, Scott C, Robb L, Metcalf D: Genetic influences determining progenitor cell mobilization and leukocytosis induced by granulocyte colony-stimulating factor. Blood 89:2736, 1997.
7. Simmons PJ, Leavesley DI, Levesque J-P, Swart BW, Haylock DN, To LB, Ashman LK, and Juttner CA: The mobilization of primitive hemopoietic progenitors into the peripheral blood. Stem Cells 12 (Suppl 1):187, 1994.
8. Lander ES and Schork NJ: Genetic dissection of complex traits. Science 265:2037, 1994.
9. Lander ES and Kruglyak L: Genetic dissection of complex traits: guidelines for interpreting and reporting linkage results. Nature Genetics 11:241, 1995.
10. Grigg AP, Roberts AW, Raunow H, Houghton S, Layton JE, Boyd AW, McGrath KM, Maher D: Optimising dose and scheduling of filgrastim (granulocyte colony-stimulating factor) for mobilization and collection of peripheral blood progenitor cells in normal volunteers. Blood 86:4437, 1995.
11. Dreger P, Haferlach T, Eckstein V, Jacobs S, Suttorp M, Löffler H, Müller-Ruchholtz W, Schmitz N: G-CSF-mobilized peripheral blood progenitor cells for allogeneic transplantation: safety, kinetics of mobilization, and composition of the graft. Brit J Haematol 87:609, 1994.
12. Lane TA, Law P, Maruyama M, Young D, Burgess J, Mullen M, Mealiffe M, Terstappen LWMM, Hardwick A, Moubayed M, Oldham F, Corringham RET, Ho AD: Harvesting and enrichment of hematopoietic progenitor cells mobilized into the peripheral blood of normal donors by granulocyte-macrophage colony-stimulating factor (GM-CSF) or G-CSF: potential role in allogeneic marrow transplantation. Blood 85:275, 1995.
13. Tjønnfjord GE, Steen R, Evensen SA, Thorsby E, Egeland T: Characterization of CD34+ peripheral blood cells from health adults mobilized by recombinant human granulocyte colony-stimulating factor. Blood 84:2795, 1994.
14. Roberts AW, DeLuca E, Begley CG, Basser R, Grigg AP, Metcalf D: Broad inter-individual variations in circulating progenitor cell numbers induced by granulocyte colony-stimulating factor therapy. Stem Cells 13:512, 1995.
15. Roberts AW, Hasegawa M, Metcalf D, Foote S: Genetic influences determining in vivo responses to granulocyte colony-stimulating factor. (submitted).
16. Muller-Sieburg CE and Riblet R: Genetic control of the frequency of hematopoietic stem cells in mice: mapping of a candidate locus to chromosome 1. J Exp Med 183:1141, 1996.
17. de Haan G and Van Zant G: Intrinsic and extrinsic control of hemopoietic stem cell numbers: mapping of a stem cell gene. J Exp Med 186:529, 1997.
18. Phillips RL, Reinhart AJ, Van Zant G: Genetic control of murine hematopoietic stem cell pool sizes and cycling kinetics. Proc Natl Acad Sci USA 89:11607, 1992.
19. Van Zant G, Eldridge PW, Behringer RR, Dewey MJ: Genetic control of hematopoietic kinetics revealed by analyses of allophenic mice and stem cell suicide. Cell 35:639, 1983.

5

ROLE OF HEMATOPOIETIC GROWTH FACTORS ON THE *EX VIVO* EXPANSION OF PRIMITIVE CORD BLOOD STEM CELLS

M. Aglietta, L. Garetto, F. Sanavio, A. Severino, A. Dané, L. Gammaitoni, G. Cavalloni, E. Perissinotto, J. Ferrario, and W. Piacibello

Department of Biomedical Sciences and Human Oncology
University of Torino Medical School
Hematology/Oncology, Mauriziano Hospital, and IRCC
(Institute for Cancer Research and Treatment)
Candiolo, Torino, Italy

INTRODUCTION

Ex vivo expansion of hematopoietic cells has recently been suggested for clinical application.[1,2] It has been demonstrated that hematopoietic compartments derived from human bone marrow (BM), mobilized peripheral blood (PB) or umbilical cord blood (CB) cells can be maintained and expanded in liquid culture or stroma coculture systems by the provision of combinations of growth factors.[3–5] The potential benefits of such studies include accelerated engraftment, reduced risk of infection, smaller stem cell harvests, and improved effectiveness of genetically modified stem cells.[1,2] Although controversies remain concerning what defined populations may or may not be useful for improving in vivo hematologic recovery, ex vivo expansion can only be considered successful when progeny receptor cells retain both pluripotent differentiation and self-renewal capacities of the original stem cells.[4] In other words, are the various growth factor combinations able to amplify the late progenitor reservoir without exhausting the stem cell pool, in order to ensure long-term post-transplantation engraftment after a myeloablative conditioning regimen? Human culture systems employing Interleukin 3 (IL3) or Granulocyte-Colony Stimulating Factor (G-CSF) maintain at least some primitive cells;[6–9] however, achieving a net expansion of early cells has proven elusive, although some laboratories have reported encouraging results.[10–12] With the discovery of new growth factors, specific for early progenitor cells, investigators have re-examined the possibility of expanding stem cells in vitro. In

Molecular Biology of Hematopoiesis 6, edited by Abraham *et al.*
Kluwer Academic/Plenum Publishers, New York, 1999.

particular, three cytokines discovered in the early 1990's, c-kit ligand or Stem Cell Factor (KL), flt3 ligand (FL), and c-mpl ligand also termed Megakaryocyte Growth and Development Factor (MGDF) or Thrombopoietin (TPO), appear to have unique activities on primitive progenitor/stem cells.

Both FLT3 and c-KIT tyrosine kinase receptors belong to the type III tyrosine kinase receptor family that includes Macrophage Colony-Stimulating Factor (M-CSF) and Platelet Derived Growth Factor (PDGF) receptors. Based on studies of the expression and function on the two receptors, it is now evident that the hematologic actions of these two cytokines are predominantly restricted to the progenitor/stem cell compartment.[13]

The ligand for c-kit has an important role in the regulation of myeloid, erythroid, and lymphoid cell development and demonstrates striking synergy with other growth factors.[13–15] c-kit receptor is expressed on blast cells obtained from a high percentage of patients with Acute Myeloid Leukemia (AML) and primary AML samples proliferate in response to KL.[16,17] Although recent data suggest that earliest stem cells might express no or very low levels of c-kit, in the long-term reconstituting stem cell compartment, c-kit appear to be expressed on more and earlier stem cells than FLT3.[18,19] Finally, c-kit-deficient stem cells are defective in their ability to reconstitute hematopoiesis in myeloablated animals.[14]

FLT3 transcript has been detected in murine and human cell populations enriched for hematopoietic stem cells and progenitors and is absent in more mature cells.[20,21] FLT3 receptor is frequently expressed on the surface of acute myelogenous leukemia blasts and FL, as a single factor or in combination with other growth factors, induces significant proliferation and colony formation by clonogenic progenitors from AML samples in primary leukemias.[22] In addition, targeted disruption of the FLT3 gene led to deficiencies in primitive hematopoietic progenitors.[23] Moreover, FL stimulates the proliferation and colony formation of a subpopulation of CD34-expressing BM, CB and fetal liver cells enriched for hematopoietic progenitors ($CD34^{+}$ $CD38^{dim}$ and $CD34^{+}$ $CD38^{+}$ $CD33^{dim}$ $HLA\text{-}DR^{+}$).[24] Single-cell and small scale experiments have suggested that FL can support renewal of long-term culture-initiating cells (LTC-IC) and extended long-term culture initiating cells (ELTC-IC) from BM.[25–27]

Thrombopoietin, a recently isolated ligand of the mpl receptor, was first identified as the cellular homologue of v-mpl, the viral oncogene transduced by the murine myeloproliferative leukemia (MPL) virus.[28] Most studies initially suggested that TPO was a lineage-restricted regulator with little or no activity on nonmegakaryocytic lineages.[29] Nevertheless, in populations of mouse fetal liver enriched for primitive hematopoietic cells, a significant proportion of cells expresses c-mpl.[30] Moreover, TPO augments growth of erythroid progenitors[31] and cells of human nonmegakaryocytic leukemias can proliferate in response of this factor.[32] An elevation of CFU-GEMM has also been observed upon TPO administration in monkeys.[33] Finally, TPO as a single factor, stimulates increases of LTC-IC within 10 days[26] and c-mpl-deficient mice displayed reduced total hematopoietic progenitor cell numbers, including multipotential blast and committed progenitors of multiple lineages.[34]

Several attempts for in vitro expansion of hematopoietic progenitors have been reported in the last few years.[1,2] Relevant increases of $CD34^{+}$ cells and of committed progenitors have been obtained (Table 1). However, in most studies, cells are permanently lost from the stem cell compartment during ex vivo culture, resulting in reduction of long-term engraftment potential. In particular, several data demonstrate that the use of growth factors such as Interleukin 1 (IL1), G-CSF, and Erythropoietin

(EPO), abrogates the reconstituting ability of hematopoietic stem cells.[1,2,35] Although long-term engraftment with ex vivo expanded hematopoietic cells has been demonstrated in mice,[36] durable engraftment in man after cell number expansion or the *in vitro* manipulations remains to be adequately tested. A dynamic model of human in vivo hematopoiesis is now represented by the use of sublethally irradiated Non Obese Diabetic Severe Combined Immunodeficient mice (NOD/SCID mice).[37–44] Recently reported data show that the intravenous injection of 5- to 8-day ex vivo expanded human hematopoietic precursors in NOD/SCID mice has resulted in the engraftment of primitive human cells that proliferate and differentiate to multiple lineages in the murine bone marrow (BM) and spleen. The engrafted human cells [defined as Competitive Repopulating Units (CRU) which represent very primitive hematopoietic cells], were not only maintained, but also expanded[45] (Table 1).

CORD BLOOD

The existence of ontogeny-related differences between primitive human hematopoietic cells has been demonstrated.[46] Stem cells isolated from fetal liver and cord blood have longer telomeres than cells purified from adult BM, which suggests that stem cells isolated from these tissues have greater proliferative potential than the equivalent cells isolated from adult tissues and that the proliferative potential of most hematopoietic stem cells is limited and decreases with age.

It is well known that umbilical cord blood contains a significantly higher number of early (CRU, LTC-IC, CFU-GEMM), and committed progenitor (CFU-GM, BFU-E, CFU-Mk) cells, a higher cloning efficiency and a more rapid response to cytokine stimulation when compared with adult peripheral blood.[47] Moreover, umbilical cord blood progenitor cells also appear to be more sensitive to ex vivo expansion with both lineage-specific and lineage-nonspecific hematopoietic growth factors.[48–50] In long-term hematopoietic suspension cultures, umbilical progenitor cell production and culture life span are significantly increased compared with progenitor cells from normal adult bone marrow (BM).[51] Fourteen-day expansion of $CD34^+$ cells isolated from umbilical cord blood stimulated with Interleukin 11 (IL11)+G-CSF was significantly greater than that from stimulated $CD34^+$ cells isolated from human adult BM (80-fold)[52] Single nonadherent low density T-Lymphocyte-depleted $CD34^+$ cells expanded ex vivo with SCF, GM-CSF, G-CSF, IL2, and EPO resulted in an eightfold increase in colony formation when umbilical cord blood $CD34^+$ were compared with adult BM.[53] Additionally, single $CD34^+$ cells isolated from umbilical cord blood are capable of at least five serial in vitro replating, demonstrating a significant level of self-renewal capacity for single-sorted $CD34^+$ umbilical cord blood stem cells.[53]

EX VIVO EXPANSION OF CORD BLOOD CELLS

We have evaluated the capacity of human umbilical cord blood cells to expand during long term cultures in well-defined culture conditions.[54–56] We used stroma-free cultures. Even though the presence of bone marrow stroma layers has been shown to increase the frequency of gene transfer into primitive cells, compared with using viral supernatants, and to prevent the loss of stem cell quality during ex vivo expansion of peripheral stem cells, other reports suggest that the long term repopulating ability

Table 1. *Ex vivo* expansion of candidate stem cell population

Source	input	Stimulus	Assay	Time (days)	Expansion (fold increase)	References
BM	$CD34^{+}Thy\text{-}1^{+}Lin^{-}$ Stroma-free	IL3 + KL	$CD34^{+}Lin^{-}$	21	4.5	Tong Exp Hematol 22:753, 1994
BM	$CD34^{+}DR^{\pm}$ Rh^{dull} Stroma-free	KL + PIXY-32	$CD34^{+}DR^{\pm}$ HPP-CFU LTC-IC	21–28 21–28 4–10	5.6 2–2.5 1	Srour Blood 81:661, 1993
BM	$CD34^{+}$ Stroma-free	IL1 + IL3 + IL6 + KL	CFU-GM	7–14	50	Moore Sem in Hematol 32:153, 1995
BM	$CD34^{+}38^{-}$ Stroma-free	FL + KL + IL3 + IL6 + NGF + MIP1α + TPO + IL11 + TNFα + IL12	LTC-IC Cell count CFCs	10 10 10	30 350	Petzer Jexp Med 183:2551, 1996
BM	$CD34^{+}Thy1^{+}Lin^{-}$ Stroma	TPO + IL3 + KL Sys1	$CD34^{+}$ $CFU\text{-}M_{K}$	7–42	30–100 8	Young Blood 88:1619, 1996
BM	$CD34^{+}38^{-}$ stroma-free	FL + IL3 + IL6 + G-CSF + NGF	LTC-IC	10–21	30–50	Petzer PNAS 93:1470, 1996
BM	$CD34^{+}Thy1^{+}Lin^{-}$ Strome-free	TPO + KL + FL + IL3 + IL6 + LIF	CAFC in vivo (SRC)	6	3.2 maintenance	Luens Blood 91:1206, 1998
MPB	$CD34^{+}$ Stroma-free	IL3 + IL6 + GM-CSF + G-CSF + KL	$CD34^{+}$ $CD33^{-}$ HPP-CFU	7	2.9	Sato Blood 82: 3600, 1993
MPB	$CD34^{+}$ Stroma-free	IL1 + IL3 + IL6 + KL + Epo	LTC-IC	12	1.2	Henschler Blood 84:2898, 1994

MPB	$CD34^+DR^\pm$ Stroma	IL3 + MIP1α M2-10B4	CFC LTC-IC	14–35	Maintenance	Prosper Blood 89:3991, 1997
MPB	$CD34^+$ Dexter LTC	FL + KL + M-CSF	Cell count LTC-IC HPP-CFC	7–49	28 2.3	Gabbianelli Blood 86:1661, 1995
MPB	$CD34^+$ Stroma/stroma-free	KL + IL3 + IL6 + TGFβ + GCSF + GM-CSF + IL1 + IL11 + cell lines	CAFC CFC	7 7	10	Breems Leukemia 11:142, 1997
MPB	$CD34^+$ Stroma	IL3 + KL + IL6 + FL + TPO + TGFβ stromal lines	CAFC CFC	7	21 15	Breems Blood 91:111, 1998
CB	$CD34^+45RA^{lo}71^{lo}$ Stroma-free	IL1 + IL3 + IL6 + KL	$CD34^+$	35	31–250	Miyani Jexp Med 78:787, 1993
CB	$CD34^+$ Stroma-free	IL1 + IL3 + IL6 + KL	LTC-IC	7–14	15–20	Moore Blood Cells 20:468, 1994
CB	$CD34^+$ Stroma-free	IL1 + IL3 + IL6 + IL11 + G-CSF + KL + MIP1α	LTC-IC	7–14	2–5	Han Exp Hematol 22:273, 1994
CB	$CD34^+38^-$ Stroma	IL3 + IL6 + KL	CFC ELTC-IC	>60	Maintenance	Hao Blood 88:3306, 1996
CB	$CD34^+$ Stroma-free	KL + FL ++ IL3 + IL6 + IL11 + TPO	CFU-Mix	14	50	Ohmizono Leukemia 11:524, 1997
CB	$CD34^+38^-$ Stroma-free	FL + KL + IL3 + IL6 + G-GSF	CFC LTC-IC in vivo (CRU)	8	100 4 2	Conneally PNAS 94:9836, 1997

of cultured cells decreases using bone marrow stroma cocultures.[57–59] Moreover, it has recently been reported that cocultures of bone marrow and cord blood cells on allogeneic stroma layers for up to three weeks resulted in a decrease and final loss of the in vivo repopulating ability in NOD/SCID mice recipients. In contrast to the loss of SCID Repopulating Cells (SRC), the cultured cells frequently contained an equal or higher number of LTC-IC compared with the fresh cells.[60] The mechanisms that underlie the loss of SRC during culture on stroma is not known. It could be due to differentiation of primitive cells during cultures or changes in the expression and/or function of adhesion receptors that could affect the engraftment potential of the primitive cells.[61–63] Stroma cells produce both early and late acting growth factors as well as negative regulators of hematopoietic cell proliferation (such as TGFß), that could be the responsibles for the decreased frequency of SRC.

In an effort to induce the long term proliferation, and possibly, the self-renewal processes of primitive hematopoietic progenitors, purified $CD34^+$ CB cells were grown in 24-well plates or in tissue culture T_{25} or T_{75} flasks in quadruplicate, in the presence of medium and early-acting growth factors (FL 50 ng/ml, KL 50 ng/ml, TPO 10 U/ml), single or in various combinations that were replaced twice a week. In order to establish the usefulness of ex vivo culture for clinical settings, pooled normal human serum subsequently substituted for Fetal Calf Serum (FCS). This substitution was not detrimental to the overall results of the ex vivo cultures. Other studies are ongoing with the use of serum-free medium.[64]

In an attempt to carefully investigate the proliferative events that take place during the long term culture, different stages of hematopoiesis were monitored:

a) cell number
b) $CD34^+$ and $CD34^+CD38^{\pm}$ subpopulations
c) CFC output (i.e. CFU-GM+BFU-E+CFU-GEMM+CFU-Mk number)
d) LTC-IC output (i.e. number of CFU-GM+BFU-E+CFU-GEMM generated in 14 to 21 day methylcellulose cultures generated by tripsinized + nonadherent cells obtained after 5 weeks of stroma cocultures)
e) SRC = SCID Repopulating Cells (i.e. the ability of human cells to completely and durably engraft the marrow, spleen, and blood of sublethally irradiated NOD/SCID mice).

It was soon evident that cultures set up without exogenously added growth factors lasted only for some days and after the third week, no cells remained live. When single factors were added, the cultures persisted for up to ten weeks (FL alone). Quite different was the behavior of the liquid cultures to which growth factors were added in combinations of two or more. In particular, analysis of the total cell count and the $CD34^+$ number (Figs. 1 and 2) performed weekly during liquid cultures showed that cultures set up in the presence of a combination of FL+TPO±KL underwent a progressive, steady increase of cell production, so that after 28–30 weeks of cultures the cell output ranged from 134,000,000 and 1,200,000,000-fold the input number and $CD34^+$ output ranged from 200,000 and 2,080,000-fold the input number (with FL+TPO or FL+KL_TPO respectively). The amplification of the stem cell pool was indirectly suggested by the analysis of CFC output (Fig. 2). In the presence of FL+TPO the CFC expansion was >28,000,000-fold and in the presence of KL+FL+TPO it reached 109,000,000-fold the initial value. Committed hemopoietic progenitors are not endowed with a self-renewal potential, therefore the progressive CFC amplification had to be substained by a progressive amplification of subpopulations(s) of stem cells which

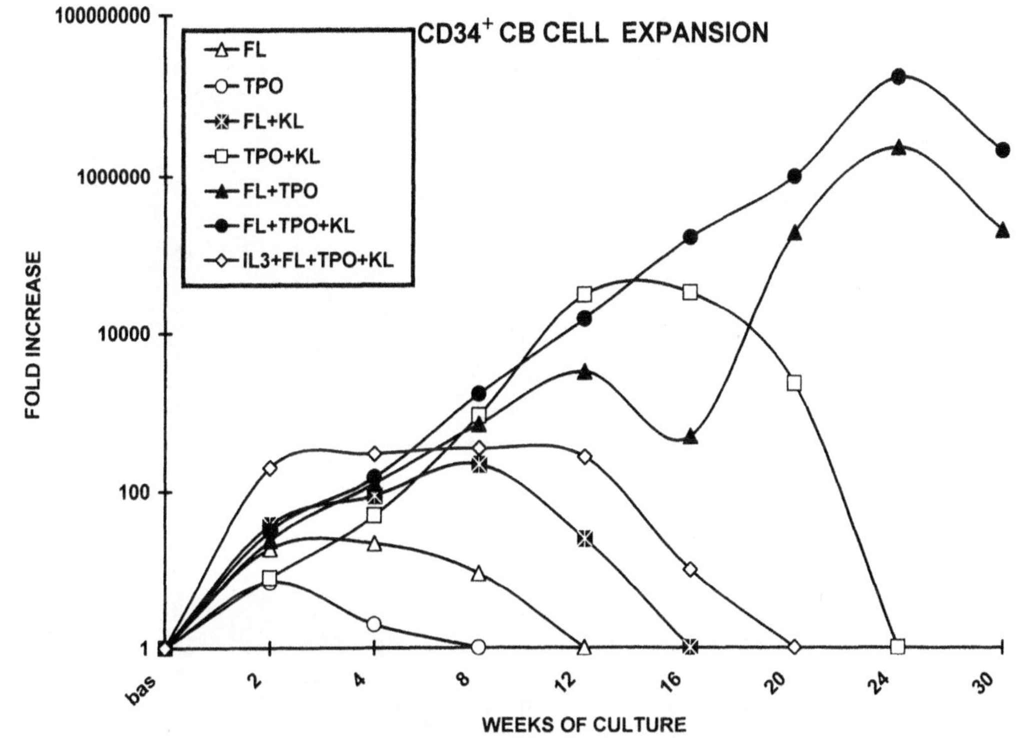

Figure 1. Expansion of hematopoietic CD34⁺ cord blood cells in stroma-free cultures stimulated by the indicated growth factors. The starting population was represented by 2×10^3 CD34⁺ CB cells in 1 ml in 24-well plates. Fresh growth factors were added twice a week. Cells were counted weekly from quadruplicate wells and the cells harvested were stained with FITC-conjugates anti CD34MoAb. The total number of each population was obtained by multiplying the percentage of each subset by the number of cells contained in each well. Results are expressed as the fold increase (compared with the starting population) and represent the mean of analyses performed on quadruplicate wells in four separate experiments.

posses self-renewal capacity, great proliferative and multilineage differentiation potential. The second, and more direct evidence derived from the LTC-IC assay. The LTC-IC is the most reliable in vitro assay capable of detecting primitive stem cells. In our hands the LTC-IC pool underwent a sustained expansion, that after 23 weeks of cultures reached 200,000-fold the initial value in the presence of FL+TPO. The addition of KL did not substantially modify the extent of the expansion.

When IL3 (10 ng/ml) was added as a supplement to any of the growth factor combinations, a progressive decrease in the cell number, CD34⁺, CFC, and LTC-IC output was seen, so that the cultures lasted only 12 to 16 weeks. These data indicate a progressive maturation of primitive stem cells at the expense of the self-renewal potential.

As a last step to asses the in vivo repopulating ability of expanded CB cells, groups of sublethally irradiated NOD/SCID mice were injected with either 10,000 and 20,000 unmanipulated cryopreserved CB cells, either the corresponding progeny of initial 10,000 or 20,000 CD34⁺ cells that were cultured for 4, 8, and 12 weeks in the presence of FL+TPO±KL. Mice that had been injected with 10,000 or 20,000 cryopreserved uncultured CD34⁺ CB cells did not show any sign of engraftment. Conversely, the expanded cells engrafted the totality of NOD/SCID mice. The levels of engraftment,

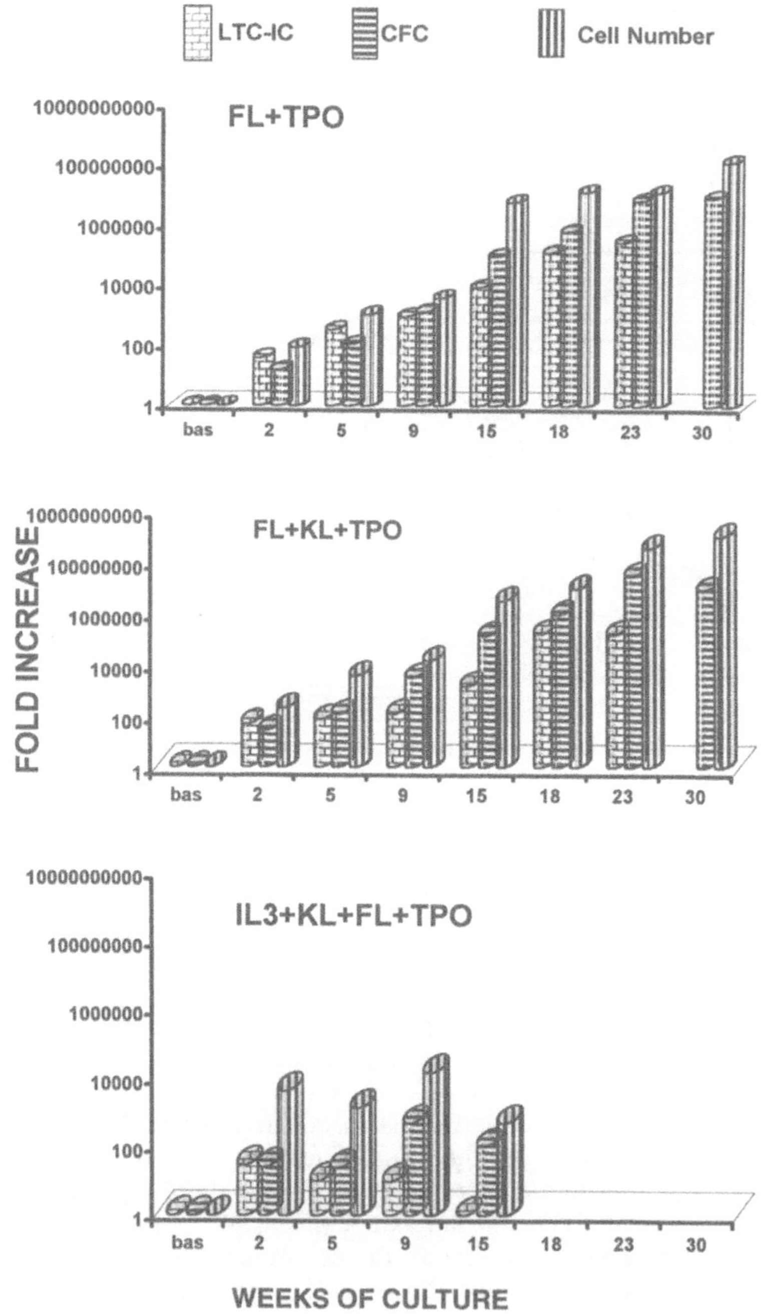

Figure 2. Production of cells and hematopoietic progenitors during long term cultures in the presence of FL+TPO, KL+FL+TPO, IL3+KL+FL+TPO. CB cells were grown for 30 weeks as described in Fig 1. Results represent the fold increase of each class of progenitors present in each well in determined periods of TLC, compared with the number of progenitors at the beginning of the cultures. Mean of four wells per point in four separate experiments.

well above that usually observed when the same number of uncultured cells were injected in the same recipients and dilution experiments, suggested that SRC were not only maintained but rather, expanded, confirming the self-renewal of primitive stem cells.

CONCLUSION

Designing optimal conditions for ex vivo expansion of human hematopoietic progenitors is based on the need for efficiency (usually assayed as the increase in the number of cells, clonogenic committed progenitors, and more primitive stem cells) and safety (mostly by minimizing exposure of the recipients to exogenous unknown proteins and infectious agents). Efficiency mostly depends on the cytokines employed and the use of the FL+TPO±KL combination is likely to have good chances to support a significant amplification of total cell and progenitor number may result in a clinical benefit, if any, it is important to consider these safety measure, even if they should result in a reduced proliferation of hematopoietic cells in culture. Finally, at least two studies[65–66] have shown that ex vivo expansion is feasible without toxicity but it has not yet been demonstrated that early reconstitution after ex vivo expanded cell transplant is shorter than that observed after unmodified transplantation and the long-term durability of reconstitution after ex vivo expanded cell transplantation has not been yet evaluated.

ACKNOWLEDGMENTS

This work was supported by fundings from the Piemonte Section of ADISCO (Associazione Italiana Donatrici di Sangue di Cordone Ombelicale); by grants to M.A. and W.P. of MURST (Ministero dell'Universita e della Ricerca Scientifica e Tecnologica), Rome and by AIRC (the Italian Association for Cancer Research), Milan, to M.A.

REFERENCES

1. Moore MAS: Expansion of myeloid stem cells in culture. Semin in Hematol 32:183,1995.
2. Emerson SG: Ex vivo expansion of hematopoietic precursors, progenitors, and stem cells: the next generation of cellular therapeutics. Blood 87:3082,1996.
3. Hylock DN, To LB, Dowse TL, Juttner CA, Simmons PJ: Ex vivo expansion and maturation of peripheral blood CD34⁺. Blood 80:1405,1992.
4. Muench MO, Firpo T, Moore MAS: Bone marrow transplantation with Interleukin 1 plus kit ligand ex vivo expanded bone marrow accelerates hematopoietic reconstitution in mice without the loss of stem cell lineage and proliferative potential. Blood 81:3643,1993.
5. Migliaccio G, Migliaccio AR, Druzin ML, Giardina PJ, Zsebo KM, Adamson SW: Long term generation of colony-forming cells in liquid culture of CD34+ cord blood cells in the presence of recombinant human stem cell factor. Blood 79:2620,1992.
6. Lansdorp PM, Drgowska W: Long term erythropoiesis from constant numbers of CD34+ cells in serum-free cultures initiated with highly purified progenitor cells from human bone marrow. J Exp Med 175:1501,1992.
7. Henschler R, Brugger W, Luft T, Frey T, Mertelsmann T, Kanz L: Maintenance of transplantation potential in ex vivo expanded CD34+ selected human peripheral blood progenitor cells. Blood 84:2898,1994.

8. Sandstrom CE, Bender JG, Papoutsakis ET, Miller WM: Effects of CD34+ selection and perfusion on ex vivo expansion of peripheral blood mononuclear cells. Blood 86:958,1995.
9. Traycoff CM, Kysak ST, Grisby S, Srour EF: Evaluation of expansion potential of cord blood and bone marrow hematopoietic progenitor cells using cell traking and limiting dilution analysis. Blood 85:2059,1995.
10. Koller MR, Emerson SG, Palsson BO: Large scale expansion of human stem and progenitor cells from bone marrow mononuclear cells in continuous perfusion cultures. Blood 2:378,1993.
11. Srour EF, Brandt JE, Briddell RA, Grigsby S, Leemhuis T, Hoffman R: Long term generation and expansion of human primitive hematopoietic progenitor cells in vitro. Blood 81:661,1993.
12. Zandstra PW, Eaves CJ, Piret JM: Expansion of hematopoietic progenitor cells in vitro. Blood 81:661,1993.
13. Lyman SD, Jacobsen SE: c-kit ligand and Flt3 ligand: stem progenitor cell factors with overlapping yet distinct activities. Blood 91:1101,1998.
14. Ogawa M, Matsuzaki Y, Nishikawa S, Hayashi, Kunisada T, Sudo T, Kina T, Nakauchi H, Nishikawa S: Expression and function of c-kit in hemopoietic progenitor cells. J Exp Med 174:63,1991.
15. Okada S, Nakauchi H, Nagayoshi K, Nishikawa S-I, Miura Y, Suda T: In vivo and in vitro stem cell function of c-kit and Sca-1 positive murine hematopoietic cells. Blood 80:3044,1992.
16. Pietsch T, Kyas U, Steffens U, Yakisan E, Hadam MR, Ludwig WD, Zsebo K, Welte K: Effects of human stem cell factor (c-kit ligand) on proliferation of myeloid leukemia cells: heterogeneity in response and synergy with other hematopoietic growth factors. Blood 80:1199,1992.
17. Carlesso N, Pregno P, Bresso P, Gallo E, Pileri A, Zsebo KM, Ferrero D: Human recombinant stem cell factor stimulates in vitro proliferation of acute myeloid leukemia cells and expands the clonogenic cell pool. Leukemia 6:642,1992.
18. Kawashima I, Zanjani ED, Almaida-Porada G, Flake AW, Zeng H, Ogawa M: CD34+ human marrow cells that express low levels of Kit protein are enriched for long-term marrow engrafting cells. Blood 87:4136,1996.
19. Rappold I, Ziegler BL, Kolher I, Marchetto S, Rosnet O, Birnbaum D, Simmons PJ, Zannettino AC, Hill B, Neu S, Knapp W, Alitalo R, Alitalo K, Ulrich A, Kanz L, Buhring HJ: Functional and phenotypic characterization of cord blood and bone marrow subsets expressing FLT3 (CD135) receptor tyrosine kinase. Blood 90:111,1997.
20. Rosnet O, Marchetto S, de La Peyriere O, Birnbaum D: Murine Flt3 gene, a gene encoding a novel tyrosine kinase receptor of the PDGFR/CSF1R family. Oncogene 6:1641,1991.
21. Matthews W, Jordan CT, Wiegand GW, Pardoll D, Lemischka IR: A receptor tyrosine kinase specific to hematopoietic stem and progenitor cell enriched populations. Cell 65:1143,1991.
22. Piacibello W, Fubini L, Sanavio F, Brizzi MF, Severino A, Garetto L, Stacchini A, Pegoraro L, Aglietta M: Effects of human Flt3 ligand on myeloid leukemia cell growth: heterogeneity in response and synergy with other hematopoietic growth factors. Blood 86:4105,1995.
23. Mackarehtschian K, Hardin JD, Moore KA, Boast S, Goff SP, Lemischka IR: Targeted disruption of the flk2/flt3 gene leads to deficiencies in primitive hematopoietic progenitors. Immunity 3:147,1995.
24. Dooley DC, Xiao M, Oppenlander BK, Plunkett JM, Lyman SD: Flt3 ligand enhances the yield of primitive cells after ex vivo cultivation of $CD34^{+}$ $CD38^{dim}$ cells and $CD34^{+}$ $CD33^{dim}$ HLA-DR+ cells. Blood 90:3903,1997.
25. Petzer AL, Hogge DE, Lansdorp PM, Reid DS, Eaves CJ: Self renewal of primitive human hematopoietic cells (long-term culture initiating cells) in vitro and their expansion in defined medium. Proc Natl Acad Sci USA 93:1470,1996.
26. Petzer AL, Zandstra PW, Piret JM, Eavees CJ: Differential cytokine effects on primitive (CD34+ $CD38^{dim}$) human hematopoietic cells: novel responses to Flt3 ligand and thrombopoietin. J Exp Med 183:2551,1996.
27. Shah AJ, Smogorzewska EM, Hannum C, Crooks GM: Flt3 ligand induces proliferation of quiescent human bone marrow CD34+ CD38- cells and maintains progenitor cells in vitro. Blood 87:3563,1996.
28. Wendling F, Penciolelli JF, Charon M, Tambourin P: Factor independent erythropoietic progenitor cells in leukemia induced by the myeloproliferative leukemia virus. Blood 73:1161,1989.
29. Kaushansky K: Thrombopoietin: the primary regulator of platelet production. Blood 86:419,1995.
30. Zeigler FC, deSauvage F, Widmer HR, Keller GA, Donahue C, Schreiber RD, Malloy B, Hass P, Eaton D, Matthews W: In vitro megakaryocytopoietic and thrombopoietic activity of c-mpl ligand (TPO) on purified murine hematopoietic stem cells. Blood 84:4045,1994.
31. Kaushansky K, Broudy VC, Grossmann A, Humes J, Lin N, Ren HP, Bailey MC, Papayannopoulou T, Forstrom JW, Sprugel KH: Thrombopoietin expands erythroid progenitors, increases red cell

production, and enhances erythroid recovery after myelosuppressive therapy. J Clin Invest 96:1683,1995.

32. Piacibello W, Sanavio F, Brizzi MF, Garetto L, Severino A, Aronica MG, Dragonetti G, Aglietta M, Pegoraro L: Megakaryocyte growth and development factor (MGDF)-induced acute leukemia cell proliferation and clonal growth is associated with functional c-mpl. Leukemia 11:531,1997.
33. Farese AM, Hunt P, Grab LB, MacVittie TJ: Combined administration of granulocyte colony stimulating factor enhances multilineage hematopoietic reconstitution in nonhuman primates after radiation-induced marrow aplasia. J Clin Invest 97:2145,1996.
34. Alexander WS, Roberts AW, Nicola NA, Li R, Metcalf D: Deficiencies in progenitor cells of multiple hematopoietic lineages and defective megakaryocytopoiesis in mice lacking the thrombopoietin receptor c-mpl. Blood 87:2162,1996.
35. Yonemura Y, Ku H, Hirayima F, Souza LM, Ogawa M: Interleukin 3 or Interleukin 1 abrogates the reconstituting ability of hematopoietic stem cells. Proc Natl Acad Sci USA 93:4040,1996.
36. Holyoake TL, Freshney MG, McNair L, Parker AN, McKay PJ, Steward WP, Fitzsimons E, Graham GH, Pragnell IB: Ex vivo expansion with stem cell factor and Interleukin 11 augments both short-term recovery posttransplant and the ability to serially transplant marrow. Blood 87:4589,1996.
37. Kamel-Reid S, Dick JE: Engraftment of immune-deficient mice with human hematopoietic stem cells. Science 242:1706,1988.
38. Lapidot T, Pflumio F, Doedens M, Murdoch B, Williams DE, Dick JE: Cytokine stimulation of multilineage hematopoiesis from immature human cells engrafted in SCID mice. Science 255:1137,1992.
39. Lowry PA, Shultz LD, Greiner DL, Hesselton RM, Kittler EL, Tiarks CY, Rao SS, Reilly J, Leif JH, Ramshaw H, Stewart FM, Quesenberry PJ: Improved engraftment of human cord blood cells in NOD/LtSz-scid/scid mice after irradiation or multiple-day injections into unirradiated recipients. Biol. Of Blood Marrow Transplant, 2:15,1996.
40. Cashman JD, Lapidot T, Wang JC, Doedens M, Shultz LD, Lansdorp P, Dick JE, Eaves CJ: Kinetic evidence of the regeneration of multilineage hematopoiesis from primitive cells in normal human bone marrow transplanted into immunodeficient mice. Blood 89:4307,1997.
41. Dick JE: Normal and leukemic human stem cells assayed in SCID mice. Semin Immunol 8:197,1996.
42. Larochelle A, Vormoor J, Hanenberg H, Wang JC, Bhatia M, Lapidot T, Moritz T, Murdoch B, Xiao XL, Kato I, Williams DA, Dick JE: Identification of primitive human hematopoietic cells capable of repopulating NOD/SCID mouse bone marrow: implications for gene therapy. Nat Med 2:1329,1996.
43. Bhatia M, Wang JCY, Kapp U, Bonnet D, Dick JE: Purification of primitive human hematopoietic cells capable of repopulating immune-deficient mice. Proc Natl Acad Sci USA 94:5320,1997.
44. Cashman J, Bockhold K, Hogge DE, Eaves AC, Eaves CJ: Sustained proliferation, multilineage differentiation and maintenance of primitive human hoemopoietic cells in NOD/SCID mice transplanted with human cord blood. Br J Haematol 98:1026,1997.
45. Conneally E, Cashman J, Petzer A, Eaves C: Expansion in vitro of transplantable human cord blood stem cells demonstrated using a quantitative assay of their lympho-myeloid repopulating activity in nonobese diabetic-scid/scid mice. Proc Natl Acad Sci USA 94:9836,1997.
46. Vaziri H, Dragowska W, Allsopp RC, Thomas TE, Harley CB, Lansdorp P: Evidence for mitotic clock in human hematopoietic stem cells: loss of telomeric DNA with age. Proc Natl Acad Sci USA 91:9857,1994.
47. Cairo MS, Wagner JE: Placental and/or Umbilical cord blood: an alternative source of hematopoietic stem cells for transplantation. Blood 90:4665,1997.
48. Broxmeyer HE, Gluckman E, Auerbach AD, Douglas GW, Friedman H, Cooper S, Hangoc G, Kurtzberg J, Bard J, Boyse EA: Human umbilical cord blood: a clinically useful source of transplantable hematopoietic stem/progenitor cells. Int J Cell Cloning 8:76,1990.
49. Thierry D, Hervatin F, Traineau R, Brossard Y, Stark R, Benbunan M, Gluckman E: Hematopoietic progenitor cells in cord blood. Bone Marrow Transplant 9:101,1992.
50. Traycoff CM, Kosak ST, Grigsby S, Srour EF: Evaluation of ex vivo expansion potential of cord blood and bone marrow hematopoietic progenitor cells using cell tracking and limiting dilution analysis. Blood 85:2059,1995.
51. Hows JM, Bradley BA, Marsh JC, Luft T, Coutinho L, Testa NG, Dexter TM: Growth of human umbilical cord blood in longterm haemopoietic cultures. Lancet 340:73,1992.
52. Van deVen C, Ishizawa L, Law P, Cairo MS: IL11 in combination with steel factor and G-CSF or GM-CSF significantly increases expansion of isolated CD34+ cell population from cord blood vs adult bone marrow. Exp Hematol 23:1289,1995.

53. Lu L, Xiao M, Shen R-N, Grigsby S, Broxmeyer HE: Enrichment, characterization, and responsiveness of single primitive CD34 human umbilical cord blood hematopoietic progenitors with high proliferative and replating potential. Blood 81:41,1993.
54. Piacibello W, Sanavio F, Garetto L, Severino A, Bergandi D, Ferrario J, Fagioli F, Berger M, Aglietta M: Extensive amplification and self-renewal of human primitive hematopoietic stem cells from cord blood. Blood 89:2644,1997.
55. Piacibello W, Sanavio F, Garetto L, Severino A, Dane A, Gammaitoni L, Aglietta M: Differential growth factor requirement of primitive cord blood hematopoietic stem cells from cord blood. Leukemia 12:718,1998.
56. Piacibello W, Sanavio F, Severino A, Dane A, Gammaitoni L, Fagioli F, Perissinotto E, Cavallo G, Kollet O, Lapidot T, Aglietta M: Engraftment in NOD/SCID mice of human CD34+ cord blood cells after ex vivo expansion: evidence for the amplification and self-renewal of repopulating stem cells. Blood 93:3736,1999.
57. Moore KA, Deisseroth AB, Reading CL, Williams DE, Belmont JW: Stromal support enhances cell-free retroviral vector transduction of human bone marrow long-term culture-initiating cells. Blood 79:1393,1992.
58. Nolta JA, Smogorzewska EM, Kohn DB: Analysis of optimal conditions for retroviral-mediated transduction of primitive human hematopoietic cells. Blood 86:101,1995.
59. Breems DA, Blokland EA, Siebel KE, Mayen AE, Engels LJ, Ploemacher RE: Stroma contact prevent loss of hematopoietic stem cell quality during ex vivo expansion of CD34+ mobilized peripheral blood stem cells. Blood 91:111,1998.
60. Gan OI, Murdoch B, Larochelle A, Dick JE: Differential maintenance of primitive human SCID-repopulating cells, clonogenic progenitors and long-term culture-initiating cells after incubation on human bone marrow stromal cells. Blood 90:641,1997.
61. Thalmeier K, Meissner P, Reisbach G, Falk M, Brechtel A, Dormer P: Establishment of two permanent human bone marrow stromal cell lines with longterm post irradiation feeder capacity. Blood 83:1799,1994.
62. Hurle RW, McCarthy JF, Verfaille CM: Direct adhesion to bone marrow stroma via fibronectin receptors inhibits hematopoietic progenitor proliferation. J Clin Invest 96:511,1995.
63. Verfaille CM, Catanzaro PM: Direct contact with stroma inhibits proliferation of human long-term culture initiating cells. Leukemia 10:498,1996.
64. Brown RL, Xu FS, Dusing S, Li Q, Fischer R, Patchen M: Serum-free culture conditions for cells capable of promoting long-term survival in lethally irradiated mice. Stem Cell 15:237,1997.
65. Brugger W, Heimfeld S, Berenson RJ, Mertelsmann R, Kana L: Reconstitution of hematopoiesis after high dose chemotherapy by autologous progenitor cells generated ex vivo. N Eng J Med 333:283,1995.
66. Williams SF, Lee WJ, Bender JG, Zimmerman R, Swinney P, Blake M, Carreon J, Schilling M, Smith S, Williams DE, Oldham F, Van Epps D: Selection and expansion of peripheral blood CD34+ cells in autologous stem cell transplantation for breast cancer. Blood 87:1687,1996.

6

THYMUS-INDEPENDENT T CELL DIFFERENTIATION *IN VITRO*

Massimo Sanchez, Elena Alfani, Giovannella Visconti,
Anna Maria Passarelli, Anna Rita Migliaccio,
and Giovanni Migliaccio[1]

Laboratorio di Biologia Cellulare
Istituto Superiore di Sanità
Rome and Ospedale Civile di Tivoli
Tivoli, Italy

ABSTRACT

We report that the mononuclear cells of human cord blood contain a cell population that support the differentiation of $CD34^+$ cells into $CD4^+$ or $CD8^+$ naive T-cells in serum-deprived cultures stimulated with stem cell factor and interleukin 7. $CD4^+$ or $CD8^+$ $CD45RA^+$ $TCR\alpha\beta^+$ T-cells are continuously produced in vitro over a period of 20 days under these conditions. The generation of T cells in these cultures is a dynamic process and clones of T cells expressing new T-cell receptor β-chain rearrangments are generated over time. These results pave the way to the development of very simple culture conditions for ex-vivo production of naive helper or cytotoxic T-cells which could be very useful for gene and immune therapy of human diseases.

INTRODUCTION

Ex vivo manipulation of the lymphoid compartments has been attempted to modify the T cell response to tumor or viral antigens[1,2] and to induce them to express foreign genes.[3,4] The culture techniques currently available allow expansion of pre-existing, and therefore already antigen-primed, T cell clones, but do not sustain the generation of new clones in the absence of accessory cells derived from fetal[5–7] or neonatal[8] thymus or of an adult[9] marrow-derived stromal cell line. The need for these

---

[1] Laboratorio di Biologia Cellulare, Istituto Superiore di Sanità. V.le Regina Elena, 299. 00161 Roma, Italy. Phone: 39-06-49902576; Fax: 39-06-49387143; e-mail: migliagi@ISS.IT

Molecular Biology of Hematopoiesis 6, edited by Abraham *et al.*
Kluwer Academic/Plenum Publishers, New York, 1999.

accessory cells makes it unlikely that the conditions which allow, either *in vitro* or *in vivo*, *ex novo* generation of T cells will be easily scaled up as required by therapeutic protocols.

Cord blood is an alternative and unlimited source of hematopoietic stem and progenitor cells for related and unrelated transplants.[10] Cord blood cells are capable to generate T cells in co-culture experiments[8,11] and the T cell compartment, although it remains immature for long time after the transplant, is of graft origin even in adult cord blood recipients with a compromised thymus (P. Rubinstein, New York Blood Center, personal communication). These results suggested to us that cord blood might contain all the cellular compartments necessary to generate new naive T cells *in vitro* and prompted us to search for the culture conditions which would make this possible. We describe that $CD34^+$ cells purified from cord blood generate new naive T cell clones if co-cultured with irradiated autologous mononuclear cells under serum-deprived conditions stimulated with Stem Cell factor (SCF) and interleukin 7 (IL-7).

MATERIALS AND METHODS

Cell Purification and Culture. Umbilical cord blood was obtained at the time of delivery from uncomplicated pregnancies previous informed consent from the mother. Mononuclear cells (CBM) isolated by density-cut separation ($\rho < 1.077$) (Ficoll-Paque; Pharmacia, Uppsala, Sweden) were either processed directly or cryopreserved (10% v/v DMSO in Hanks's balanced salt solution, HBSS, Sigma, St Louis, Mo). Highly purified $CD34^+$ cells were obtained from CBM with two rounds of purification. First, CBM cells were incubated with magnetic beads coated with anti-human CD34 antibody (Dynabeads M-450, Dynal, Oslo, Norway) and the $CD34^+$ cells detached from the beads by competition (Detachabead, Dynal) as described by the manufacturer. The cells eluted from the beads ($\cong$ 50% $CD34^+$ upon re-analysis) were then incubated with the phyco-erythrin labelled-HPCA-2 antibody (Becton-Dickinson, San Jose, CA) and the fluorescent cells isolated by cell sorter (Elite, Coulter-Counter, Miami, FL). An average of 10^4 cells ($\geq$98% $CD34^+$ upon reanalysis) were recovered for every 10^6 CBM. CBM were irradiated at 600 rad with GAMMA CELL 220 (Atomic Energy Canada Limited, Toronto, Canada). CBM (10^5/ml) and purified $CD34^+$ cells (2×10^3/ml) were expanded for up to 20 days in liquid cultures stimulated with SCF (10 ng/ml, a gift from Amgen, Thousand Oaks, CA), IL-3 (10 ng/ml, a gift of Genetic Institute, Boston, MA) and/or IL-7 (10 ng/ml, Collaborative Biomedical Product, BD Bedford, MT) under serum-deprived conditions as described.[12] The cultures were demipopulated and fresh culture medium and growth factors added as required to keep the cell concentration below 6×10^5 cells/ml. The number of myeloid progenitor cells (CFC) was evaluated in cultures made semisolid with methylcellulose (8% wt/v) dissolved in Iscove's modified Dulbecco's medium and stimulated with a mixture of growth factors including IL-3 (10 u/ml), SCF (10 ng/ml), Erythropoietin (1.5 u/ml, Epoetina α, Dompè Biotec, Milan, Italy), granulocyte colony-stimulating factor (10 ng/ml, Filgrastim, Dompè Biotec), and granulocyte macrophage-CSF (10 ng/ml, Molgramostim, Sandoz, Milan, Italy) as previously described.[12] The colonies were scored according to standard criteria after 14 days of incubation in a fully humidified CO_2 (5%) O_2 (5%) monitored incubator.

Immunophenotyping of Cultured Cells by Flow Cytometry. Cells were harvested by gentle aspiration, washed twice in phosphate buffer saline and incubated ($1–2 \times 10^5$ cells per tube) with antibodies against CD3, CD4, CD8, CD7, CD19, CD45, CD45RA, and CD45RO (from Sigma), CD34 (from Becton-Dickinson), CD2, CD33, CD56, and TCRαβ (from PharMigen, San Diego, CA) labeled with appropriate fluorochromes. Aliquots of the cells were also incubated with appropriate isotype matched antibodies as control for non specific binding. The labeled cells were analyzed with a Coulter Elite Cell Sorter and the gating set to exclude the autofluorescence signal typical of cultured cells.

RNA Extraction, cDNA Synthesis, and RT-PCR Analysis of the Expression of Non-germ Line Configurations of TCR Vβ Elements. RNA was extracted with the guanidine isothyocianate method (RNAFAST, Molecular System, San Diego, CA) as described by the manufacture. cDNA was prepared by reverse transcription with the Moloney reverse transcriptase and random primers (Gibco-BRL, Grand Island, NY). Individual non-germ line configuration of the TCR Vβ elements were amplified by RT-PCR using a common primer coupled with a primer specific for each of the 22 variable elements of the Vβ chain as described.[13] Up to 35 amplification cycles were performed in a PTC-100 Thermocycler (MJ Research Inc., Watertown, MA). The amplified products were separated by electrophoresis on a 1.5% agarose gel which was blotted onto Hybond N^+ membrane (Amersham Italia srl, Milan, Italy). The specificity of the amplification was proven by probing with an end-labeled (T4 polynucleotide kinase end labeling kit, Gibco BRL, and γ-^{32}P-ATP, 3000 Ci/mmol, Du Pont, Firenze, Italy) oligonucleotides specific for the constant portion of the Vβ elements.[13]

RESULTS AND DISCUSSION

As reported by many investigators,[10,12] cord blood mononuclear (CBM) cells generate myeloid cells (>99% CD33+) and myeloid progenitors (colony forming cells, CFC) in serum deprived cultures stimulated with stem cell factor (SCF) and interleukin 3 (IL-3). The CBM cells which are responsible for the expansion observed under these conditions are $CD34^+$ cells. In fact, the number of mature cells and of myeloid progenitors increases at day 20 by 302 ± 130 and 88 ± 38 fold, respectively, in serum deprived cultures of $CD34^+$cells stimulated with SCF and IL-3 (Table I). Virtually all the mature cells generated under these conditions are $CD33^+$.

CBM stimulated with SCF and IL-7, in the absence of IL-3, generated large numbers of differentiated cells (12.1 ± 7.4 fold increase [FI]) and of CFC (30.2 ± 7.5

Table I. Total number of cells and of CFC present at day 20 in serum-deprived cultures of CBM and $CD34^+$ cells*

Cells*	Growth Factors	Total Cell Number ($\times 10^6$)	Total Cell Number (FI)**	Total CFC Number ($\times 10^4$)	Total CFC Number (FI)
$CD34^+$	SCF + IL-3	3.02 ± 1.3	301.9 ± 13	10.6 ± 34.4	88.2 ± 38.4
$CD34^+$	SCF + IL-7	0.04 ± 0	4.0 ± 1.4	0.2 ± 0.04	1.9 ± 0.5
Irr. CBM	SCF + IL-7	0.03 ± 0	0.1 ± 0.1	0	n.a.§
Irr. CBM + $CD34^+$	SCF + IL-7	2.86 ± 2.1	286.5 ± 21	6.7 ± 3.2	48.7 ± 18.0
CBM	SCF + IL-7	6.05 ± 3.7	12.1 ± 7.4	6.4 ± 1.1	30.2 ± 7.5

*The cultures (5 ml) were initiated with a total number of 10^4 $CD34^+$ cells, 5×10^5 CBM or with a mixture of 10^4 $CD34^+$ cells and 5×10^5 irradiated (Irr.) CBM. The number of CFC contained at day 0 in cultures of $CD34^+$ cells and of CBM was 1498 ± 530 and 1575 ± 841, respectively. The results represent the mean (±SEM) of three experiments. **FI (Fold increase) is defined as the ratio between the values at day 20 and at day 0. §n.a., not applicable.

FI) by day 20 (Table I). In this case, more than 81% of the total nucleated cells gate in the lymphocyte area and 82% of those are CD4 or CD8 single positive cells (Fig. 1) which also express CD7, CD2, CD3, TCRαβ, and CD45RA (data not shown). SCF and IL-7 induce a significant increase of total nucleated cells (4.0 ± 1.4 FI) and of CFC (1.9 ± 0.5 FI) by day 20 also in cultures of purified $CD34^+$ cells (Table I). However, in this case, the expansion is generally lower than the expansion observed in the corresponding culture of CBM cells and the differentiated cells are mainly myeloid. These results indicate that SCF and IL-7 specifically enhance the numbers of T cells detectable in serum-deprived cultures of CBM but are not sufficient to sustain T cell differentiation in cultures of highly purified neonatal $CD34^+$ cells.

Purified $CD4^+$ and $CD8^+$ cells did not grow under those conditions (data not shown). Therefore CBM might contain an accessory cell population which induce $CD34^+$cells to generate T cells in cultures stimulated with SCF and IL-7. In order to test this possibility, highly purified $CD34^+$ cells (2×10^3/ml per 5 ml of culture) were co-cultured with irradiated autologous CBM (10^5/ml per 5 ml of culture). Also in this case, high numbers of nucleated cells (287 ± 215 FI) and CFC (48.7 ± 18.0 FI) were observed at day 20. Although the majority of the nucleated cells observed are still myeloid (93% $CD33^+$, not shown), 7% of the cells are in the low side and forward scatter window and a high percentage (82.4%) of those are $CD4^+$ or $CD8^+$ single positive T-cells (Fig. 1).

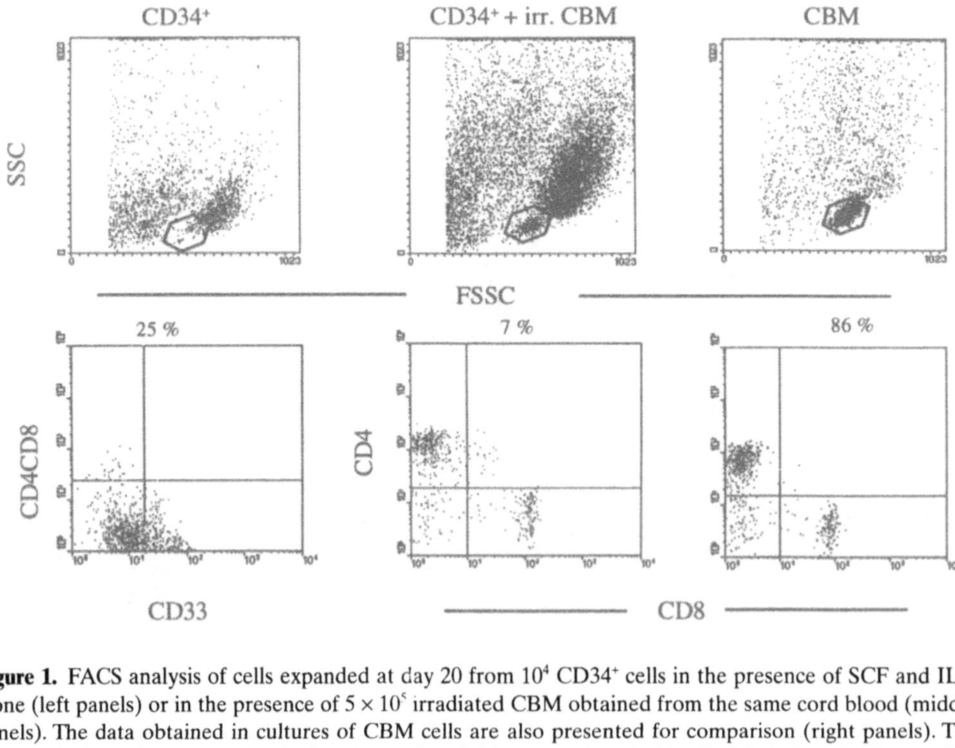

Figure 1. FACS analysis of cells expanded at day 20 from 10^4 $CD34^+$ cells in the presence of SCF and IL-7 alone (left panels) or in the presence of 5×10^5 irradiated CBM obtained from the same cord blood (middle panels). The data obtained in cultures of CBM cells are also presented for comparison (right panels). The top panels show the gate set for the low forward and side scatter lymphoid cells (circled area). The number at the bottom of the panels indicate the percent of the cells, over the total cell population, within that gate. The panels on the bottom, present the CD4 and CD8 staining of the low forward and side scatter cells selected on the top. In the case of $CD34^+$ cells cultured with SCF and IL-7 alone, the number of CD4/CD8 positive cells is so low, that the cells were also labeled with anti CD33 as control.

These results indicate that irradiate CBM cells, which are unable to proliferate, are capable to induce highly purified CD34⁺ cells to generate T cells in serum-deprived cultures stimulated with SCF and IL-7.

To characterize the T-cells clones generated *in vitro*, we have analyzed at day 0 and at day 13 and 19 of culture the expression of the non-germ line configurations for 22 different elements of the Vβ family of the T-cell Receptor (TCR)[13] β chain. Fig. 2 presents the data relative to three representative elements. CBM expressed at day 0 all the 22 Vβ elements analyzed. Very few Vβ elements were expressed with more than 1–3 different non-germ line configurations (Fig. 2 and data not shown). The non-germ line configurations expressed at day 0 varied between individual cord blood units. CBM cells expressed in culture a dynamic pattern of TCR rearrangments with some non-germ line configurations disappearing (Fig. 2, $V\beta_{5.1}$ and $V\beta_{6.1}$ at day 13 and 19) and others appearing (Fig. 2, $V_{\beta 4}$ at day 13) over time. Purified $CD34^+$ cells, on the other hand, do not express detectable levels of any of the TCR Vβ elements analyzed (data not shown). The expression of the TCR was also below detectable levels in cultures of

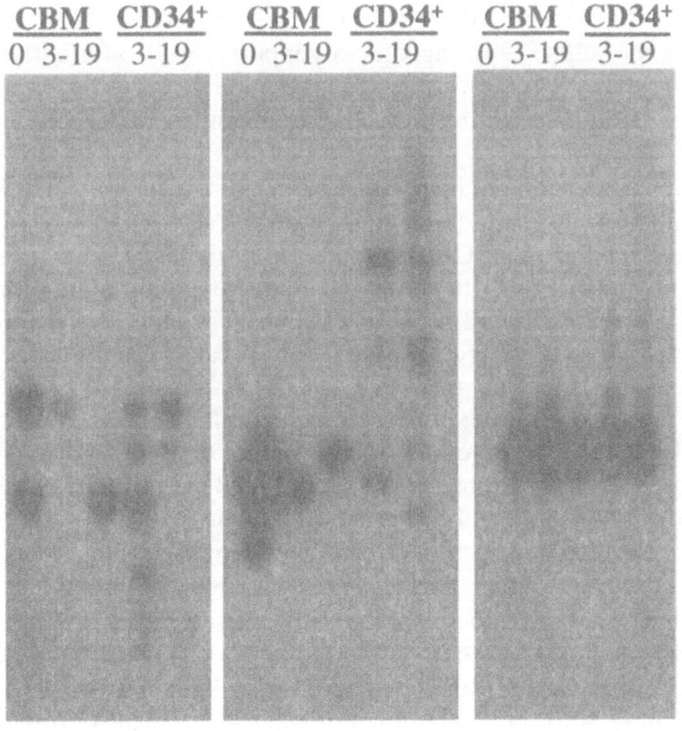

Figure 2. RT-PCR analysis of the expression of three representative (out of 22 separate elements analyzed) Vβ (β5.1, β6.1 and β4) elements of the TCR in cultures of CBM cells (CBM) and in co-cultures of $CD34^+$ cells with irradiated CBM ($CD34^+$). The number at the top of each lane indicate the day of culture.

$CD34^+$ cells stimulated with SCF and IL-7 (data not shown). However, when $CD34^+$ cells where co-cultured with irradiated CBM, numerous Vβ elements were expressed by day 13 and 19 (Fig. 2A). Interestingly, the majority of the non-germ line configurations expressed at day 13 were not expressed at day 19 and viceversa. This fact suggests that different T cell clones predominate in culture at different times and that new clones are generated, expand and disappear over time in these cultures. It was also of interest that, although the absolute number of T cells generated at day 20 is higher in cultures of CBM than in cultures of $CD34^+$ cells (Table I), much more rearrangments are expressed in cultures of purified $CD34^+$ cells than in the corresponding cultures of CBM (the cultures of CBM cells contain only half the number of $CD34^+$ cells than the cultures of purified $CD34^+$ cells). This fact may indicate that irradiated CBM are more effective than untreated CBM in sustaining differentiation of distinct T cell clones from $CD34^+$ cells. This is more so if it is taken into account that irradiated CBM do not survive in significant numbers by the end of the culture and, therefore, cannot sustain T cell proliferation for quite as long as untreated CBM.

These results prove the presence in CBM of accessory cells which are able to support *in vitro* differentiation of $CD34^+$ cells into T-cells. It would be interesting to know the phenotype of the accessory CBM cells and the factor(s) they produce which sustains T cell differentiation. T cell differentiation similar to that described here has been reported in organ cultures with explants of fetal thymus and bone marrow.[5,9] In the adult, cells capable to support T cell differentiation are present only in the thymus.[14,15] It is possible that the putative accessory cord blood cell is a microenvironment precursor in the process of migrating from the bone marrow to the thymus.

The data described here have several clinical implications. First, our data predict that adult patients transplanted with cord blood will recover an healthy immune system although it may require that CBM colonize and establish a functional microenvironment in the thymus. Second, they may facilitate gene and immunotherapy of cancer and viral diseases, in fact cord blood is easily collectable, there is no limitation to its availability as long as there are births, it is safe because can be tested for a variety of infectious diseases. Therefore, there is no problem to collect enough CBM to be used as accessory cells in co-cultures scaled up for clinical use. CBM could be used to obtain *in vitro* clinically relevant amounts of both helper or cytotoxic naive T-cells. *Ex vivo* expanded naive T cells could be given to cord blood recipients to accelerate the recovery of their immune functions. Furthermore, strategies could be designed to modify the conditions which allow production of naive cytotoxic T-cells to obtain selection of epitopes which could have an higher efficiency to recognize tumoral epitopes. These clinical applications will be so much more realistic if irradiated CBM will be proven to be effective even when co-cultured with heterologous $CD34^+$ cells, of either neonatal or adult origin, as it is the case for the generation of human T cells in SCID/hu mice in the presence of fetal explants of human origin.[16] A fact that we found very encouraging is that the *ex vivo* generated T cells we describe do not contain Natural Killer (defined as $CD56^+$) cells and have the phenotype of naive cells which are not alloreactive and, more over, do not react against any of the culture components (data not shown).

In conclusion, we report that the mononuclear cells of human cord blood contain a cell population that support the differentiation of $CD34^+$ cells into $CD4^+$ or $CD8^+$ naive T-cells in serum-deprived cultures stimulated with stem cell factor and interleukin 7.

ACKNOWLEDGMENTS

The collaboration of the nurses of the Gynecology Department of Ospedale Civile di Tivoli, Tivoli, Italy is gratefully acknowledged. This work was supported by institutional funds from Istituto Superiore di Sanità and by the EU Biotec project "Bioreactor production of human hemopoietic cells."

REFERENCES

1. Lieberman J, Skolnik PR, Parkerson GR, Fabry JA, Landry B, Bethel J, Kagan J: Safety of autologous, ex vivo-expanded human immunodeficiency virus (HIV)-specific cytotoxic T-lymphocyte infusion in HIV-infected patients. Blood 90:2196, 1997.
2. Chang AE, Shu S: Current status of adoptive immunotherapy of cancer. Crit.Rev.Oncol.Hematol. 22:213, 1996.
3. Rosenzweig M, Marks DF, Hempel D, Johnson RP: In vitro T lymphopoiesis: a model system for stem cell gene therapy for AIDS. J.Med.Primatol. 25:192, 1996.
4. An DS, Koyanagi Y, Zhao JQ, Akkina R, Bristol G, Yamamoto N, Zack JA, Chen IS: High-efficiency transduction of human lymphoid progenitor cells and expression in differentiated T cells. J.Virol. 71:1397, 1997.
5. Freedman AR, Zhu H, Levine JD, Kalams S, Scadden DT: Generation of human T lymphocytes from bone marrow $CD34^+$ cells in vitro. Nat.Med. 2:46, 1996.
6. Krowka JF, Sarin S, Namikawa R, McCune JM, Kaneshima H: Human T cells in the SCID-hu mouse are phenotypically normal and functionally competent. J.Immunol. 146:3751, 1991.
7. Plum J, De Smedt M, Defresne M-P, Leclercq G, Vandekerckhove BA: Human CD34+ fetal liver stem cells differentiate to T cells in a mouse thymic microenvironment. Blood 84:1587, 1994.
8. Yeoman H, Clark DR, Deluca D: Development of CD4 and CD8 single positive T cells in human thymus organ culture: IL-7 promotes human T cell production by supporting immature T cells. Dev.Comp.Immunol 20:241, 1996.
9. Tagoh H, Kishi H, Okumura A, Kitagawa T, Nagata T, Mori K, Muraguchi A: Induction of recombination activating gene expression in a human lymphoid progenitor cell line: requirement of two separate signals from stromal cells and cytokines. Blood 88:4463, 1996.
10. Broxmeyer HE, Hangoc G, Cooper S, Ribeiro RC, Graves V, Yoder M, Wagner JE, Vadhan-Raj S, Benninger L, Rubinstein P, Broun ER: Growth characteristics and expansion of human umbilical cord blood and estimation of its potential for transplantation in adults. Proc.Natl.Acad.Sci.U.S.A. 89:4109, 1992.
11. Blom B, Res P, Noteboom E, Weijer K, Spits H: Prethymic CD34+ progenitors capable of developing into T cells are not committed to the T cell lineage. J.Immunol. 158:3571, 1997.
12. Migliaccio G, Migliaccio AR, Druzin ML, Giardina PJ, Zsebo KM, Adamson JW: Long-term generation of colony-forming cells in liquid culture of $CD34^+$ cord blood cells in the presence of recombinant human stem cell factor. Blood 79:2620, 1992.
13. Choi Y, Kotzin B, Herron L, Callahan J, Marrack P, Kappler J: Interaction of *Staphylococcus aureus* toxin "superantigens" with human T-cells. Proc.Natl.Acad.Sci.U.S.A. 86:8941, 1989.
14. Peault B: In-vitro models of stroma-dependent lymphopoiesis. Semin.Immunol. 7:169, 1995.
15. Anderson G, Moore NC, Owen JJT, Jenkinson EJ: Cellular interactions in thymocyte development. Annu.Rev.Immunol. 14:73, 1996.
16. Yurasov S, Kollmann TR, Kim A, Raker CA, Hachamovitch M, Wong Staal F, Goldstein H: Severe combined immunodeficiency mice engrafted with human T cells, B Cells, and myeloid cells after transplantation with human fetal bone marrow or liver cells and implanted with human fetal thymus: a model for studying human gene therapy. Blood 89:1800, 1997.

7

IMMUNOGENETICS PROBLEMS IN CORD BLOOD TRANSPLANTATION

Francesca Poli, Barbara Diomelli, Nadia Ghidoli,
Cristina Tettamanti, Barbara Macchi, Mario Scalamogna,
and Girolamo Sirchia

Human cord blood (CB) is a rich source of haematopoietic stem and progenitor cells;[1,2] it can be harvested without risk to the newborn or the mother, and is currently used for bone marrow reconstitution.[1–5]

Due to the supposed partial tolerance to alloantigens of neonatal lymphocytes, placental blood transplants may require less stringent HLA matching. However, the optimal level of HLA compatibility in CB transplantation is still a matter of investigation[6] and a fully reliable definition of HLA Class I and Class II antigens/alleles is needed. Cord blood typing by standard serology presents some difficulties related to the quality and to the volume of samples.

A high background of dead cells, particularly occurring when samples are processed in the laboratory several hours after parturition, makes HLA typing unreliable unless resorting to DNA analysis. Furthermore, sera crossreactivity and the contamination with immature erythroblasts and myeloid cells[7] add other technical drawbacks.

In our setting, 14% of HLA-A,B typings of CB specimens have to be repeated[8] using trays from different suppliers and performing maternal typing to help to confirm donor's haplotypes. This implies additional costs and an overloading of work.

Molecular methods have been developed to achieve specific and efficient typing from very little material and, because of these characteristics, are well-suited for CB typing.[9] However, whereas serology for HLA Class II has been replaced by DNA methods in most of the laboratories, on the contrary for HLA Class I genomic typing is still in its infancy.[10,11]

Another aspect which is a matter of concern in CB transplantation is the level of contamination with maternal DNA which could be responsible for severe graft versus host disease (GVHD) in immunosuppressed recipients.[12,13]

While chimerism with maternal lymphocytes has been demonstrated in more than 25% of children with Severe Combined Immunodeficiency Disease,[14] many studies suggest that maternal blood cells in newborn circulation is uncommon[4,15–18] and,

---

From Centro Trasfusionale e di Immunologia dei Trapianti and Servizio per il Prelievo e Conservazione di Organi e Tessuti. Ospedale Maggiore Policlinico—Istituto di Ricovero e Cura a Carattere Scientifico, Via Francesco Sforza 35, 20122 Milano, Italy.

Molecular Biology of Hematopoiesis 6, edited by Abraham *et al.*
Kluwer Academic/Plenum Publishers, New York, 1999.

when the contamination is documented, its magnitude varies due to the different sensitivity of the methods employed.[4,19–23]

In this paper we report the experience of the Milano Cord Blood Bank on:

1. comparison between serological HLA Class I typing and PCR-SSP[24] results;
2. contamination of maternal DNA of CB units stored in our bank.

1. COMPARISON BETWEEN SEROLOGICAL HLA CLASS I TYPING AND PCR-SSP RESULTS

Since 1992, in our laboratory PCR-SSP is the choice method for urgent HLA Class II typing and showed to offer, over standard lymphocytotoxicity, a series of advantages: higher accuracy, consistency of results, independency from reagents difficult to obtain (monospecific or rare antisera), reduction of the volume of blood required, independency from cell viability, and antigen expression on the cell surface, lower cost. Moreover, PCR-SSP, with respect to serology, is equally fast and so simple that personnel can be easily trained.

For HLA Class I, many of the advantages clearly shown for HLA Class II have still to be demonstrated. In view of the relevance of HLA matching in CB transplantation,[6] we evaluated the capacity of PCR-SSP in solving problems in HLA-A, B typing for both broad and split HLA antigen definition. With this aim we have compared serology with PCR-SSP in 100 CB with doubtful or unreliable HLA-A,B typing, in the wake of what had been done years ago for HLA Class II.[25,26]

HLA-A,B typing was done on T lymphocytes isolated from 1 ml of CB samples using IM beads (Class I Dynabeads, Dynal A.S., Oslo).

All 100 cord units were typed by micro-lymphocytotoxicity using locally assembled typing trays and commercially available trays (One Lambda Inc., Canoga Park, Ca or Pel-Freez, Brown Deer, WI). PCR-SSP was carried out with primers synthesized in our laboratory (Gene Assembler Plus, Pharmacia, LKB, Uppsala) following the protocol described in the Reference Manual of the 12th International Histocompatibility Workshop.

PCR-SSP was successful in amplifying HLA alleles in all 100 CB samples.

At A and B loci 2 alleles were amplified in 80 and 81 samples respectively.

For 19 specimens, no serological conclusion was drawn due to high mortality of cell suspension while PCR-SSP allowed the definition of a clear typing.

Forty-six typing results (56.7%) were discrepant with the 2 methods (serology and PCR-SSP). HLA typing was considered discrepant also in the case of inability to define a split.

Among the discrepancies, both at A and B locus, for 4 out of the 7 samples in which serology detected a single antigen, a second allele was identified by DNA. In 6 other cases it was necessary to utilize information from serology to define HLA-A,B typing (samples 486, 489, 1076, 1372, 1686, and 1653). Finally, for samples 842 and 1103 it was impossible to exclude or attribute the antigen/allele B67; for sample 557, PCR-SSP could not discriminate between B*1503 and B*4802.

As expected, PCR-SSP turned out to be very useful in solving difficulties in HLA typing; moreover PCR-SSP made it possible to type samples with uninterpretable serology without resorting to new samples of viable cells taken from the CB units; furthermore it allowed HLA-A,B typing of CB collected at the week-end making unnecessary emergency laboratory work.

2. CONTAMINATION OF MATERNAL DNA OF CB UNITS STORED IN OUR BANK

Among the questions to be answered regarding CB use for bone marrow transplantation, one concerns the level of contamination with maternal lymphocytes that could be responsible for severe graft-versus-host disease (GVHD) in immunosuppressed recipients.

In this study, we evaluated the contamination with maternal DNA in 60 consecutive CB units suitable for transplantation, collected in 3 hospitals under a common collection protocol.[27]

To this purpose we used Polymerase Chain Reaction (PCR) amplification of minisatellite sequences followed by chemiluminescent detection. This method does not require radioactive isotopes and allows the detection of maternal cells within CB with a high sensitivity.

500 ml of CB were allotted to DNA extraction for HLA typing and minisatellite analysis.

In all cases, 2 to 10 ml of the mother's EDTA whole blood were also collected, the day following the delivery by standard venipuncture.

Apolipoprotein B gene (ApoB)[28] and D1S80[29] minisatellite analysis with a highly sensitive chemiluminescent method was carried out as described elsewhere.[30]

For the 60 cases examined, ApoB was informative for 27 units, whereas for the remaining samples D1S80 revealed useful.

Noninherited maternal specific allele was detected in 6 out of the 27 CB samples examined with ApoB; with D1S80, 5 samples were found contaminated out of the 33 informative. The level of contamination ranged from 1:100 to 1:2500.

We were able to detect the allele of the diluted DNA up to the 1:2500 dilution which correspond to the presence of 8/20,000 cells (number of cells tested in each PCR reaction).

We could also detect the child paternal allele in 3 out of the 30 mothers whose newborn was heterozygous at the loci examined.

Figure 1 shows an example of ApoB minisatellite analysis of a non contaminated and contaminated cord sample with the respective mothers.

Figure 2 reports the bands corresponding to the artificial mixes 1:500, 1:1000 and 1:2500.

3. CONCLUDING REMARKS

We have analyzed with PCR-SSP 100 consecutive CB units stored in the Milano CB Bank for which serological typing gave doubtful or unreliable results.

The results show that incorrect assignment of HLA-A,B antigens mostly occurred within the same crossreactive group. It remains to confirm these data with family studies or with sequencing methods applied to the genes involved.

However, there is a significant drawback related to PCR-SSP method. In fact, for a single HLA-A,B typing it is necessary to set up numerous PCR reactions. This is a disadvantage when handling a large amount of samples. Moreover, for the definition of certain alleles (B5, B15, and B21 subtypes), with the set of primer mixes employed in this study, it is indispensable to infer some information from serology. Obviously, increasing the number of primers, it would be possible to better define all the sequenced

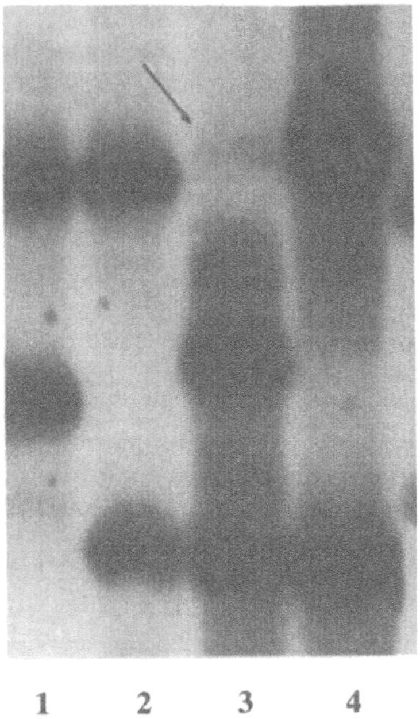

Figure 1. Lane 1 shows a noncontaminated CB sample examined with ApoB: lane 2, the respective mother. Lane 3 shows a contaminated CB sample; lane 4, the mother. Arrow, the band resulting from maternal DNA contamination.

alleles. However, the large number of PCR reactions needed would make HLA typing with this technique cumbersome and expensive. It would be more practical to use additional primer mixes only when appropriate for further subtyping.

PCR-SSOP seems to be more suitable for the screening of a great number of samples: this is our project for the near future.

As far as contamination of CB units with maternal DNA is concerned, our method was able to detect maternal contamination from as few as 2×104 nucleated

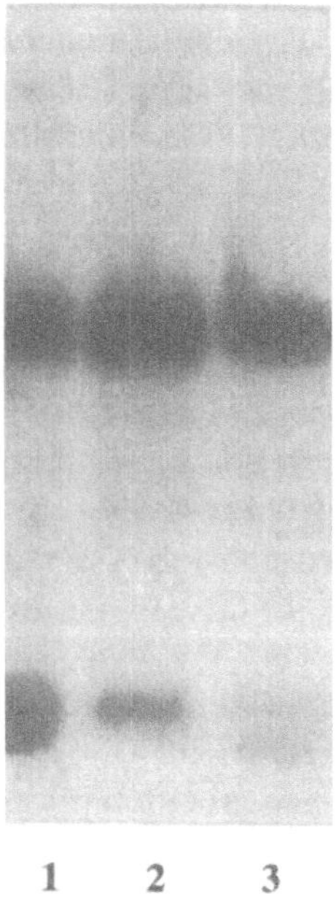

Figure 2. Sensitivity of the system used to detect maternal DNA in CB samples. Bands corresponding to the artificial mixes (standards) 1:500, 1:1000, and 1:2500 are shown (lanes 1, 2, and 3, respectively).

cells with a sensitivity of 0.04%, comparable to that reported by Petit,[31] without resorting to radioactivity.

This kind of analysis is reliable and needs only a small amount of DNA (already extracted for DRB1 typing). However, the clinical relevance of the presence of maternal DNA remains to be established. In our opinion, due to the costs, the information on maternal cell contamination should be eventually included within the quality control tests performed before releasing a unit.

REFERENCES

1. Broxmeyer HE, Douglas GW, Hangoc G, Cooper S, Bard J, English D, Arny M, Thomas L, Boyse EA: Human umbilical cord blood as a potential source of transplantable hematopoietic stem/progenitor cells. Proc Natl Acad Sci USA 86:3828, 1989.
2. Broxmeyer HE, Gluckman E, Auerbach A, Douglas GW, Friedman H, Cooper S, Hangoc G, Kurtzberg J, Bard J, Boyse EA: Human umbilical cord blood: a clinically usefull source of transplantable hematopoietic stem/progenitor cells. Int J Cell Cloning 8 (Suppl. 1):76, 1990.
3. Gluckman E, Broxmeyer HE, Adamson JW, Stevens CE: Hematopoietic reconstitution in a patient with Fanconi's anemia by means of umbilical-cord blood from an HLA-identical sibling. N Engl J Med 321:1174, 1989.
4. Rubinstein P, Rosenfield RE, Adamson JV, Stevens CE: Stored placental blood for unrelated bone marrow reconstitution. Blood 81:1679, 1993.
5. Apperley JF: Meeting report: Umbilical cord progenitor cell transplantation. Bone Marrow Transplant 14:187, 1994.
6. Gluckman E, Rocha V, Boyer-Chammard A, Locatelli F, Arcese W, Pasquini R, Ortega J, Souillet G, Ferreira E, Laporte JP, Fernandez M, Chastang C: Outcome of cord-blood transplantation from related and unrelated donors. New Engl J Med 337:373–381, 1997.
7. Kogler G, Enczmann J, Knipper A, Wernet P: HLA typing strategies of cord blood (CB) for unrelated stem cell banking within Eurocord. Hum Immunol 47(1–2) (abstr.), 1996.
8. Poli F, Scalamogna M, Crespiatico L, Macchi B, Mistò R, Nocco A, Rossini G, Scarpino C, Sioli V, Sirchia G: Comparison of serological and molecular typing for HLA-A and B on cord blood lymphocytes. Tissue Antigens 51:67, 1988.
9. Trachtenberg EA, Erlich HA. DNA-based typing for cord blood stem cell transplantation. J Hematother 5:295–300, 1996.
10. Sadler A, Petronzelli F, Krausa P, Marsh SG, Guttridge MG, Browning MJ, Bodmer JG: Low resolution DNA typing for HLA-B using sequence specific primers in allele-specific or group-specific arms/PCR. Tissue Antigens 44:148–54, 1994.
11. Bunce M, O'Neill CM, Barnardo MCNM, Krausa P, Browing MJ, Morris PJ, Welsh KI: Phototyping: Comprehensive DNA typing for HLA-A, B, C, DRB1, DRB3, DRB4, DRB5, and DQB1 by PCR with 144 primer mixes utilizing sequence-specific primers (PCR-SSP). Tissue Antigens 46:355–67, 1995.
12. Linch DC, Brent L: Can cord blood be used? Nature 340, 1989.
13. Nathan DG: The beneficence of neonatal hematopoiesis. N Engl J Med 321:1190, 1989.
14. Kadowaki JI, Zuelzer WE, Brough AJ: XX/XY lymphoid chimerism in congenital immunological deficiency syndrome with thymic alymphoplasia. Lancet 2:1152, 1965.
15. Wagner JE, Broxmeyer HE, Byrd RL, Zehnbauer B, Schmeckpeper B, Shah N, Griffin C, Emanuel PD, Zuckerman KS, Cooper S, Carow C, Bias W, Santos GW: Transplantation of umbilical cord blood after myeloablative therapy: analysis of engrafment. Blood 79:1874, 1992.
16. Hunziker RD, Gambel P, Wegmann TG: Placenta as a selective barrier to cellular traffic. J Immunol 2:1152, 1984.
17. Schroder J: Passage of leukocytes from mother to fetus. Scand J Immunol 3:369, 1974.
18. Olding LB, Oldstone MBA: Thymus-derived peripheral lymphocytes from human newborns inhibit division of their mothers lymphocytes. J Immunol 116:682, 1976.
19. Wernet P, Kogler G, Someville T: The rapid detection of the quantity (genotype) and cell lineage (immunophenotype) of contaminating maternal white cells in cord blood samples by fluorescence in situ hybridization combined with confocal laser scanning microscopy. Blood Cells 20:296, 1994.

20. Socié G, Gluckman E, Carosella E, Brosard Y, Lafon C, Brison O: Search of maternal cells in umbilical cord blood by polymerase chain reaction of two minisatellite sequences. Blood 83:340, 1994.
21. Hall JM, Lingenfelter P, Adam SL, Lasser D, Hansen JA, Bean MA: Detection of maternal cells in human umbilical cord blood using fluorescence in situ hybridization. Blood 86:2829, 1995.
22. Sociè G, Petit T, Gluckman E, Carosella ED, Brosard Y, Brison O: A highly sensitive polymerase chain reaction method reveals the ubiquitous presence of maternal cells in human umbilical cord blood. Blood (Suppl. 1), 86:392a (abstr), 1995.
23. Scaradavou A, Carrier C, Mollen N, P Rubinstein: Detection of maternal DNA in placental/umbilical cord blood by locus specific amplification of the non inherited maternal HLA gene. Blood 88:1494, 1996.
24. Olerup O, Zetterquist H: HLA-DR typing by PCR amplification with sequence-specific primers (PCR-SSP) in 2 hours: an alternative to serological DR typing in clinical practice including donor-recipient matching in cadaveric transplantation. Tissue Antigens 1992:225–35, 39.
25. Bidwell J, Bidwell E, Savage DA, Middleton D, Klouda P, Bradley B: A DNA-RELP typing system that positively identifies serological well-difined and ill-defined HLA-DR and DQ alleles, including DRw10. Transplantation 45:640–6, 1988.
26. Tiercy JM, Goumaz C, Mach B, Jeannet M: Application of HLA-DR oligotyping to 110 kidney transplant patients with doubtful serological typing. Transplantation 51:1110–14, 1991.
27. Lazzari L, Corsini C, Curioni C, Lecchi L, Scalamogna M, Rebulla P, Sirchia G: The Milan Cord Blood Bank and the Italian Cord Blood Network. J Hematother 5:117, 1996.
28. Boerwinkle B, Xiong W, Fourest E, Chan L: Rapid typing of tandemly repeated hypervariable loci by the polymerase chain reaction: application of apolipoprotein B 3'Hypervariable region. Proc Natl Acad Sci USA: 86:212, 1989.
29. Budowle B, Chakraborty R, Giusti AM, Eisenberg AJ, Allen RC: Analysis of the VNTR locus D1S80 by the PCR followed by high resolution PAGE. Am J Hum Genet 48:137, 1991.
30. Poli F, Crespiatico L, Lecchi L, Sirchia G, Scalamogna M, Sirchia SM, Garagiola I, Pedranzini L: Highly sensitive chemiluminescent method for detection of maternal cell contamination in human cord blood stored for allotransplantation: the experience of the Milano Cord Blood Bank. Blood 89:3061, 1997.
31. Petit T, Raynal B, Sociè G, Landman-Parker J, Bourhis J-H, Gluckman E, Pico J, Brison O: Higly sensitive polymerase chain reaction methods show the freqent survival of residual recipient multipotent progenitors after non- T-cell-depleted bone marrow transplantation. Blood 84:3575, 1994.

8

PRESENCE OF B CELL PROGENITORS IN THE THYMUS

Muneo Inaba, Shin-Ichiro Mori, Akira Sugihara, Biao Feng, Tomoki Ito, Kayo Inaba, and Susumu Ikehara

From First Department of Pathology
Kansai Medical University
Osaka 570-8506, and Department of Zoology
Faculty of Science
Kyoto University
Kyoto 606-8502, Japan

SUMMARY

We characterize B cell progenitors in the thymus. Though mature thymic B cells are $sIg^+/B220^+/CD43^-$, B cell progenitors in the thymus were found to be $sIg^-/B220^{med}/CD43^+$. Most progenitors showed rearranged Ig D-J and V-D-J patterns when DNA from sorted B cell progenitors was amplified by PCR, blotted and probed. When B cell progenitors were purified from the thymus and injected intra-thymically into Igh-allotype disparate recipients, B cells bearing donor-type Igh6 were detected in the thymus but not in the periphery. Thymic B cells generated from these progenitors were $CD5^+$ cells, though there was a broad range of expression of CD5 molecules. Furthermore, B cells purified from the thymus that had been inoculated with B cell progenitors were able to differentiate into antibody-forming cells under the influence of CD40 ligand plus IL-10, and secreted IgM with donor Igh6-allotype. We have thus confirmed the existence of B cell progenitors in the thymus and also characterized the surface markers and their functions.

INTRODUCTION

As has been reported, a small number of B cells exist in the thymus of normal mice and humans.[1-3] Previously we have demonstrated that more than 50% of thymic

Correspondence: Susumu Ikehara, MD, PhD, First Department of Pathology, Kansai Medical University, 10–15 Fumizono-cho, Moriguchi City, Osaka 570-8506, Japan.

Molecular Biology of Hematopoiesis 6, edited by Abraham *et al.*
Kluwer Academic/Plenum Publishers, New York, 1999.

B cells possess CD5 molecules,[1] and that thymic B cells play an important role as self-antigen-presenting cells in the negative selection process in the thymus.[4] We have also found that thymic B cells show a low reactivity to usual B cell-stimulants,[5] although they can be stimulated with MHC class II-restricted T blasts[6] or by CD40 ligand in the presence of IL-10.[7] Thymic B cells thus have unique features. We have recently found that thymic B cells are derived from the bone marrow.[8] However, the precise differentiation pathway of thymic B cells has not been clarified. It has been reported that CD43 can discriminate B cell progenitors from mature B cells in the developmental pathway of B cells; B cell progenitors but not mature B cells in the bone marrow possess CD43.[9] On the basis of the expressions of CD43, B220, and surface Ig (sIg), we here clarify the presence of B cell progenitors in the thymus, and examine their surface markers and capacity to differentiate into mature B cells in the thymic environment.

MATERIALS AND METHODS

Preparation of Thymic Progenitor or Mature B Cells

To obtain thymic B cell progenitors, the thymus cells from adult mice (8–12 wks of age) were treated with a mixture of mAbs against CD4, CD8, Mac-1, and erythrocyte-lineage cells (TER119), then incubated with sheep anti-rat IgG conjugated Dynabeads® M-450 (Dynal A.S., Oslo, Norway). Cells binding with these mAbs were depleted in the magnetic field, and the residual cells were stained with PE-anti-B220 and FITC-anti-CD43 mAbs. The stained cells were applied to a cell sorter (FACStar®, Becton Dickinson & Co.) and the $B220^{med}/CD43^{+}$ cells were separated as B cell progenitors. $B220^{high}/CD43^{-}$ cells were simultaneously purified as mature thymic B cells.

Antibodies and Flowcytometric Analyses

The surface phenotypes of the thymic B cell progenitors were analyzed after 3-color-staining as follows: i) the cells were stained with mAb against ICAM-1, LFA-1, CD44, MHC class I, MHC class II, CD4, CD8, CD45, Mac-1, HSA or normal rat IgG2a (used as an isotype-matched control), followed by ii) FITC-rat anti Ig mAb, then iii) blocked by normal rat Ig (Caltag, San Francisco, CA), iv) stained with PE-anti-B220 mAb, and, v) with biotinylated anti-CD43 mAb, and vi) finally with Av-cychrome. After staining, the cells were analyzed using a FACScan® with appropriate gate settings.

Intrathymic Injection of Thymic B Cell Progenitors

Thymic B cell progenitors prepared from C3H/He mice ($H\text{-}2^{k}$, $Igh6^{a}$) were injected (1.5×10^{4} cells) into the left thymic lobe of MHC-compatible, Igh-disparate newborn B10.BR mice ($H\text{-}2^{k}$, $Igh6^{b}$). The right lobe remained uninjected as a control. Four to seven weeks later, the thymus cells were prepared and stained with FITC-anti-$Igh6^{a}$, PE-anti-B220 and biotinylated anti-CD5 mAbs to detect donor-derived mature $CD5^{+}$ and $CD5^{-}$ B cells. FITC-anti-$Igh6^{b}$ mAb was used as a control.

PCR Analyses of Ig Heavy Chain Rearrangement in Thymic B Cell Progenitors

The rearrangement of Ig heavy chain loci in the thymic B cell progenitors was analyzed using a combination of the PCR and Southern blot techniques, as reported previously.[10,11] Thymic progenitor or mature B cells were purified using a FACStar, and the DNA in each population was prepared following the standard protocol.

RESULTS

Presence of B Cell Progenitors in the Thymus

It has been thought that the expression of B220 and CD43 is a marker of B cell progenitors in the bone marrow,[9] and we have used this criterion to check for the presence of B cell progenitors in the thymus. T cell-depleted thymic cells from C3H/He mice ($Igh6^a$) were triple-stained with anti-B220, anti-CD43, and anti-$Igh6^a$ (IgM^a) mAbs. There were two main populations in the thymic B220-bearing B cells (approximately 1.3% of the total thymic cells) after gating out the T cells ($B220^-/CD43^+$ cells). One was the $sIg^+/B220^{high}/CD43^-$ cell population. These are thought to be mature thymic B cells. The other was a newly discovered $B220^{med}/CD43^+$ cell population lacking sIgM (0.7% of the total thymic cells). To clarify the characteristics and developmental activity of these "putative" B cell progenitors, we first investigated other surface markers and compared them with those of mature thymic B cells.

Surface Markers of Thymic B Cell Progenitors

The expression of surface markers on thymic B cell progenitors was examined after triple-staining with anti-B220, anti-CD43, and other mAbs. The expression of MHC class I, CD45, ICAM-1, and LFA-1 on "putative" B cell progenitors was comparable to mature thymic B cells, and the levels of MHC class II and HSA were lower on the B cell progenitors than on the mature thymic B cells. Both populations were lacking T cell (CD4 and CD8) and macrophage (Mac-1) markers. CD44 was expressed on neither progenitor nor mature thymic B cells.

Evidence for Intra-Thymic Differentiation of Thymic B Cell Progenitors

Next, we attempted to examine whether thymic B cell progenitors differentiate into mature B cells in the thymus. $B220^{med}/CD43^+$ cells from C3H/He mice ($Igh6^a$) were purified using a FACStar, and 1.5×10^4 $B220^{med}/CD43^+$ cells were injected into the thymic lobe of neonatal B10.BR ($Igh6^b$) recipients. Seven weeks later, we detected mature thymic B cells of donor origin by mAb against donor IgM allotype ($Igh6^a$). A small number of mature thymic B cells of donor origin ($B220^+/Igh6^{a+}$) were observed in the cells from the injected thymic lobe, but not in the cells from the uninjected thymic lobe. There were approximately five times the number of $B220^+/Igh6^{a+}$ cells as injected B cell progenitors. Therefore, sIg^- thymic "progenitor" B cells appear to proliferate and differentiate into sIg^+ cells in the thymus. Furthermore, these mature B cells purified from the thymus had the potential to generate Ig-secreting cells in the presence of CD40L + IL10.[7]

It is noted that no mature B cells bearing donor Igh6^a were detected in the periphery (the spleen or lymph nodes) 7 wks after intrathymic injection. The expression of CD5 molecules on these mature thymic B cells of donor origin was next examined by three-color fluorocytometric analyses. The CD5$^+$ cells were detected in the Igh6^{a+} cells, though there was a broad range (low to moderate) of expression of CD5 molecules and the level was as detected in the freshly isolated mature thymic B cells.

Rearrangement of Immunoglobulin Heavy Chain Loci in Thymic B Cells

The Ig gene rearrangement of sIg$^-$/B220^{med}/CD43$^+$ thymic B cell progenitors was next examined using the PCR technique according to the method of Ehlich et al.[10] The population containing the thymic B cell progenitors showed a rearranged Ig D-J gene configuration as thymic mature B cells (sIg$^+$/B220^{high}/CD43$^-$) and control splenic B cells when PCR products were identified by the J_H4 probe. When PCR was carried out using a mixture of 5′ V_H gene primers (V_H7183 genes, V_HQ52 genes, and V_HJ558 genes) to examine the V-D-J rearrangement, the rearranged PCR product was detected also by the J_H2 oligo both in the thymic mature B cells and thymic B cell progenitors. The positive band representing the V-D-J rearrangement in the thymic B cell progenitors was observed. The band at 280 bp corresponds to a rearrangement of V_HQ52 genes to J_H2, and the band at 470 bp corresponds to a rearrangement of V_H7183 or V_HJ558 genes to J_H2. No V-D-J rearrangement pattern was detected in the negative control P815 mastcytoma cells. Therefore, the population identified by the surface phenotypes sIg$^-$/B220^{med}/CD43$^+$ actually contained B cell progenitors (B cell-lineage cells) exhibiting rearranged Ig D-J and V-D-J gene configurations.

DISCUSSION

Recently, it has been reported that Lin$^-$CD4^{low} precursors purified from the thymus differentiate into T cells and dendritic cells after the injection of donor type cells into the thymus, while donor type B cells are detected only when Lin$^-$CD4^{low} precursors have been injected intravenously.[12,13] Furthermore, Pèault et al. reported in vitro development of B cells and macrophages from the fetus (gestation day 12–15) using a cell line (bone marrow stromal cell line) that supports pre-B and myeloid cell differentiation.[14] These findings indicate that the early precursor/progenitor cells, which have the potential to differentiate into B cells, T cells, dendritic cells, and myeloid cells, exist in the adult or fetal thymus,[13,14] although the intermediate stage in the developmental pathway of thymic B cells is still unclear.

In this report, we have found, in the thymus, the presence of B cell progenitors with phenotype of sIg$^-$/B220^{med}/CD43$^+$ and a rearranged Ig gene configuration but lacking sIg. We have also found that these progenitors can proliferate and differentiate into mature B cells in the thymic environment and also into Ig-secreting cells in the presence of the appropriate stimuli.[7]

We have previously shown that there are two populations, CD5$^+$ and CD5$^-$ cells,[1] in the mature thymic B cells, and that both populations are derived from bone marrow cells when tested in allogeneic bone marrow transplantation.[8] This is the case when B220^{med}/CD43$^+$ cells were used as progenitors. Mainly CD5$^+$ cells with donor Igh6-allotype are detected, though the level of CD5 molecules is broad (low to moderate), being derived from B220^{med}/CD43$^+$ B cell progenitors. At present, we do not know whether these populations with this broad frequency on CD5 expression are derived from common or different progenitor B cells (with the same immunophenotype). However, it should be noted that thymic $CD5^+$ B cells are derived from this progenitor B cell population.

B cell development in the bone marrow has been extensively studied on the basis of the differential expression of surface molecules and the Ig gene configuration. The developmental pathway of B cells is known to be controlled by cytokines and stromal cells.[15]

Recently, it has been shown that $CD4^-CD8^-CD3^-/TcR^-$ triple-negative thymocytes, which have the potential to generate mature T cells, can differentiate into sIg^+ mature B cells on the selected thymic stromal cell line (θ1 stroma),[16] demonstrating that some stroma cells can directly support B cell differentiation. This is in accordance with our experiments where stroma cells can support the B cell differentiation. Hardy et al. reported the existence of pre-pro-B and pro-B cells in the bone marrow, based on not only the differential expression of CD43, BP-1, and HSA but also the Ig gene configuration in relation to the developmental stages.[9] Though it is not known whether the developmental pathway of B cells in the bone marrow is the same as in the thymus, we believe that many thymic B cell progenitors ($sIg^-/B220^{med}/CD43^+$) are in the pro-B stage with increasing D-J and V-D-J rearrangements. The pattern of surface phenotypes on our B cell progenitors also suggests this possibility. The low expression of HSA is similar to the level between Fr. A and Fr. B, but not identical to Fr. A, as reported by Hardy, et al.[9] Therefore, our assumption that thymic $B220^{med}/CD43^+$ cells are in the pro-B stage is in accordance with Hardy's report not only as regards the Ig gene level but also the expression of HSA. The slight discrepancy in the expression of HSA might be attributable to the organs where pro-B cells exist. Therefore, from these findings, we conclude that our $B220^{med}/CD43^+$ cells are in, at least, the pro-B stage.

We thank Dr. Hideo Yagita (Junten-doh Medical Univ.) for CD40L-CHO cells, Mr. Fujio Ishida and Mr. Eiichi Ohtsuki (Research Center of Kansai Medical University) for flow cytometry and image-analyzing studies, and Ms. Keiko Ando for preparing the manuscript.

ACKNOWLEDGMENT

Supported by a grant from the Ministry of Education, Science and Culture, a grant for Experimental Models for Intractable Diseases from the Ministry of Health and Welfare of Japan and a grant from the Japanese Private School Promotion Foundation.

REFERENCES

1. Miyama-Inaba M, Kuma S-I, Inaba K, Ogata H, Iwai H, Yasumizu R, Muramatu S, Steinman RM, Ikehara S: Unusual phenotype of B cells in the thymus of normal mice. J Exp Med 168:811, 1988.
2. Kumamoto T, Inaba M, Imamura Y, Nango K-I, Adachi Y, Soe Than, Inaba K, Kagawa K, Ikehara S: Characterization of B cells in human thymus. Immunobiolgy 183:88, 1991.
3. Spencer J, Choy M, Hussel T, Papadaki L, Kingston JP, Isaacson PP: Properties of human thymic B cells. Immunology 75:596, 1992.
4. Inaba M, Inaba K, Hosono M, Kumamoto T, Ishida T, Muramatu S, Masuda T, Ikehara S: Distinct mechanism of neonatal tolerance induced by dendritic cells and thymic B cells. J Exp Med 173:549, 1991.

5. Fukuba Y, Inaba M, Taketani S, Hitoshi Y, Adachi Y, Tokunaga R, Ikehara S: Functional analyses of thymic B cells. Immunobiology 190:150, 1994.
6. Inaba M, Inaba K, Adachi Y, Nango K, Ogata H, Muramatu S, Ikehara S: Functional analyses of thymic CD5[+] B cells. Responsiveness to major histocompatibility complex class II-restricted T blasts but not to lipopolysaccharide or anti-IgM plus interleukin 4. J Exp Med 171:321, 1990.
7. Inaba M, Inaba K, Fukuba Y, Mori S, Haruna H, Doi H, Adachi Y, Iwai H, Hosaka H, Hisha H, Yagita H, Ikehara S: Activation of thymic B cells by signals of CD40 molecules plus interleukin 10. Eur J Immunol 25:1244, 1994.
8. Soe Than, Inaba M, Inaba K, Fukuba Y, Adachi Y, Ikehara S: Origin of thymic and peritoneal Ly-1[+] B cells. Eur J Immunol. 22:1299, 1992.
9. Hardy RR, Carmack CE, Shinton SA, Kemp JD, Hayakawa K: Resolution and characterization of pro-B and prepro-B cell stages in normal mouse bone marrow. J Exp Med 173:1213, 1991.
10. Ehlich A, Schaal S, Gu H, Kitamura D, Muller W, Rajewsky K: Immuno-globulin heavy and light chain rearrange independently at early stages of B cell development. Cell 72:695, 1993.
11. Taketani S, Inazawa J, Abe T, Furukawa T, Kohno H, Tokunaga R, Nishimura K, Inokuchi H: The human Protoporphyrinogen Oxidase Gene (PPOX): Organization and location to chromosome 1. Genomics 29:698, 1995.
12. Wu L, Scollay R, Egerton M Pearse M, Spangurde GJ, Shortman K: CD4 expressed on earliest T-lineage precursor cells in the adult murine thymus. Nature 349:71, 1991.
13. Ardavin C, Wu L, Li C, Shortman K: Thymic dendritic cells and T cells develop simultaneously in the thymus from a common precursor population. Nature 362:761, 1993.
14. Péault B, Khazaal I, Weissman IL: In vitro development of B cells and macrophages from early mouse fetal thymocytes. Eur J Immunol 24:781, 1994.
15. Era T, Nishikawa S, Sudo T, Fu-Ho W, Ogawa M, Kunisada T, Hayashi SI, Nishikawa SI: How B-precursor cells are driven to cycle. Immunol Rev 137:35, 1994.
16. Montecino-Rodriguez E, Johnson A, Dorshkind K: Thymic stromal cells can support B cell differentiation from intrathymic precursors. J Immunol 156:963, 1996.

9

DE NOVO AND SECONDARY AML CELLS DIFFER IN THEIR KINETIC RESPONSE TO HGF PRIMING

M. A. Smith, C. J. Pallister,* R. W. Luxton,* C. R. J. Singer, and J. G. Smith*

Department of Haematology
Royal United Hospital NHS Trust, Bath
UK and Faculty of Applied Sciences
University of West of England, Bristol, UK*

ABSTRACT

This study has assessed the effect of haemopoietic growth factor (HGF) priming on the S-phase activity of cells from patients with *de novo* AML (n = 33) or AML secondary to prodromal myelodysplasia (n = 15). An immunofluorescent ligand binding assay was used to determine the occurrence of receptors for HGFs under scrutiny, namely, G-CSF, GM-CSF, IL-3, and SCF. In order to determine whether *in vitro* kinetic or receptor data might correlate with clinical outcome, data from 36 AML samples was subjected to multivariate discriminant analysis using a quadratic discriminant function.

MATERIALS AND METHODS

Marrow was obtained from 48 patients, 33 with *de novo* AML and 15 having progressed to AML from MDS, all with blast burdens >90%. Normal marrow samples were obtained from 17 individuals with non-Hodgkins lymphoma undergoing marrow examination for staging purposes. These samples were uninvolved and free of reactive features.

Light density (SG < 1.077) marrow cells (LDMCs) were depleted of T-cells and

---

Address for correspondence: M.A. Smith, Bone Marrow Laboratory, Department of Haematology, Royal United NHS Trust, Combe Park, Bath BA1 3NG, UK, Tel: (0)1225 824733, Fax: (0)1225 461044

Molecular Biology of Hematopoiesis 6, edited by Abraham *et al.*
Kluwer Academic/Plenum Publishers, New York, 1999.

monocytes using complement mediated monoclonal antibody cytolysis of CD4, CD8, and CD14 cells as previously described.[1]

T-cell and monocyte depleted LDMCs were cultured in QBSF 58™ serum free medium (SFM) (Sigma) at a concentration of 1.5×10^6/ml for seven days either in SFM alone as a negative control or with G + GM-CSF (100 U/ml each), IL-3 (10 ng/ml), SCF (50 ng/ml) or combinations of these HGFs. SCF was a kind gift of Amgen Limited. Following priming in suspension, cells were washed twice and aliquoted for flow cytometric studies.

The percentage of normal and AML T-cell and monocyte depleted LDMC's in S-phase following priming ± HGFs was determined by flow cytometric analysis of DNA content using Becton Dickinson's Cycle Test Plus ™DNA reagent kit as recommended by the manufacturer. Because of variation in individual S-phase activities, it was decided to standardise results by using an "S-phase recruitment ratio" (SPRR) obtained by dividing the resultant S-phase percentage values following HGF priming by the seven day SFM incubation control result for individual patients. Further analyses used mean absolute S-phase values.

Percentages of normal and AML T-cell and monocyte depleted LDMC's bearing surface receptors for G-CSF, GM-CSF, IL-3 and SCF were quantitatively assessed by immunofluorescent ligand binding assays using Fluorokine™ kits by R&D Systems Europe Ltd., according to the manufacturer's instructions. Only receptors functionally capable of binding their respective ligand were detected.

Multivariate discriminant analysis using a quadratic discriminant function was used to correlate in vitro data with clinical outcome. Significance values were obtained using the Mann Whitney U test for non-parametric in vitro data.

RESULTS

Comparison of Percentages of *De Novo* AML, Secondary AML, and Normal Cells Bearing Surface Receptors for Haemopoietic Growth Factors

Figure 1 shows relative percentages of cells from *de novo* and secondary AML T-cell and monocyte depleted LDMCs bearing HGF receptors. Compared with cells from normal samples significantly greater percentages of *de novo* AML cells bear receptors for G-CSF, GM-CSF, IL-3, and SCF ($p = 0.011$, <0.001, <0.001, and <0.001 respectively). Percentages of secondary AML cells bearing receptors for these HGFs except GM-CSF were significantly greater than for normal cells ($p = 0.033$, 0.346, <0.001, and 0.011 respectively). When the two groups of AML samples were compared, significantly greater percentages of *de novo* AMLs were found to carry receptors for GM-CSF and SCF than secondary AMLs ($p = 0.013$, and 0.016 respectively). No significant differences in G-CSF, GM-CSF, IL-3, and SCF relative mean receptor densities were observed when the groups were compared (results not shown).

Comparison of S-phase Recruitment Ratios by Haemopoietic Growth Factors of *De Novo* AML, Secondary AML, and Normal Cells

Kinetic recruitment ratios for *de novo* and secondary AML and normal samples assessed following seven days incubation in SFM ± G + GM-CSF, IL-3, SCF, G + GM-CSF + SCF or IL-3 + SCF is presented in Fig. 2.

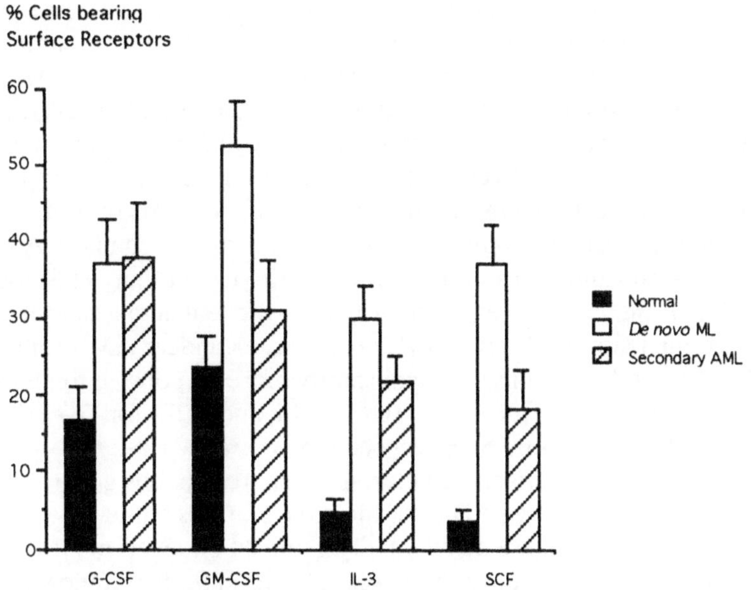

Figure 1. Comparison of percentages of T-cell and monocyte depleted cells from *de novo* AML (n = 20), secondary AML (n = 15), and normal samples (n = 17), bearing surface receptors for G-CSF, GM-CSF, IL-3, and SCF presented as mean percentages ±1 SE for each group.

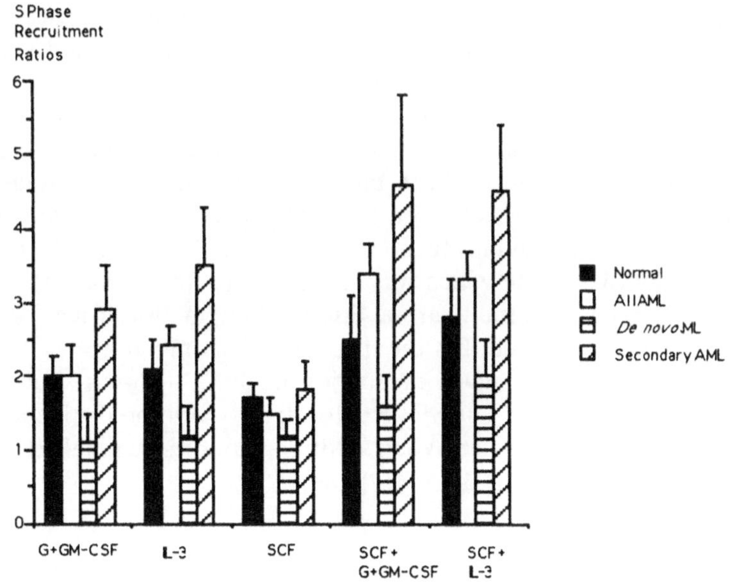

Figure 2. Comparison of S-phase recruitment ratios for normal (n = 11), *de novo* (n = 16), and secondary AML (n = 15) samples. S-phase recruitment ratios obtained by dividing percentage S-phase values following priming by respective SFM control values, are presented for T-cell and monocyte depleted samples following seven days incubation in SFM + HGFs as mean values ± 1SE.

Comparison of all AML samples with normal samples revealed no significant differences in SPRR. However, normal samples showed significantly greater recruitment into S-phase than *de novo* AML samples following priming with G + GM-CSF, IL-3 or G + GM-CSF + SCF ($p = 0.003$, 0.007, and 0.021 respectively). However, neither SCF alone nor IL-3 + SCF priming resulted in significantly different recruitment compared with similar priming of normal cells. Of all the priming permutations, IL-3 + SCF in combination was the most effective stimulator of *de novo* AML resulting in the greatest increase in cell cycling. Comparison of normal and secondary AML samples revealed no significant difference in response to priming with any HGF permutation, presumably due to the large standard error bars in the leukaemia samples.

Finally, comparison of SPRRs for *de novo* and secondary AML samples demonstrated that the secondary group was significantly more responsive to all priming permutations except SCF alone ($p = 0.007$, 0.007, 0.014, and 0.011 for G + GM-CSF, IL-3, G + GM-CSF + SCF, and IL-3 + SCF respectively).

SPRRs facilitate comparison of heterogeneous results but give no indication of absolute levels of S-phase activity. During the course of this study it became apparent that percentage baseline S-phase activity had an impact on levels of AML recruitment by HGFs. Comparison of absolute percentages of cells in S-phase following incubation in SFM ± HGFs for normal, *de novo* and secondary AML samples showed that the IL-3 + SCF priming cocktail was the most potent stimulator of S-phase activity irrespective of sample type. All normal samples were reasonably consistent with mean kinetic activity of 6.67 ± 0.9% with SFM rising to a maximum of 16.3 ± 2.3% with IL-3 + SCF which represents a 2.4 fold increase in SPRR. By contrast, both AML groups fell into distinct categories determined by high or low levels of baseline activity as illustrated in Table 1. *De novo* and secondary samples with low baseline S-phase values (<10%) showed significantly greater recruitment by SCF + IL-3 than those with high (>10%) levels ($p = 0.021$ and 0.019 respectively). However with *de novo* AML the difference in ratios was obtunded. Comparison of low with high baseline SPRRs in each disease category were 1.42 fold in *de novo* AML but 3.75 fold in secondary disease. Although the IL-3 + SCF combination was chosen to highlight these observations (as this was the most effective priming cocktail) the observation held true for all the other priming permutations except SCF alone which was ineffective (data not shown).

Results analysed using a multivariate discriminant analysis using a quadratic discriminant function demonstrated that the three best discriminators of CR were a) low baseline S-phase activity after seven incubation in SFM, b) low recruitment into S-phase by SCF + IL-3 c) and low G-CSF receptor density with predictive importance of these variables ranked in that order. When all three variables were taken together they permitted prediction of CR in 34/36 AML patients.

Table 1. SFM S-phase and S-phase recruitment ratios

	Normal n = 11	1° AML n = 22 (low) n = 11 (high)	2° AML n = 10 (low) n = 5 (high)
S phase post 7 days in SFM	6.67 ± 0.9	<10%: 5.0 ± 0.7 >10%: 16.7 ± 1.6	<10%: 2.5 ± 0.7 >10%: 23.0 ± 2.3
S phase recruitment with IL-3 + SCF	2.8 ± 0.5	<10%: 2.0 ± 0.1 >10%: 1.4 ± 0.1	<10%: 6.0 ± 1.1 >10%: 1.6 ± 0.2

DISCUSSION

To date several clinical studies have examined the use of HGFs in the supportive setting post chemotherapy or investigated their role in priming before and during treatment. A recent review of such trials suggests that administration of HGFs as a supportive measure post chemotherapy is often beneficial but no conclusive findings with regard to complete remission (CR) rate and duration have emerged when HGFs are used as part of a priming strategy.[2]

For this *in vitro* study G-CSF and GM-CSF in combination were used as priming agents as both are of clinically available. However, in order to target more primitive progenitors, IL-3 and SCF were included the latter because of its stimulatory effects at or near the level of the multipotent stem cell. In addition to exerting an effect directly on early progenitors, SCF is known to synergise strongly with G-CSF, GM-CSF, and IL-3. Exploitation of this synergistic property may make SCF a potential candidate for inclusion in priming regimens. Previous *in vitro* work by our group has demonstrated variable sensitivity of AML progenitors to Ara-C mediated cytotoxity.[3,4] We demonstrated that priming of AML cells with G-CSF, GM-CSF, IL-3, and SCF alone or in combination did not necessarily result in enhanced cytotoxic effect. Indeed, 15% of AML samples remained resistant to HGF enhanced Ara-C mediated cytotoxicity with the remainder exhibiting a variable range of priming responsivity.

This present study shows that although a greater percentage of both *de novo* and secondary AML cells possessed receptors for all HGFs studied compared with normals, this did not necessarily translate into increased S-phase. It was apparent that the baseline level of S-phase activity of AML cells governed HGF recruitment in both AML groups, with low baseline samples recruiting significantly more effectively than their high baseline counterparts. Presumably cells already in cycle, possibly via autocrine or other mechanisms, have limited capacity for further recruitment.

Several differences between *de novo* and secondary AML are well known with the overall response rate of secondary AML to initial induction chemotherapy being lower than in *de novo* leukaemia. Our finding that low baseline secondary AML samples recruited considerably more efficiently than those from low baseline *de novo* samples may be a feature of a fundamental biological difference between the two groups.

Bearing in mind that caution must be exercised when extrapolating from *in vitro* to *in vivo* situations, three variables which were apparently linked with CR were low baseline S-phase activity, low responsivity to IL-3 + SCF priming and low G-CSF receptor density. Conceivably, these factors could define a malignant cell which is less likely to proliferate in an autocrine fashion and thus be less able to recover from the effects of chemotherapy when autologous HGFs are attempting to drive the recovery of normal haemopoiesis. This predictive model was re-enforced by the finding that 10/11 patients with *de novo* and 5/5 patents with secondary disease who had baseline S-phase values >10% failed to achieve remission with the first course of chemotherapy.

In summary, several points have emerged from this study which may be of relevance to the success or failure of an HGF priming strategy in patient management. The data might suggest defective intracellular signalling in approximately one third of AML samples whether from *de novo* or secondary AML patients as the presence of growth factor receptors did not result in S-phase recruitment. It may be possible to identify those patients who will benefit from HGF therapy, whether priming or supportive, by performing *in vitro* testing to characterise the kinetic response of their presenting leukaemia population and use HGFs at the time of chemotherapy in those patients with low baseline S-phase activity who respond *in vitro*. If priming is to be successful, then the most efficacious priming cocktails should include SCF. Supportive HGF therapy would be reserved for patients with poor SPRRs. Finally, kinetic and HGF receptor factors may influence clinical outcome in AML regardless of whether HGFs have been used as part of the patient's treatment.

ACKNOWLEDGMENT

The authors are grateful for support from the Bath Leukaemia Research and Bone Marrow Transplant Fund and from Amgen UK Ltd.

REFERENCES

1. Smith JG, Seenan AK, Smith MA, Galloway E, Lesko MJ, Lucie NP, Robertson MRI, Rowan RM: Cyclical neutropenia and T8 lymphocyte mediated stimulation of granulopoiesis. British Journal of Haematology 60:481, 1985.
2. Buchner T, Hiddermann W, Wotmann B, Zuhldorf M, Rottmann R, Innig G, Maschmeier G, Ludwig WD, Sauerland MC, Heincke A: Hematopoietic growth factors in acute myeloid leukemia: supportive and priming effects. Seminars in Oncology 24:(1):124, 1997.
3. Smith MA, Singer CRJ, Pallister CJ, Smith JG: The effect of haemopoietic growth factors on the cell cycle of AML progenitors and their sensitivity to cytosine arabinoside "in vitro". British Journal of Haematology 90:767, 1995.
4. Smith MA, Smith JG, Pallister CJ, Singer CRJ: Haemopoietic growth factors, the cell cycle of acute myeloblastic leukaemia progenitors and sensitivity to cytosine arabinoside. Leukemia and Lymphoma 23:467, 1996.

10

MOLECULAR DIAGNOSIS OF PHILADELPHIA-POSITIVE LEUKEMIAS

Cedrik Haškovec, and Jaroslav Polák

Department of Molecular Genetics
Institute of Hematology and Blood Transfusion
U nemocnice 1, 128 20 Prague
Czech Republic

The **Philadelphia (Ph) chromosome** was the first chromosomal abnormality associated with a specific malignant disease in humans, namely chronic myelogenous leukemia (CML).[1] In 1973 the Ph chromosome was identified as one partner in a reciprocal chromosomal translocation between chromosomes 9 and 22—t(9;22)(q34;q11).[2] The translocation results in the formation of two hybrid genes—BCR/ABL on the Ph chromosome and ABL/BCR on 9q+. The **BCR/ABL** gene seems to be the main of (probably three)[3] molecular events leading to a development of CML.

More than 95% of patients with clinically acceptable **CML** have the BCR/ABL gene in their leukemic cells and this gene is thus a molecular hallmark of CML. This gene was found in all Philadelphia positive (Ph+) cases and we can find it also in about 5% of total CML which are without the Ph chromosome—Ph negative CML, where t(9;22) is masked by further chromosomal translocation(s). These cases are clinically indistinguishable from those with the Ph chromosome. The BCR/ABL gene is not exclusive to CML and it is also present, according to the age of the patients, in 4 to 45% of acute lymphoblastic leukemias (Ph+ **ALL**), where it is a negative prognostic factor. This gene is rarely observed in **AML** (in about 1% cases) and exceptionally in multiple **myeloma** and B **lymphoma**.[4] There are contradictory views on the presence of the BCR/ABL gene in **essential thrombocytemia** (ET). Some authors reported the presence of this gene in ET patients,[5] others take these patients as a clinical variant of CML, i.e. CML with thrombocytemic onset.[6] There are also a few cases of **myelodysplastic syndrome** (MDS) with the Ph chromosome and a rearranged BCR gene.[7] Nevertheless, it seems that in these cases the Ph chromosome is a late acquisition during the progression of the disease and that the malignant Ph+ clone is secondarily evolved.

Almost all (>99%) of BCR/ABL-positive CML patients have breakpoints in a restricted area called the **major breakpoint region** (M bcr)—Fig. 1. Breaking in the

Molecular Biology of Hematopoiesis 6, edited by Abraham *et al.*
Kluwer Academic/Plenum Publishers, New York, 1999.

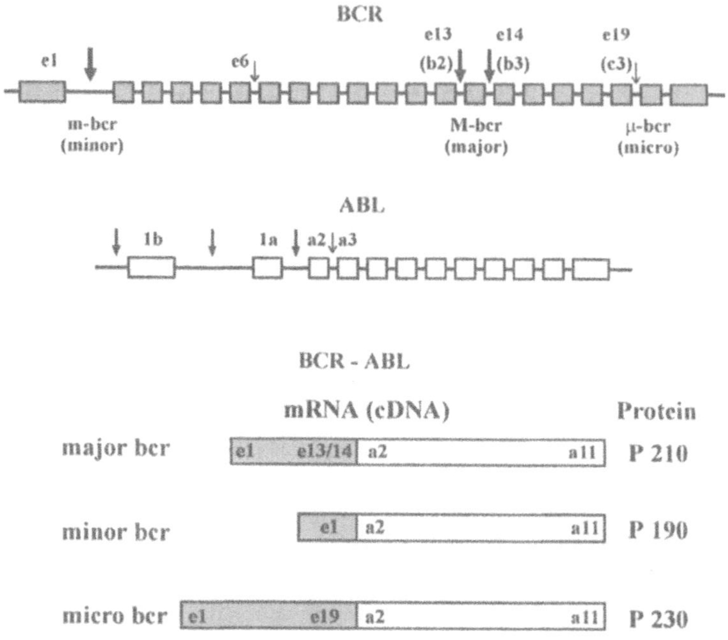

Figure 1. Simplified scheme of the BCR and ABL genes with indicated breakpoints. Possible BCR/ABL transcripts and proteins originating from individual breaks on the BCR and ABL genes.

major bcr gives two mRNA transcripts with junctions of the BCR and ABL genes: e13-a2 or e14-a2 according to a new nomenclature (b2-a2 and b3-a2 according to the old one). These mRNA transcripts encode a protein of 210kD molecular mass (P210). The BCR/ABL gene with major bcr was found also in about 50% of adult Ph+ ALL and 25–40% childhood Ph+ ALL. Very rare is this type of the BCR/ABL gene in AML and possibly in ET—see the above discussion. Some authors[8] consider, at least some, if not all, of the Ph+ ALL with the major bcr BCR/ABL gene and P210 as beginning lymphoid blast crisis of CML with an uncertain (clinically silent) chronic phase.

The second possible breakpoint region called **minor (m) bcr** in the BCR/ABL gene, which is typical for childhood Ph+ ALL, (60–75% of Ph+ cases), is located in the long first intron at the beginning of the BCR/ABL gene—Fig. 1. The transcript with minor bcr has e1-a2 junction encoding a protein of 190kD (P190). The minor bcr BCR/ABL gene was found also in about 50% of Ph+ ALL in adults and very rarely in AML (all M4 phenotype), in B cell lymphoma and multiple myeloma.[4] While this type of gene is also very rare (around 0.5%) in CML (mainly cases with monocytosis) as a sole BCR/ABL gene, it is coexpressed in certain amounts in most (>90%) CML besides the major bcr BCR/ABL gene.[9,10] The explanation of this phenomenon is probably an alternative splicing or missplicing of the major bcr BCR/ABL gene, although four cases of CML in lymphoid blast crisis with both minor and major bcr translocation were observed.[11]

Very rare are breakpoints in the third breakpoint region—**micro (μ) bcr**. In this case joining of the exon e19 (c3 according to the old classification) of the BCR gene with the exon a2 of the ABL gene is accomplished and e19(c3)-a2 transcript is formed—Fig. 1. This BCR/ABL gene codes for a protein of 230kD (P230). Only a few cases of CML have so far been published, mostly cases with a mild course of the disease,

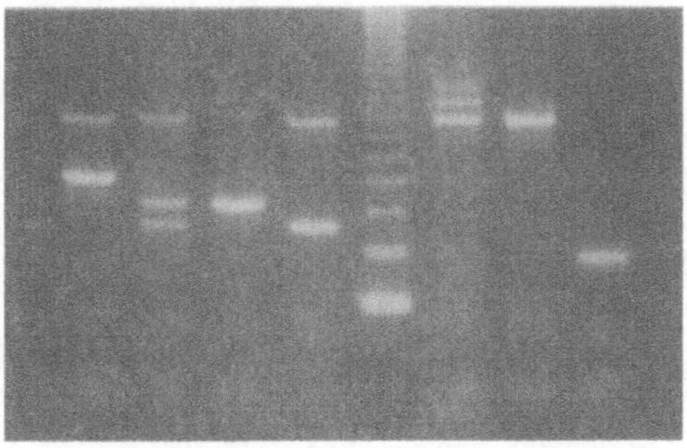

Figure 2. Results of multiplex RT-PCR amplification of the BCR-ABL and the control BCR gene, originating from the major bcr breakpoints (lanes 2, 3, 4, and 8), minor bcr (lane 1), micro bcr (lane 6), 123 bp molecular marker (lane 5) and negative control (lane 7).

some of them with thrombocytemia[12,13,14,15] classified also as Ph+ chronic neutrophilic leukemia (CNL).[13] Also one cases of "classical" CML was recently reported[16] and another case, not yet reported, was found in our laboratory. One case of a patient in myeloid blast crisis (BC) of CML was published by Wada et al.[17] and one patient diagnosed as AML (possibly also a myeloid BC CML) in our institute[18]—the result of the RT-PCR is in Fig. 2. Attempts to detect the P230 BCR/ABL protein were unsuccessful in most micro bcr cases and it was detected only in primary leukemic cells in one of our patients[18] and in a cell line derived of the patient of Wada et al.[17]

Very rare (<0.5% CML) are also cases with breakpoints in the 2nd intron of the ABL gene providing junctions of the **ABL exon a3** with either exon e13(b2), or e14(b3), or e1 of the BCR gene and formation e13-a3, e14-a3 or e1-a3 transcripts of the BCR/ABL gene—Fig. 1. Two cases of a classical CML,[19,20] one case of a myeloid bp CML[21] (the result of the RT-PCR is in Fig. 2) and five cases of ALL[20,22,23,24] with junction of the exon a3 with either the major or minor bcr BCR gene have so far been reported. The low incidence of these cases is probably due to a relatively short intron (0.6 kb) in which breaking occurs in the referred cases, in comparison with about 300 kb long region in which the vast majority of breakpoints in the ABL gene take place—Fig. 1. The existence of CML and BCR/ABL+ ALL cases without exon a2 in the BCR/ABL gene, which encodes a part of the SH 3 domain of the ABL gene, points to the SH 3 domain in the BCR/ABL protein being not necessary for the development of these diseases. These cases, though very rare, also point to the necessity to use the ABL exon a3 primer (instead of the commonly used a2 primer) in the RT-PCR—see below—Fig. 3.

There are some **other** possible breakpoints in the BCR gene and **functional BCR/ABL** junctions of the BCR/ABL gene. The following prerequisites need to be fulfilled for other functional junctions of the BCR/ABL gene: i. Breaks in the BCR and ABL genes have to produce "in frame" junction (i.e. not shifting a reading frame and not creating a stop codon) and leading to a functional BCR/ABL protein. (E.g. e12(b1)-a2 junction, which codes for a stop codon and which was found in addition to the classical major bcr e13(b2)-a2 junction,[25] does not probably have any clinical significance.) ii. The BCR/ABL protein originating from the translocation should contain the SH 2 domain and other "downstream" domains in the ABL part. (The presence of the SH 2 and other domains was shown to be necessary for the transforming activity of the ABL gene.) iii. As to the BCR gene, it seems that the presence of only the first exon e1 in the BCR/ABL gene is essential and sufficient for activation of the ABL protein. Thus the BCR part of the BCR/ABL gene provides more possibilities for functional junc-

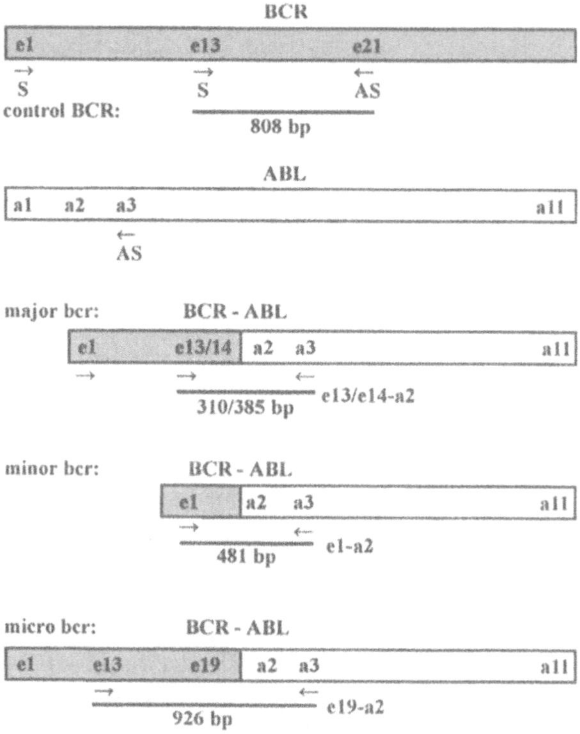

Figure 3. Position of primers and the size of the PCR products of multiplex RT-PCR of the individual variants of the BCR/ABL gene and the control BCR gene (according to Cross et al.[28])

tions with the ABL gene. According to Hochhaus et al.[26] the following theoretical possibilities exist: e6-a2, e6-a3, e12(b1)-a3, e19(c3)-a3, e20(c4)-a2, and e20(c4)-a3. One of these variants, i.e. e6-a2, and the production of a functional protein (P185) has been documented in one CML patient.[26] Other variants created also by insertions or deletions in a junctional site may produce other possibilities. One such example was recently reported in a CML patient.[27] A functional junction of the exons e2 and a2, which is normally nonfunctional, was made possible by insertion of 21 bp sequence from the intron of the ABL gene into mRNA at the junction site and making thus the junction functional.

There are several methods of detection of the Ph chromosome and the BCR/ABL gene or its expression. The fluorescence in situ hybridization (FISH) detecting the presence of the gene at a chromosomal level and the reverse transcription-polymerase chain reaction (RT-PCR) detecting an expression of the gene, are both the most suitable methods for diagnostic purposes. These methods do not require many cells, as does, e.g. assay the restriction fragment length polymorphism (RFLP), do not require dividing cells, like classical karyotyping, and they can reveal Philadelphia translocation even in cases in which the translocation is masked by subsequent translocations. As to the RT-PCR method, we successfully use the **multiplex RT-PCR** according to Cross et al.[28] Two pairs of primers are used: one pair of sense primers specific for exon e1 and e13 of the BCR gene and two antisense primers, one specific for the e21 of the BCR gene and another for the exon a3 of the ABL gene—Fig. 3. With the e13 and e21 primer we can amplify the control part of the BCR gene. With the e13 BCR primer and the a3 ABL one we can detect major as well as micro bcr BCR/ABL gene (Fig. 3). For the major bcr translocation we obtain 310 and 385 bp PCR product for e13-a2 and e14-a2 junctions, respectively—Fig. 2. (The theoretically possible e1-a3 PCR product is only exceptionally visible because of its length and its competition with preferentially amplified shorter PCR products.) Also due to a competition between the BCR/ABL and the control BCR gene amplification, the shorter BCR/ABL fragments are preferentially amplified and the control BCR should not have to be visible, as is demonstrated in one CML patient in Fig 2. In some cases, most probably due to an alternative splicing, both bands of the major bcr amplification are detectable as is seen in one CML case in Fig. 2. Combination of e1 and a3 primers gives the 481 bp PCR product for minor bcr BCR/ABL gene and the micro bcr translocation (e19-a2 junction) provides the PCR product of 926 bp—Fig. 2. Thus a combination of these four primers enables us to detect not only all so far known junctions of the BCR/ABL gene with a simultaneous amplification of the control gene, but offers also other theoretically possibilities (like e6-a2[26]) mentioned above. The sensitivity of the test is 10^{-3}–10^{-4} for diagnostic patients and 10^{-4}–10^{-5} for the BCR/ABL-positive cell line MOLM 7. The sensitivity of the test is high enough for routine diagnostic tests.

ACKNOWLEDGMENT

A study of the patients was supported by the grant of IGA MH CR no. 3563-3. The authors are grateful to Prof. R. Brdička for his critical reading of the manuscript.

REFERENCES

1. Nowell PC, Hungerford DA (1960) A minute chromosome in human CML. *Science*, **132**, 1498.
2. Rowley JD (1973) A new consistent chromosomal abnormality in chronic myelogenous leukemia identified by Quinacrine fluorescence and Giemsa staining. *Nature*, **243**, 290–293.
3. Vickers M (1996) Estimation of the number of mutations necessary to cause chronic myeloid leukemia from epidemiological data. *British Journal of Haematology*, **94**, 1–4.
4. Melo JV (1996) The diversity of BCR/ABL fusion proteins and their relationship to leukemia phenotype. *Blood*, **88**, 2375–2384.
5. Blickstein D, Aviram A, Luboshitz J, Prokocimer M, Stark P, Bairey O, Sulkers J, Shaklai M (1997) BCR-ABL transcripts in bone marrow aspirates of Philadelphia-negative essential thrombocythemia patients. Clinical presentation. *Blood*, **90**, 2768–2771.
6. Marasca R, Luppi M, Zucchini P, Longo G, Torelli G, Emilia G (1998) Might essential thrombocytemia carry Ph anomaly? Blood, **91**, 3084.
7. Verhoef G, Meeus P, Stul M, Mecucci C, Cassiman JJ, van den Berghe H, Boogaerts M (1992) Cytogenetic and molecular studies of the Philadelphia translocation in myelodysplastic syndromes. *Cancer Genetics and Cytogenetics*, **59**, 161–166.
8. Saglio G, Pane F, Martinelli, Guerrasio A (1997) BCR/ABL transcripts and leukemia phenotype: an unsolved puzzle. *Leukemia and Lymphoma*, **26**, 281–286.
9. Saglio G, Pane F, Gottardi E, Frigeri F, Buonaiuto MR, Guerrasio A, DeMicheli D, Parziale A, Fornaci MN, Martinelli G, Salvatore F (1996) Consistent amounts of acute leukemia-associated P190 BCR/ABL transcripts are expressed by chronic myelogenous leukemia patients at diagnosis. *Blood*, **87**, 1075–1080.
10. van Rhee F, Hochhaus A, Ling F, Melo JV, Goldman JM, Cross CP (1996) p190 BCR-ABL mRNA is expressed at low levels in p210-positive chronic myeloid and acute lymphoblastic leukemias. *Blood*, **87**, 5213–5217.

11. Tanaka M, Yamazaki Y, Hattori M, Tsuhita K, Utsumi M, Yoshida S (1996) The dual expression of major bcr/abl chimeric mRNA in blast crisis of chronic myelogenous leukemia. *Leukemia Resarch*, **20**, 575–580.
12. Saglio G, Guerrasio A, Rosso C, Zaccaria A, Tassinari A, Serra A, Rege-Cambrin G, Mazza U, Gavosto F (1990) New type of Bcr/Abl junction in Philadelphia chromosome-positive chronic myelogenous leukemia. *Blood*, **76**, 1819–1824.
13. Pane F, Frigeri F, Sindona M, Luciano L, Ferrara F, Cimino R, Meloni G, Saglio G, Salvatore F, Rotoli B (1996) Neutrophilic-chronic myeloid leukemia: A distinct disease with a specific molecular marker (BCR/ABL with C3/A2 junction). *Blood*, **88**, 2410–2414. (Erratum (1997) *Blood*, **89**, 4244.)
14. Yamagata T, Mitani K, Kanda Y, Yazaki Y, Hirai H (1996) Elevated platelet count features the variant type of BCR/ABL junction in chronic myelogenous leukemia. *British Journal of Haematology*, **94**, 370–372.
15. Mittre H, Leymarie P, Macro M, Leporrie M (1997) A new case of chronic myeloid leukemia with c3/a2 BCR/ABL junction. It is really a distinct disease? *Blood*, **89**, 4239–4240.
16. Wilson G, Frost L, Goodeve A, Vandenberghe E, Peake I, Reilly J (1997) BCR-ABL transcript with an e19a2 (c3a2) junction in classical chronic myeloid leukemia. *Blood*, **89**, 3064.
17. Wada H, Mizutani S, Nishimura J, Usuki Y, Kohsakai M, Komai M, Kaneko H, Sakamoto S, Delia D, Kanamaru A, Kakishita E (1995) Establishment and molecular characterization of a novel leukemic cell line with Philadelphia chromosome expressing p230 BCR/ABL fusion protein. *Cancer Research*, **55**, 3192–3196.
18. Haškovec C, Ponzeto C, Polák J, Maritano D, Zemanová Z, Serra A, Michalová K, Klamová H, Čermák J, Saglio G (1998) P230 BCR/ABL protein may be associated with an acute leukemia phenotype. *British Journal of Haematology*, submitted for publication.
19. van der Plas DC, Soekarman D, van Gent AM, Grosveld G, Hagemeijer A (1991) Bcr-abl mRNA lacking abl exon a2 detected by polymerase chain reaction in a chronic myelogeneous leukemia patient. *Leukemia*, **5**, 457–461.
20. Iwata S, Mizutani S, Nakazawa S, Yata J (1994) Heterogeneity of the breakpoint in the ABL gene in cases with BCR/ABL transcript lacking ABL exon a2. *Leukemia*, **8**, 1696–1702.
21. Polák J, Zemanová Z, Michalová K, Klamová H, Čermák J, Haškovec C (1998) A new case of chronic myeloid leukemia (CML) in myeloid blast crisis with an atypical (b3/a3) junction of the BCR/ABL gene. *Leukemia*, **12**, 250.
22. Soekarman D, van Denderen J, Hoefsloot L, Moret M, Meeuwsen T, van Baal J, Hagemeijer A, Groseveld G (1990) A novel variant of the bcr-abl fusion product in Philadelphia chromosome-positive acute lymphoblastic leukemia. *Leukemia*, **4**, 397–403.
23. Inukai T, Sugita K, Suzuki T, Ijima K, Goi K, Tezuka T, Kojika S, Hatakeyama K, Kagami K, Mori T, Okazaki T, Mizutani S, Nakazawa S (1993) A novel 203 kD aberrant BCR-ABL product in a girl with Philadelphia chromosome positive acute lymphoblastic leukemia. *British Journal of Haematology*, **85**, 823–825.
24. Tuszynski A, Dhut s, Young BD, Lister TA, Rohatiner AZ, Amas JAL, Chaplin T, Dorey E, Gibbons B (1993) Detection and significance of bcr-abl mRNA transcripts and fusion proteins in Philadelphia-positive adult acute lymphoblastic leukemia. *Leukemia*, **7**, 1504–1508.
25. Inokuchi K, Futaki M, Hanawa H, Tanosaki S, Yamaguchi H, Iwakiri R, Nomura T, Dan K (1997) Heterogeneous expression of bcr-abl fusion mRNA in a patient with Philadelphia-chromosome-positive acute lymphoblastic leukaemia. *British Journal of Haematology*, **97**, 837–840.
26. Hochhaus A, Reiter A, Skladny H, Melo JV, Sick C, Berger U, Guo JQ, Arlinghaus RB, Hehlman R, Goldman JM, Cross NCP (1996) A novel BCR-ABL fusion gene (e6a2) in a patient with Philadelphia chromosome-negative chronic myelogenous leukemia. *Blood*, **88**, 2236–2240.
27. Okamoto K, Karasawa M, Sakai H, Ogura H, Morita K, Naruse T (1997) A novel acute lymphoid leukemia type BCR/ABL transcript in chronic myelogenous leukemia. *British Journal of Haematology*, **96**, 611–613.
28. Cross NCP, Melo JV, Feng L, Goldman JM (1994) An optimised multiplex polymerase chain reaction (PCR) for detection of BCR-ABL fusion RNA in haematological disorders. *Leukaemia*, **8**, 186–189.

11

DIFFERENTIAL ENHANCEMENT OF GRAFT-VERSUS-HOST (GVH) AND GRAFT-VERSUS-LEUKEMIA (GVL) REACTIONS BY INTERLEUKIN-12 (IL-12)

M. Schmitt, Y. Miyahara, X. Gu, K. Mukae, K. Satoh, E. Nakayama, L. Bergmann, and H. Shiku

2nd Department of Internal Medicine
Mie University School of Medicine, Tsu, Japan
Tsu, Japan
Department of Parasitology and Immunology
Okayama University, Okayama, Japan
3rd Department of Internal Medicine
University of Ulm, Germany

GVH and GVL reactions were studied in a bone marrow transplantation (BMT) model of major histocompatibility complex matched, but minor histocompatibility antigen mismatched mouse strains (DBA/2 –> BALB/c). IL-12 is known to be a potent inducer of both T lymphocytes and natural killer (NK) cells. As these two cell populations are considered to be effector cells for the GVH reaction, we examined whether the administration of IL-12 results in the aggravation of this reaction. In parallel, we assessed the outcome of the GVL reaction by using the syngeneic cell line RL♂1. When administered subcutaneously to the recipient mice at the time of BMT and two days later, a dose of 200 ng of IL-12 was lethal in all of the mice receiving allogeneic transplants, but in none of the recipients of syngeneic BMT. In contrast, administration of only 25 ng of IL-12 did not result in the aggravation of the GVH, but in a 25-fold increase of the GVL reaction. Effector cells of GVH were activated $CD4^+$ T lymphocytes, whereas GVL was mediated by $CD4^+$ and $CD8^+$ T cells. NK cells were not involved in any reaction.

Address correspondence to: Michael Schmitt, MD, 3rd Department of Internal Medicine, University of Ulm, 89081 Ulm, Germany; e-mail: michael.schmitt@medizin.uni-ulm.de.

Molecular Biology of Hematopoiesis 6, edited by Abraham *et al.*
Kluwer Academic/Plenum Publishers, New York, 1999.

In summary, IL-12 revealed to be an important tool to enhance the GVL without the aggravation of the GVH reaction, opening new immunotherapeutic options for BMT strategies.

INTRODUCTION

In murine models BMT models, GVH reaction occurs rather in allogeneic than in syngeneic transplantations. This reaction is characterized by loss of anti-host CTL development, elimination of host lymphocytes, profound immunodeficiency in the acute form, B cell hyperactivity, autoantibody formation and by lupuslike disease in the chronic form.

As for most models established so far, allogeneic transplantation means transplantation across the major histocompatibility complex (MHC) barrier, it hardly meets the clinical situation where MHC-mismatches are rather avoided. Therefore our group has been working on a transplantation from DBA/2 to BALB/c mice which are MHC identical ($H\text{-}2^d$).

Several strategies have been developed to minimize GVHD like immunosuppression, T cell depletion and donor lymphocyte infusion.[1-3] On the other hand it has been a long-standing observation that freedom from relapse of leukemia after BMT is achieved to a higher degree in cases where GVHD of clinical grade I or II occurs than in cases with no GVHD.[4,5] This has been termed graft-versus-leukemia (GVL) effect.[6] It remains to be elucidated to which extend both reactions differ and, therefore might be enhanced differentially for therapeutic use.[7-9]

The severity of GVH can be influenced by cytokines like interferon gamma (IFNγ) and interleukin 2, activating T lymphocytes and natural killer (NK) cells.[8-12] Also interleukin 12 (IL-12) is know to be a potent inducer of both T lymphocytes and NK cells. Therefore we aimed at defining the differential role of this novel cytokine for GVH and GVL in our system.

MATERIALS AND METHODS

Mice. 8 to 12 week old female BALB/c ($H\text{-}2^d$), CB6F1 (BALB/c × C57/BL6, $H\text{-}2^{d/b}$), and DBA/2 ($H\text{-}2^b$) mice were purchased from SLC (Shizuoka, Japan) and kept in autoclaved cages covered by filter caps and given autoclaved water (pH 3) and food at the Laboratory Animal Center of Mie University School of Medicine.

Cells. RL♂1 cells are radiation induced leukemia cells of BALB/c origin.[13]

Irradiation. Mice were fixed in 50 ml Nunc test tubes which were further inserted into a polystyrol box to assure the application of an equal radiation dose to each animal. Using a linear accelerator (Clinac 2300, Varian, Palo Alto, CA), BALB/c mice were given a dose of 7.5 Gy.

Monoclonal antibodies. Antibodies were produced by female BALB/c nude mice injected with hybridoma cells producing anti-CD4 (GK1.5), anti-CD8 (Lyt 2.1 or Lyt 2.2), anti-INFγ (R46A2), B220 (RA3-J6B2).[14] Titers of harvested ascites fluid were checked using FACS analysis (Beckton-Dickinson).[15]

IL-12. IL-12 was kindly donated by Genetic Institute, Cambridge, MA. The bioactivity was 5×10^6 U/mg. If not otherwise indicated, the mice were given IL-12 subcutaneously every other day.

BMT. For bone marrow transplantation, cells from 10 to 15 donor mice were collected. After centrifugation, they were resuspended in RPMI. B220, GK 1.5 and Lyt 2.1 each were added. After 45 min of incubation on ice, cells were washed twice and subjected to incubation with 1:10 diluted rabbit complement at 37°C for 30 minutes. Afterwards, they were adjusted in RPMI and kept on ice till administration. 5×10^6 of these depleted bone marrow cells were then injected into the tail veins of mice about 6 hours after irradiation. The efficacy of the T cell depletion was 99% as assessed by FACS analysis of bone marrow cells.

RESULTS

Early Death of Allo-BMT Mice Is IL-12 Dose Dependent

IL-12 was administered subcutaneously to DBA/2 transplanted BALB/c recipient mice every other day, starting at the day of BMT. A dose of 200 ng IL-12 per injection resulted in 100% mortality within 4 to 7 days after BMT, whereas the recipients tolerated a dose of 25 ng per injection (Fig. 1).

Rejection of Leukemic Cells at the Time of BMT

Allogeneic BMT recipient BALB/c mice were challenged at the time of BMT with RL♂1, radiation induced leukemia cells of BALB/c origin. They rejected up to 2×10^6 cells surviving for over 4 months, which is ten times the lethal cell dose for naive BALB/c mice (Fig. 2). The administration of 25 ng of IL-12 to recipient BALB/c mice of allogeneic transplantation enabled these mice to reject up to 5×10^6, whereas in the absence of IL-12, the mice died within 15 days of tumor challenge (Fig. 2).

Effector Cells for the Aggravation of GVHD by IL-12

DBA/2 transplanted BALB/c mice given 200 ng IL-12 subcutaneously every other day, died characteristically within 4 days. To determine the effector cells for the aggravation of GVHD by IL-12, mice were given a panel of monoclonal antibodies (moAb)

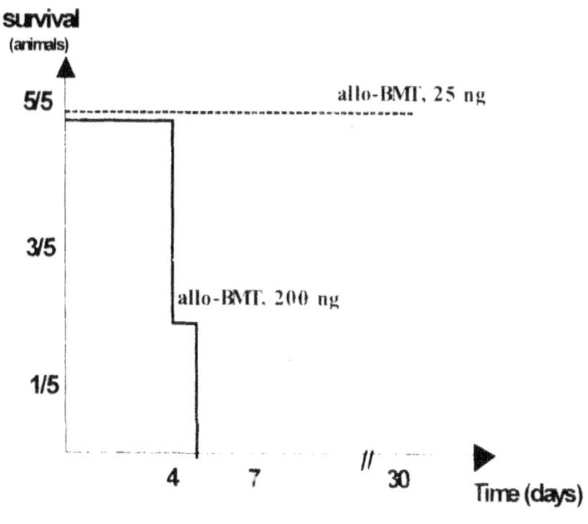

Figure 1. DBA/2 mice were transplanted to BALB/c mice. Every other day 25 or 200 ng of IL-12 were administered subcutaneously. The survival of the animals was observed.

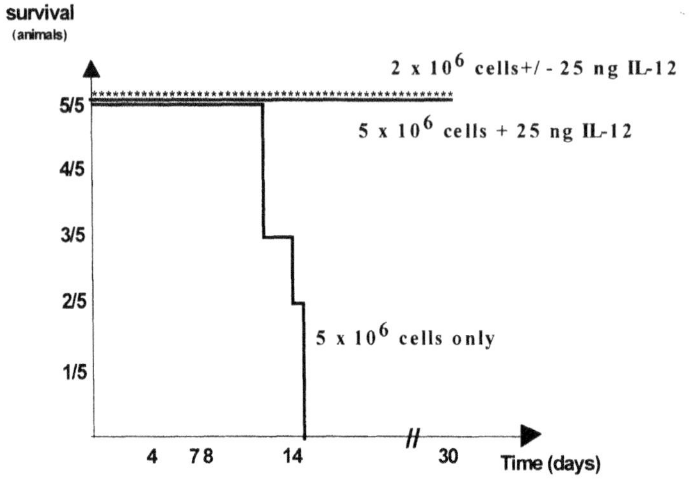

Figure 2. DBA/2 mice were transplanted to BALB/c mice. Recipient mice were challenged with RL♂1 and optionally administered 25 ng IL-12. Transplanted mice can reject 2 million leukemic cells without IL-12 treatment. In contrast, IL-12 administration is necessary for the rejection of as much as 5 million leukemic cells.

intravenously after irradiation. 2 hours later, 200 ng of IL-12 were administered subcutaneously. Another 4 hours later, T cell depleted bone marrow cells were administered. Administration of IL-12 injections was repeated on day 2 and day 4 after BMT, administration of moAb on day 3. Blocking of $CD4^+$ cells alone or $CD4^+$ and $CD8^+$ cells in DBA/2 transplanted BALB/c resulted in recovery from GVHD in 100% of these recipients, whereas blocking of $CD8^+$ cells alone failed to do so (Fig. 3). However when following the same administration schedule as described above, and when CB6F1 recipients were given allogeneic transplantation from BDF1 mice and 200 ng of IL-12 every other day, they could not be rescued by injection of anti-NK1.1 moAb. Administration of moAb against IFNγ prevented IL-12 treated allo-BMT mice from death (data not shown).

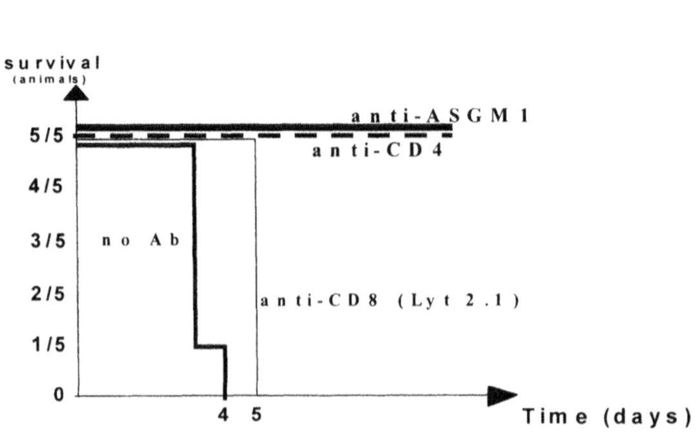

Figure 3. Allo-BMT from DBA/2 to BALB/c mice was performed. IL-12 was applied at a dose of 200 ng s.c. every other day. Recipient mice were administered antibodies against CD4, CD8, and ASGM1. Only anti-CD4 and anti-ASGM1 administration rescued mice from lethal GVH reaction, whereas anti-CD8 administration was inefficient.

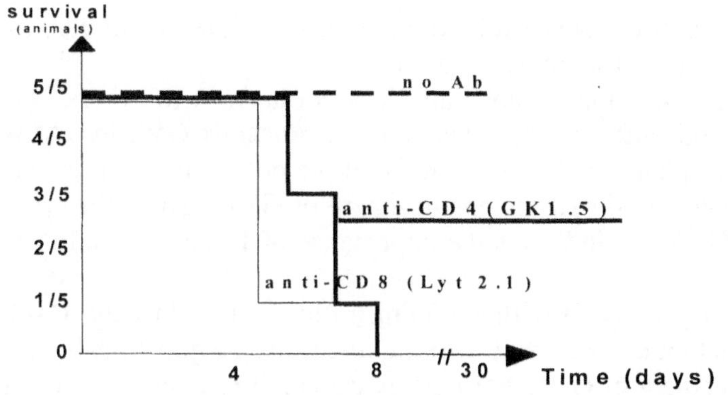

Figure 4. Allo-BMT from DBA/2 to BALB/c mice was performed. Mice were challenged with 5 million RL♂1 cells and IL-12 was applied at a dose of 25 ng s.c. every other day. Recipient mice were administered antibodies against CD4 and CD8. Both anti-CD4 and anti-CD8 administration abolished the IL-12 enhanced GVL reaction.

Effector Cells for GVL

BALB/c recipients of an allo-BMT from DBA/2 were challenged with 5×10^6 RL♂1 cells, administered 25 ng IL-12 subcutaneously every other day, and also given moAbs at the time of BMT. A control group of mice did not receive any moAbs, and survived confirming the earlier experiments (see above). The blocking of $CD4^+$ or $CD8^+$ cells resulted in a rather early death of all mice in this group (Fig. 4). In addition, the inhibiting of interferon gamma had no adverse effect on the rejection of leukemia cells (data not shown).

DISCUSSION

The differential enhancement of GVH and GVL reactions first of all requires the dissection of these overlaying processes using potentially the same effector cells and mechanisms. In our model, we clearly demonstrated that $CD4^+$ T cells are involved in both GVH (Fig. 3) and GVL (Fig. 4), whereas $CD8^+$ T cells play a role only in GVL (Fig. 4). Moreover, the effector cells of the GVH were $ASGM1^+$ (Fig. 3), and the production of Th2 cytokines (data not shown) such as IL-4 was enhanced at the serum level which suggests an activated state of the $CD4^+$ T cells. As the enhancement of GVH is blocked on the other hand by antibodies against the type 1 cytokine IFNγ, T helper cells of both type 1 and type 2 seem to be involved.

Effector cells of GVH were $ASGM1^+$, but in contrast to others,[12] we did not detect any contribution of NK cells to the GVH. We tested this possibility in a F1 system where DBA/2 and BALB/c mice were backcrossed to $NK1.1^+$ B6 mice (data not shown), whereas other groups[12] used a moAb which blocks only a subpopulation of NK cells in BALB/c mice.

As $CD8^+$ cells in our system could not be demonstrated to be GVH effector cells, we assume that macrophages played an important role as blocking of IFNγ prevented mice from death from GVH. In our system, the IFNγ secretion by DBA/2 derived $CD4^+$

donor T cells was not dependent on the presence of $CD8^+$ donor cells as we used T cell depleted (TCD) cells for transplantation.

In contrast to some groups[12] and in accordance with others,[9] we demonstrated that IL-12 is sufficiently capable to enhance significantly GVL in BALB/c mice receiving DBA/2 transplants, which show the multiple potential of this cytokine.

We demonstrated that the enhancement of GVL against the syngeneic leukemia cell line RL♂1 clearly indicates the importance of T cells whose function is activated by IL-12.

In summary, we can say that administration of IL-12 to allo-BMT recipients can enhance the activities of T cell subsets. Our results suggest that IL-12 has a potential for the enhancement of GVL. For the clinical use, it remains to be elucidated whether there might be a similar therapeutic window for the IL-12 in humans as there is in mice. Furthermore, IL-12 might not be necessarily given as a recombinant protein by subcutaneous injection, but more specifically as a gene transfected into hematopoetic cells and/or in concert with peptides of leukemia associated antigens, as e.g. pRL1a[16] for RL♂1. Such experiments are ongoing in our laboratory.

ACKNOWLEDGMENT

We would like to thank Mr. Okumura for expert assistance with the irradiation of the animals.

Supported by grants of the Ministry of Education, Science, Sports, and Culture of Japan. M.S. was awarded a Science and Technology Fellowship (STF10) by the European Commission.

REFERENCES

1. Storb R, Deeg HJ, Pepe M, Appelbaum F, Anasetti C, Beatty P, Beisinger W, Berenson R, Buckerner CD, Clift R, Doney K, Longton G, Hansen J, Hill R, Loughran Tjr., Martin P, Singer J, Sanders J, Stewart P, Sullivan K, Witherspoon R, Thomas ED. Methotrexate and cyclosporine versus cyclosporine alone for prophylaxis of graft-versus-host disease in patients given HLA-identical marrow grafts for leukemia: long-term follow-up of a controlled trial. Blood 73:1729, 1989.
2. Martin PJ, Hansen JA, Buckner CD, Sanders JE, Deeg HJ, Stewart P, Appelbaum FR, Clift R, Fefer A, Witherspoon RP, Kennedy MS, Sullivan KM, Flournoy N, Storb R, Thomas ED. Effects of in vitro depletion of T cells in HLA-identical allogeneic marrow grafts. Blood 66:664, 1985.
3. Kolb HJ, Schattenberg A, Goldman JM, Hertenstein B, Jacobsen N, Arcese W, Ljungman P, Ferrant A, Verdonck L, Niederwieser D. Graft-versus-leukemia effect of donor lymphocyte transfusions in marrow grafted patients. European Group for Blood and Marrow Transplantation Working Party Chronic Leukemia. Blood 86:2041, 1995.
4. Sullivan K, Weiden P, Storb R. Influence of acute and chronic graft-versus-host disease on relapse and survival after bone marrow transplantation from HLA-identical siblings as treatment of acute and chronic leukemia. Blood 73:1720, 1989.
5. Ringden O, Hermans J, Labopin M, Appeley J, Gorin NC Gratwohl A. The highest leukemia-free survival after allogeneic bone marrow transplantation is seen in patients with grade I graft-versus-host disease. Acute and chronic leukemia working parties of the European group for blood and marrow transplantation (EBMT). Leuk Lymphoma 24:171, 1996.
6. Horowitz MM, Gale RP, Sondel PM. Graft-versus-leukemia reactions following bone marrow transplantations in humans. Blood 75:555, 1990.
7. Johnson BD, Truitt RL. A decrease in graft-versus-host disease without loss of graft-versus-leukemia reactivity after MHC-matched bone marrow transplantation by selective depletion of donor NK cells in vivo. Transplantation 54:104, 1992.

8. Jiang YZ, Barrett AJ, Goldmann JM, Marrondis DA. Association of NK cell immune recovery with a graft-versus-leukemia effect independent of graft-versus-host disease following allogeneic bone marrow transplantation. Ann Hematol 74:1, 1997.
9. Yang YG, Segio JJ, Pearson DA, Szot GL, Shimizu A, and Sykes M. Interleukin-12 preserves the graft-versus-leukemia effect of allogeneic CD8 T cells while inhibiting CD4-dependent graft-versus-host disease in mice. Blood 90:4652, 1997.
10. Sykes M, Szot GL, Nguyen PL, Pearson DA. Interleukin-12 inhibits murine graft-versus-host disease. Blood 88:1113, 1995.
11. Zeis M, Uharek L, Glass B, Gaska T, Steinmann J, Gassmann W, Löffler H, Müller-Ruchholz W. Allogeneic NK cells as potent antileukemic effector cells after allogeneic bone marrow transplantation in mice. Transplantation 5:1734, 1995.
12. Uharek L, Zeis M, Glass B, Gassmann W, Schmitz N, Müller-Ruchholz W. High lytic activity against human leukemia cells after activation of allogeneic NK cells by IL-2 and IL-12. Leukemia 10:1758, 1996.
13. Nakayama E, Shiku H, Takahashi T, Oettgen HF, Old LJ. Definition of a unique cell surface antigen of mouse leukemia RL♂1 by cell-mediated cytotoxicity. Proc Natl Acad Sci. USA 76:3486, 1979.
14. Schmitt M, Ikeda H, Nagata Y, Gu X, Wang L, Kuribayashi K, Shiku H. Involvement of T-cell subsets natural killer (NK) cells in the growth suppression of murine fibrosarcoma cells transfected with interleukin-12 (*IL-12*) genes. Int. J. Cancer 72:505-511, 1997.
15. Gu X, Schmitt M, Hiasa A, Nagata Y, Ikeda H, Sasaki Y, Akiyoshi K, Sunamoto J, Nakamura H, Kuribayashi K, Shiku H. A novel vaccine of cholesterol-bearing polysaccharide coated oncoprotein induces both cellular and humoral immune responses against HER2 overexpressing murine sarcomas. Cancer Res 58:3385, 1998.
16. Uenaka A, Ono T, Akisawa T, Wada H, Yasuda T, Nakayama E. Identification of a unique antigen peptide pRL1 on BALB/c RL♂1 leukemia recognized by cytotoxic T lymphocytes and its relation to the *akt* oncogene. J Exp Med 180:1599, 1994.

12

EXPRESSION OF C-ERB B-2/HER-2 IN PATIENTS WITH METASTATIC BREAST CANCER UNDERGOING HIGH-DOSE CHEMOTHERAPY AND AUTOLOGOUS BLOOD STEM CELL SUPPORT

M. Bewick,[1] T. Chadderton,[1] M. Conlon,[1] R. Lafrenie,[1,3] and S. Glück[1,2,3]

Northeastern Ontario Regional Cancer Centre[1]
Sudbury, Ontario
Department of Medicine
University of Ottawa,[2] Ottawa, Ontario, Laurentian University[3]
Sudbury, Ontario

Keywords: Oncoprotein, EGFR superfamily, Plasma levels, Advanced breast cancer, Prognostic factors, Retrospective analysis.

ABSTRACT

Purpose: C-erbB-2/HER-2 (designated HER-2) is overexpressed in both primary and metastatic breast cancer and has been shown to predict for poor prognosis. We investigated the expression of HER-2 in patients with metastatic breast cancer undergoing high-dose chemotherapy (HDCT) with autologous blood stem cell (ABSC) support and correlated the presence (positive) or absence (negative) of HER-2 overexpression in these patients to response to treatment, progression free survival (PFS), and overall survival (OS).

Materials and Methods: The level of HER-2 expression was analyzed in 57 patients with metastatic breast cancer undergoing HDCT with ABSC support. Plasma from peripheral blood was taken at three different time points during the course of

Correspondence to: Stefan Glück, MD, PhD, Director of Research and Autologous Stem Cell Transplantation, Associate Professor of Medicine, University of Ottawa, 41 Ramsey Lake Road, Sudbury, Ontario. P3E 5J1, Canada, Phone: (705) 523-7312, Fax: (705) 523-7326, Email: sgluck@cancercare.on.ca

Molecular Biology of Hematopoiesis 6, edited by Abraham *et al.*
Kluwer Academic/Plenum Publishers, New York, 1999.

treatment and was analyzed using an enzyme immunoassay (EIA) to detect circulating levels of the extracellular portion of HER-2.

Results: HER-2 levels were found to be elevated (>0.2 U/mg protein) in 27/57 (47.4%) patients at one or more time points during treatment. The level of HER-2 varied during course of treatment. Following induction chemotherapy (ICT), 5 patients who were negative at the initial measurement, showed overexpression of HER-2. Three patients overexpressed HER-2 only at the post HDCT/ABSC measurement. Response to treatment was similar in patients independent of plasma HER-2 levels. Overexpression of HER-2 was associated with a significantly shorter PFS ($p = 0.004$) and OS ($p = 0.003$) following HDCT/ABSC.

Conclusion: Overexpression of HER-2, measured by EIA in patient plasma, predicts a shorter PFS and OS in patients with metastatic breast cancer treated with HDCT and ABSC support. These data also suggest that optimal timing or several measurements of circulating HER-2 levels using EIA are necessary prior to HDCT to determine overexpression since levels may vary during the course of treatment.

INTRODUCTION

The proto-oncogene c-erbB-2/neu HER-2 resides on chromosome 17 and encodes a 185 kDa transmembrane glycoprotein (designated HER-2) with intracellular tyrosine kinase activity.[1] HER-2 is a member of the EGF growth factor receptor family which consists of four closely related family members: the EGF receptor (EGFR, ERBB1), HER-2 (ERBB2/NEU), HER-3 (ERBB3), and HER-4 (ERBB4).[2] Both homodimers and heterodimers are formed between the various family members mediating a complex system of signal transduction affecting both cell proliferation and growth.[2–6]

The importance of these growth factor receptors on the growth and regulation of breast cancer has been clearly demonstrated. Overexpression and/or amplification of HER-2 has been found in greater than 20%–30% of primary breast tumours and to varying degrees in other cancers (ovarian, gastric, colorectal, lung, salivary, bladder, pancreas, endometrial, cervical, oral, and prostate).[7–12] Both HER-2 overexpression and gene amplification have been often found to correlate with poorer survival, particularly in patients with breast cancer with involvement of lymph nodes.[7–14,47]

Although HER-2 oncoprotein overexpression correlates with a reduction in patient survival, only a few studies have evaluated the incidence of HER-2 overexpression in patients with metastatic breast cancer (MBC) and examined whether increased levels of this protein indicates a poorer prognosis and/or reduced responsiveness to systemic treatment.[14–24] In these studies, 30–50% of patients with MBC have been shown to have elevated levels of circulating (plasma, serum) extracellular domain of HER-2.[14–19] Overexpression has been shown to correlate with an increased tumour burden and reduced survival.[14–16,18,22–25]

Since EIA is a relatively simple assay commonly used in clinical medicine, the possibility that elevated levels of extracellular HER-2 could indicate poor prognosis in patients with MBC is of particular interest.

High-dose chemotherapy (HDCT) followed by autologous blood stem cell (ABSC) support is rapidly becoming the preferred treatment for patients with MBC, high risk Stage II and Stage III breast cancer in the United States.[26] Little is known about the incidence of HER-2 gene overexpression in patients with MBC undergoing this treatment and whether increased levels of this protein indicates a poor prognosis and/or a reduced responsiveness to this treatment.

In this retrospective study, we have examined the incidence of overexpression and changes in the levels of HER-2 protein in the plasma of 57 patients with advanced breast cancer treated with HDCT and ABSC support. Results were analyzed and correlated to response to treatment, progression free survival (PFS), and overall survival (OS).

MATERIALS AND METHODS

1. Patients

Peripheral blood and ABSC specimens were collected during the course of treatment of 57 patients with metastatic breast cancer (MBC) enroled into clinical trials using HDCT and ABSC transplantation. Patients were between 18 and 55 years of age and had either hormone receptor negative tumours or had failed at least one hormone treatment regimen for MBC. The median age was 42.5 years (range, 26–55 years). The WHO performance status was 0, 1 or 2. The majority of patients have a performance status of 0 and 1. The initial breast cancer presentation consisted of 5 patients with stage I, 23 patients with stage II, 4 patients with stage IIIA and 2 patients with stage IIIB. The remaining patients presented with metastatic disease (stage IV). 22 patients had bony metastasis, 17 patients metastasis in the lymph nodes, 15 patients had viceral metastasis (liver and/or lung). Patients with CNS metastasis were excluded from this study. The number of metastatic sites were 1 or 2 in the majority of patients, only 15 patients had 3 or more sites involved with metastasis. Most of our patients (29) had previously adjuvant chemotherapy consisting of either Cyclophosphamide (CTX), Methotrexate, 5-Fluorouracil (CMF) or anthracycline containing regimen. Thirty-two had previously received tamoxifen.

The study was approved by institutional review board and the patients were enroled into the clinical trial when they fulfilled the entry criteria and signed an informed consent.

All patients received treatment initially consisting of two cycles induction chemotherapy (ICT). This consisted of CTX, 500 mg/m^2 body surface area (BSA) intravenously (IV); Doxorubicin (DOX), 50 mg/m^2 BSA IV or Epirubicin (EPI), 60 mg/m^2 BSA IV; 5-FU, 500 mg/m^2 BSA IV. Patients whose tumours did not progress after these two cycles continued on the study which included further ICT using dose-escalation of CTX to 2000 mg/m^2 BSA IV with the same doses of DOX or EPI and 5-FU. CD34+ cells were mobilized with either rhGM-CSF or rhG-CSF at 5 or 10 μg/kg, body weight (BW) starting one day after ICT. An apheresis procedure (AP) was performed on average for 4 consecutive days after the rebound phase was reached (when WBC values increased to >2.5/nl) using a Fenwal CS 3000 Plus (Baxter Healthcare Co., Deerfield, IL, USA) or COBE (COBE Laboratories, Inc., Lakewood, CO, USA) blood cell separator. A total blood volume of 10 litres per AP was processed at a flow rate of 60–70 ml/min. AP collections were continued until a minimum of 2×10^6 CD34+ cells/kg BW was obtained as determined by standard flow cytometric evaluation.

High-dose chemotherapy consisted of CTX (6 g/m^2 BSA IV), Mitoxantrone (70 mg/m^2 BSA IV), and either Carboplatin (800 mg/m^2 BSA IV) or Vinblastine (12 mg/m^2 BSA IV), all administered in 4 equal doses over 4 days. Three patients did not receive HDCT due to disease progression after ICT. Four patients received a second regimen of HDCT with CTX and Mitoxantrone and Carboplatin. One patient received a second HDCT with Thiotepa (500 mg/m^2 BSA IV) and CTX (6 g/m^2 BSA IV). CD34+ cells were re-infused from cryopreserved AP containing greater than 2×10^6 CD34+/kg BW that were pooled and washed according to the method established by *Glück* et al.[27]

2. HER-2 Plasma Levels

HER-2 plasma levels were determined using the HER-2 serum EIA kit (Chiron Diagnostics, formerly Triton Diagnostics, Alameda, CA, USA) according to the manufacturer's instructions. This kit uses a monoclonal antibody-based immunoenzymatic assay to quantitate the shed HER-2 fragment in plasma.

Blood samples obtained during the course of treatment were collected in ethylene-diaminetetraacetic acid (EDTA) vacutainer tubes. Plasma was collected by centrifugation at 400 g for 10 minutes. The plasma was removed, frozen in aliquots and stored at –70°C until required for assay. Plasma HER-2 levels were examined at three different time points: 1) at initial enrolment to the clinical program; 2) at the time of first or second apheresis (following ICT with rhGM-CSF or rhG-CSF); 3) approximately 1 month after HDCT and ABSC reinfusion. Prior to assay, frozen plasma samples were thawed gradually and centrifuged at 8000 rpm for 5 minutes in eppendorf tubes. Protein concentration was measured by the Bradford protein dye method,[28] using bovine serum albumin as the standard.

Measurements of HER-2 levels in plasma of 7 healthy women resulted in a mean concentration of 13.8 U/ml ± 2.3 S.D. To further offset any variation due to patient plasma differences during treatment, data was expressed as U/mg plasma protein or 0.134 U/mg protein ± 0.02 S.D. The cut-point for positivity (i.e. overexpression) was defined as 21 U/ml or 0.2 U/mg protein (using the mean plus three times the standard deviation calculation from *Isola* et al.[24]). Plasma samples expressing HER-2 levels below this cut-point were defined as "HER-2 negative" and those samples equal to or above this cut-point were defined as "HER-2 positive." More significant differences between the HER-2 positive and negative patient populations were found when the data were calculated and analyzed using the 0.2 U/mg protein cut-point and all data was subsequently evaluated using this cut-point.

PATIENT EVALUATION AND STATISTICAL ANALYSIS

Progression free survival (PFS) was defined as the time from study entry to documented signs of recurrence or progression of disease. Overall survival (OS) was defined as the time from study entry to death due to MBC. Survival curves were calculated using the Kaplan-Meier method. Differences between survival times were analyzed by the Log-rank, Wilcoxon, and Cox tests for survivorship functions and hazard ratios were calculated by Cox's proportional hazards method for univariate analysis.

All patients had a physical and radiological evaluation based on WHO criteria at various times during the course of treatment and subsequently every 2–3 months for 2 years and then every 6 months until evidence of relapse.

Patients were considered to have achieved a complete response (CR) if no evidence of disease was found for at least 4 weeks. Partial response (PR) was defined as a reduction in measurable disease volume to less than 50% of prestudy size for at least 4 weeks. Stable disease (SD) was defined as no significant change (no more than 25% above or 50% below prestudy size) in evaluable disease for at least four weeks. Progressive disease (PD) was defined as a 25% or more increase in the minimum size of one or more measurable lesions or the presence of new metastatic lesions.

Only patients with CR or PR were considered to have responded to treatment. Fisher's exact test was used to test for response difference according to HER-2 expression levels.

RESULTS

1. HER-2 Levels during Treatment

The data shown are derived from plasma samples of 57 patients with MBC treated with HDCT and ABSC support. HER-2 plasma levels in these samples were determined using the c-erbB-2 serum EIA kit (Chiron Diagnostics, formerly Triton Diagnostics Alameda, CA.). Plasma HER-2 levels were examined at three time points: 1) At initial enrolment to the clinical trial; 2) Following ICT, at the time of apheresis; 3) At a 4–6 week restaging after HDCT and ABSC support.

Table 1 shows the number of samples overexpressing HER-2 at each serial measurement. The overall percentage of patients that overexpress HER-2 is within the range of that found in the few studies that have examined overexpression in MBC; i.e. 20–57% of patients with MBC have elevated HER-2 plasma concentrations.[11–17] It should be noted that although this study examined HER-2 expression at three time points during the course of treatment of patients with metastatic breast cancer undergoing HDCT and ABSCT, for some patients not all three plasma samples were available for measurement. Overall, it was found that 27 of the 57 patients examined over-expressed HER-2 (Table 1) at one or more of the time-points examined. The HER-2 EIA results of these HER-2 positive patients were then examined at each of the time-points. At the first time point, plasma samples from 25 of the 27 patients that were found to be HER-2 positive in this study were available for analysis. 16 of these 25 patients (64%) were HER-2 positive at this time-point. At the second time-point, samples from all 27 HER-2 positive patients were available for analysis. 21 of these 27 patients (77.8%) over-expressed HER-2 at this time-point. This included 5 patients that were negative at the initial measurement. At the third time-point (after HDCT and ABSCT), plasma samples from 24 of the 27 overall HER-2 positive patients were available for assessment. 13 of these 24 patients (54.2%) overexpressed HER-2. This

Table 1. Plasma HER-2 overexpression in peripheral blood

	Initial		At Apheresis		Post Transplant		Overall Positive	
	No.	%	No.	%	No.	%	No.	%
Variable	16	32.7	21	37.5	14	29.2	27	47.4
Mean	0.89		0.59		0.33		0.61	
Range	0.22–4.65		0.20–3.18		0.21–0.55		0.20–4.65	
Total	49		56		47		57	

No. designates number of HER-2 positive; % designates percent HER-2 positive.
The lower limit for the amount of overexpressed protein is 0.2 units/mg total protein. Overall positive plasma HER-2 refers to patients with HER-2 levels ≥ 0.2 U/mg protein at any one or more of the three time points.

included 3 patients who were negative at the two previous measurements. Plasma samples from all three time-points were available from 24 of the 27 overall HER-2 positive patients. Only 8 of these 24 patients (33.3%) over-expressed HER-2 at all 3 time points.

The mean expression values of secreted HER-2 decreased during the course of treatment in 17 of 22 evaluated patients who were HER-2 positive initially and/or at time of apheresis. Eleven of these 17 values decreased to below the cut-point of 0.2 U/mg protein after HDCT and ABSC re-infusion.

2. Survival and Clinical Course Analysis

Patients that overexpressed levels of HER-2 (>0.2 U/mg) on at least one measurement were considered HER-2 positive. Twenty-four HER-2 negative patients responded to chemotherapy (17 PR and 7 CR) and 21 HER-2 positive patients responded to treatment (17 PR and 4 CR). The total number of nonresponders (PD +SD) was 5 for both HER-2 positive and HER-2 negative patients. There was no significant difference in the response to chemotherapy between HER-2 positive and HER-2 negative patients (Fisher's exact test, data not shown).

OS and PFS analysis using the Kaplan-Meier method on the basis of patient HER-2 expression levels are shown in Table 2. The median OS for all patients on this study was found to be 25.1 months; and the median PFS was 10.4 months.

The difference in OS was highly significant between patients with HER-2 negative and HER-2 positive plasma using the Log-rank test ($p = 0.003$), Wilcoxon ($p = 0.0018$), and Cox tests ($p = 0.0040$). The overall median survival for HER-2 positive patients was 17.2 months and for HER-2 negative patients, it was 28.9 months. The relative risk or hazard ratio was 2.5. This means that on average that HER-2 positive patients die at 2.5 times the rate of HER-2 negative patients.

Table 2. Overall survival and progression free survival results according to HER-2 overexpression

Plasma HER-2 Expression	Median Overall Survival	Median Progression Free Survival
(+)HER-2 (n = 27)	17.2 months	7.5 months
(C.I.)[a]	(8.2, 19.9)	(5.5, 10.4)
(−)HER-2 (n = 30)	28.9 months	14.3 months
(C.I.)[a]	(25.1, indeterminate)	(8.3, 24.6)
Hazard ratio (relative risk)	2.5 (95% C.I. = 1.3, 4.7)	2.3 (1.3, 4.1)
p*[b]	0.0030	0.0040
p**[c]	0.0018	0.0070
p***[d]	0.0040	0.0051

(+) HER-2 and (−) HER-2 designates those patients with ≥0.2 U/mg protein at one or more time point or below 0.2 U/mg protein respectively.

[a]C.I. 95% Confidence Intervals.

[b]*Log-rank test of survivorship. p values are derived from a comparison of survival functions for (+) HER-2 and (−) HER-2 patients.

[c]**Wilcoxon estimate of survivorship.

[d]***Cox estimate of survivorship.

The hazard ratio designates a relative risk of death or relapse between HER-2 positive and HER-2 negative patients.

Differences in the duration of PFS was also highly significant using the Log-rank test (0.004), Wilcoxon (0.007), and Cox tests (0.005) (see Table 2) between HER-2 plasma positive and negative patients. The median PFS of HER-2 plasma positive patients was 7.5 months compared with 14.3 months for HER-2 negative patients. The relative risk or hazard ratio was 2.3. Disease progression in HER-2 positive patients occurs at 2.3 times the rate of HER-2 negative patients.

Figure 1 show the OS and PFS curves respectively, of patients with plasma positive and negative HER-2 expression.

There were no significant differences in OS and PFS between the group of patients who initially overexpressed HER-2 (at the first measurement) and that continued to overexpress HER-2 (6 patients) following HDCT/ABSC support and those patients that initially overexpressed HER-2 and then became HER-2 negative following HDCT/ABSC support (11 patients) (data not shown).

An independent analysis at each of the three serial measurements for HER-2 expression was performed. Analysis of the PFS showed that only the hazard ratio at the second measurement (at the time of apheresis) was significant for those patients which overexpressed HER-2 and those that did not (hazard ratio: 2.04, 95% CI: 1.14–3.65, $p = 0.017$).

Analysis of OS at each time point showed significant differences in hazard ratio between the HER-2 positive and HER-2 negative group at this second measurement (hazard ratio: 2.66, 95% CI: 1.40–5.05, $p = 0.003$) and also at the initial measurement (hazard ratio: 2.25, 95% CI: 1.15–4.43, $p = 0.018$).

DISCUSSION

A simple, quantitative EIA procedure was used to examine HER-2 overexpression in patient plasma samples. No clear cutoff value has been established in a variety of studies which have examined overexpression using this procedure and previous studies have individually determined the cut-point for HER-2 overexpression in patient serum and plasma in various ways. *Isola* et al.[24] examined HER-2 levels in preoperative and follow-up serum samples in patients with breast cancer using the EIA assay manufactured by Triton Diagnostics. Serum levels in 113 patients with benign breast disease had a mean of 5.7 ± 4.8 S.D. U/ml and the cut-point for positivity was defined as 20 U/ml (using the mean plus three times the standard deviation calculation). *Revillion* et al.[17] using the HER-2 plasma assay manufactured by Triton determined a cutoff value of 27 U/ml from healthy women ($n = 30$). *Volas* et al.[20] selected a cut-point of 30 U/ml also based on serum results of healthy controls also using the EIA kit by Triton Diagnostics. *Kandl* et al.[14] measured the HER-2 antigen fragment in the serum of 79 women with advanced breast cancer. They found that a serum concentration of >10 U/ml had a significant impact on OS from the time of diagnosis of metastatic disease.

We determined the cut-point for HER-2 overexpression using plasma from normal blood samples to be 21 U/ml. After normalizing for plasma protein, a cut-point of 0.2 U/mg plasma protein was obtained. Any variations due to patient differences during treatment were therefore offset. Protein in plasma samples was found to vary considerably particularly during ICT (range 45.6 mg/ml to 125.7 mg/ml) using the Bradford protein determination method.[28] More significant discrimination between various HER-2 expressing populations was found using data analyzed in relation to total plasma protein.

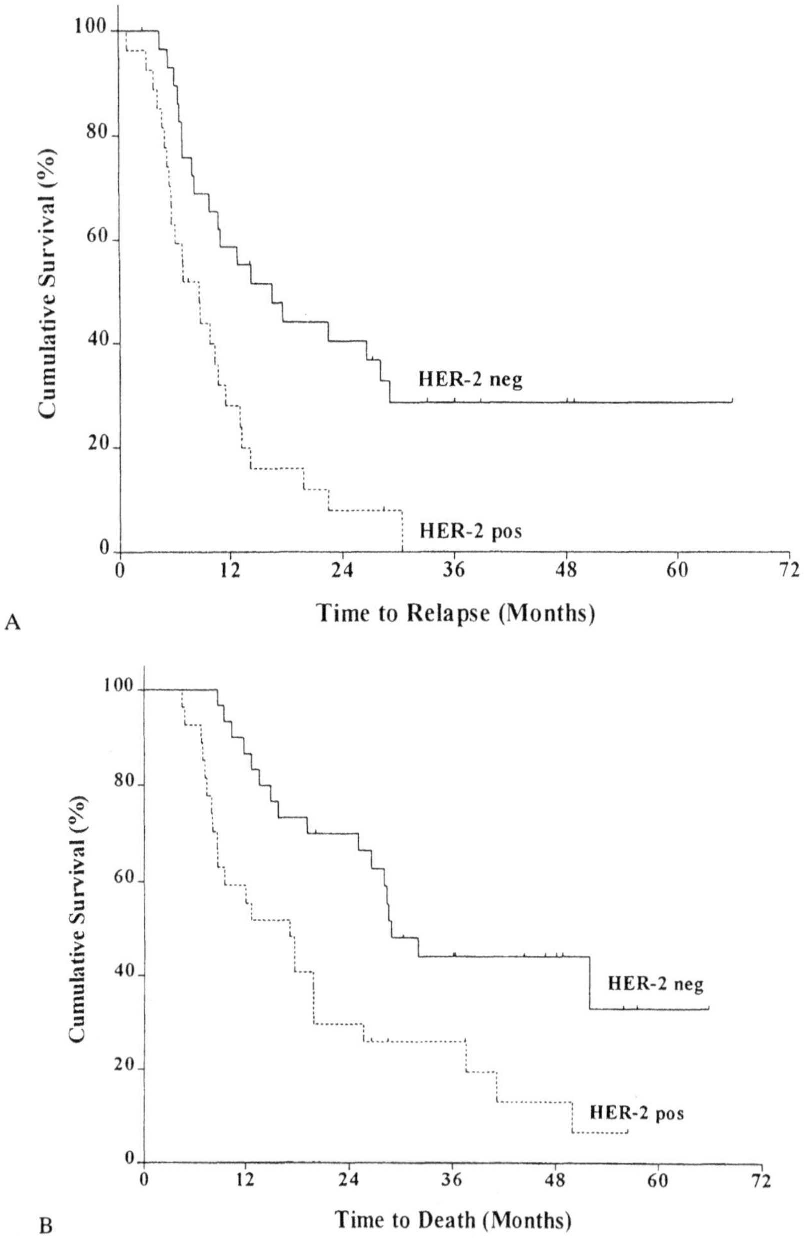

Figure 1. Kaplan-Meir survival plots according to the levels of circulating HER-2 fragment in plasma of 57 patients with MBC undergoing HDCT with ABSC support. Levels of HER-2 were determined using an enzyme immunoassay. HER-2 pos represents patients with overexpression of plasma HER-2 ($\geq$0.2 U/mg protein) and HER-2 neg represents patients with control levels of plasma HER-2. Tick marks represent censored patients. Figure 1A Kaplan-Meir plot of progression free survival. Figure 1B Kaplan-Meir plot of overall survival.

Serial changes in HER-2 overexpression were measured in this study. These results show that changes in HER-2 levels can occur during treatment and that patients can have both negative and positive HER-2 levels during the course of treatment. In 3 of the 27 HER-2 overexpressing patients, HER-2 was only detected at the post HDCT and ABSC transplantation measurement. Five patients that were negative at the initial measurement were found to overexpress HER-2 at the second measurement after ICT at apheresis. *Revillion* et al., using the same EIA method, also observed that 5 of 23 initially HER-2 negative advanced breast cancer patients became HER-2 positive (overexpressing) during treatment with chemotherapy.[17] Serial changes in HER-2 plasma levels were also observed by *Isola* et al.[24] In patients that overexpressed HER-2 prior to HDCT, there were no significant differences in OS and PFS between those patients which lost HER-2 overexpression (11/17) or maintained overexpression (6/17) post HDCT (data not shown). This again is similar to studies which have shown that changes in HER-2 overexpression do not correlate with clinical outcome.[17–20]

It has been shown that chemotherapy and cytokines used to mobilize haematopoietic stem cells into blood can also mobilize tumour cells and the observed increase in the number of patients with circulating HER-2 after ICT may be related to this effect.[29,30] In this study, if only an initial measurement of HER-2 had been determined, 8 patients who were actually positive for HER-2 overexpression would have been excluded. Therefore, during treatment more than one measurement (or a determination of the optimal timing of measurement) with this method may be necessary in order to evaluate HER-2 expression. An analysis of PFS and OS at each time-point in relation to HER-2 overexpression showed that for both PFS and OS, the hazard ratio at the second measurement (at apheresis, following ICT with rhG-CSF or rhGM-CSF growth factor) showed significant survival differences and therefore, this may be the optimum time to evaluate HER-2 overexpression prior to HDCT.

In this study, we found that 47.4% of patients with MBC overexpress HER-2. These results are consistent with the few studies that have examined overexpression in MBC, ie. 20–57% of patients with MBC have elevated HER-2 plasma/serum concentration.[14–20,22–25] These HER-2 positive patients had a significantly decreased OS and PFS following HDCT and ABSC transplantation (Table 2 and Fig. 1). Therefore, the presence of elevated plasma HER-2 may be an important prognostic factor, that could be used to select more effective and additional therapies for these patients (e.g. directly targeted treatments using immunologic therapy).

It is still not certain whether patients with elevated levels of HER-2 would benefit more from specific and/or increased dosages of chemotherapeutic agents as suggested previously by some clinical studies.[31–34] Results from a study by *Muss* et al.[31] suggest that patients with node positive early breast cancer who overexpress HER-2 may benefit from higher doses of adjuvant chemotherapy particularly when using doxorubicin. Results of this retrospective study indicate that increased dosages of drugs used in HDCT do not significantly improve response and survival in patients with MBC who overexpress HER-2. (Four of the 5 patients which received an additional HDCT treatment were HER-2 positive).

HER-2 overexpression has been observed to result in resistance to a wide variety of drugs both *in vitro* and clinically. Correlations have been found between HER-2 overexpression and resistance of breast tumour cells *in vitro* to a wide variety of agents, including alkylating agents such as cis-platinum and CTX,[35] tumour necrosis factor[36] and natural killer cells.[37] Several clinical studies have also shown an association between HER-2 overexpression and drug resistance to alkylating agents, anthracyclines and CMF like regimens.[31,38–40] Additional studies have shown a correlation of HER-2 overexpression with tamoxifen[41–44] and hormone therapy[25,45] resistance in breast cancer. Alternatively, the poor prognosis associated with HER-2 overexpression may occur as a result of increased proliferation and metastasis as indicated by other studies.[13,24,46] In another recent study, HER-2 overexpression was described as having both prognostic and predictive value with regard to adjuvant therapy in patients with lymph node positive breast cancer.[47] Other biomarkers, e.g. estrogen receptor, high S-phase fraction, p53 mutation may also contribute estimating survivals. This together with our finding could be of potential clinical relevance by making choices to alternative or more aggressive treatments available to selected patients.

In conclusion, in this retrospective study, we identified a potential prognostic factor relating to the overexpression of HER-2 in plasma using an EIA method, in women with MBC undergoing HDCT and ABSC transplantation. We feel that multivariate analysis of HER-2 expression would not be useful in this small retrospective study but would provide valuable information when examined in conjunction with a larger prospective study. A larger prospective study currently underway would confirm whether or not this biomarker is an independent prognostic factor.

ACKNOWLEDGMENTS

The authors would like to thank Dr. C. Germond, Dr. P. Lopez, Dr. P. Cano, Dr. M. Dorreen, Dr. G. Knight, and Dr. C. Sinoff of the Medical Oncology Department and Nursing staff of the Northeastern Ontario Regional Cancer Centre for their contribution and dedication to their clinical work. Ms. Nadia Krane for her excellent secretarial support.

Part of this paper was presented at the 39th Annual ASH Meeting, San Diego, California, December 1997 and also at the 20th Annual San Antonio Breast Cancer Symposium, San Antonio, Texas, December 1997.

This research was supported by grants from the Northern Cancer Research Foundation, Sudbury, Ontario; Northern Ontario Heritage Fund, Sault Ste. Marie, Ontario, Canada.

REFERENCES

1. Maguire, HC and Green, MI. The neu (c-erbB-2) oncogene. Semin Oncol., *16*: 148–155, 1989.
2. Earp, HP, Dawson, TL, Li, X, et al. Heterodimerization and functional interaction between EGF receptor family members: A new signalling paradigm with implications for breast cancer research. Breast Cancer Res Treat., *35*: 115–132, 1995.
3. Reese, DM and Slamon, DJ. HER-2/neu signal transduction in human breast and ovarian cancer. Stem Cells, *15*: 1–8, 1997.
4. Sliwkowski, MX, Schaefer, G, Akita, RW, et al. Coexpression of erbB2 and erbB3 proteins reconstitutes a high affinity receptor for heregulin. J Biol Chem., *269*: 14661–14665, 1994.
5. Wallasch, C, Weiss, FU, Niederfellner, G, et al. Heregulin-dependent regulation of HER2/neu oncogenic signalling by heterodimerization with HER3. EMBO Journal, *14*: 4267–4275, 1995.
6. Lewis, GD, Lofgren, JA, McMurtrey, AE, et al. Growth regulation of human breast and ovarian tumour cells by heregulin: Evidence for the requirement of erbB2 as a critical component in mediating heregulin responsiveness. Cancer Res., *56*: 1457–1465, 1996.
7. De Potter, CR and Schelfhout, A-M. The neu protein and breast cancer. Virchows Archiv., *426*: 107–115, 1995.
8. Singleton, TP and Strickler, JG. Clinical and Pathological Significance of the c-erbB-2 (HER-2/neu) oncogene. Pathol Annu., *27*: 165–190, 1992.
9. Perren, TJ. C-erbB-2 oncogene as a prognostic marker in breast cancer. Br J Cancer, *63*: 328–332, 1991.
10. Hynes, NE and Stern, DF. The biology of erbB-2/neu/HER-2 and its role in cancer. Biochim Biophys Acta., *1198*: 165–184, 1994.

11. Slamon, DJ, Godolphin, W, Jones, LA, et al. Studies of the HER2/neu proto-oncogene in human breast and ovarian cancer. Science, *24*: 707–712, 1989.
12. Slamon, DJ. Human Breast Cancer: Correlation of relapse and survival with amplification of the HER-2/neu Oncogene. Science, *235*: 177–182, 1989.
13. Kallioniemi, OP, Holli, K, Visakorpi, T, et al. Association of c-erbB-2 protein over-expression with high rate of cell proliferation, increased risk of visceral metastasis and poor long-term survival in breast cancer. Int J Cancer, *49*: 650–655, 1991.
14. Kandl, HL, Seymour, L, and Bezwoda, WR. Soluble c-erbB-2 fragment in serum correlates with disease stage and predicts for shortened survival in patients with early- stage and advanced breast cancer. Br J Cancer, *70*: 739–742, 1994.
15. Narita, T, Funahashi, H, Satoh, Y, et al. C-erbB-2 protein in the sera of breast cancer patients. Breast Cancer Res Treat., *24*: 97–102, 1992.
16. Kynast, B, Binder, L, Marx, D, et al. Determination of the c-erbB2 translational product p185 in serum of breast cancer patients. J Cancer Clin Oncol., *119*: 249–252, 1993.
17. Revillion, F, Hebbar, M, Bonneterre, J, et al. Plasma c-erbB2 concentrations in relation to Chemotherapy in Breast Cancer Patients. European Journal of Cancer, *32A*: 231–234, 1996.
18. Molina, R, Filella, X, Zanon, G, et al. C-erbB-2 oncoprotein in the sera and tissue of patients with breast cancer. Utility in prognosis. Anticancer Res., *16*: 2295–2300, 1996.
19. Anderson, TI, Paus, E, Nesland, JM, et al. Detection of c-erbB-2 related protein in sera from breast cancer patients. Relationship to ERBB2 gene amplification and c-erbB-2 overexpression in tumour. Acta Oncol., *34*: 499–504, 1995.
20. Volas, GH, Leitzel, K, Teramoto, Y, et al. Serial serum c-erbB-2 levels in patients with breast carcinoma. Cancer, *78*: 267–272, 1996.
21. Hubbard, AL, Doris, CP, Thompson, AM, et al. Critical determination of the frequency of c-erbB-2 amplification in breast cancer. Br J Cancer, *70*: 434–439, 1994.
22. Fehm, T, Maimonis, P, Wetz, S, et al. Influence of circulating c-erbB-2 serum protein on response to adjuvant chemotherapy in node-positive breast cancer patients. Breast Cancer Res Treat., *43*: 87–95, 1997.
23. Molina, R, Jo, J, Zanon, G, et al. Utility of C-erbB-2 in tissue and in serum in the early diagnosis of recurrence in breast cancer patients: comparison with carcinoembryonic antigen and CA 15.3. Br J Cancer, *74*: 1126–1131, 1996.
24. Isola, J, Holli, K, Oksa, H, et al. Elevated erbB-2 oncoprotein levels in preoperative and follow-up serum samples define an aggressive disease course in patients with breast cancer. Cancer, *73*: 652–658, 1994.
25. Leitzel, K, Teramoto, Y, Konrad, K, et al. Elevated serum c-erbB-2 antigen levels and decreased response to hormone therapy of breast cancer. J Clin Oncol., *13*: 1129–1135, 1995.
26. Antman, KH, Rowling, PA, Vaughan, WP, et al. High-dose chemotherapy with autologous haematopoietic stem-cell support for breast cancer in North America. J Clin Oncol., *15*: 1870–1879, 1997.
27. Glück, S, Porter, K, Chadderton, T, et al. Depletion of DMSO (Dimethylsulfoxide) prior to autografting after high-dose chemotherapy for metastatic breast cancer. Blood, *82*(10)1: 430a, 1993.
28. Bradford, MM. A rapid and sensitive method for the quantitation of microgram quantities of protein utilizing the principle of protein-dye binding. Analytical Biochem., *72*: 248–254, 1976.
29. Brugger, W, Bross, KJ, Glatt, M, et al. Mobilization of tumour cells and haematopoietic progenitor cells into peripheral blood of patients with solid tumours. Blood, *83*: 636–640, 1994.
30. Shpall, EJ and Jones, RB. Release of tumour cells from bone marrow. Blood, *83*: 623–625, 1994.
31. Muss, HB, Thor, A, Berry, DA, et al. c-erbB-2 expression and response to adjuvant therapy in women with node-positive early breast cancer. N Engl J Med., *330*: 1260–1266, 1994.
32. Allred, DC, Clark, GM, Tandon, AK, et al. HER-2/neu in node negative breast cancer: prognostic significance of overexpression influenced by the presence of in situ carcinoma. J Clin Oncol., *10*: 599–605, 1992.
33. Gusterson, BA, Gelber, RD, Goldhirsch, A, et al. Prognostic importance of c-erbB-2 expression in breast cancer. International (Ludwig) Breast Cancer Study Group. J Clin Oncol., *10*: 1049–1056, 1992.
34. Wright, C, Caims, J, Cantwell, BJ, et al. Response to mitoxantrone in advanced breast cancer: correlation with expression of C-erbB-2 protein and glutathione S-transferase. Br J Cancer, *65*: 271–274, 1992.
35. Hancock, MC, Langton, BC, Chan, T, et al. A monoclonal antibody against c-erbB-2 protein enhances the cytotoxicity of cis-diamine dichloroplatinum against human breast cancer and ovarian tumour cell lines. Cancer Res., *51*: 4575–4580, 1991.

36. Hudziak, RM, Schlissinger, J and Ullrich, A. Increased expression of the putative growth factor receptor p185 (HER-2) causes transformation and tumorigenesis of NIH3T3 cells. Proc. Natl Acad Sci USA, *84*: 7159–716, 1987.
37. Wiltschke, C, Tyl, E, Steininger, A, et al. Increased NK-cell activity correlates with low or negative expression of the HER-2/neu oncogene in breast cancer patients. Proc AACR, *32*: 203, 1991.
38. Têtu, B and Brisson, J. Prognostic significance of HER-2/neu oncoprotein expression in node- positive breast cancer. Cancer, *73*: 2359–2365, 1994.
39. Giai, M, Roagna, R, Ponzone, R, et al. Prognostic and predictive relevance of c-erbB-2 and ras expression in node positive and negative breast cancer. Anticancer Res., *14*: 1441–1450, 1994.
40. Stål, O, Sullivan, S, Wingren, S, et al. C-erbB-2 expression and benefit from adjuvant chemotherapy and radiotherapy of breast cancer. Eur J Cancer, *31A*: 2185–2190, 1995.
41. Nicholson, S, Wright, C, Sainsburg, RC, et al. Epidermal growth factor receptor (EGFR) as a marker for poor prognosis in node negative breast cancer patients: neu and tamoxifen failure. J Steroid Biochem Mol Biol., *37*: 811–814, 1990.
42. Benz, CC, Scott, GK, Sarup, JC, et al. Estrogen-dependent, tamoxifen-resistant tumorigenic growth of MCF-7 cells transfected with HER2/*neu*. Breast Cancer Res Treat., *24*: 85–95, 1993.
43. Witters, LM, Kumar, R, Chinchilli, VM, et al. Enhanced activity of the combination of tamoxifen plus HER-2-neu antibody. Breast Cancer Res Treat., *42*: 1–5, 1997.
44. Carlomagno, C, Perrone, F, Gallo, C, et al. C-erb-B-2 overexpression decreases the benefit of adjuvant tamoxifen in early-stage breast cancer without auxiliary lymph node metastases. J Clin Oncol., *14*: 2702–2708, 1996.
45. Yamauchi, H, O'Neill, A, Gelman, R, et al. Prediction of response to antiestrogen therapy in advanced breast cancer patients by pretreatment circulating levels of extracellular domain of the HER-2/c-neu protein. J of Clin Onc., *15*: 2518–2525, 1997.
46. Borg, AB, Baldetorp, M, Ferno, D, et al. ErbB2 amplification in breast cancer with a high rate of proliferation. Oncogene, *6*: 137–143, 1991.
47. Sjögren, S., Inganäs, M, Lindgren, A, et al. Prognostic and predictive value of c-erB-2 overexpression in primary breast cancer, alone and in combination with other prognostic markers. J of Clin Onc., *16*: 462–469, 1998.

13

CLINICAL FACTORS CONTRIBUTING TO THE PACE OF ENGRAFTMENT AFTER ALLOGENEIC AND AUTOLOGOUS STEM CELL TRANSPLANTATION

Multivariate Analyses

Gerald J. Elfenbein*

* Division of Bone Marrow Transplantation
Department of Internal Medicine
University of South Florida College of Medicine
Tampa, Florida 33612, USA

Previously, we have shown that a number of clinical factors contributed to the pace of recovery of granulocytes (>500/uL) and platelets (>20,000/uL) after high dose therapy when followed by autologous bone marrow transplantation or when followed by autologous peripheral blood cell transplantation (where hematopoietic progenitor cells had been "mobilized" by a variety of regimens). Recently, we evaluated two additional patient populations to determine whether such clinical factors remained important determinants regulating recovery of blood counts after hematopoietic stem cell transplantation. The first population was 164 consecutive recipients of allogeneic, unstimulated, unmanipulated bone marrow from healthy donors. The second population was 276 consecutive recipients of autologous peripheral blood cells collected after mobilization with cyclophosphamide based chemotherapy and granulocyte-colony stimulating factor. Using time dependent, censored data, multivariate analyses, we found a number of clinical factors made significant contributions to engraftment (two tailed values of $P < 0.05$). After allogeneic bone marrow transplant: 1) for granulocyte recovery—patient diagnosis, degree of HLA match between donor and patient, high dose regimen, granulocyte-colony stimulating factor use after transplant, and acute

---

Please direct correspondence to: Gerald J. Elfenbein, M.D., F.A.C.P., Roger Williams Medical Center, 825 Chalkstone Avenue, Providence, RI 02908 USA; Telephone: (401) 456-6565; FAX: (401) 456-6793

Molecular Biology of Hematopoiesis 6, edited by Abraham *et al.*
Kluwer Academic/Plenum Publishers, New York, 1999.

graft-versus-host disease prophylaxis all made independent contributions and 2) for platelet recovery—only patient gender contributed to the pace of engraftment. After autologous peripheral blood cell transplant: 1) for granulocyte recovery—only the number of CD34+ cells per kilogram weight made a contribution but 2) for platelet recovery—in addition to CD34+ cell dose, high dose regimen made an independent contribution. We conclude that, even for relatively uniform populations of stem cells (healthy, unstimulated, unmanipulated, allogeneic bone marrow, and chemotherapy/growth factor mobilized autologous peripheral blood cells), clinical factors as well as graft factors (e.g., CD34+ cell content) make independent contributions to the pace of engraftment. Engraftment appears to be sensitive to modulation by a number of clinical factors.

INTRODUCTION

After hematopoietic stem cell transplant (SCT), the day that the absolute granulocyte (AGC) exceeds 500/uL and the day the platelet count (PLT) exceeds 20,000/uL are important clinical milestones as they signal the end of the periods of high risk of life-threatening infection and "spontaneous" bleeding, respectively. Understandably, then, any insight into the clinical factors which control the pace of engraftment after SCT is very important for it may provide information about hematopoietic regulatory mechanisms which may, in turn, provide rational and valuable avenues for clinical manipulations to shorten aplasia after SCT.

Previously, we have reported results concerning AGC recovery from studies of patients receiving autologous bone marrow transplants (BMT) that showed: 1) granulocyte-colony stimulating factor (G-CSF), when given after autologous BMT, hastened engraftment more rapidly than did granulocyte-macrophage colony stimulating factor (GM-CSF) which improved the pace of engraftment as compared to autologous BMT after which no hematopoietic growth factor was given,[1] 2) the high dose regimen itself made a significant impact on the pace of engraftment (recovery of counts after mitoxantrone and thio-TEPA (MITT) was slower than after ifosfamide, carboplatin, and etoposide (ICE))[2], and 3) G-CSF stimulated bone marrow engrafted faster than did unstimulated bone marrow after autologous BMT.[3] Further, we have reported results concerning AGC recovery from studies of patients receiving autologous peripheral blood cell transplants (PBCT) that showed: 1) patients with multiple myeloma engrafted more rapidly than did patients with lymphomas after busulfan and cyclophosphamide (second version, BUCY2),[4] 2) PBC collected after cyclophosphamide based chemotherapy and G-CSF mobilization engrafted more rapidly than did those collected after mobilization with G-CSF alone which, in turn, engrafted faster than those collected after GM-CSF,[5,6] 3) the pace of engraftment was affected by the high dose regimen itself (engraftment after the MITT regimen was slower than after either the ICE regimen or the BUCY2 regimen which had similar paces of engraftment,[5] and 4) the content of PBC collections determine the pace of engraftment in that low CD34+ cell content was associated with a higher probability of late engraftment.[7] Finally, we have reported data that showed the dependence of PLT recovery upon patient diagnosis,[4] stem cell mobilization regimen,[5,6] and high dose regimen[6] as well as upon CD34+ cell content of the graft.[7]

With these results in mind, we looked into two entirely different populations of patients to determine if the observations that clinical factors contributed to engraftment were generalizable and to determine what clinical factors may condition the pace of engraftment for these two populations. Crucial to these analyses is the fact that each population was as homogeneous as possible with respect to the stem cells that were transplanted; this was not the case for the studies mentioned above.[1–7]

METHODS AND PATIENTS

The analyses reported herein are retrospective, multivariate and time dependent with censoring (Cox's proportional hazards method). The endpoints of the studies were the first of three consecutive days that AGC > 500/uL and the first of three consecutive days that PLT > 20,000/uL without platelet transfusion support. Data were abstracted from patient records. All patients whose data were evaluated had given written informed consent to participate in clinical trials that were approved and reviewed by the Institutional Review Board of the University of South Florida. Clinical factors entered into the analyses were deemed to make independent contributions to the pace of engraftment if two tailed values of $P < 0.05$.

There were two discrete patient populations studied. The first population received healthy donor BMT. 175 consecutive patients were evaluated but 11 were ineligible for further analysis because they received either syngeneic BMT, T-cell depleted allogeneic BMT, G-CSF primed (mobilized) allogeneic BMT or G-CSF mobilized (primed) PBCT. Thus, there were 164 consecutive evaluable patients available for analysis who received allogeneic, unstimulated, unmanipulated BMT. The second population received autologous PBCT. 278 consecutive patients receiving PBC collected after cyclophosphamide based chemotherapy followed by G-CSF were evaluated but 2 were excluded from further analysis because of missing data about the graft (number of CD34+ cells). Thus, there were 276 consecutive evaluable patients available for analysis who received chemotherapy/growth factor mobilized PBCT. In so far as we could determine, each population was uniform with respect to its source of stem cells and the stimulation status of its stem cells. On the other hand, the populations were otherwise heterogeneous. For example, allogeneic BMT patients may or may not have received growth factors after transplant and bleeding patients were transfused to PLT > 50,00/uL until bleeding ceased. For these latter patients, the day of PLT > 20,000/uL may be artifactually late.

RESULTS

The results of the multivariate analyses are shown in Table I. Listed are the significant clinical factors for both AGC and PLT recovery and their respective two tailed values of P. For allogeneic BMT recipients, 5 clinical factors contributed to the pace of AGC recovery (patient diagnosis, degree of HLA match between donor and patient, high dose regimen, use of growth factors after BMT, and acute graft-versus-host disease (GVHD) prophylaxis) but only one (patient gender) appeared to contribute to the pace of PLT recovery. For autologous PBCT recipients, as expected, the dose of CD34+ cells in the graft (number per kilogram body weight) was significant for both AGC and PLT recovery. Additionally, the high dose regimen was a significant determinant of PLT recovery.

Having completed the analyses of AGC and PLT recovery for allogeneic BMT and autologous PBCT separately, we performed a mini-meta-analysis by joining the two patient population databases together to determine if there were common themes

Table I. Determinants of engraftment after stem cell transplantation from multivariate analyses

Type of transplant	Recovery regulated	Clinical factor	2 tailed p value	Engraftment may be delayed if:
Allogeneic BMT (N = 164)	AGC	Patient diagnosis	0.00012	Other than multiple myeloma
"	AGC	Degree of HLA match	0.033	No genotypic HLA identity
"	AGC	High dose regimen	0.00093	Other than BUCY2 or TBI based
"	AGC	Growth factor	0.012	None is used
"	AGC	Acute GVHD prophylaxis	0.00052	Methotrexate included
"	PLT	Patient gender	0.022	Recipient is male
Autologous PBCT (N = 276)	AGC	CD34+ cell dose	<0.000001	<2 × 10exp6 per kg
"	PLT	High dose regimen	0.000052	MITT, TNT, or BUCY2 with or without VP16
"	PLT	CD34+ cell dose	<0.000001	<7 × 10exp6 per kg

TBI = total body irradiation; TNT = Taxol(R), Novantrone(R), and thioTEPA; and VP16 = etoposide.

across the patient populations (grand N = 440). The results of this effort are shown in Table II. Four clinical factors were determined to be significant for AGC and PLT recovery. Interestingly, 3 of the 4 factors were identical for both AGC and PLT recovery (disease stage, stem cell priming method, and acute GVHD prophylaxis regimen). The fourth factor for AGC recovery was the use of hematopoietic growth factors after SCT and for PLT recovery was the high dose regimen. Figures 1 and 2 show the effects

Table II. Determinants of engraftment for all 440 allogeneic BMT and autologous PBCT recipients from multivariate analyses in a mini-meta-analysis (two tailed values of P)

Clinical factor	Day to AGC > 500/μL	Day to PLT > 20,000/μL
BMT sequence number	0.33	0.70
Patient age	0.53	0.21
Patient gender	0.80	0.44
Patient diagnosis	0.46	0.58
Disease stage	0.045	0.0054
Stem cell priming	0.0025	0.015
Stem cell source (BM or PB)	0.44	0.75
Degree of HLA match	0.54	0.097
High dose regimen	0.41	0.0012
Growth factor use	0.0046	0.50
CD34+ cell selection	0.81	0.18
Acute GVHD prophylaxis	0.0019	0.0082

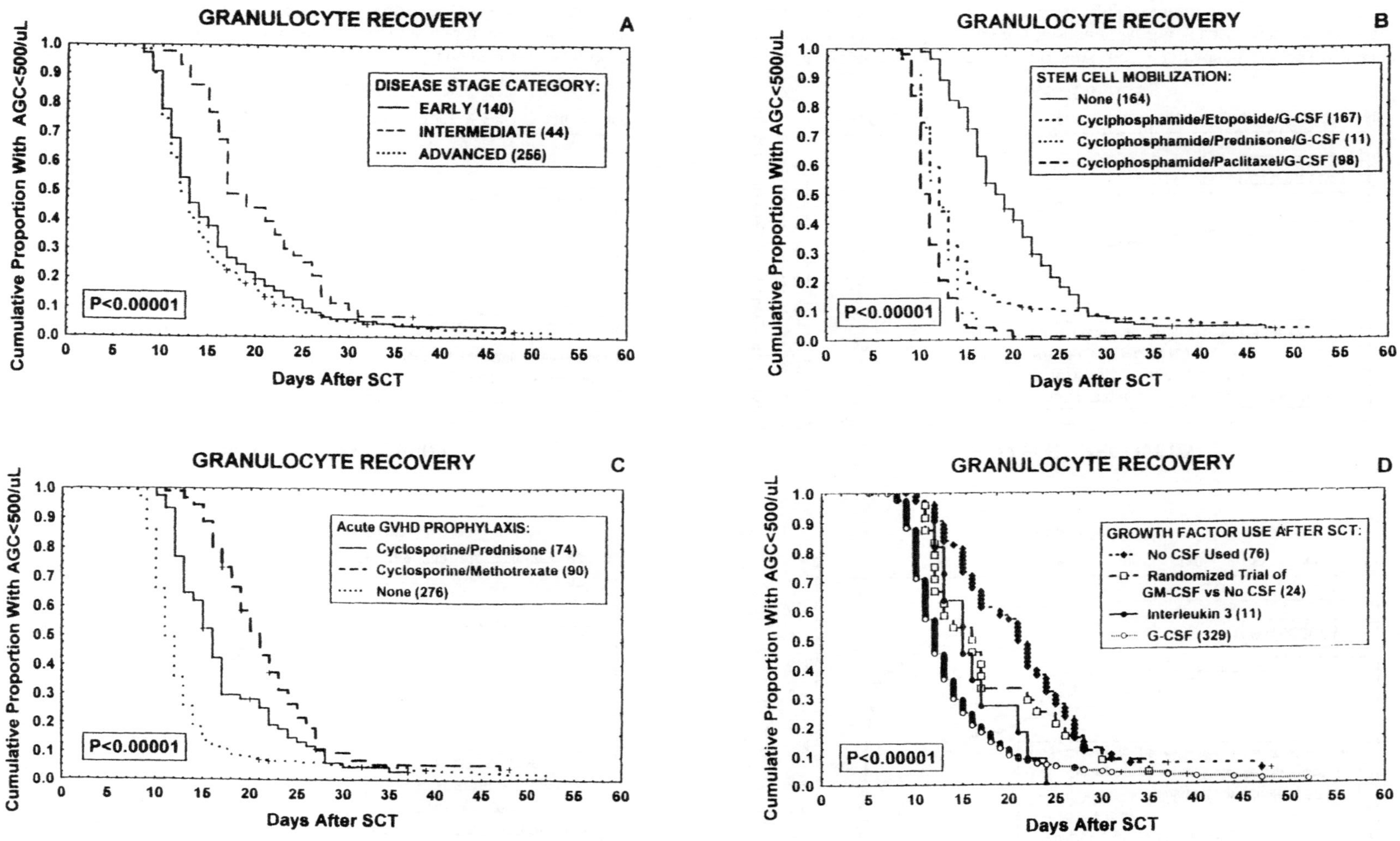

Figure 1. Granulocyte recovery. All plots are of the Kaplan-Meier type showing the percentage of the at risk population who have not yet reached the engraftment endpoint of AGC > 500/ul. See text for details and abbreviations. Abbreviation not defined in text is GM-CSF = granulocyte-macrophage colony stimulating factor.

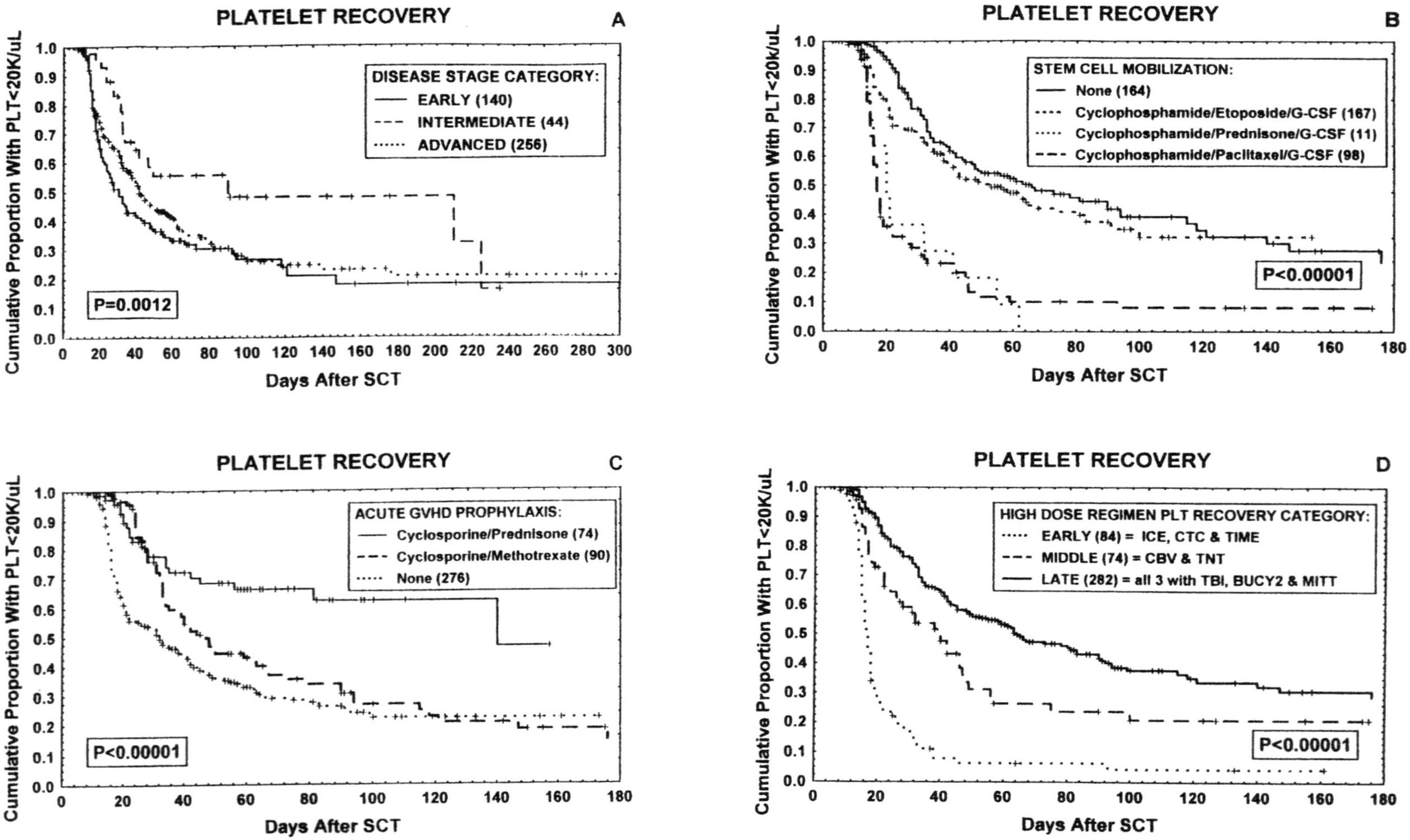

Figure 2. Platelet recovery. All plots are of the Kaplan-Meier type showing the percentage of the at risk population who have not yet reached the engraftment endpoint of PLT > 20,000/ul. See text for details and abbreviations. Abbreviations not defined in text are CTC = cyclophosphamide, thio-TEPA, and carboplatin; TIME = topotecan, ifosfamide, mesna, and etoposide; CBV = cyclophosphamide, BCNU, and etoposide (VP-16); and TBI = total body irradiation.

of the significant clinical factors in a univariate manner (with two tailed values of P coming from the log rank test).

DISCUSSION

The purpose of these analyses was to confirm and extend our previous findings that measurable/evaluable clinical factors contributed to the pace of engraftment of not just granulocytes but platelets as well in two homogeneous but quite distinct and dissimilar patient populations. Five independent contributors to engraftment were found for granulocytes after allogeneic BMT. Only the use of hematopoietic growth factors and the acute GVHD prophylaxis regimen were anticipated in advance to contribute to recovery of the AGC where mechanisms are relatively easy to implicate (proliferation of granulocyte precursors, on the one hand, and cytotoxicity to myeloid progenitor cells, on the other hand). With respect to patient diagnosis, degree of HLA match, and high dose regimen, the mechanism(s) whereby these clinical factors exert their influence(s) is (are) as yet unclear but it is attractive to hypothesize that their impact is upon the hematopoietic microenvironment any reduction of function of which could lead to delayed engraftment.[2,7] At this point in time, how patient gender exerts its influence on PLT recovery after allogeneic BMT is also unclear but one attractive hypothesis is that gender is merely a surrogate for marrow cell dose per kilogram body weight of the recipient as women on the average are lighter in weight than are men. Taken together, these observations indicate that, despite transplanting normal, healthy bone marrow collected from donors in the "resting" state (i.e., unstimulated *in vivo*) and without manipulation *ex vivo* (e.g., T-cell depletion), engraftment is modulated by a number of clinical factors and not simply determined just by marrow cell dose (except perhaps in the instance of PLT recovery).

For cyclophosphamide based chemotherapy and G-CSF mobilized PBC, we see the expected relationship of CD34+ cell content of the graft and the pace of engraftment, i.e., for low CD34+ cell content, the probability of late engraftment rises. Whereas, in the previous studies mentioned above,[1-7] there was considerable variability in the stem cell mobilization (priming) regimen and a number of clinical factors contributing to engraftment were identified; in this analysis of relatively homogeneous PBC collected from a variety of patients with respect to diagnoses and high dose regimens, we were still able to identify high dose regimen as a determinant of PLT recovery. This confirms the basic concept that clinical factors other than just CD34+ cell content of PBC collections determine engraftment as well.

The mini-meta-analysis confirms the roles of stem cell priming/mobilization regimen and acute GVHD prophylaxis regimen in regulating the pace of recovery for both AGC and PLT but with some interesting twists. Additionally, disease stage appears important in affecting both AGC and PLT recovery; growth factor use is significant for AGC recovery; and high dose regimen is a significant determinant of PLT recovery. Why disease stage (which was not a significant contributor in the separate, multivariate analyses for allogeneic BMT and autologous PBCT) is now significant for both AGC and PLT recovery (Figs. 1a and 2a) is unclear; however, only intermediate stage disease patients had delayed engraftment suggesting that we might be seeing something other than a dose-response effect with slower and slower engraftment for more and more advanced disease perhaps due to cumulative toxicity to the marrow microenvironment by chemotherapy delivered prior to high dose therapy. Because no autologous PBCT patient was categorized as having intermediate stage disease (only early and advanced classifications were used), it is possible that intermediate stage disease is a surrogate for allogeneic BMT and degree of HLA match between donor and recipient (P = 0.097 in the multivariate analysis; see Table II). As expected, all categories of chemotherapy/growth factor mobilized PBC produced earlier (and similar) AGC recovery than resting BM cells (Fig. 1b). However, even though the differences in PBC mobilization regimens were relatively minor (i.e., cyclophosphamide with either etoposide, paclitaxel, or prednisone followed by G-CSF), not all PBC mobilization regimens produced the same pace of recovery of PLT (i.e., cyclophosphamide/etoposide/G-CSF mobilized PBC produced no more rapid recovery than did resting BM cells (Fig. 2b) both of which produced slower PLT recovery than did cyclophosphamide/paclitaxel or prednisone/G-CSF mobilized PBC). Why is unclear at the present time.

On the other hand, that growth factor use after transplant affected only AGC recovery (Fig. 1d) is not too difficult to understand as only G-CSF, GM-CSF and interleukin 3 were used for allogeneic BMT recipients; G-CSF was routinely used after autologous PBCT. The impact of high dose regimen upon PLT recovery (Fig. 2d) is, also, not too difficult to understand 1) if thrombopoiesis is more dependent upon the microenvironment than is granulopoiesis and/or 2) if the process of PLT recovery is more marginal than that of AGC recovery (Fig. 1c). This latter possibility is supported by the higher threshold level of CD34+ cells below which the probability of late recovery for PLT rises as compared to the threshold for late AGC recovery (see Table I). Last, as expected, methotrexate, when included in acute GVHD prophylaxis, delayed AGC recovery. Quite unexpected, however, PLT recovery was later for patients receiving acute GVHD prophylaxis including prednisone (Fig. 2c). This observation is provocative and there is, as yet, no suitable hypothesis to explain this paradox.

One last comment needs to be made from observations concerning 6 of the 11 healthy allogeneic marrow donor transplants that were excluded from these analyses. These 6 healthy donors had stem cells collected after 3 days of G-CSF therapy. The recipients of G-CSF primed healthy donor stem cells had more rapid recovery of AGC (median day 11 versus median day 18; log rank, two tailed P = 0.18) and much more rapid recovery of PLT (median day 27 versus median day 66: log rank, two tailed P = 0.0081) after allogeneic SCT (whether the cells were collected from the bone marrow or the peripheral blood) than did recipients of unprimed allogeneic BMT. That the AGC recovery time difference is not statistically significant while the PLT recovery time difference is statistically significant is a reflection of the lesser magnitude of the percent reduction in recovery times (39% versus 59%, respectively) and the small sample size of the G-CSF primed group *vis a vis* the unprimed group (6 versus 164). Nonetheless, we believe that the shortened time to recovery of AGC with G-CSF primed allogeneic stem cells is biologically and clinically meaningful. Thus, there may be a role for priming healthy donors with G-CSF before allogeneic bone marrow or PBC collection to speed up engraftment.

REFERENCES

1. Fields KK, Elfenbein GJ, Perkins JB, Janssen WE, Ballester OF, Hiemenz JW, Zorsky PE, Kronish LE, Foody MC: High-dose ifosfamide/carboplatin/etoposide: Maximum tolerable doses, toxicities, and hematopoietic recovery after autologous stem cell reinfusion. Seminars in Oncology 21 (suppl 12):86–92, 1994.

2. Elfenbein GJ, Janssen WE, Perkins JB: Relative contributions of marrow microenvironment, growth factors, and stem cells to hematopoiesis *in vivo* in man: Review of results from autologous stem cell transplant trials and laboratory studies at the Moffitt Cancer Center, in Sackstein R, Janssen WE, Elfenbein GJ (eds): Bone Marrow Transplantation: Foundations for the 21st Century; Annals of the New York Academy of Sciences, Vol 770, New York, NY, The New York Academy of Sciences, 1995, pp 315–338.
3. Janssen WE, Smilee RC, Elfenbein GJ: A prospective randomized trial comparing blood vs. marrow derived stem cells for hematopoietic replacement following high dose chemotherapy. Journal of Hematotherapy 4:139–140, 1995.
4. Ballester OF, Agaliotis DP, Hiemenz JW, Janssen WE, Fields KK, Zorsky PE, Goldstein SC, Perkins JB, Elfenbein GJ: Phase I-II study of high dose busulfan and cyclophosphamide followed by autologous peripheral blood stem cell transplantation for hematological malignancies: Toxicities and hematopoietic recovery. Bone Marrow Transplantation 18:9–14, 1996.
5. Janssen WE and Elfenbein GJ: Mobilization of peripheral blood stem cells: Are all regimens created equal? in Levitt D, Mertelsmann R (eds): Hematopoietic Stem Cells: Biology and Therapeutic Applications. New York, NY, Marcel Dekker, 1995, pp 403–419.
6. Janssen WE, Elfenbein GJ, Fields KK, Hiemenz JW, Zorsky PE, Ballester OF, Goldstein SC, Smilee R, Kronish L, Beach B, LeParc G: Comparison of cell collections and rates of post-transplant granulocyte recovery when G-CSF and GM-CSF are used as mobilizers of peripheral blood stem cells for autotransplantation, in Dicke KA, Keating A (eds): Autologous Bone Marrow Transplantation, Proceedings of the 7th International Symposium. Arlington, TX, 1995, pp 527–539.
7. Elfenbein GJ, Janssen WE, Perkins JB, Partyka JS, Fields KK: Mathematical modeling of human hematopoiesis: Lessons learned from the bedside following autologous peripheral blood stem cell transplants. in Dicke KA, Keating A (eds): Autologous Blood and Marrow Transplantation, Proceedings of the 8th International Symposium. Charlottesville, VA, Carden Jennings, 1997, pp 443–466.

14

CLINICAL RELEVANCE OF BONE MARROW HISTOLOGIC EXAMINATION IN HIV-INFECTED PATIENTS

Silvia Cantoni, Anna Maria Nosari, Pier Luigi Oreste,° Alberto Volonterio,* Maria Cristina Moioli,* Liliana Caggese,* and Enrica Morra

Department of Hematology
* Infectious Diseases
° Oncology and Pathology
Niguarda Ca'Granda Hospital
Milan, Italy

ABSTRACT

Infection with HIV is frequently associated with hematological abnormalities whose origin is multifactorial: HIV indirect action on marrow microenvironment; drug-related myelotoxicity; neoplastic and infectious diseases localized to the bone marrow (BM).

To assess the usefulness of BM biopsy in the diagnostic evaluation of HIV+ patients and the prevalence of BM abnormalities, 612 biopsies—350 *in vita*, 262 *post mortem*—performed between 1981 and 1996 were reviewed. Indications to *in vita* biopsies were: FUO (71%), lymphoma staging (7%), cytopenias (22%).

Of the 612 BM biopsies performed, 70 (11.5%) yielded specific results: 30 (5%) were positive for neoplastic (NHL 19, HD 7, MM 2, B-ALL 2) and 40 (6.5%) for infectious diseases (mycobacteria 28, granuloma 4, toxoplasma 2, *cryptococcus* 3, *pneumocystis* 1, leishmania 1, CMV 1).

Dysplastic changes were found in the remaining 88.5% biopsies. BM hyper-normocellularity was found in 60% of specimens; hypoplasia (40%) was usually associated with myelotoxic drugs administration and mycobacterial infection.

Corresponding Author: Silvia Cantoni, Department of Hematology, Niguarda Ca'Granda Hospital, Piazza Ospedale Maggiore, 3 Milan, Italy, FAX: 39 2 6444 2033

Molecular Biology of Hematopoiesis 6, edited by Abraham *et al.*
Kluwer Academic/Plenum Publishers, New York, 1999.

Our data show that in HIV+ pts BM biopsy is useful 1- in the evaluation of FUO, especially when mycobacterial infection or lymphoma are suspected; 2- in diagnosing and staging hematologic malignancies; 3- in evaluating cytopenias and bone marrow cellularity in view of the institution of myelotoxic therapies.

INTRODUCTION

Although a decrease of CD4+ lymphocytes is the most characteristic abnormality associated with HIV infection, anomalies of blood hematologic parameters—anemia, leucopenia, and thrombocytopenia, alone or variably combined—are also frequently encountered in HIV-positive patients.

Prevalence of cytopenias usually increases with disease progression and according to some Authors, cytopenia can be observed in over 80% of individuals with late stage infection.[1]

Moreover, hematologic malignancies, especially high grade non-Hodgkin's lymphomas (NHL), are common in these severely immunodepressed patients. Immunodepression may also result in opportunistic infections which may localise to the bone marrow and interfere with myelopoiesis.

In this setting, BM examination has been shown to be useful in the evaluation of peripheral blood cytopenias, of fever of unknown origin (FUO) and in the diagnosis and staging of hematologic malignancies.[2]

In order to assess the usefulness of bone marrow examination in the diagnostic evaluation of HIV-positive patients and the prevalence of specific bone marrow abnormalities, bone marrow biopsies performed between 1981 and 1996 at our Hospital were reviewed and constitute the object of this report.

METHODS

During the years 1981–1996 a total of 612 bone marrow biopsies were performed on HIV infected patients followed in the same time period at our Institution: 350 biopsies were performed *in vita*, while 262 were *post mortem* autoptic specimens. Patients at all stages of HIV infection were included.

In vita bone marrow biopsies were taken for the following clinical indications: evaluation of anemia, leucopenia, thrombocytopenia or any combination of these conditions; evaluation of fever of unknown origin, coupled with marrow microbiological culture; exclusion of marrow's malignant infiltration. More than one indication to bone marrow biopsy might have been present.

Biopsies were taken from the posterior iliac crest and processed as usual.

The sections were stained with Haematoxylin and Eosin, Giemsa and PAS for morphologic studies. Special stains for reticulin (Gomori's stain), acid-fast bacilli (Ziehl-Neelsen) fungi (Grocott's stain) were also employed. *In situ* hybridization for CMV and monoclonal antibodies against toxoplasma (APAAP method) were used when indicated.

Screening for NHL and HD was done with the following monoclonal antibodies: CD20, CD79a, CD3, CD4, CD45Ro, CD15, CD30, CD68.

Special immunocytochemical stainigs were also applyed when clinically indicated.

RESULTS

A total of 612 bone marrow biopsies were reviewed. *In vita* biopsies were performed in the following settings: FUO, 71% of patients; diagnosis of hematological malignancies and staging of NHL or HD, 7%; peripheral blood cytopenias, 22%.

Bone marrow hypercellularity or normocellularity was found in 60% of specimens. Hypoplasia was present in the remaining 40% of biopsies and it was most often associated with concomitant administration of known myelotoxic drug regimens (antiviral and anti-protozoan drugs; zidovudine) or mycobacterial infection, independent of concomitant presence of characteristic granuloma.

Some degree of dysplasia was evident in 88.5% of biopsies. As far as erythropoiesis is concerned, dysplastic features resembled in many respects those associated with myelodisplastic syndromes of HIV-negative patients. Varying degrees of erythroid hyperplasia associated with megaloblastosis were present in 20% of specimens. Sideroblasts were very rarely seen. Megakaryocytes almost always showed prominent dysplastic changes, plurilobulated nuclea and very scant cytoplasm being the most frequently encountered anomalies. Contrary to HIV-negative myelodysplastic syndromes, micromegacariocytes were very rarely seen. Myelopoiesis was the least affected. No characteristic morphologic abnormalities were noted; the percentage of blast cells was always within normal range and no Auer rods were seen.

Plasma cells were increased in 50% of patients, typically localizing in a perivascular position. No aggregates or abnormal morphology were noted and plasmocytosis never exceeded 10% of total nucleated cells—with the only exception of the two myeloma patients.

A B- and T-cell derived lymphoid infiltrate, often organized in small size, interstitial nodules or aggregates, was found in the majority of specimens (60%).

Of the 612 biopsies performed, 70 (11.5%) yielded definite diagnostic findings: 30 (5%) were positive for involvement of bone marrow by hematologic malignancies; 40 (6.5%) were positive for infection. Details on diagnostic findings are given in Table I.

Our study group comprised 57 patients with systemic NHL. Among these patients,

Table I. Diagnostic findings in 612 bone marrow biopsies from HIV-positive patients

	In vita		*Post-mortem*	
biopsies (total)	350	57%	262	43%
non-Hodgkin's lymphoma	13	3.7%	6	2.3%
Hodgkin's disease	6	1.7%	1	0.4%
multiple myeloma	1	0.3%	1	0.4%
B-ALL	1	0.3%	1	0.4%
mycobacterial infection	16	4.6%	12	4.6%
granuloma	4	1%	—	—
Cryptococcus	1	0.3%	2	0.8%
Toxoplasma gondii	—	—	2	0.8%
Leishmania	1	0.3%	—	—
Pneumocystis carinii	—	—	1	0.4%
Cytomegalovirus	1	0.3%	—	—

B-ALL: B-acute lymphoblastic leukemia

Table II. Results of 54 bone marrow biopsies performed *in vita* on a group of 57 patients with HIV-related systemic non-Hodgkin's lymphoma

Histology	# of patients	Negative	Positive	Not performed
peripheral T cell lymphoma	2	2	—	—
low-grade B cell lymphoma	3	—	3	—
MALT lymphoma	3	2	—	1
intermediate-grade lymphoma	6	4	2	—
high-grade lymphoma	32	24	7	1
ALCL	11	3	7	1
total	57	35 (65%)	19 (35%)	

MALT: mucosal associated lymphoid tissue; ALCL: anaplastic large cell lymphoma

39 *in vita* biopsies and 15 autoptic specimens were obtained; 3 patients refused the procedure. Histologic results are shown in Table II. Among MALT lymphomas, the 2 high-grade ones were bone marrow negative, while the one patient with a low-grade MALT refused biopsy. In 5 of 39 *in vita* procedures (13%) undertaken for evaluation of FUO, a positive bone marrow biopsy was the initial site of diagnosis of NHL; in the remaining 87% of patients, biopsy was performed for staging purposes.

A bone marrow biopsy was performed in all 7 HD patients included in our study group. In the 6 patients in whom biopsy was taken *in vita*, the bone marrow was the initial site of diagnosis in 4 cases, while in 2 patients biopsy was performed for staging purposes. In one patient, HD was discovered at autopsy and the bone marrow specimen was found positive.

Thirty-six specimens displayed morphologically recognizable micro-organisms; in four cases, only granuloma suggestive of mycobacterial infection were found (Table I). Mycobacteria were the most commonly observed micro-organism in our series.

DISCUSSION

The bone marrow is often involved during the course of HIV infection and bone marrow examination is frequently performed for evaluation of patients with cytopenias and FUO, in the latter case to obtain marrow cultures and identify infectious and neoplastic diseases.

In our experience, indications to submitting the patient to this procedure have changed through the years. Up until 1987, *in vita* biopsies were preferentially done in patients with isolated or concomitant cytopenias, mainly because pathogenetic mechanisms underlying these hematologic alterations were largely unknown. After recognition of the high incidence of NHL among HIV-positive patients and their inclusion as AIDS defining events, the large majority of later biopsies were performed in patients with FUO in order to differentiate infections—especially atypical mycobacteriosis—from lymphomas. This explains the relatively low number (77/350) of *in vita* biopsies performed for the evaluation of cytopenias in our series.

Our data on cellularity show a higher prevalence of hypocellular bone marrow specimens compared to other reported series.[2,3] This is probably due to the high number of autoptic specimens in our series which show marrow morphology of end stage AIDS, when hypocellularity is very frequent both as a result of advanced disease and widespread use of myelotoxic drugs.[4]

No strict, direct correlation between peripheral blood cellularity and bone marrow cellularity or, more precisely, bone marrow reserve, has been found.[2] Therefore we believe that bone marrow biopsy is still useful in those patients with cytopenias who are candidate to administration of potentially myelotoxic drugs or in whom immune-mediate destruction of blood cells (e.g. immune-mediate thrombocytopenia) is suspected on clinical grounds.

The overall data of dysplasia in our series do not substantially differ from previous reports.[2-5] Aspecific dysplastic changes were frequent. Contrary to myelodysplasia of HIV-negative patients, micromegakaryocytes were very rare and no tendency to blastic transformation was observed.

As already reported, reactive plasmocytosis and lymphoid infiltrates were common, also in the context of a dysplastic marrow. However, infiltration patterns and use of monoclonal antibodies make differentiation from malignant lymphoid proliferation straightforward.

In our experience, evaluation of FUO still represents the main indication to bone marrow biopsy, coupled with marrow cultures. As already reported, even in acutely ill patients, bone marrow evaluation is a low complication rate procedure which provided evidence of infection in up to 42% of patients tested.[6] Moreover, morphologically recognizable micro-organisms seen on biopsy, may provide diagnostic results more rapidly than blood cultures and, at least in the case of mycobacterial infections, with similar sensitivities in detecting the micro-organism.[6]

Atypical mycobacteria were the most frequent micro-organism found in our series. Moreover, the finding of a hypocellular marrow was frequently associated with mycobacterial infection at other sites. Therefore we believe that the finding of a hypocellular marrow in HIV patients with FUO makes the search for mycobacterial infection mandatory.

In the presence of FUO, it is also necessary to rule out malignancies and especially hematologic ones, considering the high incidence of lymphomas among HIV-positive patients. Bone marrow biopsy is especially useful in febrile patients without enlarged lymph nodes susceptible to be biopsied. A positive bone marrow biopsy can represent, although seldom, the initial site of diagnosis of NHL and of HD: in our series, in 5/39 patients (13%) with subsequent diagnosis of systemic NHL and in 4/6 patients (66.5%) with subsequent diagnosis of HD who underwent bone marrow biopsy for FUO, the bone marrow was the diagnostic site.

Peculiar extra-nodal involvement is characteristic of HIV-related NHL. However, as far as bone marrow involvement is concerned, their biological behaviour is diverse. Similarly to the general population, low-grade NHL showed marrow involvement while high-grade MALT lymphomas were all biopsy negative. On the contrary, intermediate- and high-grade NHL colonize the bone marrow more frequently than in HIV-negative patients; in our series 16 of 49 patients (32.5%) with a diagnosis of intermediate-high grade lymphoma who underwent biopsy.

In our experience bone marrow biopsy, although invasive, still plays a role in the evaluation of HIV-positive patients. Clearly correct timing of the procedure and selection of patients are important in order to obtain significative diagnostic results.

REFERENCES

1. Scadden DI, Zon LI, Groopman JE: Pathophysiology and management of HIV associated hematologic disorders. Blood 4:1455, 1989.
2. Ciaudo M, Doco-Lecompte T, Guettier C, d'Agay MF, David F, Rioual N, Buu-Hoi A, Molina JM, Kazatchkine M, Modai J, Lecompte T: Revised indications for bone marrow examinations in HIV-infected patients. Eur J Haematol 53:168, 1994.
3. Holland KH, Spivak JL: The haematological manifestations of acquired immune deficiency syndrome. Baillière's Clinical Haematology 3:103,1990.
4. Bain BJ: The haematological features of HIV infection. Br J Haematol 99:1, 1997.
5. Mir N, Costello C, Luckit J, Lindley R: HIV-disease and bone marrow changes: a study of 60 cases. Eur J Haematol 42:339,1989.
6. Keiser P, Rademacher S, Smith JW: Utility of bone marrow culture and biopsy in the diagnosis of disseminated infections in AIDS. Am J Hematol 56:1, 1997.

15

MOLECULAR BIOLOGY OF CHILDHOOD SOLID TUMORS

Promises Maintained and Promises Postponed

Carlo Dominici[1,2] and Per Kogner[2]

[1] Department of Pediatrics
La Sapienza University
I-00161 Rome, Italy
[2] Childhood Cancer Research Unit
Department of Woman and Child Health
Karolinska Hospital and Institute
S-171 76 Stockholm, Sweden

1. INTRODUCTION

More than twenty years ago, the approach to tumor biology changed dramatically after the identification of genes that were shown to be specifically affected in human cancers, the cellular proto-oncogenes. By understanding their functions and by subsequently recognizing that, basically, cancer (either sporadic or inherited) is a genetic disease deriving from an accumulation of mutations able to promote clonal selection of tumor cell populations with increasingly aggressive phenotypes, new methods of study developed.[1,2] Nowdays, the extensive application of the techniques of molecular biology to cancer research has led to the characterization of genes whose structural and/or functional alterations are thought to be responsible for producing and modulating the different phases of tumor development and progression.[3] The great majority of these alterations take place in somatic cells and are therefore found in tumor cells alone. On the other hand, about 1% of all human tumors arise in individuals affected with inherited cancer syndromes who carry a germline mutation that is thus found in every cell of their body.[4]

Understanding the critical role played by gene function disruptions in human tumors has allowed findings obtained from molecular genetics to be integrated into a wide range of clinical settings with potentially promising diagnostic and therapeutic

Molecular Biology of Hematopoiesis 6, edited by Abraham *et al.*
Kluwer Academic/Plenum Publishers, New York, 1999.

approaches, e.g., establishment of diagnosis, staging, assessment of risk class, monitoring of response to treatment;[2,3,5] genetic counseling and/or testing in predisposed families;[6,7] identification of specific targets for therapy and design of new anticancer drugs;[8,9] and restoration of disrupted gene function.[10]

As far as the aforementioned issues are concerned, even brief details would be beyond the limits of this introductory chapter. Nonetheless, it is undeniable that the integration of molecular genetics into clinical management has yielded promising results, particularly in pediatric oncology. The identification of several tumor histotypes based on molecular characteristics and no longer simply on their morphology has permitted, for example, a more reliable and reproducible diagnosis.[11] It is hoped that this will enable a better evaluation of therapeutic results to be made in the future; something which appears even more likely considering that the different molecular features, for a given histotype, seem to distinguish distinct subsets of patients with different clinical outcomes.[12] Likewise, molecular investigation, consenting in some cases a more sensitive and specific identification of rare tumor cells, has enabled more accurate staging of disease at diagnosis as well as a more precise detection of minimal residual disease during treatment.[13,14] Furthermore, the assessment of risk class also including molecular markers, seems to have improved our ability to discriminate subsets of patients with different outcomes.[15,16] Finally, a few genes responsible for tumor predisposition have been identified enabling genetic testing to be included in the clinical management of those individuals with well-defined inherited cancer syndromes.[4] In many aspects, the promises of molecular biology have been maintained.

Regretfully, despite the remarkable progress briefly summarized above, no such parallel has been obtained in patient cure rates. Ongoing studies exploiting the possibility of gene alterations, which are also responsible for the genetic instability in tumor cells, as the key to tumor cell sensitivity will require time before being clinically validated. In spite of the in-depth understanding we have gained of the gene alterations underlying tumor development and progression, Ian Tannock's statement that "For the next few years, cancer management outside the context of a clinical trial will continue to be based almost entirely on surgery, radiation therapy, and systemic treatment with chemotherapy or hormones."[17] seems destined to turn out true.

2. SESSION ON "MOLECULAR BIOLOGY OF CHILDHOOD SOLID TUMORS"

This is the first time that the Symposium on Molecular Biology of Hematopoiesis dedicates, along with the traditional sessions on matters regarding hematopoiesis, a few sessions focused on both molecular genetics and clinical management of childhood solid tumors. In particular, the session on "Molecular Biology of Childhood Solid Tumors" organized and chaired by us, includes five presentations aimed at giving a brief up to date introduction to the molecular features of childhood solid tumors and to the present-day possibilities of integrating these data into clinical management. Since it is the first of such sessions, it would have been interesting to provide an update on each type of childhood solid tumor. However, since this is beyond the scope of the Symposium, and since other sessions focus on soft tissue sarcomas and retinoblastoma, the session presented here will be limited to the consideration of three histotypes only, namely: Wilms tumor, the Ewing family of tumors, and neuroblastoma.

Kathy Pritchard-Jones describes the molecular genetics of sporadic and familial Wilms tumor, the most common renal tumor in children, as well as of the related genetic syndromes. Although several genes have been implicated in Wilms tumor, only *WT1* gene has been proven to be responsible for a proportion of cases. She thus addresses the questions of what other genes are involved and how they relate to *WT1*, suggesting that, through their disregulated expression, the contribution of genes such as *WT1* and *IGF-2*—infrequently if ever mutated—may be much greater than expected.

Olivier Delattre reports on the Ewing family of tumors, the second most common type of bone tumor in children and adolescents, which may also occur in soft tissues. Presently, rather than by its microscopic appearance, the Ewing family of tumors is characterized by its molecular features, i.e., the fusion of the *EWS* gene on 22q12 with one of the ETS family members of transcription factors. Notably, different outcomes have been reported to be associated with the diverse chimeric transcripts. He also describes the possibility of using RT-PCR to detect chimeric transcripts in bone marrow and the prognostic significance of these findings in patients with otherwise localized disease.

Tatsuya Matsuo and Carol Thiele address some fundamental issues regarding the mechanisms underlying proliferation and differentiation in neuroblastoma, a tumor that arises from sympatho-adrenal precursors and represents the most frequent extracranial childhood solid neoplasm. They show that *in vitro* exposure of neuroblastoma cell line SMS-KCNR to all-*trans* RA reduces cell proliferation probably by decreasing *MYCN* gene expression, thereby affecting both p27 and cyclin/cdk kinase activities, and ultimately blocking tumor cells in G_1. The interest of these findings is emphasized by the *in vivo* use of retinoids and also by a recent report from the Children's Cancer Group demonstrating a significant improvement of outcome in children with high-risk loco-regional or metastatic neuroblastoma when treated with 13-*cis* RA.[18]

Peter Ambros and Co-workers discuss three genotypic features closely related to the *in vivo* clinico-biological behavior of neuroblastoma, i.e., the amplification of *MYCN* gene, the deletion at 1p36, and the nuclear DNA ploidy. Until recently, these characteristics were thought to result from specific events occurring during the earlier phases of neuroblastoma development and, as such, were considered to remain stable throughout all subsequent phases of tumor development and progression. However, it has been recently evidentiated in a small minority of cases, using *in situ* hybridization, that intratumoral heterogeneity for these features exists in different topographic areas and developmental stages. They therefore address the question of whether these microscopic foci of *MYCN* amplified and/or 1p deleted tumor cells are responsible for tumor progression and what the possible consequences of this heterogeneity for tumor sampling and patient management may be. They also report that a similar deletion at 1p36 can be detected in a subset of Ewing tumors.

Per Kogner and co-workers point out how and to what extent molecular assessment of tumor biology is nowdays integrated into the upfront evaluation for children with neuroblastoma and other solid tumors. Even in the light of the possibility of intratumoral heterogeneity, they discuss the advantages and drawbacks of some of the currently available approaches for carrying out such a molecular evaluation, briefly discussing the present limits of targeted therapy and gene therapy in neuroblastoma.

We hope that the papers here presented constitute a limited but somewhat coherent ensemble of some themes of current interest regarding the application of molecular biology and molecular genetics in childhood solid tumors.

REFERENCES

1. Vogelstein B, Kinzler KW. The multistep nature of cancer. Trends Genet **1993**; 9:138–141.
2. Weinberg RA. The integration of molecular genetics into cancer management. Cancer **1992**; 70:1653–1658.
3. Haber DA, Fearon ER. The promise of cancer genetics. Lancet **1998**; 351:SII1–SII8.
4. Fearon ER. Human cancer syndromes: clues to the origin and nature of cancer. Science **1997**; 278:1043–1050.
5. Sidransky D. Nucleic acid-based methods for the detection of cancer. Science **1997**; 278:1054–1058.
6. Ponder B. Genetic testing for cancer risk. Science **1997**; 278:1050–1054.
7. Nichols KE, Li FP, Haber DA, Diller L. Childhood cancer predisposition: applications of molecular testing and future implications. J Pediatr **1998**; 132:389–397.
8. Hartwell LH, Szankasi P, Roberts CJ, Murray AW, Friend SH. Integrating genetic approaches into the discovery of anticancer drugs. Science **1997**; 278:1064–1068.
9. Huang PS, Heimbrook DC. Oncogene products as therapeutic targets for cancer. Curr Opin Oncol **1997**; 9:94–100.
10. Hall SJ, Chen SH, Woo SL. The promise and reality of cancer gene therapy. Am J Hum Genet **1997**; 61:785–789.
11. Delattre O, Zucman J, Melot T, Sastre Garau X, Zucker JM, Lenoir GM, Ambros PF, Sheer D, Turc-Carel C, Triche TJ, Aurias A, Thomas G. The Ewing family of tumors: a subgroup of small-round-cell tumors defined by specific chimeric transcripts. N Engl J Med **1994**; 331:294–299.
12. Zoubek A, Dockhorn-Dworniczak B, Delattre O, Christiansen H, Niggli F, Gatterer-Menz I, Smith TL, Jürgens H, Gadner H, Kovar H. Does expression of different *EWS* chimeric transcripts define clinically distinct risk groups of Ewing tumor patients? J Clin Oncol **1996**; 14:1245–1251.
13. Miyajima Y, Kato K, Numata S, Kudo K, Horibe K. Detection of neuroblastoma cells in bone marrow and peripheral blood at diagnosis by the reverse transcriptase-polymerase chain reaction for tyrosine hydroxylase mRNA. Cancer **1995**; 75:2757–2761.
14. Miyajima Y, Horibe K, Fukuda M, Matsumoto K, Numata S, Mori H, Kato K. Sequential detection of tumor cells in the peripheral blood and bone marrow of patients with stage IV neuroblastoma by the reverse transcription-polymerase chain reaction for tyrosine hydroxylase mRNA. Cancer **1996**; 77:1214–1219.
15. Brodeur GM. Molecular basis for heterogeneity in human neuroblastomas. Eur J Cancer **1995**; 31A:505–510.
16. Castleberry RP, Pritchard J, Ambros P, Berthold F, Brodeur GM, Castel V, Cohn SL, De Bernardi B, Dicks-Mireaux C, Frappaz D, Haase GM, Haber M, Jones DR, Joshi VV, Kaneko M, Kemshead JT, Kogner P, Lee REJ, Matthay KK, Michon JM, Monclair R, Roald BR, Seeger RC, Shaw PJ, Shimada H, Shuster JJ. The International Neuroblastoma Risk Groups (INRG): a preliminary report. Eur J Cancer **1997**; 33:2113–2116.
17. Tannock IF. Conventional cancer therapy: promise broken or promise delayed? Lancet **1998**; 351:SII9–SII16.
18. Reynolds CP, Villablanca JC, Stram DO, Harris R, Seeger RC, Matthay KK. 13-cis-Retinoic acid after intensive consolidation therapy for neuroblastoma improves event-free survival: a randomized Children's Cancer Group (CCG) study. Proc Am Soc Clin Oncol **1998**; 17:2a (abstr.).

16

MOLECULAR GENETICS OF WILMS TUMOR

K. Pritchard-Jones

Institute of Cancer Research & Royal Marsden Hospital
Downs Road, Sutton
Surrey SM2 5PT, UK

1. INTRODUCTION

Wilms tumor is an embryonal kidney cancer affecting 1 in 10,000 children.[1] Its remarkable histological mimicry of structures and cell types seen during normal nephrogenesis suggests it results from mutations in genes involved in this process. Analysis of the role in nephrogenesis of the first Wilms tumour gene to be isolated, WT1, has led to Wilms tumour being considered a paradigm for the relationship of cancer to normal development.[2] However, the genetics of Wilms tumour is clearly much more complex than the original model of two mutations in a single tumour suppressor gene, derived from the work of Knudson and Strong.[3] There is currently evidence for at least ten different genes involved in this tumour type, which do not always behave as classical tumour suppressors[4] (Table 1).

2. THE WT1 GENE

WT1 is one of the few cancer predisposition genes to have a highly tissue-specific expression pattern, mainly in the developing genitourinary system and illustrates that tumorigenesis and malformation of the same organ can be pleiotropic effects of mutation in a single gene. Although WT1 mutations account for only a minority of sporadic Wilms tumours, study of this fascinating protein and its associated malformation syndromes has led to important insights into the molecular basis of nephrogenesis. Mice engineered to lack both copies of the WT1 gene ("knock out" mouse model) fail to develop either kidneys or gonads.[5] Humans with constitutional WT1 mutations are likely to have genitourinary malformation, which is a dominant phenotype with incomplete penetrance, whereas tumorigenesis requires further somatic mutations

Molecular Biology of Hematopoiesis 6, edited by Abraham *et al.*
Kluwer Academic/Plenum Publishers, New York, 1999.

Table 1. Genes/Genetic loci implicated in Wilms tumor

Locus	Genes(s)	Associations	Ref.
11p13	WT1	WAGR, DDS syndromes 5–10% sporadic WT; occasional familial WT	2, 10, 31
	WIT1	Co-deleted with adjacent WT1 gene in some WTs-? involved in regulation of WT1	32
11p15.5	Uncertain	BWS (familial and sporadic) Disregulated expression or mutation of 4 known imprinted genes (IGF2, H19, p57, KVLQT1) but no intragenic mutation in sporadic WTs	20, 33, 34
17q12–q21	Unknown	Familial Wilms tumor (some cases only)	24
19q13.3–13.4	Unknown	Familial Wilms tumor (some cases only)	25
Xq26	GPC3	Simpson-Golabi-Behmel syndrome	19
7p15	Unknown	Constitutional translocations Allele loss in sporadic WT	27
17p	p53	Rare component of Li-Fraumeni syndrome Associated with anaplasia and clonal evolution in 5% sporadic WT	29, 35
16q	Unknown	Allele loss in sporadic WT	28
1q21–q31	Unknown	Hereditary hyperthyroidism-jaw tumor syndrome	36
Unknown	Unknown	Familial Wilms tumor	26, 37
Unknown	Unknown	Perlman syndrome	38

WT1 encodes a zinc finger protein which can function as a transcription factor but which is also capable of binding RNA.[2] A complex series of isoforms are possible due to two alternative splices, an alternative initiation codon and the possibility of RNA editing (Fig. 1). The functional effects of insertion of three amino acids (KTS) in the linker between the third and fourth zinc fingers are the best understood. Both WT1+KTS and WT1–KTS are capable of binding DNA, but with differing sequence specificities and affinities.[6] WT1– KTS isoforms are found in diffuse domains of transcriptional activity in the nucleus whereas WT1–KTS isoforms are found in discrete speckles whose integrity depends on the presence of RNA and which contain some elements of the splicing machinery.[7] One possible mechanism for the functional effects of the KTS splice is to reduce DNA binding activity and favour other interactions of WT1

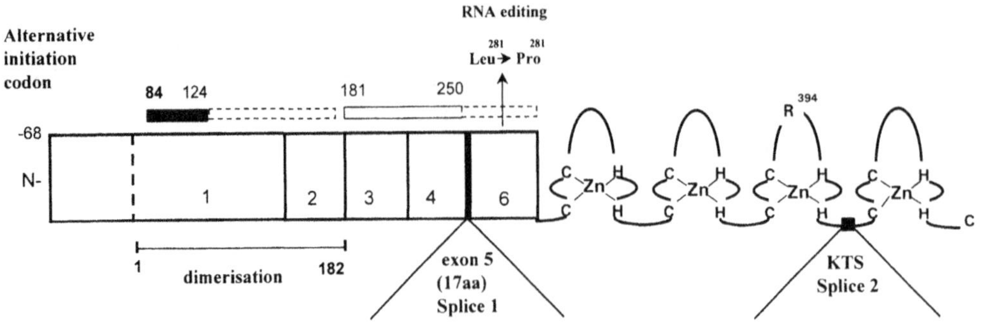

Figure 1. Schematic view of the WT1 coding region showing the four zinc fingers. Filled box, repression domain, open box, activation domain. Arginine[394] is critical for making contact with DNA and is frequently mutated in Denys-Drash syndrome.

either with other proteins, through its N-terminal domain, or with RNA, through the altered zinc finger structure. However, a recent analysis suggests that the DNA- and RNA binding domains of WT1 overlap and are mutually competitive and that WT1–KTS binds both with a greater affinity than +KTS isoforms.[8] The ratio of +:– KTS splicing is almost constant between tissue types whereas splicing of exon 5 varies and may account for tissue-specific functions of WT1.[9] Intronic mutations preventing formation of +KTS isoforms have been shown to underlie Frasier syndrome (see later).

The WT1 gene has now been analysed in almost 600 cases of sporadic Wilms tumours and is mutated in only 5–10% (reviewed in[10] and updated WT1 mutational database maintained at *http://www.umd.necker.fr*).There is a degree of genotype/phenotype correlation, in that if one selects tumours with stromal predominant histology, often manifest as differentiation towards a muscle phenotype, they are more likely to harbour a WT1 mutation.[11] Approximately two thirds of WT1 mutations in sporadic Wilms tumour produce a truncated protein with a disrupted DNA binding domain and in over 70% the wild type allele is lost, consistent with WT1 acting as a recessive tumour suppressor gene. Some bilateral cases also follow the two-hit model, with germline mutation of one WT1 allele followed by mutation of the second allele in the tumours. However, there is also evidence that mutant WT1 may sometimes have either dominant or dominant-negative properties.[12] Also, in some tumours there is evidence that events at other loci cooperate with mutation of only a single WT1 allele. This suggests that other WT-predisposition genes are interacting in a pathway that also involves WT1.

3. GENETIC PREDISPOSITION TO WILMS TUMOR

Study of migrants indicates that the 3–4 fold international variation in incidence of Wilms tumour relates to ethnicity rather than geography, which suggests the involvement of genetic factors.[13] Approximately 1–2% of WTs are familial and a further 2% occur as part of recognised Wilms tumour-predisposing congenital malformation syndromes, which usually result from *de novo* mutations. The proportion of the remaining 95% of apparently sporadic Wilms tumours which are due to a constitutional mutation in a predisposition gene is at present unclear, although up to 10% are associated with isolated hemihypertrophy or genitourinary malformation. Recent follow-up studies of offspring of survivors have provided some empirical estimates of heritability, suggesting this is less than 5%.[14]

3.1. Syndromes Involving WT1

Three congenital malformation syndromes involve the WT1 gene. The first to be described was the WAGR syndrome (*W*ilms tumour, *a*niridia, *g*enito-urinary malformation, and mental *r*etardation), which was instrumental in cloning the WT1 gene. These children have complete loss of one WT1 allele on large interstitial deletions and their 30–50% risk of developing Wilms tumour is usually accompanied by mutation of the remaining WT1 allele.[2] Two rare syndromes usually presenting as severe early onset nephrotic syndrome, involve intragenic WT1 mutations. Denys-Drash Syndrome (DDS) was originally described as a triad of male pseudohermaphroditism, early onset nephrotic syndrome and Wilms tumour but should now be expanded to include patients with only two of these three features and normal females with the characteristic nephropathy (diffuse mesangial sclerosis).[15] Frasier syndrome, a combination of gonadal dysgenesis, nephrotic syndrome, and predisposition to gonadoblastoma rather than Wilms tumour, has considerable phenotypic overlap with DDS.[16] DDS involves mainly missense mutations in zinc fingers 2 or 3 of WT1, affecting amino acids critical for the stability of DNA binding.[10] By contrast, children with Frasier syndrome have intronic mutations which prevent normal splicing of the KTS insert between zinc fingers 3 & 4, showing that isoform imbalance affects normal development.[17] Whether this is directly tumorigenic is less certain as there is an increased risk in a streak gonad whatever the genetic lesion. Since some DDS mutations result in a truncated WT1 protein rather than an amino acid substitution, it has been suggested that the constitutionally mutant protein behaves as a dominant negative or antimorph, somehow interfering with the function of the remaining wild type allele.[12] Direct evidence for this comes from transfection studies showing that the N-terminal domain of WT1 can alter the normal subnuclear localisation pattern of wildtype WT1.[18] Comparing WAGR, Denys-Drash and Frasier syndromes shows that the presence of a mutant form or an isoform imbalance of the WT1 protein has a more deleterious effect on genitourinary development than does reduced dosage through complete absence of one allele. In all three syndromes, children with an XY genotype are more likely to have external genital abnormalities than genotypic females. The converse of this is that all phenotypic females with early onset nephrotic syndrome should be karyotyped and analysed for signs of either DDS or Frasier syndromes.

3.2. Other Wilms Tumor Predisposition Syndromes

The second WT locus, at 11p15, was initially identified as a region of specific allele loss in sporadic Wilms tumours but was subsequently shown to contain the gene or genes responsible for Beckwith-Wiedemann syndrome (BWS).[2] This syndrome is characterised by pre and postnatal overgrowth which may be asymmetrical leading to hemihypertrophy and is accompanied by an approximately 8% risk of malignant tumours, of which half are Wilms. The expressivity of the syndrome is highly variable and there is phenotypic overlap with other overgrowth syndromes such as the Simpson-Golabi-Behmel (SGB) syndrome for which the gene (glypican 3, GPC3) has recently been cloned at Xq26.[19] Even within the BWS locus at 11p15, there is evidence for the existence of more than one WT gene. The rare translocations associated with BWS fall into 2 clusters separated by more than 4 Mb. The cyclin dependent kinase inhibitor, p57, has recently been shown to be mutated in 10% of BWS cases, none of whom developed WT, although one had neuroblastoma.[20] Disregulation of imprinted genes in the 11p15 region, such as insulin like growth factor 2 and H19 may also be involved in both BWS and its associated embryonal tumours. It is therefore intriguing that there is tentative evidence that GPC3 encodes an IGF binding protein, suggesting that both syndromes may result from disruption of a common pathway. A contribution of prenatal growth factor excess to the development of Wilms tumour is also suggested by the finding of significantly increased birth weight in patients with both sporadic and BWS-associated WTs.[21] Other very rare syndromes (Perlman, hyperparathyroidism-jaw tumour syndrome) have an ill-defined risk of Wilms tumour due to the small number of affected cases. Some other better characterised syndromes show weak association with Wilms tumour, including trisomy 18, Sotos' syndrome, Klippel-Trenaunay-Weber syndrome, Bloom's syndrome, and neurofibromatosis type I.[22] The strength of the association is difficult to quantify and, in the case of NF1, may be due to relative underdiagnosis of NF1 in the normal childhood population. Although the cancer risk in Bloom's syndrome is high (~40%), this is most commonly leukaemia or lymphoma, with only two out of 100 cancers in the registry being Wilms tumour.[23] The NF1 and Bloom's syndrome genes have been isolated, encoding a GAP-protein and DNA helicase respectively, but no mutational analysis has been reported yet in Wilms tumours. Wilms tumour is a rare component of the Li-Fraumeni syndrome, although p53 mutation is more often associated with anaplasia in sporadic tumours.

3.3. Familial Wilms Tumor

Familial Wilms tumour is rare and is not usually associated with either congenital malformation or other tumour types. A few small pedigrees have been shown to involve transmission of WT1 mutation but the larger pedigrees in which linkage analysis is possible do not involve genes on either chromosome 11 or other sites of allele loss in sporadic Wilms tumour. Recently, one pedigree with 7 cases of WT has localised a gene for familial WT to chromosome 17q.[24] The lack of allele loss in the tumours suggests that FWT1 may be a weakly dominant oncogene rather than a classical tumour suppressor. Five other largish pedigrees have been shown not to involve this FWT1 gene locus but have been tentatively linked to 19q.[25] Certainly, familial WT is heterogeneous at both a clinical and genetic level. Bilateral disease, presence of metastases and age at diagnosis tend to cluster in specific families. Furthermore, chromosomes 17 and 19 can be excluded as the site of the predisposing gene in several smaller pedigrees[25,26,37] (& Rahman, personal communication). The contribution of FWT1 and FWT2 to apparently sporadic Wilms tumour is potentially significant, given that they both appear to have relatively low penetrance. However, their absolute involvement must await their identification and analysis in sporadic tumours. There also appears to be a WT-predisposing gene at chromosome 7p, where constitutional translocations have been described in more than one patient.[27]

4. GENETIC CHANGES IN SPORADIC WILMS TUMORS

Allele loss studies have been invaluable in localising tumour-predisposition genes. Only one third of sporadic Wilms tumours show allele loss involving chromosome 11p, and in one third of these it is confined to the BWS locus at 11p15 rather than the WT1 locus at 11p13.[2] Whether the approximately 20% of Wilms tumours with 11p allele loss but no WT1 mutation involve the multiple BWS genes requires further investigations. Allele loss on other chromosomes is infrequent in Wilms tumours. Approximately 20% show 16q loss and this is suggested to be a poor prognostic feature.[28] 7p is also suggested as a site of consistent allele loss by smaller studies. Mutation of the p53 gene is rare, occurring in less than 5% of tumours and is associated with anaplasia and a poor prognosis.[29] In some tumours, mutation of either WT1 or p53 is associated with clonal evolution rather than tumour initiation, resulting in islands of muscle differentiation or anaplasia, respectively.[29,30]

5. CONCLUSIONS

Several genes are involved in Wilms tumour development and some or all of them may contribute to the rare heritable form of the disease. So far, only one gene, *WT1*, has molecular proof of its identity as a Wilms tumour gene but it is involved in only a small minority of tumours. There are no easy clues as to what proportion of the remaining 90% of sporadic Wilms tumours can be attributed to each of the nine or more candidate WT loci. It is likely that most Wilms tumours are due to somatic mutations in one or more of these genes and that progression to frank tumour may require further events. We are on the verge of identifying many of these WT genes. The next step will be to see if they fit into a single or related developmental pathway(s) or act independently; a challenge that may lead to therapeutic and preventative insights.

ACKNOWLEDGMENTS

I thank Mrs J Neil for secretarial assistance. Dr. Kathy Pritchard-Jones is supported by the Cancer Research Campaign.

REFERENCES

1. Birch J, Breslow N: Epidemiological features of Wilms tumour. Hematol/Oncol Clinics N Am 9:1157, 1995.
2. Hastie ND: The genetics of Wilms' tumor—a case of disrupted development. Ann Rev Genet 28:523, 1994.
3. Knudson AG, Strong LC: Mutation and cancer: a model for Wilms tumour of the kidney. J Natl Cancer Inst 48:313, 1972.
4. Pritchard-Jones K: Molecular genetic pathways to Wilms tumor. Crit Rev Oncog 8:1, 1997.
5. Kreidberg JA, Sariola H, Loring JM, Maeda M, Pelletier J, Housman D, Jaenisch R: WT-1 is required for early kidney development. Cell 74:679, 1993.
6. Nakagama H, Heinrich G, Pelletier J, Housman DE: Sequence and structural requirements for high-affinity DNA binding by the WT1 gene product. Mol Cell Biol 15:1489, 1995.
7. Larsson SH, Charlieu JP, Miyagawa K, Engelkamp D, Rassoulzadegan M, Ross A, Cuzin F, van HV, Hastie ND: Subnuclear localization of WT1 in splicing or transcription factor domains is regulated by alternative splicing. Cell 81:391, 1995.
8. Bardeesy N, Pelletier J: Overlapping RNA and DNA binding domains of the wt1 tumor suppressor gene product. Nucleic Acids Res 26:1784, 1998.
9. Renshaw J, King-Underwood L, Pritchard-Jones K: Differential Splicing of Exon 5 of the Wilms tumour (WT1) Gene. Genes Chromosomes & Cancer 19:256, 1997.
10. Little M, Wells C: A clinical overview of WT1 gene mutations. Hum Mutation 9:209, 1997.
11. Schumacher V, Schneider S, Figge A, Wildhardt G, Harms D, Schmidt D, Weirich A, Ludwig R, RoyerPokora B: Correlation of germ-line mutations and two-hit inactivation of the WT1 gene with Wilms tumors of stromal-predominant histology. Proc Natl Acad Sci USA 94:3972, 1997.
12. Little MH, Williamson KA, Mannens M, Kelsey A, Gosden C, Hastie ND, van HV: Evidence that WT1 mutations in Denys-Drash syndrome patients may act in a dominant-negative fashion. Hum Mol Genet 2:259, 1993.
13. Stiller CA, Parkin DM: International variation in the incidence of childhood renal tumours. Brit J Cancer 62:1026, 1990.
14. Hawkins MM, Winter DL, Burton HS, Potok MH: Heritability of Wilms' tumor. J Natl Cancer Inst 87:1323, 1995.
15. Jadresic J, Wadey RB, Buckle B, Barratt TM, Mitchell CD, Cowell JK: Molecular analysis of chromosome region 11p13 in patients with Drash syndrome. Hum Genet 86:497, 1991.
16. Moorthy AV, Chesney RW, Lubinsky: Chronic renal failure and XY gonadal dysgenesis: "Frasier" syndrome—a commentary on reported cases. Am J Med Genet Suppl 3:297, 1987.
17. Barbaux S, Niaudet P, Gubler MC, Grunfeld JP, Jaubert F, Kuttenn F, Fekete CN, Souleyreau-Therville N, Thibaud E, Fellous M, McElreavey K: Donor splice-site mutations in WT1 are responsible for Frasier syndrome. Nature Genet 17:467, 1997.

18. Englert C, Vidal M, Maheswaran S, Ge Y, Ezzell RM, Isselbacher KJ, Haber DA: Truncated WT1 mutants alter the subnuclear localization of the wild- type protein. Proc Natl Acad Sci USA 92:11960, 1995.
19. Pilia G, HughesBenzie RM, MacKenzie A, Baybayan P, Chen EY, Huber R, Neri G, Cao A, Forabosco A, Schlessinger D: Mutations in GPC3, a glypican gene, cause the Simpson-Golabi-Behmel overgrowth syndrome. Nature Genet 12:241, 1996.
20. O'Keefe D, Dao D, Zhao R, Sanderson D, Warburton L, Weiss A, Anyane-Yeboa, Tycko B: Coding mutations in p57KIP2 are present in some cases of Beckwith-Wiedemann syndrome but are rare or absent in Wilms tumours. Am J Hum Genet 61:295, 1997.
21. Leisenring WM, Breslow NE, Evans IE, Beckwith JB, Coppes MJ, Grundy P: Increased birth weights of National Wilms tumour study patients suggest a growth factor excess. Cancer Res 54:4680, 1994.
22. Clericuzio CL: Clinical phenotypes and Wilms tumor. Med Pediatr Oncol 21:182, 1993.
23. German J: Bloom's syndrome. XX. The first 100 cancers. Cancer Genet Cytogenet 93:100, 1997.
24. Rahman N, Arbour L, Tonin P, Renshaw J, Pelletier J, Baruchel S, Pritchard-Jones K, Stratton MR, Narod SA: Evidence for a familial Wilms' tumour gene (FWT1) on chromosome 17q12–q21. Nature Genet 13:461, 1996.
25. McDonald JM, Douglass EC, Fisher R, Geiser CF, Krill CE, Strong LC, Virshup D, Huff V: Linkage of familial Wilms' tumor predisposition to chromosome 19 and a two-locus model for the etiology of familial tumors. Cancer Res 58:1387, 1998.
26. Rahman N, Abidi F, Ford D, Arbour L, Tonin P, & 19 others . . . & Narod S, Schwartz C, Pritchard-Jones K, Stratton MR: Confirmation of *FWT1* as a Wilms' tumour susceptibility gene and phenotypic characteristics of Wilms' tumour attributable to *FWT1*. Hum Genet 103:547–556, 1998.
27. Reynolds PA, Powlesland RM, Keen TJ, Inglehearn CF, Cunningham AF, Green ED, Brown KW: Localization of a novel t(1;7) translocation associated with Wilms' tumor predisposition and skeletal abnormalities. Genes Chromosom Cancer 17:151, 1996.
28. Grundy PE, Telzerow PE, Breslow N, Moksness J, Huff V, Paterson MC: Loss of heterozygosity for chromosomes 16q and 1p in Wilms' tumors predicts an adverse outcome. Cancer Res 54:2331, 1994.
29. Bardeesy N, Beckwith JB, Pelletier J: Clonal expansion and attenuated apoptosis in Wilms' tumors are associated with p53 gene mutations. Cancer Res 55:215, 1995.
30. Miyagawa K, Kent J, Moore A, Charlieu JP, Little MH, Williamson KA, Kelsey A, Brown KW, Hassam S, Briner J, Hayashi Y, Hirai H, Yazaki Y, van Heyningen V, Hastie ND: Loss of WT1 function leads to ectopic myogenesis in Wilms' tumour [letter]. Nature Genet 18:15, 1998.
31. Kaplinsky C, Ghahremani M, Frishberg Y, Rechavi G, Pelletier J: Familial Wilms' tumor associated with a WT1 zinc finger mutation. Genomics 38:451, 1996.
32. Malik KT, Wallace JI, Ivins SM, Brown KW: Identification of an antisense WT1 promoter in intron 1: implications for WT1 gene regulation. Oncogene 11:1589, 1995.
33. Lee MP, Hu RJ, Johnson LA, Feinberg AP: Human KVLQT1 gene shows tissue-specific imprinting and encompasses Beckwith-Wiedemann syndrome chromosomal rearrangements. Nature Genet 15:181, 1997.
34. Steenman MJ, Rainier S, Dobry CJ, Grundy P, Horon IL, Feinberg AP: Loss of imprinting of IGF2 is linked to reduced expression and abnormal methylation of H19 in Wilms' tumour. Nature Genet 7:433, 1994.
35. Hartley AL, Birch JM, Tricker K, Wallace SA, Kelsey AM, Harris M, Jones PH: Wilms' tumor in the Li-Fraumeni cancer family syndrome. Cancer Genet Cytogenet 67:133, 1993.
36. Szabo J, Heath B, Hill VM, Jackson CE, Zarbo RJ, Maette LE, Chew SL, Besser GM, Thakker RV, Huff V, Lepperts MF, Heath III H: Hereditary hyperparathyroidism-Jaw tumour syndrome: The endocrine tumour gene HRPT2 maps to chromosome 1q21–q31. Am J Hum Genet 56:944, 1995.
37. Huff V, Amos CI, Douglass EC, Fisher R, Geiser CF, Krill CE, Li FP, Strong LC, McDonald JM: Evidence for genetic heterogeneity in familial Wilms' tumor. Cancer Res 57:1859, 1997.
38. Greenberg F, Copeland K, Gresik M: Expanding the spectrum of Perlman syndrome. Am J Med Genet 29:773, 1988.

17

GENETIC HETEROGENEITY IN EWING TUMORS AND NEUROBLASTOMAS

P. F. Ambros, I. M. Ambros, G. Amann, C. M. Hattinger, R. Ladenstein, R. Kerbl, S. Rumpler, A. Luegmayr, and H. Gadner

CCRI, Children's Cancer Research Institute
St. Anna Kinderspital
Kinderspitalgasse 6, A-1090 Vienna (P.F.A.; I.M.A.; C.M.H.; S.R.; A.L.;H.G.)
Inst. f. Clinical Pathology, Vienna, Austria (G.A.)
St. Anna Kinderspital, Vienna, Austria (R.L.; H.G.)
Univ. Children's Hosp. Graz, Austria, (R.K.)

1. OVERVIEW

The aim of this study was to investigate genetic changes occurring in Ewing Tumors (ETs) and Neuroblastomas (NBs). We analyzed whether the same genetic alteration, i.e. the deletion of genetic material at chromosomal region 1p36, occurs in both tumor types at a comparable rate and whether these aberrations are acquired during tumor progression. Consecutive tumor samples from the same patients at different stages of the disease and topographically different samples from single tumors were studied. Furthermore, we focused on the detection of the status of the *MYCN* oncogene in NBs. Deletions at 1p36 are found in ETs and NBs at a comparable frequency (approx. 20%) and account for an unfavorable prognosis in localized disease.[1,2,3] In ETs, the loss of 1p36 material was frequently observable in a sub-population of cells and is therefore regarded as a later event in the tumor development. In NBs, it seemed more likely that the aberrations found in a subgroup of tumors (i.e. del1p, *MYCN* amplification, and gain of 17q) are very early on initial events as virtually all tumor cells within an individual tumor were thought to bear the respective genetic aberra-

**Correspondence to*: Peter F. Ambros Ph.D., CCRI, St. Anna Kinderspital, Kinderspitalgasse 6 A-1090 Vienna, Tel. ++43-1-40470, FAX: ++43-1-4087230, e-mail: ambros@ccri.univie.ac.at

Molecular Biology of Hematopoiesis 6, edited by Abraham *et al.*
Kluwer Academic/Plenum Publishers, New York, 1999.

tion. This view is supported by the results obtained so far.[4] However, recent data indicate that deletions at 1p36 and also *MYCN* amplifications can be acquired during tumor progression (Ambros et al. in press), because sub-populations of cells showing these aberrations interspersed between cells not showing these aberrations exist in evolutionary early tumors.

2. DELETIONS AT 1P36

Deletions at the short arm of chromosome 1 have been found in a wide variety of tumors.[5,6,7] Many of these tumors are of neuroectodermal origin such as NB,[8] melanoma,[9] pheochromocytoma[10] medullary thyroid carcinoma,[11] or Merkel cell carcinoma.[12] Loss of heterozygosity or total loss of one copy of the sub-terminal segment of chromosome 1 has also been demonstrated in other tumor types such as breast cancer,[13] colon cancer,[14] and pediatric germ cell tumors.[15] Interestingly, aberrations involving a deletion at 1p36 occur at different stages of the disease depending on the type of tumor. In melanoma, loss of heterozygosity (LOH) at the distal 1p appears to be a late event in tumor progression.[9] Observations in NB indicate that deletions at 1p36 usually represent very early or even initial events, and in a minor percentage of NBs the deletion at 1p36 is acquired during tumor progression (Ambros et al. in press).

2.1. Deletions at 1p36 in Ewing Tumors

Ewing tumors have a common genetic marker.[16,17,18] The reciprocal translocation t(11;22)(q24;q12) or aberrations involving 22q12 have been seen in more than 90% of the cases analyzed by classical cytogenetics.[19] Recently, the molecular events underlying the rearrangements of 22q12 in ETs have been elucidated. The *EWS* gene at 22q12 is fused with a gene of the *ETS* proto-oncogene family, either the *FLI1* gene at 11q24,[20,21] or with other members of this gene family situated on different chromosomes. Besides these constant aberrations, facultative numerical and structural aberrations have been observed in ETs, including gains of chromosomes 2, 5, 7, 8, 9, and 12, the non-reciprocal translocation t(1;16)(q12;q11.2),[22,23,24,25] and deletions at the short arm of chromosome 1.[3,7,23,26]

2.1.1. Intratumoral Heterogeneity in Ewing Tumors. Tumor samples from 58 patients were analyzed using classical cytogenetics and/or in situ hybridization (ISH) techniques.[3] Losses of genetic material at 1p36 were observed in 18% (9/51) of the cases. In localized tumors and in metastases, we frequently found this aberration in subpopulations of cells. Thus we can conclude that deletions at 1p36 are acquired in the evolution of ETs. The del1p, which is frequently presented with additional material on the tip of this chromosome possibly leading to the deletion, was found in combination with one or two intact chromosomes 1. The high percentage of tumor cells with intact chromosomes 1 besides cells with deletion of 1p36 most likely obscures the picture and explains the difficulty to visualize this aberration by PCR or Southern blot based LOH studies.

To assess the possible prognostic impact of the del1p in localized ETs, we correlated the genetic data with the relapse-free survival, the presence of metastatic disease at diagnosis, with the location of the primary tumor and with age and sex of the patients. Interestingly, within the group of patients with localized disease, deletions of DNA at the short arm of chromosome 1, even when present in only a subpopulation of cells identified a subgroup of patients (6/22) with unfavorable clinical outcomes.[3] However, it is noteworthy that this aberration has no influence on the outcome of patients with disseminated disease, which might be explained by the different treatment regimens used for localized and disseminated diseases.

2.2. Deletions at 1p36 in Neuroblastomas

In contrast to ETs, no single genetic defect that is common to all NBs is known so far. Two main groups can be defined on the basis of the DNA content of these tumors, *i.e.* diploid (2n) and near-triploid (3n) NBs.[27] In both tumor types identical genetic aberrations are described, although the frequency of these facultative genetic aberrations is much higher in the diploid group as compared to the near-triploid group. The deletion at 1p36 and gain of chromosomal 17q material are the most frequently observed aberrations.

2.2.1. Genetic Heterogeneity in Neuroblastomas. Using fluorescence in situ hybridization (FISH) techniques to unambiguously detect chromosome 1p aberrations also in small foci, two localized and one stage 4 tumors with heterogeneity for the deletion at 1p36 were identified. Furthermore, acquisition of deletions at 1p36 could be verified in four tumors. However, in two cases, a bifocal tumorigenesis could not be excluded, as the DNA content of the primary and recurrence was different. In addition, we observed two simultaneously occurring primaries in one patient, one displaying a 1p deletion, and the other with intact chromosome regions 1p36.3 (Ambros et al. unpublished observation).

3. *MYCN* AMPLIFICATION IN NEUROBLASTOMAS

Amplification of the *MYCN* oncogene is found in 20 to 25% of all NBs analyzed. Up to now, the data shown by Brodeur and coworkers[4] indicated that the genetic constellation of neuroblastic tumors is stable and does not alter in the initial tumor growth or during tumor progression. The authors found a similar copy number in the primary tumors and the recurrences or in the metastases. No indication for the existence of *MYCN* amplified sub-populations of cells in a non-amplified tumor or an acquired *MYCN* amplification was found. These data were generated by Southern blotting, thus giving an average figure of the *MYCN* copy number. As DNA extraction of a larger piece of tumor is needed to perform this technique, focal clusters of cells carrying a gene amplification can be "diluted" by the majority of cells not showing this aberration and can therefore be missed. Another important fact is that those essential pioneer studies were performed before the NB mass screening was introduced in the US. The urinary mass screening or the use of ultrasonography, however, enable the detection of tumors which are probably early in their development. Furthermore, the detection of genetic aberrations in sub-areas is facilitated by in situ hybridization techniques on touch preparations and on histological sections.

3.1. Genetic Heterogeneity in Neuroblastomas

We were able to analyze tumors found by NB mass screening, performed at a time when a relatively high proportion of genetically unfavorable tumors are detected.[28] Three tumors (one stage 4s and two localized NBs) were identified which showed intratumoural heterogeneity for *MYCN* amplification (Ambros et al. in press). The FISH techniques, using a *MYCN* probe and a chromosome 2 specific reference probe, was applied on thin sections of paraffin embedded tumor tissues. Foci of cells with up to 20fold amplification of the *MYCN* oncogene were embedded in a majority of non amplified tumor cells. This pattern resembled the finding in metastatic prostatic carcinomas where *MYCC* amplifications were found in a focal pattern.[29]

4. CONCLUSION

Our data show that chromosome 1p deletions occur in a subpopulation of ET cells and that this aberration, even if present only in a sub-population of cells, is very likely responsible for an unfavorable outcome of patients with localized tumors. In contrast to the current view, we could demonstrate that intratumoral heterogeneity of del1p and *MYCN* amplification and acquisition of these aberrations is also found in NBs. Up to now, we can only speculate about the frequency of these focal aberrations as only a limited number of tumors were analyzed in such detail. The detection of such cell clones is important, because the deletion at 1p36 and/or *MYCN* amplification are most likely responsible for aggressive local growth and development of metastases and thus *MYCN* status is already used for the stratification of the therapy.

ACKNOWLEDGMENT

This study was supported by the Forschungsinstitut für krebskranke Kinder im St. Anna Kinderspital.

REFERENCES

1. Ambros PF, Ambros IM, Strehl S, Bauer S, Luegmayr A, Kovar H, et al. Regression and progression in neuroblastoma. Does genetics predict tumour behaviour? Eur.J.Cancer 1995; 31A:510–515.
2. Caron H, van Sluis P, de KJ, Bokkerink J, Egeler M, Laureys G, et al. Allelic loss of chromosome 1p as a predictor of unfavorable outcome in patients with neuroblastoma. N.Engl.J.Med. 1996; 334: 225–230.
3. Hattinger CM, Rumpler S, Strehl S, Ambros IM, Zoubek A, Pötschger U, Gadner H, Ambros PF. Prognostic impact of deletions at 1p36 and numerical aberrations in Ewing Tumors. Genes Chromosomes Cancer 1999; 24:243–254.
4. Brodeur GM, Hayes FA, Green AA, Casper JT, Wasson J, Wallach S, et al. Consistent N-myc copy number in simultaneous or consecutive neuroblastoma samples from sixty individual patients. Cancer Res. 1987; 47:4248–4253.
5. Douglass EC, Green AA, Hayes FA, Etcubanas E, Horowitz M, Wilimas JA. Chromosome 1 abnormalities: a common feature of pediatric solid tumors. J.Natl.Cancer Inst. 1985; 75:51–54.
6. Atkin NB. Chromosome 1 aberrations in cancer. Cancer Genet.Cytogenet. 1986; 21:279–285.
7. Mitelman F. Catalog of chromosome aberrations in cancer. 5th Edition. 1994.
8. Brodeur GM, Sekhon GS, Goldstein MN. Chromosomal aberrations in human neuroblastomas. Cancer 40, 2256–2263. 1977.
9. Dracopoli NC, Harnett P, Bale SJ, Stanger BZ, Tucker MA, Housman DE, et al. Loss of alleles from the distal short arm of chromosome 1 occurs late in melanoma tumor progression. Proc.Natl.Acad.Sci.U.S.A. 1989; 86:4614–4618.
10. Mathew CG, Smith BA, Thorpe K, Wong Z, Royle NJ, Jeffreys AJ, et al. Deletion of genes on chromosome 1 in endocrine neoplasia. Nature 1987; 328:524–526.

11. Moley JF, Brother MB, Fong CT, White PS, Baylin SB, Nelkin B, et al. Consistent association of 1p loss of heterozygosity with pheochromocytomas from patients with multiple endocrine neoplasia type 2 syndromes. Cancer Res. 1992; 52:770–774.
12. Harnett PR, Kearsley JH, Hayward NK, Dracopoli NC, Kefford RF. Loss of allelic heterozygosity on distal chromosome 1p in Merkel cell carcinoma. A marker of neural crest origins? Cancer Genet. Cytogenet. 1991; 54:109–113.
13. Devilee P, van Vliet M, Bardoel A, Kievits T, Kuipers DN, Pearson PL, et al. Frequent somatic imbalance of marker alleles for chromosome 1 in human primary breast carcinoma. Cancer Res. 1991; 51:1020–1025.
14. Leister I, Weith A, Bruderlein S, Cziepluch C, Kangwanpong D, Schlag P, et al. Human colorectal cancer: high frequency of deletions at chromosome 1p35. Cancer Res. 1990; 50:7232–7235.
15. Stock C, Ambros IM, Lion T, Haas OA, Zoubek A, Gadner H, Ambros PF. Detection of numerical and structural chromosome abnormalities in pediatric germ cell tumors by means of interphase cytogenetics. Genes, Chromosomes and Cancer 1994; 11:40–50.
16. Aurias A, Rimbaut C, Buffe D, Dubousset J, Mazabraud A. Chromosomal translocations in Ewing's sarcoma. N Engl J Med 1983; 309:496–497.
17. Iselius L, Lindsten J, Aurias A, Fraccaro M, Bastard C, Bottelli AM, et al. The 11q;22q translocation: a collaborative study of 20 new cases and analysis of 110 families. Hum.Genet. 1983; 64:343–355.
18. Turc-Carel C, Philip I, Berger MP, Philip T, Lenoir GM. Chromosomal translocations in Ewing's sarcoma. N Engl J Med 1983; 309:497–498.
19. Turc CC, Aurias A, Mugneret F, Lizard S, Sidaner I, Volk C, et al. Chromosomes in Ewing's sarcoma. I. An evaluation of 85 cases of remarkable consistency of t(11;22)(q24;q12). Cancer Genet.Cytogenet. 1988; 32:229–238.
20. Delattre O, Zucman J, Plougastel B, Desmaze C, Melot T, Peter M, et al. Gene fusion with an ETS DNA-binding domain caused by chromosome translocation in human tumours. Nature 1992; 359:162–165.
21. Zucman J, Delattre O, Desmaze C, Plougastel B, Joubert I, Melot T, et al. Cloning and characterization of the Ewing's sarcoma and peripheral neuroepithelioma t(11;22) translocation breakpoints. Genes Chromosomes Cancer 1992; 5:271–277.
22. Mugneret F, Lizard S, Aurias A, Turc CC. Chromosomes in Ewing's sarcoma. II. Nonrandom additional changes, trisomy 8, and der(16)t(1;16). Cancer Genet.Cytogenet. 1988; 32:239–245.
23. Douglass EC, Green AA, Hayes FA, Etcubanas E, Horowitz M, Wilimas JA. Chromosome 1 abnormalities: a common feature of pediatric solid tumors. J.Natl.Cancer Inst. 1985; 75:51–54.
24. Hattinger CM, Rumpler S, Ambros IM, Strehl S, Lion T, Zoubek A, et al. Demonstration of the translocation der(16)t(1;16)(q12;q11.2) in interphase nuclei of Ewing tumors. Genes Chromosomes Cancer 1996; 17:141–150.
25. Armengol G, Tarkkanen M, Virolainen M, Forus A, Valle J, Bohling T, et al. Recurrent gains of 1q, 8 and 12 in the Ewing family of tumours by comparative genomic hybridization. Br. J Cancer 1997; 75:1403–1409.
26. Sainati L, Leszl A, Montaldi A, Ninfo V, Basso G. Is the deletion of the short arm of chromosome 1 a prognostic factor in pediatric peripheral primitive neuroepithelioma (PNET)? [letter]. Med.Pediatr.Oncol. 1996; 26:143–144.
27. Look AT, Hayes FA, Shuster JJ, Douglass EC, Castleberry RP, Bowman LC, et al. Clinical relevance of tumor cell ploidy and N-myc gene amplification in childhood neuroblastoma: a Pediatric Oncology Group study. J.Clin.Oncol. 1991; 9:581–591.
28. Kerbl R, Urban CE, Ladenstein R, Ambros IM, Spuller E, Mutz I, et al. Neuroblastoma screening in infants postponed after the sixth month of age: a trial to reduce "overdiagnosis" and to detect cases with "unfavorable" biologic features. Med Pediatr. Oncol. 1997; 29:1–10.
29. Jenkins RB, Qian J, Lieber MM, Bostwick DG. Detection of c-myc oncogene amplification and chromosomal anomalies in metastatic prostatic carcinoma by fluorescence in situ hybridization. Cancer Res. 1997; 57:524–531.

18

MECHANISMS OF RETINOID INHIBITION OF CELL PROLIFERATION IN NEUROBLASTOMA

Tatsuya Matsuo and Carol J. Thiele

Cell & Molecular Biology Section
Pediatric Oncology Branch
National Cancer Institute
10 Center Dr. MSC-1928
Bethesda MD, 20892-1928

ABSTRACT

Cyclin/cdk kinases stimulate progression through the G_1 phase of the cell cycle while members of the CIP/KIP gene family function to inhibit cdk kinase activity and cell cycle progression. To study the mechanism of retinoid induced decrease in NB cell growth, the expression and activity of cyclin/cdk kinases, and their inhibitors was evaluated in KCNR NB cells arrested in G_1 by treated with all trans-retinoic acid (RA). After 2 days in 5 uM RA and coincident with the arrest of cells in G_1, there is a 10 and 5-fold decrease in CDK2 and CDK4 kinase activity, respectively. The decrease in kinase activity occurs despite the fact that the protein levels of the components of the cyclin-cdk kinases are unchanged during this time. RA stimulates a 5-fold increase in $p27^{KIP}$ steady-state levels and a comparable increase in $p27^{KIP}$ complexed to cdk2 and cdk4 compared to untreated cells. The levels of other CIP/KIP genes are unchanged. RA does not stimulate an increase in $p27^{KIP}$mRNA suggesting an increase in the stability of $p27^{KIP}$ or other post-translational modifications. The mechanism by which RA increases $p27^{KIP}$ may be by decreasing N-myc levels. We find that over-expression of N-myc leads to decreases in $p27^{KIP}$ and increases in cyclin/cdk kinases activity and an increase in cells in G_1. Increases in $p27^{KIP}$ are important in inhibiting the growth of NB, since infection of NB with adenoviruses containing $p27^{KIP}$ arrested cells in G_1 and decreased cyclin/cdk kinase activity. These data suggest a model in which the RA induced arrest of NB cells is mediated by a decrease in N-myc levels leading to a increase in p27 that binds and inhibits cyclin/cdk kinase activity causing an arrest of cells in G_1.

Molecular Biology of Hematopoiesis 6, edited by Abraham *et al.*
Kluwer Academic/Plenum Publishers, New York, 1999.

1. INTRODUCTION

Amplificaiton of N-myc is a marker of Neuroblastoma tumors that have a particularly poor prognosis. Our findings that retinoids cause a decrease in N-myc that is accompanied by inhibition of cell proliferation and induction of differentiation[1] have led to clinical studies investigating the efficacy of retinoids in the treatment of neuroblastoma. Arrest of cell cycle, induction of differentiation or cell death are among the mechanisms by which retinoids control tumor cell growth. In neuroblastoma, we have found that retinoids inhibit cell proliferation by decreasing DNA synthesis and causing an accumulation of cell in the G1 phase of the cell cycle.

The transition of cells through the G1 phase of the cell cycle is mediated by sets of protein complexes that phosphorylate key proteins needed to transit the G1 restriction point and initiate DNA synthesis. Cyclins D, E, and A form complexes with cdk2, 4, or 6 to phosphorylate proteins such as the tumor suppressor gene RB. Phosphorylation of RB prevents binding and sequestration of the E2F transcription factors. E2Fs stimulate transcription of genes important in DNA synthesis such as DNA polymerase α. The activity of the G1 cyclin dependent kinases are regulated in part by two families of proteins called cyclin dependent kinase inhibitors (CKIs). CKI include members of the INK4 (p15, 16, 18, & 19) and the Cip/Kip (p21, 27, 57) gene families and these proteins bind to cyclin/Cdk complexes and inhibit their kinase activity leaving Rb underphosphorylated and in an active state, capable of binding pocket proteins such as E2F.[2]

To study the molecular mechanism by which retinoids regulate growth in NB cells we evaluated the expression and activity of components of the G1 cyclin dependent kinases and G1 CKIs.

2. MATERIALS AND METHODS

2.1. Cell Culture

The NB cell line SMS-KCNR (KCNR) was cultured as described previously[1] and treated with indicated concentrations of all-trans-retinoic acid (Sigma, St. Louis, MO) or 95% ethanol control solvent for up to 120hrs. Cells were treated with virus for 24 hours, washed and cultured for indicated periods of time.

2.2. Antibodies

The following antibodies were used in this study: against Cdk2, M2-G; (Santa Cruz Biotech., Santa Cruz, CA, sc-163-G); p27, C-19-G (Santa Cruz Biotech., sc-528G); N-Myc, C-19 (Santa Cruz Biotech., sc-791); Rb, G3-245 (Pharmingen, San Diego, CA, 14001A). Horseradish peroxidase (HRP) linked secondary antibodies were purchased from Amersham (Arlington Heights, IL).

2.3. Immunoprecipitation and Immunoblot

Protein extraction, immunoprecipitation, and immunoblot were performed as previously described.[3,4]

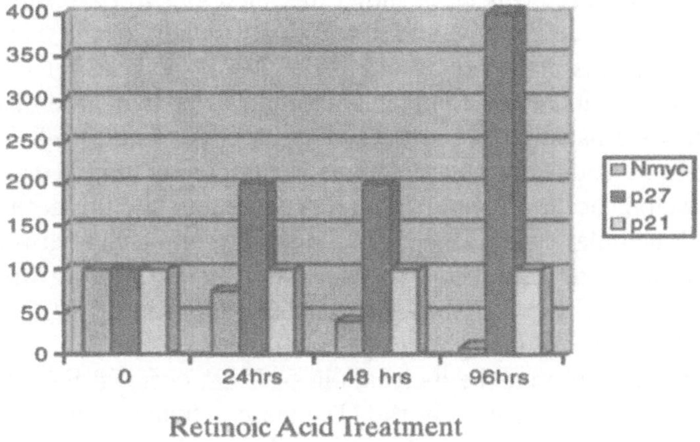

Figure 1. Inverse relationship between p27 and N-myc levels.

3. RESULTS AND DISCUSSION

Figure 1 shows the relative steady-state levels of p67 N-myc, p27 and p21 levels as a percent of levels at initiation of treatment after various treatments with all trans-retinoic acid. Significant changes in the steady-state levels of the cyclins did not change over the first 48 hrs of RA treatment and are not shown. Consistent with our previous results showing a decrease in N-myc mRNA levels, the steady-state levels of p67 N-myc within 24 hrs of RA treatment. Co-incident with this decrease in N-myc is an increase in p27 levels. The increase in p27 but not p21. The increase in p27 is not accompanied by an increase in p27 mRNA indicating that in NB, retinoids do not directly regulate transcription of p27 or p21.[4]

The increase in p27 steady-state levels indicates that this CKI may be critical in inhibiting the cyclin dependent kinase activity. To determine if p27 could bind to cyclin dependent kinases, these cdk2, cdk4, and cyclin E were immunoprecipitated, and a portion of the immunoprecipitate assessed for kinase activity while another portion of the immunoprecipitate was electrophoresed on SDS-PAGE, blotted and probed for p27 that was co-immunoprecipitated or bound to these molecules. Table I shows the kinetics of maximum p27 bound to cyclin dependent kinases (initial value = 0; maximal bound = 100) and the kinase activity (initial kinase activity = 100; background level = 0) derived from densitometric analysis of appropriately exposed Western blots and *in*

Table I. Relationship of bound p27 and specific kinase activity

Time (RA)	0 hrs	24 hrs	48 hrs	72 hrs	96 hrs
p27 + cyclinE	1*	10	20	80	100
E dep. kinase	100^	100	30	10	0
p27 + cdk2	1*	90	100	100	100
cdk2 dep. kinase	100^	90	0	0	0
p27 + cdk4	1*	50	100	100	100
cdk4 dep. kinase	100^	100	20	0	0

*numbers represent the relative amount of p27 bound with 100% being maximum amount bound; ^numbers represent the % kinase activity compared to controls which are 100% at t_0.

vitro kinase activity blots.[4] This table shows that increases in p27 bound to the cyclin kinase complexes precedes inactivation of the kinase activity of the specific complexes. For example, p27 is maximally bound to cdk2 within 24–48 hrs after RA treatment at which time the kinase activity is completely inhibited. Maximal binding of p27 to cdk4 occurs at 48–72 hrs. This is followed by inactivation of cdk4 kinase 48–72 hrs after RA treatment and by 72 hours the kinase activity is completely inhibited. The majority of p27 bound to cyclinE occurs 72–96 hrs after RA treatment and it is at this time that the kinase activity is completely abolished. The temporal relationship between increases in p27 binding to cyclin/cdk complexes and decreased kinase activity strongly supports the hypothesis that it is the p27 that is specifically inhibiting the G1 dependent cdk kinases during RA induced neuroblastoma cell growth.

These data indicate that the increases in p27 that occur in RA treated NB cells are key in mediating the decrease in the G1 kinase activities. Supporting this data is our finding that increasing p27 levels in NB cells by adenovirus infection causes an increase in cells in G1 and a decrease in G1 cyclin dependent kinase activity (T. Matsuo unpublished data). The mechanism by which RA causes the increase in the steady-state levels of p27 is unknown. However studies have shown that there is inverse relationship between c-myc and p27 and that c-myc can overcome p27 inhibition of cell cycle progression.[5] As c-myc and N-myc are structurally and functionally related it is not unexpected that N-myc may specifically antagonize the activity of p27 in NB cells as well. Our data is consistent with a model in which the RA induced decrease in N-myc would "liberate" p27 allowing it to bind to G1 cyclin dependent kinases, inhibit their activity and arrest the cells in the G1 phase of the cell cycle.

REFERENCES

1. Thiele CJ, Reynolds CP, Israel MA, Decreased espression of N-myc precedes retinoic acid-induced differnetiation of human neuroblastoma. Nature 313:404–406, 1985
2. Sherr CJ, Roberts JM, Inhibitors of mammalian G1 cyclin-dependent kinases. Genes Dev. 9:1149–1163, 1995
3. Gaetano C, Matsuo T, Thiele CJ, Identification and characterizatio of a retinoic acid-regulated human homologue of the unc-33-like phosphoprotein gene (hUlip) from neuroblastoma cells. J. Biol. Chem. 272:12195–12201, 1997
4. Matsuo T, Thiele CJ, p27: a key mediator of retinoic acid induced growth arrest in the SMS-KCNR human neuroblastoma cell line. Oncogene 16:3337–3343, 1998
5. Vlach J, Hennecke S, Alevizopoulos K, Conti D, Amati B, Growth-arrest by the cyclin-dependent kinase inhibitor p27 Kip1 ia abrogated by c-myc. EMBO J 15:6595–6604, 1996

19

HIGH-DOSE CHEMOTHERAPY FOR BREAST CANCER IN EUROPE

From Fashion To Randomization

Giovanni Rosti,* Patrizia Ferrante, Maurizio Marangolo, John Crown, Henry Roche, Patrice Viens, and Alois Gratwohl**

EBMT Solid Tumors Working Party
* Chairman—Department of Oncology and Hematology Ravenna Italy
and ** Hematology University of Basel
Switzerland

In the last few years high-dose chemotherapy expanded throughout Europe as a treatment option for breast carcinoma patients. In 1993, 400 patients were reported during the annual European survey, while in 1996 more than 2100 patients received this treatment modality, and we may foresee that at least 2500 women underwent this procedure in 1998. From the European Group for Blood and Marrow Transplantation (EBMT) Registry data base more than 95% of the high-dose programs have been conducted with the support of Peripheral Blood Progenitor Cells (PBPC) last year, with a minority of patients receiving Autologous Bone Marrow Transplantation (ABMT) as hematopoietic support. This data are slightly different from those recently reported by the Autologous Bone Marrow Transplantation Registry of North America,[1] indicating that 25% of the patients received ABMT, alone or in combination with PBPC (data referring to 1995). Patients selected for high-dose chemotherapy programs are equally divided into metastatic and adjuvant ones, with 10% with inflammatory breast carcinoma.

There has been a tremendous decrease in the number of patients receiving this treatment modality for metastatic disease in a refractory phase of their disease, only 2% in 1996, compared to 59% in 1989. Man age of the patients is 45 ± 8 years with a median of 45 years (range 24–65).

Corresponding author: Giovanni Rosti Department of Oncology and Hematology, Ospedale Civile Viale Randi 5 48100 Ravenna (Italy), Phone +39 335 471029 Fax +39 0 544285246

Molecular Biology of Hematopoiesis 6, edited by Abraham *et al.*
Kluwer Academic/Plenum Publishers, New York, 1999.

Treatment related mortality declined from 22% in 1989 to 7% in 1991 to 2% in 1996 and the reasons for this trend are probably due to a better patients' selection as well as to better knowledge of the procedures by many groups, and probably to the introduction of PBPC transplantation.

Out of 2839 patients in the data base, the vast majority received a single transplantation, while 382 had a double graft and 59 a triple transplantation. Moreover from a survey conducted by the EBMT Solid Tumors Working Party in late 1995 at least 30% of the patients were treated according to protocols.[2]

ADJUVANT THERAPY AND INFLAMMATORY BREAST CANCER

So far 524 patients with adjuvant high-dose chemotherapy are in the data base. The mean number of involved axillary nodes is 15 ± 7 (median 14—range 0–42). Eighty-three percent of the procedures were supported by PBPC, and 3% by ABMT + PBPC. Hematopoietic growth factors were used after chemotherapy in 80% of the courses. In 477 patients full data on hematopoietic recovery are available: mean number of days to reach PMN > 50 ul is 11 ± 4 for patients receiving PBPC and 16 ± 6 for those with ABMT as hematologic support ($p < 0.01$). A statistically significant difference was observed also in platelets recovery (>50,000 ul): 16 ± 9 days with PBPC versus 22 ± 12 days for ABMT.

At a median follow-up of 28 months event-free survival is 60% (mean follow-up is 36 months) and so far no difference is observed between the cohort of patients with >9 or <9 positive nodes (Fig. 1).

An interesting result comes out from the event-free survival analysis of a cohort of patients bearing a dismal prognosis (>20 positive axillary nodes): 50% of 63 patients are alive with no disease at 27 months (Fig. 2). Three hundred fifty patients are evalu-

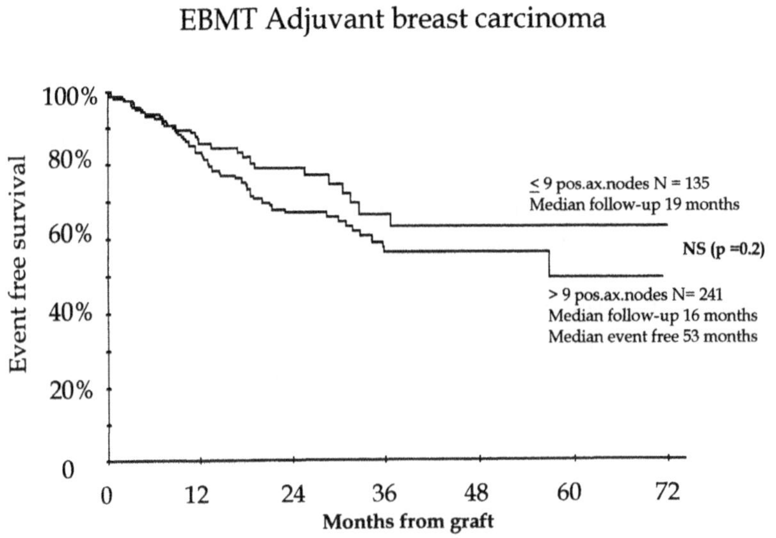

Figure 1. Event free survival for patients with high-risk (<9 and >9 positive axillary nodes) operable breast carcinoma receiving high-dose chemotherapy (EBMT data base).

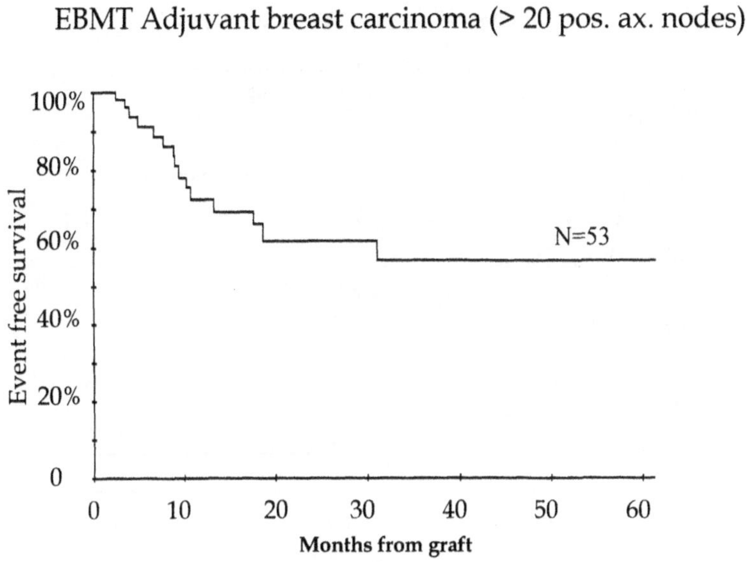

Figure 2. Event free survival for patietns with >20 positive axillary nodes receiving high-dose adjuvant chemotherapy (EBMT data base).

able for inflammatory breast carcinoma, with a median follow-up of 26 months and with nearly a 40% disease-free survival at 5 years.

METASTATIC DISEASE

Within the EBMT Registry data base 1293 patients with metastatic disease at the time of transplantation are included. As already shown by the North American Registry[1] the best results in terms of event-free survival are achieved in those patients receiving high-dose chemotherapy at the time of complete remission (i.e. late intensification policy): median event-free survival is 40% at 38 months for transplantation in CR versus for 18 months for those receiving intensification in partial remission, tability of the disease or progression (Fig. 3).

ONGOING TRIALS

In order to switch from phase II studies which have extensively been presented in the recent literature with sometimes excellent results as in the adjuvant[3,4] as well as in the metastatic[5] settings, many investigators began to set up clinical randomized phase III trials comparing high-dose chemotherapy versus standard-dose treatment. In the field of adjuvant therapy, a dozen of large trials are ongoing in Europe and the accrual for the early ones will be closing this year. Table 1 shows the main characteristics of these studies. From the table some differences emerge in comparing European and US studies. In Europe, the majority of the trials accrue patients with a reduced number of positive lymph nodes (4 compared to 10 in the US ones). Nearly all studies use an anthracyclin containing regimen in the standard arm, but no taxanes so far are employed, and except for the study IBCSG 15–95 (see Table) the mobilization regimen

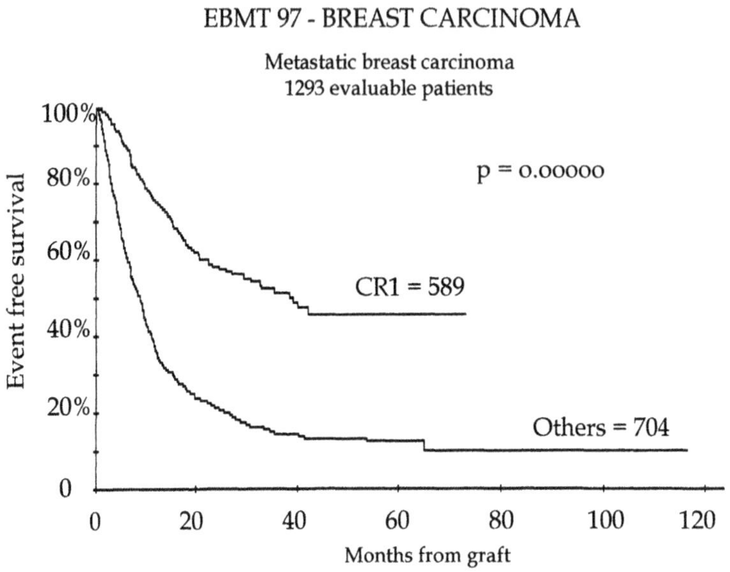

Figure 3. Event free survival for patients with metastatic disease receiving high-dose cheomtherapy in first complete remission (CR1) or in other disease status (partial remission, stable or progressive disease).

is based on the association of standard chemotherapy plus hematopoietic growth factors (G-CSF). The first results will be available by the year 2000 or 2001.

In metastatic disease there are only two major ongoing studies, one from the PEGASE 03 which compares in patients up to 60 years with metastatic disease receiving 4 courses of FEC (with Epirubicin at the dose of 100 mg/sqm) no further therapy or intensification with Cyclophosphamide 6 gr/sqm and Thiotepa 800 mg/sqm. This study is going to accrue for another year to reach the target of 180 randomized patients. In March 1998 another trial was launched on behalf of a joint venture between EBMT and the EBDIS (European Breast Dose Intensity Studies) Group.

Table 1. European ongoing high-dose chemotherapy phase III randomized studies for patients with high-risk operable breast cancer

Major European on-going adjuvant studies in breast cancer (July 1999)

Group	# nodes	Standard	High-dose	# pts	Mobilizing
Italian	≥4	EPI → CMF	HDS	386*	EPI 120 mg/sqm
Anglo-Celtic	≥4	ADM → CMF	ADM + CT	560	CTX 4 g/sqm
Scandinavian	≥5 or >8	FEC	FEC + STAMP V	525*	FEC
Netherlands	≥4	FEC	FEC + STAMP V	850*	FEC
EBMT/PEGASE	≥8	FEC	FEC + CMA	314*	FEC
German	≥10	EC → CMF	EC → CMT	230	FEC
IBCSG 15–95	≥4	EC → CMT	HD-EC	230	Filgrastim

HDS = high-dose sequential.
STAMP V= Cyclophosphamide, Thiotepa, Carboplatin.
CMA = Cyclophosphamide, Mitoxantrone, Alkeran.
CMT = Cyclophosphamide, Mitoxantrone, Thiotepa.
HD-EC = High-dose Epirubicin/Cyclophosphamide.
*studies closed.

This trial is comparing, in patients up to 60 years, the combination of Docetaxel 75 mg/sqm and Doxorubicin 50 mg/sqm for four courses followed by CMF four courses versus the same Docetaxel-Doxorubicin combination for three courses followed by double transplantation with VIC (Etoposide 1200 mg/sqm, Ifosfamide 12 gm/sqm, and Carboplatin AUC 18) and Cyclophosphamide (6 gr/sqm plus Thiotepa 800 mg/sqm). A total of 264 patients are going to be accrued in a 24-month period.

CONCLUSIONS

High-dose chemotherapy or breast carcinoma seems to proceed in the proper way from sometimes "wild" phase two studies to a more organized network of well designed phase III randomized trials.

In Europe in 1999 this relatively high-technology, high-cost procedure is not considered as the standard therapy for breast carcinoma patients in any disease status[6]. Anyway the policy of authorization for high-dose chemotherapy in solid tumors is different from country to country (i.e. in the Netherlands the Government is taking care of the Dutch National Study only, while in Italy it is decided on a local basis by local health authorities). The cost of an autologous procedures varies from country to country: from 18,000 US dollars in Spain (non-complicated autograft) to 40,000 US dollars in Italy (official reimbursement from the State, despite complications or not). We think that with the European Union becoming a reality, probably in Europe we are going to achieve a common agreement and consensus also on the regulations and reimbursement policies within countries.

REFERENCES

1. Antman KH, Rowlings PA, Vaughan WA, et al. High-dose chemotherapy with autologous hematopoietic stem-cell support for breast cancer in North America. J Clin Oncol 15:1302, 1997.
2. Rosti G, Ferrante P, Philip T, Marangolo M. High-dose chemotherapy (HDC) for solid tumors in Europe: a survey by the European Group for Blood and Marrow Transplantation (EBMT). J Clin Oncol 14:968a, 1996.
3. Peters WP, Ross M, Vredenburgh JJ, Meisenberg B, Marks LB, Winer E, Kurtzberg J, Bast RC, Jr., Jones R, Shpall EJ, Wu K, Rosner G, Gilbert C, Mathias B, Coniglio D, Petros W, Henderson IC, Norton L, Weiss RB, Budman D, Hurd D. High-dose chemotherapy and autologous bone marrow support as consolidation after standard-dose adjuvant therapy for high-risk primary breast cancer. J Clin Oncol 11:1131, 1993.
4. Gianni AM, Siena S, Bregni M, Di Nicola M, Orefice S, Cusumano F, Salvadori B, Luini A, Greco M, Zucali R, Rilke F, Zambetti M, Valagussa P, Bonadonna G. Efficacy, toxicity, and applicability of high-dose sequential chemotherapy as adjuvant treatment in operable breast cancer with 10 or more involved axillary nodes: five-year results. J Clin Oncol 15:2312, 1997.
5. Ayash LF, Wheeler C, Fairclough D, Schwartz G, Reich E, Warren D, Schipper L, Antman KH, Frei E, III, Elias AD. Prognostic factors for prolonged progression-free survival with high-dose chemotherapy for advanced breast cancer. J Clin Oncol 13:2043, 1995.
6. Goldman JM, Schmitz N, Niethammer D, Gratwohl A. Allogeneic and autologous transplantation for haematological diseases, solid tumours and immune disoredrs: current practice in Europe in 1998. Accreditation Sub-Committee of the European Group for Blood and Marrow Transplantation. Bone Marrow Transplant 21:1–7, 1998.

20

COMBINATION OF CARBOPLATIN AND ETOPOSIDE IN THE MANAGEMENT OF INTRAOCULAR RETINOBLASTOMA

M. A. Castello,[1] A. Schiavetti,[1] G. Ragni,[1] T. Hadjistilianou,[2] A. Clerico,[1] C. Cappelli,[1] E. Properzi,[1] and R. Frezzotti[2]

Dept of Paediatrics University "La Sapienza", Rome[1]
Institute of Ophthalmological Sciences
University of Siena, Italy[2]

ABSTRACT

The combination of carboplatin and etoposide has been tested in intraocular retinoblastoma (IORB), in order to avoid enucleation or radiotherapy and to permit local therapy.

Seventeen patients with intraocular RB have been treated with Carboplatin (1000 mg/sqm) and Etoposide (300 mg/sqm) on day 1, at intervals of 21–28 days and reduced by 1/3 in patients weighing less than 10–12 kg. Patients younger than 6 months received carboplatin at a dose of 25 mg/kg and etoposide at a dose of 7 mg/kg.

The disease-free survival was 91.7% with an average follow-up of 21.3 months.

After 2–4 courses of chemotherapy tumor regression was observed in 13 patients (76.5%). All patients received Argon-Laser therapy, after chemotherapy. A complete response has been observed in 52.9% of patients. Seven patients relapsed, three of whom received further courses of chemotherapy with carboplatin+etoposide obtaining a complete response.

- In 76.5% of the cases, the administration of carboplatin+etoposide resulted in a mass shrinkage
- In 31.2% of the cases neither enucleation nor radiotherapy were administered
- carboplatin+etoposide was effective also in relapsed patients
- Longer follow-up is needed on account of the risk of a second cancer

Molecular Biology of Hematopoiesis 6, edited by Abraham *et al.*
Kluwer Academic/Plenum Publishers, New York, 1999.

INTRODUCTION

Investigations have been carried out at the Paediatric Department of Rome University "La Sapienza", in collaboration with the Institute of Ophthalmological Sciences of Siena to establish the efficacy of the CBDCA+E association also in IORB. The aim of this pilot study is to obtain shrinkage of the lesions, in order to avoid the cosmetic and visual consequences of enucleation and the potential severe long-term consequences of radiotherapy as well as to reduce retinal scarring after local therapy.

The results are reported here of primary chemotherapy with CBDCA+E in 17 children with IORB.

PATIENTS AND METHODS

Between January 1990 and February 1998, 17 patients (7 males, 10 females) with IORB have been treated with the CBDCA+E association. Chemotherapy consisted of CBDCA (1000 mg/sqm) and E (300 mg/sqm) on day 1, at intervals of 21–28 days, being reduced by 1/3 in patients weighing less than 12 kg. Patients younger than 6 months received CBDCA at the dose of 25 mg/kg and E at the dose of 7 mg/kg.

The eligibility criteria for chemotherapy were: patients with unilateral or bilateral IORB, diagnosed by ophthalmoscopic examinations, as follows: tumors larger than 4 disk diameters (dd) with or without retinal detachment, localized in zone I (an oval described by points one foveal disk distance from a line connecting the centre of the fovea and the centre of the disk) or II (the remainder of the retina, see: Proposed International Classification, Fig. 1); any size tumors located in zone I; multifocal tumors; no extensive vitreous seeding and/or severe glaucoma.

All patients underwent ophthalmoscopic evaluation after each chemotherapy cycle before being submitted to argon-laser treatment. Non responders have been treated differently.

Minimal haematological requirements for chemotherapy were: WBC >2000/mmc, ANC >1000/mmc, platelet count <100,000/mmc, bilirubin <1,5 mg%, creatinine clearance >70 cc/min.

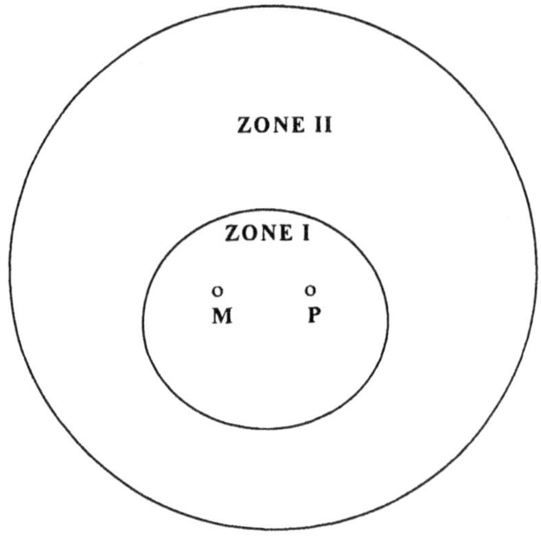

Figure 1. Proposed International Classification of Intraocular Retinoblastoma
Zones of Disease:
Zone I—An oval described by points one foveal disc distance from a line connecting the centre of the fovea and the centre of the disc. M: macula, P: papilla.
Zone II—The remainder of the retina.

Seventeen patients (7 males, 10 females) were enrolled in this pilot study; mean age at diagnosis was 5.7 months, range 0–27 months. Five were pretreated with external beam radiotherapy, and two received also chemotherapy (one of whom with carboplatin). Fifteen had bilateral IORB, whilst 2 had unilateral IORB. Of a total of 32 eyes affected, 11 have been enucleated on account of extensive vitreous seeding (6 cases), severe glaucoma (3 cases), or optic nerve invasions with the cut end free of disease (2 cases). In 21 affected eyes, tumoral involvement was in zone I in 8 eyes, in zone II in 4 eyes, in zones I, and II in 9 eyes.

The response to chemotherapy detected by ophthalmoscopic evaluation was based on the following regression patterns: type 0: Shrinkage of the tumor until the complete disappearance of the tumor; type 1: White calcified mass that resembles a cottage cheese; type 2: Pink translucent mass that lacks calcium and resembles a fish-flesh; 3: Combination of types 1 and 2 (Fig. 2a,b).

RESULTS

The disease-free survival was 91.7%, with a mean follow-up of 21.3 months (range 6–78 months).

All patients received chemotherapy, each course followed by ophthalmoscopic evaluation and argon laser therapy. These patients received a mean of 6.6 courses of chemotherapy (range 3–7) with CBDCA+E. After 2–4 courses of chemotherapy, regression patterns (types 0, 1, 2, and 3) have been observed in 13 patients (76.5%). All received argon-laser therapy and a complete response was observed in 52.9% of patients.

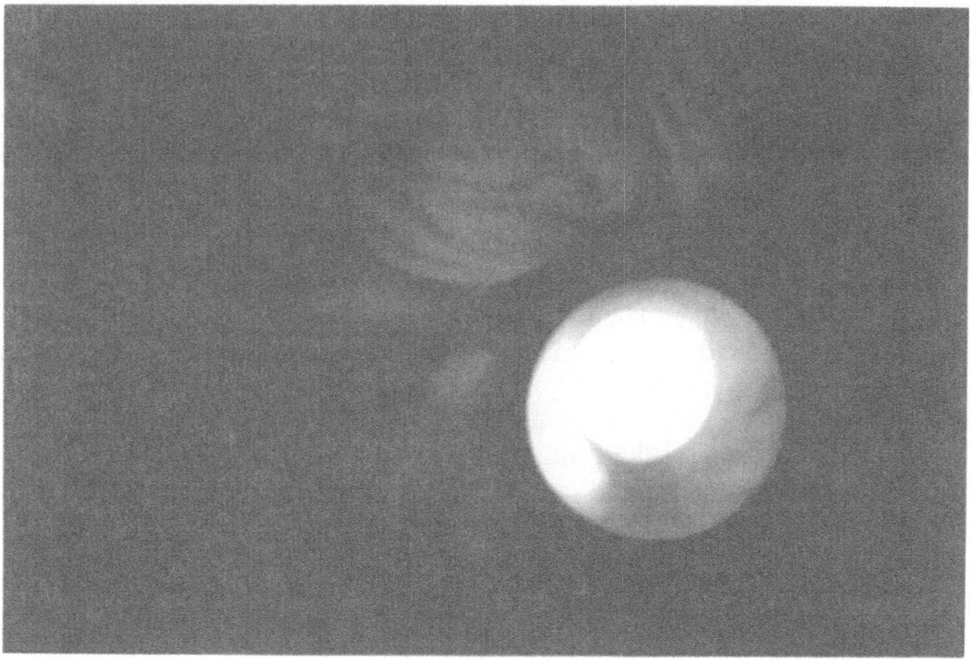

Figure 2a. Retinoblastoma balloon like. At onset.

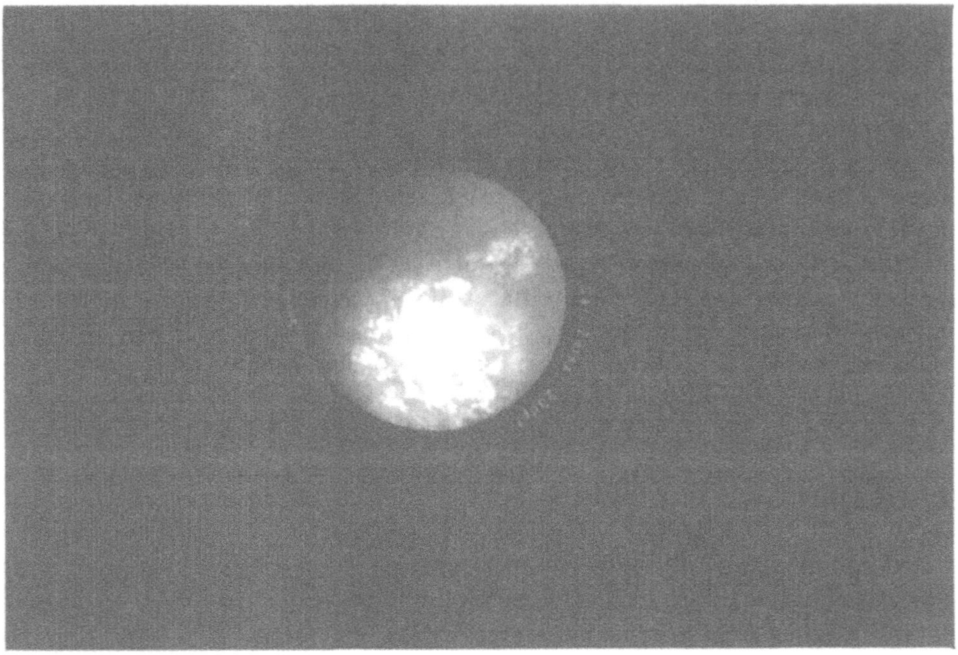

Figure 2b. Complete remission after 4 courses of chemotherapy with carboplatin and etoposide and argon-laser therapy.

Six patients showed no relapse and received neither RT nor enucleation after a mean follow-up of 23 months (range 6–78 months). Five had a bilateral IORB, 3 of whom enucleated at diagnosis, 3 patients (2 bilateral, 1 unilateral) were not submitted to enucleation.

Seven patients relapsed 4–12 months after the end of treatment.

Three of them received further courses of chemotherapy (2–6 courses) and argon laser therapy which led to a second complete response (follow-up: 10 months, range 6–18 months). One patient was submitted to enucleation at diagnosis, the other two (one bilateral and one unilateral) were not enucleated.

The remaining four received local radiotherapy for retinal haemorrhage and for extensive disease (3 cases). Two of these patients relapsed again and enucleation was necessary. These four patients had bilateral IORB.

In four patients, the disease progressed under treatment (after 2 courses). All had bilateral IORB, 3 cases had familial disease, one had 13 chromosomal deletion. Two were pretreated with radiotherapy and one also with platinum derivatives. Two patients died of the disease after radiotherapy and enucleation, one was lost to follow-up and one received radiotherapy and at present fares well.

Toxicity was manageable and was considered acceptable also for further treatment in the relapsed patients.

DISCUSSION AND CONCLUSIONS

Since 1990, we have attempted to extend treatment with CBDCA+E in the management of intraocular tumors, in order to obtain shrinkage of the masses, thus allowing more conservative treatment on the remaining tumoral tissue.

The aim of our study was to avoid enucleation and radiotherapy as first line therapy in tumors eligible for this kind of treatment. Radiotherapy is associated with important side effects such as: sicca syndrome, corneal ulceration, perforation, cataracts developing after 18 months, radiation retinopathy, optic nerve infarction, and orbital growth problems.

The use of chemotherapy with argon-laser therapy has allowed us to obtain complete remission in 9 cases (56.25%), and thus avoid radiotherapy and its side effects. The use of radiotherapy has been restricted to selected relapsed cases or to non responders.

Five previously untreated patients (31.2%), 3 bilateral and 2 unilateral were not submitted to enucleation and received only chemotherapy and argon-laser therapy. Of particular interest is the complete response obtained in three relapsed patients (3 cases) on the same treatment, two of whom treated for second and third relapses (appearing between 3 and 12 months after the end of the treatment-periods).

The efficacy and tolerable toxicity of this treatment lead to optimism regarding a more conservative approach in the management of unilateral and bilateral IORB.

A longer follow-up is needed to evaluate possible late effects of the therapy especially in patients with a constitutional abnormality.

REFERENCES

1. Reese AG, Merriam GL, Martin HE: Treatment of bilateral retinoblastoma by irradiation and surgery. Report on 15 years results. Am J Ophthalmol 32:175, 1949.
2. Ellesworth RM: The practical management of retinoblastoma. Trans Am Ophthalmol Soc 67: 462–534, 1969.
3. Morgan G: Diffuse infiltrating retinoblastoma. Br J Ophthalmol 55: 600–606, 1971.
4. Kingstone JE, Hungerford JL: Retinoblastoma. In: Plowman PN, Pinkerton CK (eds) Pediatric Oncology. London, Chapman & Hall Medical, p. 268–290, 1992.
5. Shields JA, Shields CL: Current management of retinoblastoma. Mayo Clin Proc 69:50–56, 1994.
6. Donaldson SS, Egberth PR, Newsham I, Cavenee WK: Retinoblastoma. In: Pizzo P, Poplack D (eds) Philadelphia, Lippincott Raven p.699, 1997.
7. Abramson DH, Ellesworth RM, Rozakis GW: Cryotherapy for retinoblastoma. Arch Ophthalmol 100:1253, 1982.
8. Abramson DH: Cryotherapy in retinoblastoma. In: Jacobiec, ed. Technique in ophthalmic surgery. Philadelphia, WB Saunders, p.433, 1984.
9. Lemerle J, Bloch-Michel E: Chemotherapy in retinoblastoma: preliminary report and warning. In: Lommatzsch PK, Blody FC (eds) Intraocular tumors. New York, Springer-Verlag, p.541, 1983.
10. Pratt CB, Crom DB, Howard C: The use of chemotherapy for extraocular retinoblastoma. Med Pediatr Oncol 13:330–333, 1985.
11. Castello MA, Clerico A, Deb G et al.: High dose Carboplatin in combination with Etoposide for childhood brain tumors. Am J Pediatr Hematol Oncol 12:297–300, 1990.
12. Lee White: Chemotherapy for retinoblastoma: where do we go from here ? Ophthalmic Pediatr and Genetics 12(3):115–130, 1991.
13. Doz F, Quintana E, Michon J: High dose chemotherapy using VP 16, carboplatin and cyclophosphamide with autologous bone marrow transplantation (ABMT) rescue in 9 children with extraocular retinoblastoma. Proc Am Soc Clin Oncol 11:371, 1992 (abstr 1282).
14. Doz F, Neuenschwander S, Plantaz D, Courbon B, Gentet JC, Bouffet E, Masseri V, Vannier JP, Mechinaud F, Desjardins L, Vielh P, Zucker JM: Etoposide and carboplatin in extraocular retinoblastoma: a study by the Société Française d'Oncologie Pédiatrique. J Clin Oncol 13 (4):902–909, 1995.
15. Shields JA, Augsberger JJ: Current approaches to diagnosis and management of retinoblastoma. Surv Ophthalmol 25:347–371, 1981.
16. Sanders BM, Draper GJ, Kingston JE: Retinoblastoma in Great Britain, 1969–80. Incidence, treatment, and survival. Br J Ophthalmol 72:516–583, 1988.
17. Malogolowkin M, Steele D, Deegan W, J.Ortega, and L.Murphee Systemic chemotherapy (CRX) for intraocular retinoblastoma (IO-RB). Proc Am Soc Clin Oncol 14:449, 1995 (abstr 1437).

18. Seregard S, Kock E, Trampe E: Intravitreal chemotherapy for recurrent retinoblastoma in an only eye. Br J Ophthalmol 79:194, 1995.
19. Castello MA, Schiavetti A, Ragni G, Hadjistilianou T, Clerico A, Properzi E, Frezzotti R: Chemotherapy with etoposide and carboplatin in intraocular retinoblastoma. Med Pediatr Oncol 27(4):257, 1996.
20. Castello MA, Schiavetti A, Ragni G, Hadjistilianou T, Clerico A, Properzi E, Cappelli C, Frezzotti R: Primary chemotherapy in intraocular retinoblastoma with carboplatin and etoposide. Med Pediatr Oncol 29(5):414, 1997.
21. Shields CL, De Potter P, Himelstein BP, Shields JA, Meadows AT, Maris JM: Chemoreduction in the initial management of intraocular retinoblastoma. Arch Ophthalmol. 114:1330–1338, 1996.
22. Doz F, Desjardins L, Quintana E, Levi C, Zucker JM: Efficacy and toxicity of neoadjuvant chemotherapy using etoposide and carboplatin in 20 patients with intraocular retinoblastoma. Med Pediatr Oncol 27(4):257, 1996.
23. Greenwald MJ, Strauss LC: Treatment of intraocular retinoblastoma with carboplatin and etoposide chemotherapy. Ophthalmology 103(12):1989, 1996.

HYPER-IL-6 (H-IL-6), A FUSION PROTEIN OF SOLUBLE IL-6 RECEPTOR (sIL-6R), AND INTERLEUKIN-6 (IL-6), ACTS SYNERGISTIC WITH THROMBOPOIETIN (TPO) AND STEM CELL FACTOR (SCF) IN EXPANDING MEGAKARYOCYTE PROGENITORS FROM HUMAN CD34^{++}/CD90^{+} CELL

Josef Birkmann, *Stefan Rose-John, Manfred Smetak, Thomas Sievert, and Walter M. Gallmeier

Inst. f. Med. Oncology & Hematology
Med. Clinic 5; Nürnberg, Germany
*Med. Clinic I, University of Mainz, Germany

It has been shown that signalings from c-kit and gp130, the signal-transducing receptor component of the IL6 receptor, act synergistic for the ex-vivo expansion of multipotential hematopoietic progenitors. A similar synergistic effect has been demonstrated for signalings from c-kit, c-mpl, and flt3. While c-kit is activated by stem cell factor (SCF), gp130 can be activated by the complex of soluble IL-6 receptor (sIL-6R) and interleukin-6 (IL-6). Recently, a bioactive designer cytokine, H-IL-6, a fusion protein consisting of sIL-6R and IL-6 linked by a flexible peptide chain has been shown to expand human hematopoietic colony-forming cells. We tested the activity of H-IL-6 alone and in combination with TPO, SCF, and the flt3 ligand (FL) on mobilized purified human primitive CD34++/CD90+ progenitor cells in a 7 day serum-free liquid culture system. Apart from a moderate expansion of CD34+ cells (2.5-fold to 3.5-fold) the most striking effects of H-IL-6 were observed in megakaryopoiesis. In contrast to TPO, H-IL-6 alone was not sufficient for survival or proliferation of CD34++/CD90+ progenitor cells. Yet the combination of H-IL-6 and TPO expanded CD41+ cells to the same extent as did the combination of TPO + SCF (20-fold). A maximum yield of megakaryocyte progenitors (more than 50-fold) was obtained with the combination of H-IL-6, TPO, and SCF.

Molecular Biology of Hematopoiesis 6, edited by Abraham *et al.*
Kluwer Academic/Plenum Publishers, New York, 1999.

A combination of sIL-6R and IL-6 in the presence of SCF was found to stimulate expansion of human cord blood progenitor cells.[1] We and others have shown that TPO, the ligand of c-mpl, is an early-acting cytokine able to expand $CD34^+$ hematopoietic progenitor cells together with SCF.[2–8] Similar synergistic effects were demonstrated for signalings from c-kit, c-mpl and flt3.[9] The receptor protein gp130 can be activated by the complex of IL-6 (50 ng/ml) and high doses of IL-6R (>1000 ng/ml). Hyper-IL-6 (H-IL-6) is a semi-synthetic fusion protein consisting of sIL-6R and IL-6 linked by a flexible peptide chain.[10] This fusion protein H-IL-6 been shown to stimulate gp130 much more effectively than the instable IL-6/sIL-6R complex and to expand human CFU-GM in a dose-dependent manner. This effect of H-IL-6 on committed human hematopoietic progenitor cells prompted the question whether H-IL-6 was able to stimulate early $CD90^+/CD34^{++}$ progenitors and whether additive or synergistic effects with other early-acting cytokines could be observed.

The following experiments were carried out to determine the role of H-IL-6, alone and in combination with TPO, SCF, and FL, with respect to survival, proliferation and differentiation induction in human early progenitor cell populations, to evaluate the potential of these cytokine combinations on the expansion of $CD34^+$ cells and on the expansion of megakaryocytic progenitors.

PATIENTS AND METHODS

Aliquots of leukapheresis samples taken from 5 patients, (1 Hodgkin's Disease, 1 Myeloma, 2 NHL, 1 Ewing's Sarcoma) were processed in a 2-step procedure for the purification of progenitor cells.

In the 1st step, $CD34^+$ cells were highly enriched by immunomagnetic positive selection using the CD34 Progenitor Cell Isolation Kit (Miltenyi Biotec GmbH, Bergisch Gladbach, Germany) according to the manufacturer's instructions. Briefly, after blocking non-specific binding with human polyclonal IgG, the cells were incubated at 4°C with a class-II anti-CD34 monoclonal antibody (QBEND/10, mouse IgG1) coupled to colloidal superparamagnetic microbeads and then run over a MiniMAC'S™ separation column. During this process, the $CD34^+$ cells coated with the monoclonal antibody attached to the microbeads are retained through a strong magnetic field, while the negative cells are allowed to pass through the column. After removing the column from the magnet, the $CD34^+$ cells are eluted. Reanalysis of the enriched cell fraction, performed with a class-III anti-CD34 antibody, (HPCA-2, clone 8G12, Becton Dickinson, Heidelberg, Germany), revealed a purity of 88.2% to 98%. The recovery rate was 52%–75%, and the trypan-blue dye exclusion test showed a viability of 82%–92%. Re-analysis of the enriched $CD34^+$ cells showed a 35%–40% co-expression of CD90.

In the 2nd step, these $CD34^+$ cells were separated into a $CD34^{++}/CD90^+$ and a $CD34^{++}/CD90^-$ population using an EPICS Elite fluorescence-activated cell sorter (Coulter Corp., Hialeah FL). The recovery rate of this procedure is 40%–50%; both viability and purity of the final cell product were >97%. As $CD34^{++}/CD90^+$ cells comprise only a very small proportion of the nucleated cells in the separation product (<0.5%), and as >75% of cells are lost during enrichment and washing procedures, roughly 1 to 1.5 billion cells from the separation product were needed for one complete assay.

Cells were cultured in 96 well round-bottom microtiter plates at a density of 10,000 cells/well. We used serum-free culture medium, IMDM supplemented with L-glutamine, penicillin/streptomycin, and "BIT", i.e. bovine serum albumin, 50 mg/ml, bovine insulin, 50 µg/ml, and human transferrin, 1 mg/ml (StemCell Technologies Inc, Vancouver, BC).

Cytokines were added at the following concentrations:

H-IL-6: 10 ng/ml, TPO: 2.5 ng/ml, SCF: 20 ng/ml, and FL: 50 ng/ml. All single, double, and triple cytokine combinations and the 4-cytokine combination were tested.

On day 4, 50% of the medium was replaced and fresh cytokines were added. On day 7, survival, proliferation and differentiation were evaluated by

- screening each well with the inverted microscope for gross morphology,
- counting the cell number from each well (Neubauer chamber),
- determining the proportion of viable cells by trypan blue exclusion,
- preparing a cytospin from an aliquot of each assay,
- using all other cells for 3-color flow cytometry analysis,
- calculating proliferation rates and absolute cell numbers for each marker.

All assays were carried out in quadruplicates.

RESULTS

1. Expansion of CD34$^+$ Cells

In contrast to TPO, H-IL-6 alone was not sufficient for survival or proliferation of human CD34^{++} progenitor cells. This applies both for the CD90$^+$ and the CD90$^-$ subpopulations.

When H-IL-6 was combined with TPO, slightly more CD34$^+$ cells were found on day 7 than with TPO alone. The combination of H-IL-6 with SCF or FL yielded only about 10% of the absolute counts of CD34$^+$ cells compared to the combination of TPO with SCF or FL (Fig. 1).

The triple combinations containing H-IL-6 yielded a 1.5-fold to 2.5-fold net expansion of CD34$^+$ cells. With all four cytokines a 3.5-fold expansion was obtained. Yet maximum expansion of CD34$^+$ cells (4.5-fold) was reached with the combination TPO + FL + SCF without H-IL-6 (Fig. 1).

2. Expansion of Megakaryocyte Progenitors

Several cytokines with thrombopoietic activity like IL-6 or IL-11 act via stimulation of gp130, the target receptor of H-IL-6. On the other hand, TPO is not only an early-acting cytokine, but also a megakaryocyte growth and development factor. We therefore evaluated the effects of the chosen cytokine combinations with respect to generation of CD41$^+$ megakaryocyte progenitors.

While H-IL-6 alone was not able to generate CD41$^+$ cells, the combination of H-IL-6 and TPO displayed a strong synergistic activity very similar to the combination of SCF and TPO. Both combinations yielded the same amount of CD41$^+$ cells, i.e. about a 15-fold expansion from day 0, but subpopulation analysis showed that H-IL-6 in the presence of TPO generated by far more mature CD41$^+$/CD42b$^+$ megakaryocyte progenitors (Fig. 2).

These findings were confirmed by morphology.

The combinations of H-IL-6 with FL or SCF without TPO did not expand CD41$^+$ cells; the triple combination H-IL-6 + SCF + FL showed about the same effect as TPO + FL.

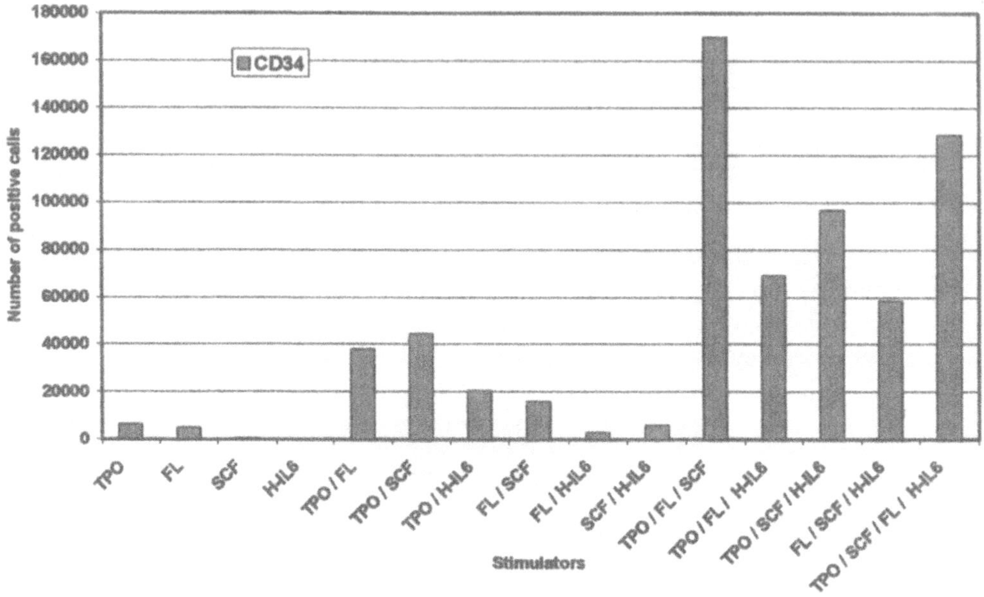

Figure 1. Effects of different cytokine combinations on the expansion of CD34⁺ cells (number of CD34⁺ cells on day 0 = 40,000).

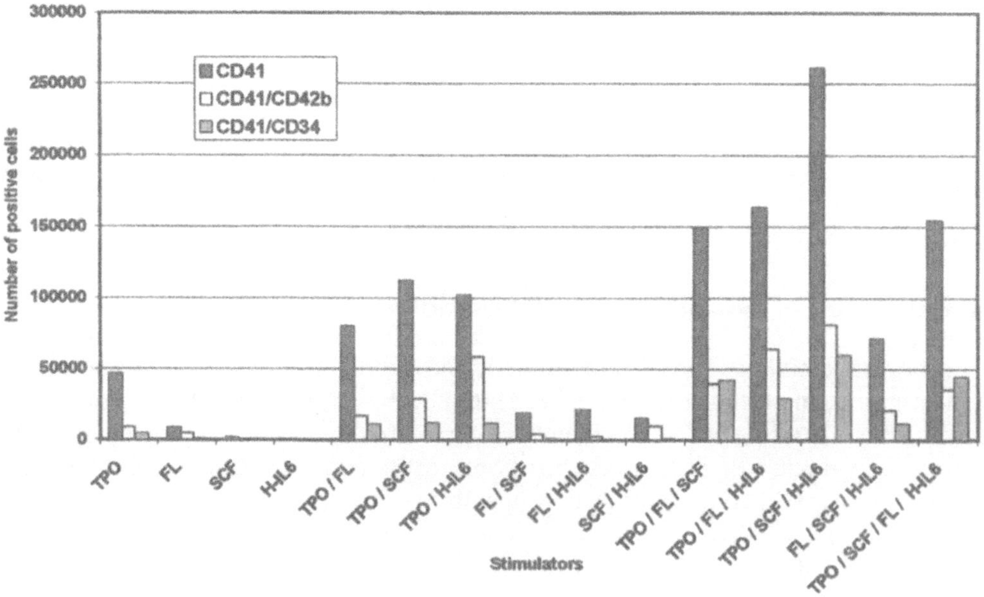

Figure 2. Effects of different cytokine combinations on the expansion of CD41⁺ cells and subpopulations of CD41⁺ cells (number of CD41⁺ cells on day 0 = 5000).

Addition of H-IL-6 to the "thrombopoietic" mixture of TPO + SCF generated more than 260,000 CD41$^+$ cells by day 7, corresponding to 50-fold expansion of CD41$^+$ cells. Notably, this combination yielded both a high number of immature CD41$^+$/CD34$^+$ progenitors and an even higher amount of more mature CD41$^+$/CD42b$^+$ committed progenitors (Fig. 2).

The fact that CD41$^+$ cells were most efficiently expanded by H-IL-6 + TPO + SCF also applies for the CD34^{++}/CD90$^-$ target population.

DISCUSSION

Stimulation by H-IL-6 alone is not sufficient for survival of early CD34$^+$ subpopulations co-expressing the CD90 molecule. This may be due to the fact that gp130 expression is very low on this subpopulation or that signalings from gp130 alone cannot maintain survival or proliferation in these cells. The combination with TPO + H-IL-6 yields about 50% of the initial CD34$^+$ cell counts, while SCF + H-IL-6 yield only about 12%, suggesting that additional signals from c-mpl are more likely to promote survival than signals from c-kit.

When both c-mpl and c-kit are stimulated together with gp130 (TPO + SCF + H-IL-6) a 2.5-fold expansion of CD34$^+$ cells is observed. Further addition of FL, i.e. stimulation of flt3, increases CD34$^+$ cell expansion to 3.5-fold. On the other hand, the combination of TPO, FL, and SCF without H-IL-6 is even more effective in increasing CD34$^+$ cells (4.5-fold), suggesting that *signalings from c-mpl, c-kit, and flt3* are most important in this subpopulation to expand the immature phenotype. These findings are consistent with the data recently published by Solar et al.[11] who demonstrated that the hematopoietic repopulating ability both of murine and of human early progenitor cells resides in the c-mpl$^+$ fraction.

The combinations of H-IL-6 with TPO showed a synergistic effect with respect to expansion of CD41$^+$ megakaryocyte progenitors, which was as potent as the combination of TPO + SCF (about 20-fold) which we had published earlier.[2] The addition of H-IL-6 to these thrombopoietic cytokines TPO and SCF produced a more than 50-fold expansion of CD41$^+$ cells. Subpopulation analysis showed that in the CD41$^+$ cells generated by stimulation with H-IL-6-containing combinations the number of more mature megakaryocytic progenitors co-expressing CD42b exceeds the number of more immature CD41$^+$ cells still co-expressing CD34.

At least 2 cytokines with a well-known thrombopoietic activity, IL-6, and IL-11, act through signalings of the gp130 receptor. The thrombopoietic activity of thrombopoietin is mediated by the c-mpl receptor. Our data suggest that signalings from gp130 triggered by a potent stimulator like H-IL-6 together with signalings from c-mpl and c-kit drive immature CD90$^+$ hematopoietic progenitor cells into the cell cycle and strongly commit them to the megakaryocytic lineage. Thus, the bioactive designer cytokine H-IL-6 may play an important role in the ex vivo generation and expansion of immature and mature megakaryocytic progenitors, i.e. the cell compartments which are most important for rapid thrombocyte engraftment.

ACKNOWLEDGMENT

The authors thank Mrs. Anja Effmert, Mrs. Elisabeth Holzmann, Mrs. Manuela Stüwe, and Mrs. Andrea Weissenberger for their excellent technical assistance. We are grateful to Dr. Jakob Bogenberger (Amgen Inc., Thousand Oaks, USA) and Prof. Karl Welte (Hannover Medical School, Hannover, Germany) for providing the recombinant thrombopoietin, and to ICC (Ismaning, Germany) for the SCF. This work was supported by Deutsche Krebshilfe, Bonn, Germany.

REFERENCES

1. Sui XW, Kohichiro T, Tanaka R, Tajima S, Muraoka K, Ebihara Y, Ikebuchi K, Yasukawa K, Taga T, Kishimoto T, Nakahata T: gp130 and c-Kit signalings synergize for ex vivo expansion of human primitive hemopoietic progenitor cells. PNAS 92:2859, 1995.
2. Birkmann J, Oez S, Smetak M, Kaiser G, Kappauf H, Gallmeier WM: Effects of Recombinant Human Thrombopoietin Alone and in Combination with Erythropoietin and Early-Acting Cytokines on Human Mobilized Purified CD34+ Progenitor Cells Cultured in Serum-Depleted Medium. Stem Cells 15:18, 1997.
3. Borge OJ, Ramsfjell V, Veiby OP, Murphy MJ, Lok S, Jacobsen SEW: Thrombopoietin, but not erythropoietin promotes viability and inhibits apoptosis of multipotent murine hematopoietic progenitor cells in vitro. Blood 88:2859, 1996.
4. Ebihara Y, Tsuji K, Lyman SD, Sui XW, Yoshida M, Muraoka K, Yamada K, Tanaka R, Nakahata T: Synergistic action of Flt3 and gp130 signalings in human hematopoiesis. Blood 90(11):4363, 1997.
5. Kaushansky K, Broudy VC, Grossmann A, Humes J, Lin N, Ren HP, Bailey MC, Papayannopoulou T, Forstrom JW, Sprugel KH: Thrombopoietin expands erythroid progenitors, increases red cell production, and enhances erythroid recovery after myelosuppressive therapy. J Clin Invest 96:1683, 1995.
6. Kobayashi M, Laver JH, Kato T, Miyazaki H, Ogawa M: Thrombopoietin supports proliferation of human primitive hematopoietic cells in synergy with steel factor and/or interleukin-3. Blood 88:429, 1996.
7. Rasko JEJ, OFlaherty E, Begley CG: Mpl ligand (MGDF) alone and in combination with stem cell factor (SCF) promotes proliferation and survival of human megakaryocyte, erythroid, and granulocyte/macrophage progenitors. Stem Cells 15:33, 1997.
8. Young JC, Bruno E, Luens KM, Wu S, Backer M, Murray LJ: Thrombopoietin stimulates megakaryocytopoiesis, myelopoiesis, and expansion of CD34(+) progenitor cells from single CD34(+)Thy-1(+)Lin(−) primitive progenitor cells. Blood 88:1619, 1996.
9. Ramsfjell V, Borge OJ, Cui L, Jacobsen SEW: Thrombopoietin directly and potently stimulates multilineage growth and progenitor cell expansion from primitive (CD34(+)CD38(−)) human bone marrow progenitor cells—Distinct and key interactions with the ligands for c-kit and flt3, and inhibitory Ef-Fects of TGF-beta and TNF-alpha. J Immunol 158(11):5169, 1997.
10. Fischer M, Goldschmitt J, Peschel C, Brakenhoff JPG, Kallen KJ, Wollmer A, Grotzinger J, Rose-John, S: A bioactive designer cytokine for human hematopoietic progenitor cell expansion. Nat Biotechnol 15(2):142, 1997.
11. Solar GP, Kerr WG, Zeigler FC, Hess D, Donahue C, de Sauvage FJ, Eaton DL: Role of c-mpl in Early Hematopoiesis. Blood 92:4, 1998.

22

TOWARDS VACCINATION WITH DEFINED TUMOR ANTIGENS?

P. van der Bruggen, P. Chaux, S. Mandruzzato,[#] B. Van den Eynde, P. G. Coulie,* M. Marchand, and T. Boon

Ludwig Institute for Cancer Research
Brussels Branch
74 avenue Hippocrate
B-1200 Brussels, Belgium
[#]Department of Oncology and Surgical Sciences
University of Padova
64 via Gattamelata
I-35128 Padova, Italy
*Cellular Genetics Unit
Université Catholique de Louvain
74 avenue Hippocrate
B-1200 Brussels, Belgium

1. SUMMARY

Following the identification of the first tumor antigens recognized by CTL on melanoma cells, concern was expressed that tumors of other histological types might be less likely to express such antigens. However, the results obtained now with other types of tumors suggest that there is no fundamental difference in this respect between melanoma and other tumors. As new antigens are discovered, an increasing proportion are found to be the result of mutations, several of which may play a role in tumoral transformation or progression. Some of these tumor antigens are currently in the early stages of clinical study. There can be little doubt that the coming years will witness a large number of clinical trials involving peptides, proteins, DNA, and recombinant defective viruses, while increasingly sensitive tools are developed to measure the induction of CTL responses in immunized patients.

2. EXISTENCE OF MOUSE TUMOR REJECTION ANTIGENS

In 1943, Ludwik Gross induced sarcomas in mice by applying methylcholanthrene to their skin.[1] Such tumors could be transplanted to grow progressively on mice belong-

Molecular Biology of Hematopoiesis 6, edited by Abraham *et al.*
Kluwer Academic/Plenum Publishers, New York, 1999.

ing to the same inbred strain as the original tumor bearer. However, when transplanted mice were cured of their tumor by surgery and reinoculated subsequently with the same tumor, the tumor usually failed to grow. This observation was the first evidence suggesting that tumors can carry specific transplantation antigens that cause their rejection in immunized hosts. This work was later extended by, among others, Prehn and Main,[2] who revealed the great diversity of antigens in methylcholanthrene-induced tumors.

In the early 1970s, these observations led to a great deal of optimism regarding the prospects of cancer immunotherapy. The theory of the "immune surveillance of cancer" was put forward by Burnet[3] who proposed that most tumors are rejected by the immune system, the cancerous disease being the result of an occasional failure of this surveillance. To many, this implied that non-specific boosting of the immune defences of cancer patients, such as that achieved with BCG, might result in cures. But neither the theory, nor these clinical implications, received much confirmation. Mouse experiments have demonstrated that the cells of the immune system that constitute the specific component of tumor rejection responses and that carry long term memory are the T lymphocytes.[4–6] However, it is very likely that, in addition to cytolytic T lymphocytes (CTL), tumor rejection involves several other effector cells, such as natural killer cells and macrophages.

It has been demonstrated that most animal tumors express tumor antigens that can constitute targets for rejection responses, but that these tumors rarely elicit immune responses without first being manipulated to improve their immunogenicity.[7,8]

3. HUMAN TUMOR REJECTION ANTIGENS

A large number of independent studies have demonstrated that lymphocytes of cancer patients can be restimulated *in vitro* to produce CTL that show specificity for the tumor cells.[9–13] Autologous mixed lymphocyte-tumor cell cultures (MLTC) have been carried out with peripheral blood cells of cancer patients as a source of lymphocytes. Many such studies have involved melanoma, but anti-tumor CTL populations have also been obtained against other types of tumors, such as sarcomas, head and neck carcinomas, and pancreatic carcinomas.[14–16] In human tumors associated with viral transformation, CTL specific for HTLV-infected cells were obtained from T cell leukaemia patients.[17,18] However, no CTL directed against Epstein-Barr virus components have been observed in patients carrying EBV-positive tumors. In the latter, the expression of all EBNA products are repressed with the exception of EBNA-1, for which no CTL has ever been isolated. This is in sharp contrast with EBV-induced lymphoma that occurs in immunosuppressed patients and expresses many other EBV antigens that can be recognized by CTL.[19–21] Another approach to obtain anti-tumor CTL is the use of TILs. When melanoma metastases are cultured in medium containing IL-2, $CD8^+$ T lymphocytes will proliferate. These lymphocytes exert significant cytolytic activity against the autologous tumor.[22–24]

CTL recognize small peptides of 8–11 aminoacids long that are embedded in the groove of MHC class I molecules. These short peptides are generated from larger cellular proteins- or viral proteins- by degradation in the cell by proteolytic enzymes, like the proteasome.[25,26] This complex contains several units and the presence of some of these is dependent of IFN-γ. The small peptides are actively transported to the endoplasmic reticulum by the peptide transporter heterodimer TAP, where these peptides will bind to MHC class I molecules with the help of chaperones.[27]

Three methods have been used to identify the peptides presented to tumor-specific T cells. The first is a genetic approach based on the transfection of recombinant DNA libraries into cells expressing the MHC presenting molecule, in order to isolate the gene encoding the antigen.[28] Once the gene has been isolated, the antigenic peptide is deduced from the sequence of the putative protein. The second method is a biochemical purification of peptides eluted from the MHC class I molecules of the tumor cells. Tumor peptides are fractionated by HPLC, and the different fractions are then tested for their ability to sensitize target cells for lysis by relevant CTL, with the positive fraction being further purified and then, sequenced. Peptide sequencing usually requires sophisticated mass spectrometry methods.[29] When post-translational modification of the protein occurs, the sequence of the peptide can not be deduced from the sequence of the protein and the biochemical approach is then the only way to identify the natural epitope presented at the cell surface.[30] The third method of identification is the reverse of the first two and has been used to identify antigens encoded by proteins known to be overexpressed or mutated in tumor cells. Candidate peptides carrying consensus anchor motives for a certain HLA are located, tested for HLA binding, and loaded on antigen-presenting cells that are then used to stimulate lymphocytes *in vitro*.[31,32] This straightforward approach was greeted with enthusiasm several years ago, but has not proved to be very productive for the isolation of human antigens so far. The major drawback, it seems, is that many peptide-specific CTL do not recognize HLA-matched tumor cells expressing the protein endogenously. At least three different reasons may explain this paradox. In some cases, the CTL activated *in vitro* with a high concentration of peptide may have a too low affinity to recognize the low amounts of peptide naturally displayed at the cell surface.[33] Other CTL, obtained by stimulation with synthetic peptides that were not extensively purified, may in fact be directed against contaminants, such as the protective groups incorporated during peptide.[34] Lastly, the peptide may not be displayed at the cell surface because the processing machinery generates a different set of peptides.

The list of tumor antigens, recognized by T cells, that have been identified at the genetic level is growing. Based on the pattern of expression of the parent protein, the antigens can be classified into six major groups.

3.1. Tumor-Specific Shared Antigens

Tumor-specific shared antigens are encoded by genes that are silent on normal cells and are activated in many tumors belonging to different histological types. Thus, the activation of specific genes in tumors can generate tumor rejection antigens that are shared by many tumors. Prototype antigens of this group are those encoded by gene *P1A* in the mouse[35] and by the *MAGE* genes in humans.[36]

The activation of the human *MAGE* genes usually results from a demethylation of their promoter that correlates with a non-specific demethylation throughout the genome.[37,38] Human antigens shared by tumors expressing the genes and bearing the appropriate MHC type represent promising targets for cancer immunotherapy. *BAGE* and *GAGE*, two new families of genes, with similar expression profiles to the *MAGE*-type genes, have been identified and are listed in Tables I and II. Expression has been detected in substantial fractions of human tumors and the list of histological types of

Table I. List of tumor antigens recognized by T cells and their encoding genes

gene/protein	expression in tumors	expression in normal tissues	HLA restriction
Tumor-specific shared antigens			
MAGE-1	various	testis	A1, Cw16
MAGE-3	various	testis	A1, A2, A24, B44
MAGE-6	various	testis	Cw16
BAGE	various	testis	Cw16
GAGE-1/2	various	testis	Cw6
GAGE-3 to 6	various	testis	A29
NY-ESO-1	various	testis, ovary	A2
GnTV[a]	melanoma	none	A2
HER-2/neu[b]	breast and ovarian carcinoma	ubiquitous (low level)	A2
mucin[c]	pancreatic and breast carcinoma	lactating breast	non-MHC-restricted
Tumor antigens resulting from mutations			
MUM-1	melanoma	ubiquitous	B44
CDK4	melanoma	ubiquitous	A2
β-catenin	melanoma	ubiquitous	A24
HLA-A2[d]	renal cell carcinoma	ubiquitous	—
bcr-abl (b3a2)	chronic myeloid leukemia	ubiquitous	DR4
CASP-8	head and neck squamous cell carcinoma	ubiquitous	B35
KIAAO205	bladder tumor	ubiquitous	B44
K-ras	pancreatic adenocarcinoma	ubiquitous	B35, DR6
Differentiation antigens			
tyrosinase	melanoma	melanocytes	A2, A24, B44, DR4
Pmel17^{gp100}	melanoma	melanocytes	A2, A3
Melan-A$^{MART\text{-}1}$	melanoma	melanocytes	A2
gp75$^{TRP\text{-}1}$ [e]	melanoma	melanocytes	A31
TRP-2	melanoma	melanocytes	A31
MC1R	melanoma	melanocytes	A2
CEA	gut carcinoma	ubiquitous (low level)	A2
PSA	prostate carcinoma	prostate	A2
Antigens overexpressed in tumors			
HER-2/neu	breast and ovarian carcinoma	ubiquitous (low level)	A2
P53	various	ubiquitous (low level)	A2
PRAME[f]	various	testis ovary endometrium adrenals	A24

All the references can be found in[110].
[a]aberrant trancript of N-acetyl glucosaminyl transferase V (GnTV) that is found only in melanomas.
[b]overexpressed in 30% of breast and ovarian carcinomas, and in a lower fraction of other carcinomas.
[c]MHC-unrestricted recognition of a repeated motif that is unmasked in tumors due to mucin underglycosylation.
[d]the mutation affects the HLA-A2 gene itself.
[e]peptide translated from an alternative open reading frame of the gp75 transcript.
[f]the antigen is recognized by CTLs bearing an NK inhibitory receptor that prevents lysis of cells expressing certain HLA-C molecules.

tumors expressing *MAGE*-type genes is increasing (Table II). The only normal cells in which significant expression of genes of this category has been detected are placental trophoblasts and testicular germ,[39] which do not express MHC class I molecules and are therefore not implicated in antigen expression. Hence such antigens can be considered as strictly tumor-specific cells.[40] These conclusions are further strengthened by immunization studies carried out with the tumor antigen encoded by mouse gene *P1A*, which is also expressed in testis. Immunization of male mice produced strong P1A-specific CTL responses that did not cause testis inflammation or alteration of fertility.[41]

Table II. Expression in tumor samples of genes encoding T-cell antigens[a]

	Percentage of tumors expressing:					
Histological type	*MAGE-1*	*MAGE-3*	*BAGE*	*GAGE-1, 2*	*NY-ESO-1*	*PRAME*
Melanomas					34	
primary lesions	16	36	8	13		88
metastases	48	76	26	28		95
Nonsmall cell lung carcinomas	49	47	4	19	20	
adenocarcinomas						46
squamous cell carcinomas						78
Head and neck tumors	28	49	8	19	20	39
Bladder carcinomas	22	36	15	12	44	
Sarcomas	14	24	6	25	32	39
Mammary carcinomas	18	11	10	9	30	27
Gastric carcinomas	29				0	
Esophageal carcinomas	53	47				
Prostatic carcinomas	15	15	0	10	25	
Colorectal carcinomas	2	17	0	0	0	
Renal cell carcinomas	0	0	0	0	0	41
Ovarian carcinomas	28	17	15	31	25	
Neuroblastomas	18	54				
T cell leukemias related to HTLV-1 infection	53	11				
Acute myeloid leukemia[b]						37
Acute lymphoid leukemia[b]						16

[a]Expression was measured by RT-PCR on total RNA using primers specific for each gene. Data were compiled from results obtained at the Brussels branch of the Ludwig Institute, and from other sources:[111] for gastric carcinomas,[112] for esophageal carcinomas,[113] for ovarian carcinomas,[114] for neuroblastomas, and[115] for T cell leukemias related to HTLV-1 infection.
[b]Remarkably, all acute myeloblastic leukaemias that carry the chromosomal translocation t (8; 21), which fuses the genes AML1 and ETO, expressed PRAME at a high level[116].

Another interesting approach, named SEREX, has led to the identification of three additional *MAGE*-type gene families (reviewed in).[42] It is based on the observation that the serum of cancer patients often contains antibodies directed against intracellular proteins. A screening of phage expression libraries with such sera yielded *MAGE* sequences but also new interesting sequences, notably the *NY-ESO-1*[43] and the *SSX2* gene.[44]

Finally, representational difference analysis has been used to enrich cDNA libraries for sequences expressed in tumors and not in normal tissues. This has led to the identification of new *MAGE* sequences and also to that of gene *LAGE*, which belongs to the same family as *NY-ESO-1*.[45,46]

A few HLA-A1 melanoma patients whose tumor cells express *MAGE-1* were injected with autologous blood monocytes pulsed with a *MAGE-1* encoded peptide.[47,48] In another pilot clinical study, the MAGE-3.A1 peptide was administered to 39 *MAGE-3* positive, melanoma tumor-bearing HLA-A1 patients. Of the 25 patients who received the three monthly injections, seven showed objective regressions of metastases. However, no CTL response against the MAGE-3.A1 antigen could be detected in their blood.[49,50] In a recent study, melanoma patients were immunized, by injection in non-invaded lymph nodes, with autologous dendritic cells pulsed with either tumor lysates or a combination of synthetic peptides derived from melanocyte differentiation or MAGE proteins. Most of the patients developed a specific DTH reaction, and several showed clinical responses.[51]

3.2. Mucins

By stimulating draining lymph node cells of pancreatic cancer patients with an allogeneic pancreatic tumor cell line, $CD8^+$ CTL with α-β T cell receptor were obtained that lysed most pancreatic and breast tumor cell lines.[52,53] Remarkably, however, they functioned without MHC restriction. Pancreatic and breast mucins are both encoded by gene *MUC-1*.[54,55] These molecules have a glycosylated protein core with a repeated motif of 20 amino acids. In breast and pancreatic tumor cells, mucin is underglycosylated, and this seems to unveil epitopes of the repeated peptide motif that can be recognized both by specific antibodies and also, remarkably, by CTL. Cells transfected with a construct containing only two tandem repeats of the mucin motif stimulate CTL very effectively, presumably because they are not glycosylated at all.[56]

Immunization against unglycosylated mucins offers a very interesting prospect in immunotherapy. In a pilot clinical trial, a 105 amino-acid peptide, containing five repeated mucin epitopes, was injected with BCG to 67 breast, colon, and pancreatic cancer patients. No clinical benefit was reported. However, an increase in the mucin-specific CTL frequency was seen in the blood lymphocytes of most patients.[57]

3.3. Antigens Resulting from Mutations

Besides gene activation, gene mutation is another mechanism that can generate new potential rejection antigens. Mutations produce either new peptides that bind to class I molecules or new epitopes from peptides that were already capable of binding. Gene mutation can also lead to the expression of mouse tumor antigens presented to $CD4^+$ T lymphocytes by MHC class II molecules. A point mutation in the gene encoding the ribosomal protein L9 modifies one amino acid of an antigenic peptide recognized by a mouse CTL on an UV-induced fibrosarcoma.[58] The great majority of mouse tumor antigens identified to date result from mutations. This may be because many of the mouse tumors used for study were induced with carcinogens or radiation. However, antigens of this category have also been found on spontaneous mouse tumors and on human tumors (Table I). Mutations affecting human genes *CDK4* and β-catenin may be involved in oncogenesis, since they were found in several independent tumors and have a demonstrated effect on the activity of the encoded proteins.[59–61] The *CDK4* mutation prevents the protein from binding to its inhibitor, p16. This appears to alter the regulation of the cell cycle, favoring uncontrolled growth of the tumor cells. The β-catenin mutation results in stabilization of the protein, which favors the constitutive formation of complexes with transcription factors, such as Lef-1. Constitutive β-catenin/Lef-1 complexes may result in persistent transactivation of as yet unidentified target genes, thereby stimulating cell proliferation or inhibiting apoptosis.[61,62] Another mutation which may antagonize apoptosis was identified recently with CTL specific for a human squamous-cell carcinoma.[63] The antigen is encoded by a mutated form of the *CASP-8* gene, which codes for a protease called caspase-8, FLICE or MACHα1. This protease is required for induction of apoptosis through the Fas and TNFR1 receptors, and the ability of the altered protein to trigger apoptosis appears to be reduced relative to the normal caspase-8. Combined with the observations made with the CDK4 and β-catenin mutations, this suggests that a number of point mutations generating tumor antigens also play a role in tumoral transformation or progression.

Again, human antigens of this category were initially found on melanoma, but, recently, similar antigens have also been identified on other tumor types (Table I). Myeloid leukemias often express chimeric protein bcr-abl, which results from a t(9;22) chromosomal translocation. $CD4^+$ T cells, raised *in vitro* against a peptide centered on the fusion region of bcr-abl, were recently found to recognize HLA-DR4 leukemic blasts expressing bcr-abl.[64] This indicates that another source of tumor-specific antigens are the chimeric proteins encoded at the sites of chromosomal translocations. Since identical translocations occur frequently in certain types of leukemias, these antigens should be shared by many leukemias.

Although the tumor-specificity of antigens resulting from mutations is optimal, the fact that they are often unique prevents their use as components of generalized cancer vaccines. It is also somewhat surprising that none of the frequently mutated oncogenes or tumor-supressor genes, such as *ras* or *p53*, have been picked up as targets of CTL raised against tumor cells. However, synthetic ras and p53 mutant peptides have been used to generate CTL or rejection responses against tumors carrying the corresponding mutant genes.[65-68]

3.4. Differentiation Antigens

A large number of CTL directed against human melanoma were found to recognize not only a majority of melanomas but also normal melanocytes. The notion that such CTL recognize melanocyte differentiation antigens was confirmed when the antigens were identified at the molecular level (Table I). A good example is tyrosinase, a melanocyte protein that gives rise to different peptides that are presented either by class I or class II molecules to CD8 or CD4 T cells.[69,70] The role of CTL against such antigens in melanoma rejection is not clear, but is supported by the reported association of vitiligo with prolonged survival and spontaneous regression of melanoma.[71] In a pilot clinical study, peptides derived from the melanocytic lineage differentiation protein, gp100, were used to immunize HLA-A2 melanoma patients, with or without subsequent systemic IL-2. Few responses were observed in the group receiving peptide only, whereas 13 out of 31 patients having received peptide plus IL-2 displayed objective clinical responses.[72] A number of patients have also been injected with peptides corresponding to melanoma differentiation antigens mixed with incomplete Freund's adjuvant or GM-CSF. In these studies, peptide-specific CTL were obtained *in vitro* by restimulating blood lymphocytes from these patients with peptide-pulsed cells.[73-75] Objective tumor regressions were documented in a few patients.

Melanocytes are also present in the choroid layer of the retina, therefore the potential ophthalmic toxicity of active immunotherapy targeted against these antigens needs to be evaluated very carefully. Similar concerns about autoimmune side-effects apply to carcinoembryonic antigen (CEA), an oncofoetal protein expressed in the normal colon epithelium and in most gut carcinomas, that was shown to contain a peptide epitope recognized by T cells from immunized patients.[76]

If the tissue expressing the differentiation antigen is expendable and is resected in its entirety by the surgeon in the course of therapy, then the differentiation antigen corresponding to this tissue may be used for immunotherapy without risk from autoimmunity. This applies to the prostate-specific antigen (PSA), a glycoprotein of the prostatic epithelium. Two distinct peptide epitopes presented by HLA-A2 to specific CTL were recently described and may prove to be very useful for T cell immunotherapy in prostate cancer patients who have undergone prostatectomy, where the only PSA-expressing cells would be in metastatic deposits.[77]

A particular type of tumor-specific antigen is found in B-cell malignancies such as follicular lymphoma and multiple myeloma. All cells from the malignant clone produce the same immunoglobulin, which contains idiotypic determinants that may induce immune responses specific for the tumor cells. Clinical immunization trials require the preparation of an individual vaccine for each patient enrolled. Such a trial was performed on 41 patients with non-Hodgkin's B-cell lymphoma, who received their monoclonal immunoglobulin coupled to keyhole limpet hemocyanin, on several occasions. Half of the patients developed a specific CTL response. A significantly longer disease-free interval and overall survival was observed in this group of patients, as compared with the group without a detectable immune response.[78]

3.5. Overexpressed Tumor Antigens

A number of CTL raised against autologous tumor cells have been found to recognize antigens encoded by genes expressed in both normal and tumoral tissues, although the CTL appeared to lyse tumor cells specifically.

HER-2/neu is found at high levels in about 30% of breast and ovarian carcinomas. A peptide derived from HER-2/neu was found to be the target of lymphocytes infiltrating some HLA-A2 ovarian carcinomas.[79,80] Similarly, a CTL clone raised by *in vitro* priming against a p53 wild-type peptide was found to lyse HLA-A2 tumor cells overexpressing p53.[81,82] The concentration of p53 protein is increased in many tumor cells, principally due to an extended half-life of the mutant forms compared to the wild-type.

It is very difficult to know whether immunization against antigens such as HER-2/neu or p53 will lead to autoimmune damage in normal tissues, where the same antigens are expressed at low levels. Recognition of CTL epitopes has been shown to require expression of the encoding gene above a certain threshold.[83] In some cases, this threshold may not be reached in normal tissues. This is not easy to test, however, due to the scarcity of cell lines derived from normal tissues. In other cases, the low expression measured at the RNA level in a given type of normal tissue may in fact reflect high expression by only a subtype of cells which would then exceed the threshold. Two studies performed *in vivo* in mouse models suggest that adoptive transfer of CTL directed against a tumor antigen that is also present at a lower level in normal tissues, does not cause autoimmune toxicity. One study involved transgenic mice expressing a retroviral antigen in all their tissues.[84] The other used CTL directed against a wild-type p53 peptide, which were isolated from p53 knock-out mice and transferred into normal mice.[85] In another study, mice immunized against HER-2/neu peptides developed CTL and tumor rejection responses without apparent toxicity to normal tissues expressing HER-2/neu.[86]

3.6. Viral Antigens

Antigens derived from oncogenic viruses constitute another category of potentially useful tumor antigens. A number of viral antigens have been studied in detail on virally-induced mouse tumors and shown to be relevant for tumor rejection.[87–91] In humans, the best example is oncoprotein E7 of human papilloma virus 16, which is present in most cervical carcinomas. Tumor-specific CTL have been elicited by *in vitro* sensitization with E7 peptides presented by HLA-A2.[92]

Eight patients with late stage cervical cancer were vaccinated with a recombinant vaccinia virus expressing the E6 and E7 proteins of HPV16 and 18. No clinical side-effects were observed, nor any clinical benefit reported. Each patient mounted an anti-vaccinia antibody response. An anti-HPV antibody response was measured in three patients and an HPV-specific CTL response in only one patient.[93]

4. PERSPECTIVES WITH DEFINED ANTIGENS

In our opinion, the results of many years of anti-cancer vaccination studies can be summarized by stating that success has been too rare to be convincing, but too frequent to give up hope. It is very tempting to believe that a number of reported favorable clinical outcomes of patients injected with autologous, xenogenized or allogenic tumor cells are indeed due to an anti-tumor immune response. The identification of tumor rejection antigens and their encoding genes should reveal new possibilities to decrease the number of unknown factors in immunization studies. The expression of genes such as *MAGE* might be confirmed readily by reverse transcriptase-PCR analysis of a very small tumor sample and, moreover, the patients will be typed for HLA expression. Purified protein or peptides represent a major agent for immunization, because they can be produced on a large scale. It is possible that success will depend on more effective adjuvant formulations. Another use of protein and peptide could be the pulsing of autologous antigen presenting cells such as macrophages or dendritic cells. Such cells could then be reinfused to the patient to present the antigen. This may also be achieved with recombinant vaccinia or adenovirus constructs.[94–96] Adenovirus constructs were found quite effective in the production of CTL responses against tumor antigens in mice.[97]

Even if strong CTL responses are generated, there may be limitations to the clinical efficacy and usefulness of immunization. The anti-tumoral effectiveness of CTL in patients vaccinated with antigenic peptides may be limited by the lack of tumor-reactive CD4 T helper lymphocytes. The identification of such epitopes presented by HLA class II molecules will provide us with novel reagents for use in the development of multiple peptide vaccines and the monitoring of patients' immune responses.

This approach may be challenged by antigen-loss variants or HLA-loss variants of the tumor that escape rejection, that may appear in some patients. Antigen-loss variants may not present a major problem because most tumors seem to carry several different tumor rejection antigens. Therefore, if immunity can be raised against several of these antigens, the loss of one of them should not result in tumor escape. A more serious problem would be posed by the occurrence of HLA-loss variants. Loss of HLA expression has been observed in many instances for several types of tumors.[98–102] Here, two situations should be distinguished. In the first, a few HLA-loss variants are present among antigenic tumor cells. This situation could probably be handled by the immune system. Activated CTL not only lyse their target but also produce cytokines, such as TNF, that can kill nearby tumor cells that have lost the antigen. One interesting example involves a melanoma tumor having lost the expression of most of its HLA alleles, except HLA-A24. An antigen presented to autologous CTL on HLA-24 is encoded by a new gene, named *PRAME*, that is expressed in some normal tissues, albeit at low levels, and is overexpressed in many tumors (Table II). The anti-PRAME CTL does not recognize cells of another melanoma metastasis from the same patient, even though these cells carry the PRAME antigen. The reason for this discrepancy is that the latter tumor cells express all HLA class I molecules, and that the CTL bears natural killer inhibitory receptors (KIR) that prevent lysis of HLA class I positive cells. It is possible that such CTL expressing KIR play a role in the defense against tumor cells that have a partial loss of HLA class I expression.[103] But a very different situation might exist where large tumor masses contain cells that have all lost HLA expression, possibly because of selection brought about by a previous spontaneous immune response. It is not clear whether NK cells might be able to eliminate such tumors.

Numerous other clinical trials are being performed. For ethical reasons, these trials are currently restricted to metastatic patients with a high tumor burden. The fact that tumor regressions are occasionally observed under those conditions is remarkable; and it is promising for the success of immunotherapy in less advanced cancer patients.

Tools are being developed to measure with higher sensitivity the induction of CTL responses in immunized patients.[104] Real questions are now emerging concerning the value of the classical limiting dilution analysis (LDA). The number of precursors may be 100 times lower than those detected by other methods, such as the ones described below.[105] In the enzyme-linked immunospot (ELISPOT) assay, TNF-α or IFN-γ secreted by single T cells in response to antigen are immunochemically visualized as spots.[106,107] An alternative approach is to measure the number of interferon-producing T cells after stimulation with the antigen in the presence of Brefeldin A to avoid the release of the cytokine, permeabilization, and intracellular staining with an antibody directed against IFN-γ. These approaches bypass the need for several rounds of stimulation *in vitro*. A third possibility is the direct labelling of T cell receptors, specific for a given antigen, with soluble HLA tetramers presenting the antigenic peptide.[108] This approach was used successfully to detect anti-HIV CTL present at very high frequencies (around 5×10^{-3}) in the blood of diseased patients. If it also proves successful in the case of cancer patients who most probably have much lower frequencies, it may be a very convenient tool to monitor anti-tumor CTL.[109] An unique advantage of this method is that it may be used as a first purification step for the CTLp: blood cells may be labelled with peptide-HLA tetramer and sorted either by flow cytometry or by magnetic sorting. The enriched population may then be tested for production of cytokines in an ELISPOT assay, or stimulated *in vitro* with peptides and tested for their specificity. This combination of techniques may achieve low background, and high sensitivity and specificity.

A better understanding of the requirements for an effective anti-tumor response will probably not only rely on the clinical results, but also on the careful study of the tumor cells and of the immune responses of the immunized patients.

ACKNOWLEDGMENTS

We are grateful to Nathalie Krack and Simon Mapp for their very precious help in the preparation of the manuscript and to Rosalie Luiten for critical reading. Pascal Chaux was supported by the Association pour la Recherche sur le Cancer (ARC), France, and by a postdoctoral fellowship from the "Training and mobility of the researchers" program of the European Commission.

REFERENCES

1. Gross L: Intradermal immunization of C3H mice against a sarcoma that originated in an animal of the same line. Cancer Res. 3:326–333, 1943
2. Prehn RT, Main JM: Immunity to methylcholanthrene-induced sarcomas. J. Natl. Cancer Inst. 18:769–778, 1957
3. Burnet FM: Immunological surveillance in neoplasia. Transplant. Rev. 7:3–25, 1971
4. Rouse BT, RÖllinghoff M, Warner NL: Anti-ϑ serum-induced suppression of the cellular transfer of tumour-specific immunity to a syngeneic plasma cell tumour. Nature New Biol. 238:116–117, 1972
5. Kripke ML: Antigenicity of murine skin tumors induced by ultraviolet light. J. Natl. Cancer Inst. 53:1333–1336, 1974
6. Uyttenhove C, Maryanski J, Boon T: Escape of mouse mastocytoma P815 after nearly complete rejection is due to antigen-loss variants rather than immunosuppression. J. Exp. Med. 157:1040–1052, 1983
7. Boon T, Kellermann O: Rejection by syngeneic mice of cell variants obtained by mutagenesis of a malignant teratocarcinoma cell line. Proc. Natl. Acad. Sci. USA 74:272–275, 1977
8. Boon T, Van Pel A: Teratocarcinoma cell variants rejected by syngeneic mice : protection of mice immunized with these variants against other variants and against the original malignant cell line. Proc. Natl. Acad. Sci. USA 75:1519–1523, 1978
9. Livingston PO, Shiku H, Bean MA, Pinsky CM, Oettgen HF, Old LJ: Cell-mediated cytotoxicity for cultured autologous melanoma cells. Int. J. Cancer 24:34–44, 1979
10. Mukherji B, MacAlister TJ: Clonal analysis of cytotoxic T cell response against human melanoma. J. Exp. Med. 158:240–245, 1983
11. Knuth A, Danowski B, Oettgen HF, Old L: T-cell mediated cytotoxicity against autologous malignant melanoma : analysis with interleukin-2-dependent T-cell cultures. Proc. Natl. Acad. Sci. USA 81:3511–3515, 1984
12. Hérin M, Lemoine C, Weynants P, Vessière F, Van Pel A, Knuth A, Devos R, Boon T: Production of stable cytolytic T-cell clones directed against autologous human melanoma. Int. J. Cancer 39:390–396, 1987
13. Anichini A, Fossati G, Parmiani G: Clonal analysis of the cytolytic T-cell response to human tumors. Immunol. Today 8:385–389, 1987
14. Slovin SF, Lackman RD, Ferrone S, Kielly PE, Mastrangelo MJ: Cellular immune response to human sarcomas: cytotoxic T cell clones reactive with autologous sarcomas. I. Development, phenotype, and specificity. J. Immunol. 137:3042–3048, 1986
15. Wölfel T, Herr W, Coulie P, Schmitt U, Meyer zum Büschenfelde K-H, Knuth A: Lysis of human pancreatic adenocarcinoma cells by autologous HLA-class I-restricted cytolytic T-lymphocyte (CTL) clones. Int. J. Cancer 54:636–644, 1993
16. Yasumura S, Hirabayashi H, Schwartz DR, Toso JF, Johnson JT, Herberman RB, Whiteside TL: Human cytotoxic T-cell lines with restricted specificity for squamous cell carcinoma of the head and neck. Cancer Res. 53:1461–1468, 1993
17. Kannagi M, Sugamura K, Sato H, Okochi K, Uchino H, Hinuma Y: Establishment of human cytotoxic T cell lines specific for human adult T cell leukemia virus-bearing cells. J. Immunol. 130:2942–2946, 1983
18. Mitsuya H, Matis LA, Megson M, Bunn PA, Murray C, Mann DL, Gallo RC, Broder S: Generation of an HLA-restricted cytotoxic T cell line reactive against cultured tumor cells from a patient infected with human T cell leukemia/lymphoma virus. J. Exp. Med. 158:994–999, 1983
19. Gavioli R, De Campos-Lima P-O, Kurilla MG, Kieff E, Klein G, Masucci MG: Recognition of the EBV encoded nuclear antigens EBNA-4 and EBNA-6 by HLA-A11-restricted cytotoxic T-lymphocytes. Implications for down-regulation of HLA-A11 in Burkitt's lymphoma. Proc. Natl. Acad. Sci. USA 89:5862–5866, 1992
20. Khanna R, Burrows SR, Kurilla MG, Jacob CA, Misko IS, Sculley TB, Kieff E, Moss DJ: Localization of Epstein-Barr virus cytotoxic T cell epitopes using recombinant vaccinia: implications for vaccine development. J. Exp. Med. 176:169–176, 1992
21. Murray RJ, Kurilla MG, Brooks JM, Thomas WA, Rowe M, Kieff E, Rickinson AB: Identification of target antigens for the human cytotoxic T cell response to Epstein-Barr virus (EBV): implications for the immune control of EBV-positive malignancies. J. Exp. Med. 176:157–168, 1992
22. Itoh K, Tilden AB, Balch CM: Interleukin-2 activation of cytotoxic T-lymphocytes infiltrating into human metastatic melanomas. Cancer Res. 46:3011–3017, 1986

23. Muul LM, Spiess PJ, Director EP, Rosenberg SA: Identification of specific cytolytic immune responses against autologous tumor in humans bearing malignant melanoma. J. Immunol. 138:989–995, 1987
24. Itoh K, Platsoucas CD, Balch CM: Autologous tumor-specific cytotoxic T lymphocytes in the infiltrate of human metastatic melanomas. Activation by interleukin 2 and autologous tumor cells, and involvement of the T cell receptor. J. Exp. Med. 168:1419–1441, 1988
25. Groettrup M, Soza A, Kuckerkorn U, Kloetzel P-M: Peptide antigen production by the proteasome: complexity provides efficiency. Immunol. Today 17:429–435, 1996
26. Cerundolo V, Benham A, Braud V, Mukherjee S, Gould K, Macino B, Neefjes J, Townsend A: The proteasome-specific inhibitor lactacystin blocks presentation of cytotoxic T lymphocyte epitopes in human and murine cells. Eur. J. Immunol. 27:336–341, 1997
27. Kelly A, Powis SH, Kerr L-A, Mockridge I, Elliott T, Bastin J, Uchanska-Ziegler B, Ziegler A, Trowsdale J, Townsend A: Assembly and function of the tWo ABC transporter proteins encoded in the human major histocompatibility complex. Nature 355:641–644, 1992
28. De Plaen E, Lurquin C, Brichard V, van der Bruggen P, Renauld J-C, Coulie P, Szikora J-P, Wölfel T, Van Pel A, Boon T: Cloning of genes coding for antigens recognized by cytolytic T lymphocytes., in Lefkovits I (ed): The Immunology Methods Manual, MHC ligands, and peptide binding, Academic Press Ltd, 1997, p 692–718
29. Cox AL, Skipper J, Chen Y, Henderson RA, Darrow TL, Shabanowitz J, Engelhard VH, Hunt DF, Slingluff CL, Jr.: Identification of a peptide recognized by five melanoma-specific human cytotoxic T cell lines. Science 264:716–719, 1994
30. Skipper JCA, Hendrickson RC, Gulden PH, Brichard V, Van Pel A, Chen Y, Shabanowitz J, Wölfel T, Slingluff CL, Jr, Boon T, Hunt DF, Engelhard VH: An HLA-A2-restricted tyrosinase antigen on melanoma cells results from posttranslational modification and suggests a novel pathway for processing of membrane proteins. J. Exp. Med. 183:527–534, 1996
31. Jung S, Schluesener HJ: Human T lymphocytes recognize a peptide of single point-mutated, oncogenic ras proteins. J. Exp. Med. 173:273–276, 1991
32. Celis E, Tsai V, Crimi C, DeMars R, WentWorth PA, Chesnut RW, Grey HM, Sette A, Serra HM: Induction of anti-tumor cytotoxic T lymphocytes in normal humans using primary cultures and synthetic peptide epitopes. Proc. Natl. Acad. Sci. USA 91:2105–2109, 1994
33. Dahl AM, Beverley PCL, Stauss HJ: A synthetic peptide derived from the tumor-associated protein mdm2 can stimulate autoreactive, high avidity cytotoxic T lymphocytes that recognize naturally processed protein. J. Immunol. 157:239–246, 1996
34. Purcell AW, Chen W, Ede NJ, Gorman JJ, Fecondo JV, Jackson DC, Zhao Y, McCluskey J: Avoidance of self-reactivity results in skewed CTL responses to rare components of synthetic immunogens. J. Immunol. 160:1085–1090, 1998
35. Van den Eynde B, Lethé B, Van Pel A, De Plaen E, Boon T: The gene coding for a major tumor rejection antigen of tumor P815 is identical to the normal gene of syngeneic DBA/2 mice. J. Exp. Med. 173:1373–1384, 1991
36. De Plaen E, Arden K, Traversari C, Gaforio JJ, Szikora J-P, De Smet C, Brasseur F, van der Bruggen P, Lethé B, Lurquin C, Brasseur R, Chomez P, De Backer O, Cavenee W, Boon T: Structure, chromosomal localization, and expression of twelve genes of the MAGE family. Immunogenetics 40:360–369, 1994
37. De Smet C, De Backer O, Faraoni I, Lurquin C, Brasseur F, Boon T: The activation of human gene MAGE-1 in tumor cells is correlated with genome-wide demethylation. Proc. Natl. Acad. Sci. USA 93:7149–7153, 1996
38. Serrano A, Garcia A, Abril E, Garrido F, Ruiz-Cabello F: Methylated CpG points identified within MAGE-1 promoter are involved in gene repression. Int. J. Cancer 68:464–470, 1996
39. Takahashi K, Shichijo S, Noguchi M, Hirohata M, Itoh K: Identification of MAGE-1 and MAGE-4 proteins in spermatogonia and primary spermatocytes of testis. Cancer Res. 55:3478–3482, 1995
40. Haas GG, Jr., D'Cruz OJ, De Bault LE: Distribution of human leukocyte antigen-ABC and -D/DR antigens in the unfixed human testis. Am. J. Reprod. Immunol. Microbiol. 18:47–51, 1988
41. Uyttenhove C, Godfraind C, Lethé B, Amar-Costesec A, Renauld J-C, Gajewski TF, Duffour T, Warnier G, Boon T, Van den Eynde BJ: The expression of mouse gene *P1A* in testis does not prevent safe induction of cytolytic T cells against a P1A-encoded tumor antigen. Int. J. Cancer 70:349–356, 1997
42. Sahin U, Türeci O, Pfreundschuh M: Serological identification of human tumor antigens. Curr. Opin. Immunol. 9:709–716, 1997

43. Chen Y-T, Scanlan MJ, Sahin U, Türeci Ö, Gure AO, Tsang S, Williamson B, Stockert E, Pfreundschuh M, Old LJ: A testicular antigen aberrantly expressed in human cancers detected by autologous antibody screening. Proc. Natl. Acad. Sci. USA 94:1914–1918, 1997
44. Türeci Ö, Sahin U, Schobert I, Koslowski M, Schmitt H, Schild H-J, Stenner F, Seitz G, Rammensee H-G, Pfreundschuh M: The *SSX-2* gene, which is involved in the t(X;18) translocation of synovial sarcomas, codes for the human tumor antigen HOM-MEL-40. Cancer Res. 56:4766–4772, 1996
45. Lucas S, De Smet C, Arden KC, Viars CS, Lethé B, Lurquin C, Boon T: Identification of a new MAGE gene with tumor-specific expression by representational difference analysis. Cancer Res. 58:743–752, 1998
46. Lethé B, Lucas S, Michaux L, De Smet C, Godelaine D, Serrano A, De Plaen E, Boon T: LAGE-1 a new gene with tumor specificity. Int. J. Cancer 76:903–908, 1998
47. Mukherji B, Chakraborty NG, Yamasaki S, Okino T, Yamase H, Sporn JR, Kurtzman SK, Ergin MT, Ozols J, Meehan J, Mauri F: Induction of antigen-specific cytolytic T cells in situ in human melanoma by immunization with synthetic peptide-pulsed autologous antigen presenting cells. Proc. Natl. Acad. Sci. USA 92:8078–8082, 1995
48. Hu X, Chakraborty NG, Sporn JR, Kurtzman SH, Ergin MT, Mukherji B: Enhancement of cytolytic T lymphocyte precursor frequency in melanoma patients following immunization with the MAGE-1 peptide loaded antigen presenting cell-based vaccine. Cancer Res. 56:2479–2483, 1996
49. Marchand M, Weynants P, Rankin E, Arienti F, Belli F, Parmiani G, Cascinelli N, Bourlond A, Vanwijck R, Humblet Y, Canon J-L, Laurent C, Naeyaert J-M, Plagne R, Deraemaeker R, Knuth A, Jäger E, Brasseur F, Herman J, Coulie PG, Boon T: Tumor regression responses in melanoma patients treated with a peptide encoded by gene MAGE-3. Int. J. Cancer 63:883–885, 1995
50. Marchand M, van Baren N, Weynants P, Brichard V, Dréno B, Tessier M-H, Rankin E, Parmiani G, Arienti F, Humblet Y, Bourlond A, Vanwijck R, Liénard D, Beauduin M, Dietrich P-Y, Russo V, Kerger J, Masucci G, Jäger E, De Greve J, Atzpodien J, Brasseur F, Coulie PG, Van der Bruggen P, Boon T: Tumor regressions observed in patients with metastatic melanoma treated with an antigenic peptide encoded by gene MAGE-3 and presented by HLA-A1. Int. J. Cancer 80:219–230, 1999
51. Nestle FO, Alijagic S, Gilliet M, Sun Y, Grabbe S, Dummer R, Burg G, Schadendorf D: Vaccination of melanoma patients with peptide- or tumor lysate-pulsed dendritic cells. Nature Medicine 4:328–332, 1998
52. Barnd DL, Kerr LA, Metzgar RS, Finn OJ: Human tumor-specific cytotoxic T cell lines generated from tumor-draining lymph node infiltrate. Transplant. Proc. 20:339–341, 1988
53. Barnd DL, Lan MS, Metzgar RS, Finn OJ: Specific, major histocompatibility complex-unrestricted recognition of tumor-associated mucins by human cytotoxic T cells. Proc. Natl. Acad. Sci. USA 86:7159–7163, 1989
54. Gendler S, Taylor-Papadimitriou J, Duhig T, Rothbard J, Burchell J: A highly immunogenic region of a human polymorphic epithelial mucin expressed by carcinomas is made up of tandem repeats. J. Biol. Chem. 263:12820–12823, 1988
55. Lancaster CA, Peat N, Duhig T, Wilson D, Taylor-Papadimitriou J, Gendler SJ: Structure and expression of the human polymorphic epithelial mucin gene: an expressed VNTR unit. Biochem. Biophys. Res. Commun. 173:1019–1029, 1990
56. Magarian-Blander J, Domenech N, Finn OJ: Specific and effective T-cell recognition of cells transfected with a truncated human mucin cDNA. Ann. N.Y. Acad. Sci. 690:231–243, 1993
57. Barratt-Boyes SM: Making the most of mucin: a novel target for tumor immunotherapy. Cancer Immunol Immunother 43:142–151, 1996
58. Monach PA, Meredith SC, Siegel CT, Schreiber H: A unique tumor antigen produced by a single amino acid substitution. Immunity 2:45–59, 1995
59. Wölfel T, Hauer M, Schneider J, Serrano M, Wölfel C, Klehmann-Hieb E, De Plaen E, Hankeln T, Meyer zum Büschenfelde K-H, Beach D: A p16^{INK4a}-insensitive CDK4 mutant targeted by cytolytic T lymphocytes in a human melanoma. Science 269:1281–1284, 1995
60. Robbins PF, El-Gamil M, Li YF, Kawakami Y, Loftus D, Appella E, Rosenberg SA: A mutated β-catenin gene encodes a melanoma-specific antigen recognized by tumor infiltrating lymphocytes. J. Exp. Med. 183:1185–1192, 1996
61. Rubinfeld B, Robbins P, El-Gamil M, Albert I, Porfiri E, Polakis P: Stabilization of β-catenin by genetic defects in melanoma cell lines. Science 275:1790–1792, 1997
62. Peifer M: β-catenin as oncogene: the smoking gun. Science 275:1752–1753, 1997
63. Mandruzzato S, Brasseur F, Andry G, Boon T, van der Bruggen P: A CASP-8 mutation recognized by cytolytic T lymphocytes on a human head and neck carcinoma. J. Exp. Med. 186:785–793, 1997

64. ten Bosch GJA, Joosten AM, Kessler JH, Melief CJM, Leeksma OC: Recognition of BCR-ABL positive leukemic blasts by human CD4+ T cells elicited by primary in vitro immunization with a BCR-ABL breakpoint peptide. Blood 88:3522–3527, 1996
65. Skipper JCA, Stauss HJ: Identification of tWo cytotoxic T lymphocyte-recognized epitopes in the Ras protein. J. Exp. Med. 177:1493–1498, 1993
66. Peace DJ, Smith JW, Chen W, You S-G, Cosand WL, Blake J, Cheever MA: Lysis of ras oncogene-transformed cells by specific cytotoxic T lymphocytes elicited by primary in vitro immunization with mutated ras peptide. J. Exp. Med. 179:473–479, 1994
67. Noguchi Y, Chen Y-T, Old LJ: A mouse mutant p53 product recognized by CD4+ and CD8+ T cells. Proc. Natl. Acad. Sci. USA 91:3171–3175, 1994
68. Abrams SI, Stanziale SF, Lunin SD, Zaremba S, Schlom J: Identification of overlapping epitopes in mutant ras oncogene peptides that activate CD4+ and CD8+ T cell responses. Eur. J. Immunol. 26:435–443, 1996
69. Brichard V, Van Pel A, Wölfel T, Wölfel C, De Plaen E, Lethé B, Coulie P, Boon T: The tyrosinase gene codes for an antigen recognized by autologous cytolytic T lymphocytes on HLA-A2 melanomas. J. Exp. Med. 178:489–495, 1993
70. Topalian SL, Gonzales MI, Parkhurst M, Li YF, SouthWood S, Sette A, Rosenberg SA, Robbins PF: Melanoma-specific CD4+T cells recognize nonmutated HLA-DR-restricted tyrosinase epitopes. J. Exp. Med. 183:1965–1971, 1996
71. Rosenberg SA, White DE: Vitiligo in patients with melanoma: normal tissue antigens can be target for cancer immunotherapy. J. Immunother. 19:81–84, 1996
72. Rosenberg SA, Yang JC, Schwartzentruber DJ, Hwu P, Marincola FM, Topalian SL, Restifo NP, Dudley ME, Schwarz SL, Spiess PJ, Wunderlich JP, Parkhurst MR, Kawakami Y, Seipp CA, Einhorn JH, White DE: Immunologic and therapeutic evaluation of a synthetic peptide vaccine for the treatment of patients with metastatic melanoma. Nature Medicine 4:321–327, 1998
73. Salgaller ML, Marincola FM, Cormier JN, Rosenberg SA: Immunization against epitopes in the human melanoma antigen gp100 following patient immunization with synthetic peptides. Cancer Res. 56:4749–4757, 1996
74. Jager E, Ringhoffer M, Dienes HP, Arand M, Karbach J, Jager D, Ilsemann C, Hagedorn M, Oesch F, Knuth A: Granulocyte-macrophage-colony-stimulating factor enhances immune responses to melanoma-associated peptides in vivo. Int. J. Cancer 67:54–62, 1996
75. Cormier JN, Salgaller ML, Prevette T, Barracchini KC, Rivoltini L, Restifo NP, Rosenberg SA, Marincola FM: Enhancement of cellular immunity in melanoma patients immunized with a peptide from MART-1/Melan-A. Cancer J. Sci. Am. 3:37–44, 1997
76. Tsang KY, Zaremba S, Nieroda CA, Zhu MZ, Hamilton JM, Schlom J: Generation of human cytotoxic T cells specific for human carcinoembryonic antigen epitopes from patients immunized with recombinant vaccinia-CEA vaccine. J. Natl Cancer Inst 87:982–990, 1995
77. Correale P, Walmsley K, Nieroda C, Zaremba S, Zhu M, Schlom J, Tsang KY: In vitro generation of human cytotoxic T lymphocytes specific for peptides derived from prostate-specific antigen. J Natl Cancer Inst 89:293–300, 1997
78. Hsu FJ, Caspar CB, Czerwinski D, Kwak LW, Liles TM, Syrengelas A, Taidi-Laskowski B, Levy R: Tumor-specific idiotype vaccines in the treatment of patients with B-cell lymphoma. Long-term results of a clinical trial. Blood 89:3129–3135, 1997
79. Fisk B, Blevins TL, Wharton JT, Ioannides CG: Identification of an immunodominant peptide of HER-2/neu protooncogene recognized by ovarian tumor-specific cytotoxic T lymphocytes lines. J. Exp. Med. 181:2109–2117, 1995
80. Lustgarten J, Theobald M, Labadie C, LaFace D, Peterson P, Disis ML, Cheever MA, Sherman LA: Identification of Her-2/neu CTL epitopes using double transgenic mice expressing HLA-A2.1 and human CD.8. Hum. Immunol. 52:109–118, 1997
81. Röpke M, Hald J, Guldberg P, Zeuthen J, Norgaard L, Fugger L, Svejgaard A, Van Der Burg S, Nijman HW, Melief CJM: Spontaneous human squamous cell carcinomas are killed by a human cytotoxic T lymphocyte clone recognizing a wild-type p53-derived peptide. Proc. Natl. Acad. Sci. USA 93:14704–14707, 1996
82. Gnjatic S, Cai Z, Viguier M, Chouaib S, Guillet J-G, Choppin J: Accumulation of the p53 protein allows recognition by human CTL of a wild-type p53 epitope presented by breast carcinomas and melanomas. J. Immunol. 160:328–333, 1998
83. Lethé B, van der Bruggen P, Brasseur F, Boon T: MAGE-1 expression threshold for the lysis of melanoma cell lines by a specific CTL. Melanoma Research 7:S83–S88, 1997

84. Hu J, Kindsvogel W, Busby S, Bailey MC, Shi Y-Y, Greenberg PD: An evaluation of the potential to use tumor-associated antigens as targets for antitumor T cell therapy using transgenic mice expressing a retroviral tumor antigen in normal lymphoid tissues. J. Exp. Med. 177:1681–1690, 1993
85. Vierboom MPM, Nijman HW, Offringa R, Vandervoort EI, Vanhall T, Vandenbroek L, Fleuren GJ, Kenemans P, Kast WM, Melief CJM: Tumor eradication by wild-type p53-specific cytotoxic T lymphocytes. J. Exp. Med. 186:695–704, 1997
86. Nagata Y, Furugen R, Hiasa A, Ikeda H, Ohta N, Furukawa K, Nakamura H, Furukawa K, Kanematsu T, Shiku H: Peptides derived from a wild-type murine proto-oncogene c-erbB-2/HER2/neu can induce CTL and tumor suppression in syngeneic hosts. J. Immunol. 159:1336–1343, 1997
87. Klarnet JP, Kern DE, Okuno K, Holt C, Lilly F, Greenberg PD: FBL-reactive CD8+ cytotoxic and CD4+ helper T lymphocytes recognize distinct Friend murine leukemia virus-encoded antigens. J. Exp. Med. 169:457–467, 1989
88. Tanaka Y, Tevethia MJ, Kalderon D, Smith AE, Tevethia SS: Clustering of antigenic sites recognized by cytotoxic T lymphocyte clones in the amino terminal half of SV40 T antigen. Virology 162:427–436, 1988
89. Kast WM, Offringa R, Peters PJ, Voordouw AC, Meloen RH, van der Eb AJ, Melief CJM: Eradication of adenovirus E1-induced tumors by E1A specific cytotoxic T lymphocytes. Cell 59:603–614, 1989
90. Plata F, Langlade-Demoyen P, Abastado JP, Berbar T, Kourilsky P: Retrovirus antigens recognized by cytolytic T lymphocytes activate tumor rejection in vivo. Cell 48:231–240, 1987
91. Kast WM, Melief CJM: Fine peptide specificity of cytotoxic T lymphocytes directed against adenovirus-induced tumours and peptide-MHC binding. Int. J. Cancer Suppl. 6:90–94, 1991
92. Ressing ME, Sette A, Brandt RMP, Ruppert J, WentWorth PA, Hartman M, Oseroff C, Grey HM, Melief CJM, Kast WM: Human CTL epitopes encoded by human papillomavirus type 16 E6 and E7 identified through in vivo and in vitro immunogenicity studies of HLA-A*0201-binding peptides. J. Immunol. 154:5934–5943, 1995
93. Borysiewicz LK, Fiander A, Nimako M, Man S, Wilkinson GWG, Westmoreland D, Evans AS, Adams M, Stacey SN, Boursnell MEG, Rutherford E, Hickling JK, Inglis SC: A recombinant vaccinia virus encoding human papillomavirus types 16 and 18, E6 and E7 proteins as immunotherapy for cervical cancer. Lancet 347:1523–1527, 1996
94. Taylor J, Trimarchi C, Weinberg R, Languet B, Guillemin F, Desmettre P, Paoletti E: Efficacy studies on a canarypox-rabies recombinant virus. Vaccine 9:190–193, 1991
95. Rosenfeld MA, Yoshimura K, Trapnell BC, Yoneyama K, Rosenthal ER, Dalemans W, Fukayama M, Bargon J, Stier LE, Stratford-Perricaudet L, Perricaudet M, Guggino WB, Pavirani A, Lecocq J-P, Crystal RG: In vivo transfer of the human cystic fibrosis transmembrane conductance regulator gene to the airway epithelium. Cell 68:143–155, 1992
96. Taylor J, Weinberg R, Tartaglia J, Richardson C, Alkhatib G, Briedis D, Appel M, Norton E, Paoletti E: Nonreplicating viral vectors as potential vaccines: recombinant canarypox virus expressing measles virus fusion (F) and Hemagglutinin (HA) glycoproteins. Virology 187:321–328, 1992
97. Warnier G, Duffour M-T, Uyttenhove C, Perricaudet M, Lurquin C, Haddada H, Boon T: Induction of a cytolytic T cell response in mice with a recombinant adenovirus coding for tumor antigen P815A. Int. J. Cancer 67:303–310, 1996
98. Lopez-Nevot MA, Esteban F, Ferron A, Gutierrez J, Oliva MR, Romero C, Huelin C, Ruiz-Cabello F, Garrido F: HLA class I gene expression on human primary tumours and autologous metastases: demonstration of selective losses of HLA antigens on colorectal, gastric, and laryngeal carcinomas. Br. J. Cancer 59:221–226, 1989
99. Momburg F, Koch S: Selective loss of β2-microglobulin mRNA in human colon carcinoma. J. Exp. Med. 169:309–314, 1989
100. Momburg F, Ziegler A, Harpprecht J, Möller P, Moldenhauer G, Hämmerling GJ: Selective loss of HLA-A or HLA-B antigen expression in colon carcinoma. J. Immunol. 142:352–358, 1989
101. Smith MEF, Bodmer SG, Kelly AP, Trowsdale J, Kirkland SC, Bodmer WF: Variation in HLA expression on tumors: an escape from immune response. CSH Symp. Quant. Biol. 54:581–586, 1989
102. Cabrera T, Concha A, Ruiz-Cabello F, Garrido F: Loss of HLA heavy chain and β2-microglobulin in HLA negative tumours. Scand. J. Immunol. 34:147–152, 1991
103. Ikeda H, Lethé B, Lehmann F, Van Baren N, Baurain J-F, De Smet C, Chambost H, Vitale M, Moretta A, Boon T, Coulie PG: Characterization of an antigen that is recognized on a melanoma showing partial HLA loss by CTL expressing an NK inhibitory receptor. Immunity 6:199–208, 1997
104. McMichael AJ, O'Callaghan CA: A new look at T cells. J. Exp. Med. 187:1367–1371, 1998

105. Kuroda MJ, Schmitz JE, Barouch DH, Craiu A, Allen TM, Sette A, Watkins DI, Forman MA, Letvin NL: Analysis of gag-specific cytotoxic T lymphocytes in simian immunodeficiency virus-infected Rhesus monkeys by cell staining with a tetrameric major histocompatibility complex class I-peptide complex. J. Exp. Med. 187:1373–1381, 1998
106. Herr W, Schneider J, Lohse AW, Meyer zum Büschenfelde K-H, Wölfel T: Detection and quantification of blood-derived $CD8^+$ T lymphocytes secreting tumor necrosis factor alpha in response to HLA-A2.1-binding melanoma and viral peptide antigens. J. Immunol. Methods 191:131–142, 1996
107. Herr W, Linn B, Leister N, Wandel E, Meyer zum Büschenfelde KH, Wölfel T: The use of computer-assisted video image analysis for the quantification of $CD8^+$ T lymphocytes producing tumor necrosis factor alpha spots in response to peptide antigens. J. Immunol. Methods 203:141–152, 1997
108. Altman JD, Moss PAH, Goulder PJR, Barouch DH, McHeyzer-Williams MG, Bell JI, McMichael AJ, Davis MM: Phenotypic analysis of antigen-specific T lymphocytes. Science 274:94–96, 1996
109. Dunbar PR, Ogg GS, Chen J, Rust N, van der Bruggen P, Cerundolo V: Direct isolation, phenotyping and cloning of low-frequency antigen-specific cytotoxic T lymphocytes from peripheral blood. Curr. Biol. 8:413–416, 1998
110. Van den Eynde B, van der Bruggen P: T cell-defined tumor antigens. Curr. Opin. Immunol. 9:684–693, 1997
111. Katano M, Nakamura M, Morisaki T, Fujimoto K: Melanoma antigen-encoding gene-1 expression in invasive gastric carcinoma: correlation with stage of disease. J. Surg. Oncol. 64:195–201, 1997
112. Quillien V, Raoul JL, Heresbach D, Collet B, Toujas L, Brasseur F: Expression of MAGE genes in esophageal squamous-cell carcinoma. Anticancer Res. 17:387–391, 1997
113. Russo V, Dalerba P, Ricci A, Bonazzi C, Leone BE, Mangioni C, Allavena P, Bordignon C, Traversari C: MAGE, BAGE, and GAGE genes expression in fresh epithelial ovarian carcinomas. Int. J. Cancer 67:457–460, 1996
114. Corrias MV, Scaruffi P, Occhino M, De Bernardi B, Tonini GP, Pistoia V: Expression of MAGE-1, MAGE-3 and MART-1 genes in neuroblastoma. Int. J. Cancer 69:403–407, 1996
115. Shichijo S, Sagawa K, Brasseur F, Boon T, Itoh K: MAGE-1 gene is expressed in T-cell leukemia. Int. J. Cancer 65:709–710, 1996
116. Van Baren N, Chambost H, Ferrant A, Michaux L, Ikeda H, Millard I, Olive D, Boon T, Coulie PG: PRAME, a gene encoding an antigen recognized on a human melanoma by cytolytic T cells, is expressed in acute leukaemia cells. Br. J. Haematol. 102:1376–1379, 1998

23

MOLECULAR MONITORING OF NON-HODGKIN'S LYMPHOMAS

Alena Slavíčková, Veronika Ullmannová, Eva Benešová, and Pavel Klener

Charles University Prague, Fac. Med. 1
1st Department of Internal Medicine

ABSTRACT

The knowledge of a level of minimal residual disease (MRD) is important for further prognosis. In some cases a simple semi-quantitative approaches may contribute to a monitoring of disease activity instead of laborious precise techniques. We, using PCR-based methods, employ the limiting dilution methodology and the multiplex IgH/ras PCR in monitoring of non-Hodgkin's lymphomas (NHL). Applying the limiting dilution methodology as a simple dose-response assay for the t(14,18) translocation and for the clonal CDR3, minimal amounts of total cells initiating positive results are determined in stored diagnostic DNA samples on a single PCR run. Assuming that specific targets are diluted proportionally in dilution of total genomic DNA, samples showing lower critical concentrations of total DNA are considered as containing higher portion of cells possessing the marker and vice versa. In optimized multiplex IgH/Ras PCR co-amplifying clonal CDR3 rearrangement of immunoglobulin heavy chain gene (IgH) and the codon 61 of Hras 1 gene (ras), the amount of CDR3 product as the disease marker is related to the ras product as a standard marker of all cells. It appears that both approaches may provide clinically useful information on disease activity and treatment outcome.

The use of monitoring minimal residual disease for an assessment of treatment effectiveness and the detection of early signs of relapse in leukemia has been already reported.[1,2] In NHLs the +(14,18) is the most common chromosomal translocation and its qualitative detection is being frequently used in disease monitoring.[3–6] Further, in almost all patients with B-cell malignancies the clonal rearrangement of the

---

Corresponding author: A. Slavickova, Lab. Mol. Biol., 1st Internal Medicine, VFN, U Nemocnice 2, 128 08 Prague 2, Czech republic. Phone: +42 02/2496 3113, Fax: +42 02/29 79 32, e-mail: SLAVIKOV@mbox.cesnet.cz

Molecular Biology of Hematopoiesis 6, edited by Abraham *et al.*
Kluwer Academic/Plenum Publishers, New York, 1999.

immunoglobulin heavy chain gene (IgH), namely Complementarity Determining Region III (CDR3), is detectable and provides the disease marker as well.[7,8]

The use of limiting dilution methodology producing a calibration curve[9–11] appears to be an uncomplicated semi-quantifying approach[12] and a verification of clinical validity of this method seems to be meaningful. As for the second method, to relate the amplification of specific PCR targets to the amplification of a standard marker present in all cells might be a reliable way of quantifying, especially when it is done in a single test tube as a multiplex PCR. However, it is very important to optimizes multiplex PCR conditions to prevent any reaction from being preferential.

In this report we present results obtained on correlation of limiting dilution methodology with clinical status of HNL patient and the optimized conditions for the multiplex IgH/ras PCR enabling the detection of clonal CDR3 with good sensitivity.

MATERIALS AND METHODS

Patient Material and Diagnostic Samples

The diagnosis based on standard morphologic and immunohistochemical parameters and the updated Kiel classification are used for the typing. DNAs from bone marrow (BM) and peripheral blood (PB) from patients with B cell NHL at diagnosis and during the disease course as possible and from the *RAJI* cell line[13] are extracted by a variation of 6M NaCl technique[14] and stored at 4°C or at −20°C. Prior to PCR, the DNA concentration and integrity is estimated on 0.9% agarose gels stained with ethidium bromide, samples are diluted with sterile distilled water and denaturated.

PCR and Analyses of Amplified Products

Limiting Dilution Analyses. One—step PCR analyses of the t(14,18) and clonal CDR3 are performed in 30–33 cycles with 1–2U Rec Tac DNA polymerase/30–60 µl, 200 mM dNTPs, 0.2–2000 ng of purified DNA, and 1 µM primers of sequences as described in references.[4] The compared samples are examined in the same PCR with identical reaction mixture using identical 10% polyacrylamide (PAGE) or 3% agarose gel electrophoresis to decide on positive or negative responses to DNA amount in reaction. The specificity of t(14, 18) amplification product is verified by blot hybridisation with the digoxigenin 3′tailed *bcl-2* probe of sequence 5′GCC TGT AAC ACA GAC CC 3′. Using serial dilution of DNA in dose-response assays, the PCR sensitivities as critical amounts of total cells for preceding (N_p) and following (N_a) samples and the relative PCR—sensitivity as their ratio (N_p/N_a) are evaluated. Considering the samples with lower critical concentration of total DNA as containing higher portion of specific markers, the case of $N_p > N_a$ (e.g. the sample a containing higher portion of markers than the earlier sample p), relative PCR-sensitivity $N_p/N_a > 1$ is considered as an indicator of an increase of cells bearing the marker and *vice versa*.

Multiplex IgH/ras PCR

30–50 µl of GeneAmp^R (Perkin Elmer) reaction buffer with 0.4–0.25 mM dNTPs and primers (IDT, Coralville, IA) and the thermal profile as indicated in the Table 1 are employed. Hot starts are achieved with 2–5 U of the AmpliTaq Gold™ DNA polymerase activated 8 min at 94°C at the beginning of cycling. Amplification products are analyzed on 10% polyacrylamide gels stained with ethidium bromide. The amount of products is evaluated by gel densitometry using the Ultra Lum documentation system and the Gel-Pro *ANALYZER* 1-D gel analysis software (Media Cybernetics). In analyses of two or more samples the identical amounts of stored DNA are examined in one PCR run and an evaluation of the CDR3/ras product ratio is accomplished at exponential amplification phase and equal amount of ras products. An increase of the CDR3/ras is viewed as a sign of an increase of malignant cells and *vice versa.*

RESULTS

Correspondence of Limiting Dilution Analyses with Clinical Status

Clinically blinded laboratory estimations were completed with clinical data additionally in ten patients. The critical amounts of total cells initiating positive reaction (N) estimated on bone marrow samples varied in orders during disease and were much higher at clinically luckier periods. In all cases of regressed disease the values of critical amount of total cells obtained for late diagnostic samples were higher then those for preceding sample ($N_a > N_p$) indicating a decrease of cells bearing the marker ($N_p/N_a < 1$). At relapses or disease in progression the higher fraction of malignant cells resulted in lower total critical amount (Fig. 1).

Sensitivity of the Optimized Multiplex IgH/ras PCR

At reaction conditions summarised in the Table 1 the CDR3 products obtained with the Raji cell line in a simple IgH PCR and in the multiplex IgH/ras PCR were similar and comparable with ras products. These conditions were successful with almost all diagnostic samples.

Applications of Multiplex IgH/ras PCR

The product ratio CDR3/ras depends upon initial amount of DNA in amplification (Figs 2 and 3). Data obtained with stored diagnostic samples at identical ras yield appears to be in accordance with clinical findings, e.g. lower ratios are observed in clinically luckier periods. Results of both methods are in good accordance and the IgH/ras PCR is viewed as probably suitable for quantitative molecular monitoring of

Table 1. Conditions for the Multiplex IgH/ras PCR

Primers:		
JH:	5′ ACC TgA ggA gAC ggT gAC C 3′	0.6–1.0 μM
FR3A:	5′ ACA Cgg CC/Tg/C TgT ATT ACT gT 3′	1.2–2.0 μM
HA-61:	5′ AgA CgT gCC TgT Tgg ACA TC 3′	0.1–0.5 μM
HB-61:	5′ CgC ATg TAC Tgg TCC CgC AT 3′	0.1–0.5 μM

Reaction buffer: 2.5–1.0 mM $MgCl_2$, 2–6 U AmpliTaq Gold ™/30–50 μl, 500–250 μM dNTPs.

cycling: 94°C/10′, (94°C/45″, 58°C/90″, 72°C/120″)$_{30-33x}$, 72°C/7′

The conditions for the co-amplification of clonal CDR3 and codon 61 of H ras 1 gene which produce similar sensitivity of CDR3 and ras detection on amplification of DNA from the Raji cell line and seem to be successful with almost all diagnostic samples from NHL cases.

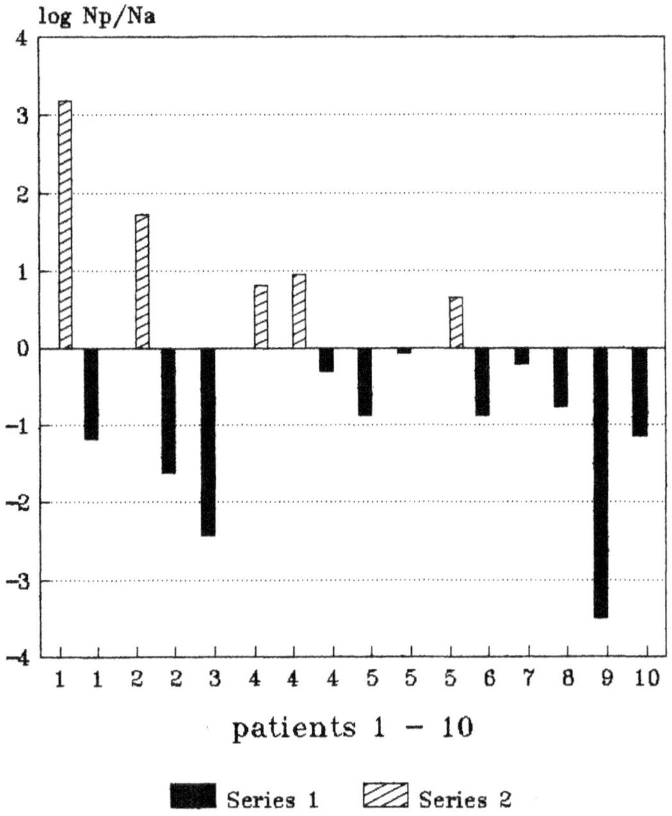

Figure 1. Correspondence of limiting dilution analyses with clinical status. The presented laboratory estimations were clinically blinded and completed with clinical data additionally in ten patients. The critical amounts of total cells initiating positive reaction (N) estimated on bone marrow samples varied in order of magnitude during disease and were much higher at clinically luckier periods. In all cases of regressed disease the values of critical amount of total cells obtained for late diagnostic samples were higher then those for preceding sample ($N_a > N_p$) indicating a decrease of cells bearing the marker ($N_p/N_a < 1$). At relapses or disease in progression the higher fraction of malignant cells resulted in lower total critical amount ($N_a < N_p$, $N_p/N_a > 1$). Series 1, regressions; series 2, progressions.

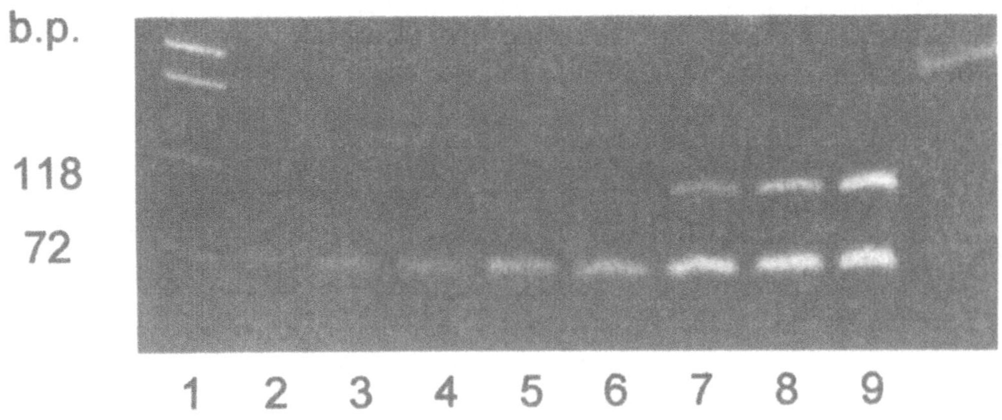

Figure 2. Limiting dilution of multiplex IgH/Ras PCR accomplished with patient's DNA. Lane 1: size marker—72, 118, 194, 234 base pairs from bellow. Lanes 2–9: 73 b. p.—Ras products, 116 b. p.—CDR3 products.

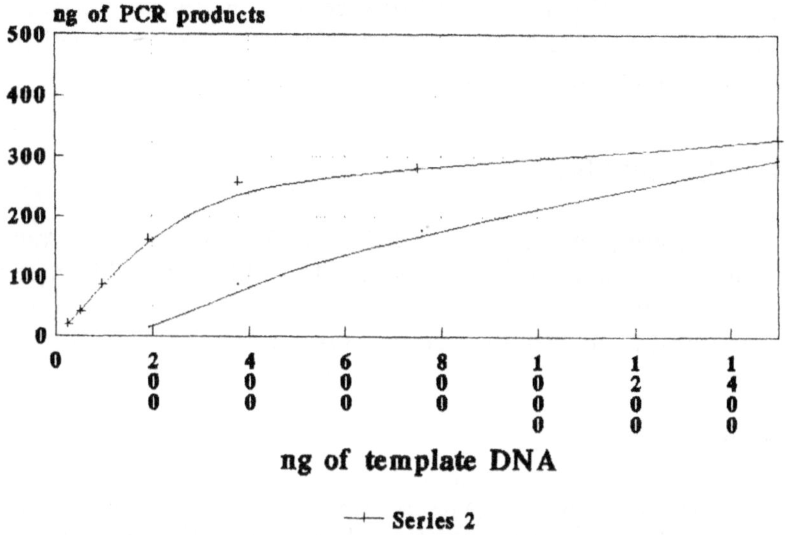

Figure 3. Quantification of multiplex IgH/Ras PCR. The numeration of results shown on the Fig. 2. Variable ratios CDR3 (series 1) : ras (series 2) depending on initial patient's template concentrations indicate that the comparison of diagnostic samples should be done only at the same ras amplification at exponential phase of PCR.

B-malignancies. In CDR3 negative cases the amplified Ras products confirm that the reaction was not anywise inhibited in the particular test tube and the use of IgH/ras PCR instead of a simple IgH PCR is helpful.

DISCUSSION

Quantitative monitoring of a malignant disease will help to identify patients at high risk of relapse. Equally, it is important to identify those patients, who will or will not require prolonged courses of chemotherapy and highly sensitive PCR—based methods are advantageous for this purpose.

Concerning the markers used in PCR techniques presented here, the translocation t(14, 18) is reported in approximately 40% European and 80% US cases of follicular lymphomas and in 30% with diffuse disease.[5] This IgH/*bcl 2* translocation usually persists for a long time and if it is present, then is it a good marker suitable for quantitative comparison in stored diagnostic samples. Employing consensus primers in clonality assays (IgH PCR), about 85% PCRs is positive,[15] 15% of it may be in benign reactive processes.[16] However, clonal changes may occur occasionally during the disease course. The sufficient sensitivity of CDR3 detection must be achieved by optimization of IgH/ras PCR. Ras oncogene[18] as a referring marker appears to be appropriate in co-amplifying with the clonal CDR3.

The quality of a sample, particularly clinical specimen, may be an important variable and always needs to be checked over after a storage. The quantity of markers vary by orders and thus the simple semi-quantitative limiting dilution approach may track the proliferating activity of malignant clone and be helpful in decisions, whether a positive PCR is initiated by DNA fragments released after chemotherapy. Naturally, the predictive value of results depends on frequency of taking samples.

The use of Q-PCR for the monitoring of MRD has already been recommended[1,17] in leukemia. The two techniques described here for non-Hodgkin's lymphomas seem to be applicable, however their value in the routine management of patients will need to be confirmed in much larger studies.

ACKNOWLEDGMENT

The authors are grateful to H. Achylisova, D. Hlouskova, E. Vodickova, and R. Zounar for their technical assistance and to Dr. J. Miller and I. Machackova for the cultivation of Raji cell line.

Supported by the grant of the Czech Ministry of Health No. 4317.

REFERENCES

1. Lion T: Clinical implications of qualitative and quantitative polymerase chain reaction analysis in the monitoring of patients with chronic myelogenous leukemia. Bone Marrow Transplant. 14: 505, 1994.
2. Tobal K, Liu Yin JA: Molecular monitoring of Minimal residual disease in acute myeloblastic leukemia with t(8, 21) by RT-PCR. Leuk. Lymph. 31: 115, 1998.
3. Crescenzi M, Seto M, Herzig GP, Weiss PO, Grifith RC, Korsmeyer SJ: Termostable DNA polymerase chain amplification of t(14, 18) chromosomal breakpoints and detection of minimal residual disease. Proc. Natl. Acad. Sci. 85: 4869, 1988.
4. Lambrechts AC, de Ruiter PE, Dorssers LCJ, van't Veer MB: Detection of residual disease in translocation t(14, 18) positive non-Hodgkin's lymphoma, usung the polymerase chain reaction: a comparison with conventional staging methods. Leukemia 6: 29, 1992.
5. Pezzella f, Ralfkiaer e, Ghatter KC, Mason DY: The 14, 18 translocatin in European cases of follicular lymphoma: comparison of Soutrhern blotting and the polymerase chain reaction. Brit. J. Haematol. 76: 58, 1990.
6. Meijerink JPP, Smetsers TFCM, Raemaekers JMM, Bogman MJJT, DeWtte T, Mensink EJBM: Quantitation of follicular non-Hodgkin's lymphoma cells carrying t(14, 18) by competitive polymerase chain reaction. Brit. J. Haematol. 84: 250, 1993.
7. Deane M, Norton JD: Detection of immunoglobulin gene rearrangement in B lymphoid malignances by polymerase chain reaction gene amplification. Brit. J. Haematol. 74: 251, 1990.
8. Trainor K, Brisco MJ, Story CJ, Morley AA: Monoclonality in B-lymphpproliferative disorders detected at the DNA level. Blood 75: 2220, 1990.
9. Brisco MJ, Tan LW, Osborn AM, Morley AA: Development of a highly sensitive assay, based on the polymerase chain reaction, for rare B-lymphocyte clones in a polyclonal population. Brit. J.Haematol 75: 163, 1990.
10. Brisco MJ, Condon J, Sykes PJ, Noeh SH, Morley AA: Detection and quantitation of neoplastic cells in acute lymphoblastic leukemia, by use of the polymerase chain reaction. Brit. J. Haematol. 79: 211, 1991.
11. Sykes PJ, Neoh SH, Brisco MJ, Hughes E, Condon J, Morley AA: Quantitation of targets for PCR by use of limiting dilution. Biotechniques 13: 444, 1992.
12. Oupenskaia MV, Johnston DA, Roberts WM, Estrov Z, Zipf TF: Accurate quantitation of residual B-precursor acute lymphoblastic leukemia by limiting dilution and PCR based detection system: A description of the method and the principles involved. Leukemia 9: 321, 1995.
13. Pulvertaft RJV: Cytology of Burkitt's tumour (African lymphoma). *Lancet* 1: 238, 1964.
14. Miller SA, Dykes DD, Polesky HF: A simple salting out procedure for extracting DNA from human nucleated cells. Nucl. Ac. Res. 1: 1215, 1988.
15. Linke B, Pytlich J, Tiemann M, Suttorp M, Parwaresch R, Hiddeman W, Kneba M: Identification and structural analysis of rearranged immunoglobulin heavy chain genes in lymphomas and leukemias. Leukemia 9: 840, 1995.
16. Orphanos V, Anagnastou D, Papadaki TH, Maniatis GM, Athanassiadou A: Detection of gene rearrangements in reactive lymphoid processes. Leukemia and Lymphoma 9: 103, 1993.
17. Muto A, Mori S, Matsushita H, Awaya N, Ueno H, Takayama N, Okamoto S, Rizaki M, Ikeda Y: Serial quantificatin of minimal residual disease of t(8, 21) acute myelogenous leukaemia with RT-competitive PCR assay. Brit. J. Haematol. 95: 1, 1996

24

CLASSIFICATION OF MALIGNANT LYMPHOMAS

State of the Art

S. A. Pileri and E. Sabattini

Service of Pathologic Anatomy/Haematopathology Unit
Institute of Haematology
S. Orsola Hospital
University of Bologna

1. REAL CLASSIFICATION: HISTORY, PRINCIPLES, AND VALIDATION STUDIES

The International Lymphoma Study Group (ILSG) (Table 1) was founded in 1991 by 19 experienced European and American haematopathologists with the scope to discuss and compare the problematics connected with the diagnosis and, above all, the classification of malignant lymphomas. After two meetings dealing with mantle cell lymphoma[1] and Hodgkin's disease,[2] in 1993 a first draft of a new lymphoma classification was discussed: this topic was considered of primary importance by all the members due to the long lasting differences and contrasts between the American and the European schools.[3-13] The ILSG reckoned that the two classifications in use (the Working Formulation and the Updated Kiel Classification) (WF, UKC)[11-13] were for different reasons[14-16] unsatisfactory and an update of either scheme difficult or even impossible.

Taking also into consideration what reported in the literature in the last decade, the ILSG thought that the best classification of lymphoid neoplasms was a mere list of all distinct clinico-pathologic entities each defined by morphology, phenotype, molecular characteristics, clinical features and, when possible, the normal counterpart. Moreover, it had to comprise all lymphoid neoplasms, including plasmacytoma, leukaemias, and Hodgkin's disease, and to differentiate the so called non-Hodgkin's lymphomas into B- and T-cell types, subdividing each of the latter into derived from lymphocyte precursor and peripheral-lymphocytes. All officially recognized entities were listed and termed "accepted" or "provisional", the latter adjective being attributed to those

Molecular Biology of Hematopoiesis 6, edited by Abraham *et al.*
Kluwer Academic/Plenum Publishers, New York, 1999.

Table 1. International lymphoma study group members

1)	Peter M. Banks, Carolinas Medical Center, Charlotte NC, USA
2)	Elias Campos, Hospital Clinic, Barcelona, Spain
3)	John K.C. Chan, Queen Elisabeth Hospital, Hong-Kong
4)	Riccardo Dalla Favera, College of Physicians and Surgeons, New York, USA
5)	Geoges Delsol, Faculty of Medicine Purpan, University Paul Sabatier, Toulouse, France
6)	Christine De Wolf-Peeters, University of Leuven, Leuven, Belgium
7)	Brunangelo Falini, University of Perugia, Perugia, Italy
8)	Kevin C. Gatter, John Radcliffe Hospital, Oxford University, Oxford, UK
9)	Philip Kluin, Leiden University Hospital, Leiden University, The Netherlands
10)	Philippe Gaulard, Hôpital Henri Mondor, Créteil, France
11)	Thomas M. Grogan, University of Arizona Medical School, Tucson, USA
12)	Peter G. Isaacson, University College London Medical School, London, UK
13)	Elaine S. Jaffe, National Cancer Institute, Bethesda, MD, USA
14)	Daniel M. Knowles, New York Hospital, Cornell University Medical Center, New York, USA
15)	Nancy Lee Harris, Massachusetts General Hospital, Harvard University, Boston, MA, USA
16)	David Y. Mason, John Radcliffe Hospital, Oxford University, Oxford, UK
17)	Shigeo Mori, Institute of Medical Science, University of Tokyo, Japan
18)	Hans-Konrad Müller-Hermelink, University of Würzburg, Würzburg, Germany
19)	Stefano A. Pileri, University of Bologna, Bologna, Italy
20)	Miguel A. Piris, Hospital Virgen de la Salud, Toledo, Spain
21)	Elisabeth Ralfkiaer, University of Copenhagen, Herlev, Denmark
22)	Harald Stein, Klinikum Benjamin Franklin, Free University of Berlin, Berlin, Germany
23)	Ih-Jen Su, National Chen Kung University, Taiwan
24)	Roger A. Warnke, Stanford University Medical School, Stanford CA, USA
25)	Lawrence M. Weiss, City of Hope National Medical Center, Duarte CA, USA

lymphomas for which convincing data were available in the literature, but lacking either consensus or enough diagnostic experience among the ILSG members.

This proposal, that was called Revised European-American Lymphoma (REAL) Classification and published in Blood September 1994[17] (Table 2), was the first to meet the largest consensus among experienced haematopathologists, to comprehend nodal and extranodal lymphomas and to avoid the indication of grades of malignancy. Regarding the latter point, the decision made by the ILSG was in accordance to what reported in the literature about the variable aggressiveness observed in different patients affected by the same lymphoma and attributable to several factors (indeces of proliferation and apoptosis, activation of oncogenes, presence of hybrid fusion genes, development of multidrug resistance, microambient, possible correlation with microorganisms, etc.).[18–35] Accordingly, it was stated that the grade of malignancy could neither be defined by response to treatment nor by pure morphology.[36,37]

As expected, the reactions to the REAL, both among clinicians and pathologists, ranged from favourable to totally adverse.[38–49] Some, in fact, claimed the lack of a validation and application process and the missing beloved grade of malignancy, while others had on the other side appreciated its biologic cut and the attention drawn on the single patient rather than on the whole category. A remarkable proposal for a worldwide validation study was made by J. Armitage, after a meeting at the National Cancer Institute/USA in March 1994, with the aim to verify the applicability and reproducibility of the REAL[17] as opposed to the WF[13] and the UKC.[11,12] At the same time, the study should have also allowed to establish how important were immunophenotyping and clinical data for the diagnosis and to show the incidence of

Table 2. Revised European-American lymphoma classification (17)

B-Cell Neoplasms
- I. Precursor B-cell neoplasm: Precursor B-lymphoblastic leukaemia/lymphoma
- II. Peripheral B-cell neoplasms
  1. B-cell chronic lymphocytic leukaemia/prolymphocytic leukaemia/small lymphocytic lymphoma
  2. Lymphoplasmacytoid lymphoma/immunocytoma
  3. Mantle cell lymphoma
  4. Follicle center lymphoma, follicular
     Provisional cytological grades: I (small cell), II (mixed small and large cells), III (large cells)
     Provisional subtype: diffuse, predominantly small cell type
  5. Marginal zone lymphoma
     Extranodal (MALT type ± monocytoid B cells)
     Provisional subtype: Nodal (± monocytoid B cells)
  6. Provisional entity: Splenic marginal zone lymphoma (± villous lymphocytes)
  7. Hairy cell leukaemia
  8. Plasmacytoma/plasma cell myeloma
  9. Diffuse Large B-cell lymphoma*
     Subtype: Primary mediastinal (thymic) B-cell lymphoma
  10. Burkitt's lymphoma
  11. Provisional entity: High-grade B-cell lymphoma, Burkitt-like*

T-Cell and Putative NK-Cell Neoplasms
- I. Precursor T-cell neoplasm: Precursor T-lymphoblastic leukaemia/lymphoma
- II. Peripheral T-cell and NK-cell neoplasms
  1. T-cell chronic lymphocytic leukaemia/prolymphocytic leukaemia
  2. Large granular lymphocyte leukaemia
     T-cell type
     NK-cell type
  3. Mycosis fungoides/Sezàry syndrome
  4. Peripheral T-cell lymphomas, unspecified*
     Provisional cytological categories: medium-sized cell, mixed medium and large cell, large cell, lymphoepithelioid
     Provisional subtype: Hepatosplenic $\gamma\delta$ T-cell lymphoma
     Provisional subtype: Subcutaneous panniculitis T-cell lymphoma
  5. Angioimmunoblastic T-cell lymphoma (AILD)
  6. Angiocentric lymphoma
  7. Intestinal T-cell lymphoma (± entreopathy associated)
  8. Adult T-cell lymphoma/leukaemia (ATL/L)
  9. Anaplastic large cell lymphoma (ALCL), CD30+, T- and null-cell types
  10. Provisional entity: Anaplastic large-cell lymphoma, Hodgkin's-like

Hodgkin's Disease
- I. Lymphocyte predominance
- II. Nodular sclerosis
- III. Mixed Cellularity
- IV. Lymphocyte depletion
- V. Provisional entity: Lymphocyte-rich classical HD

each type of lymphoma in the different geographic areas. For this purpose, from 9 Institutions (Omaha, Vancouver, Capetown, London, Bellinzona/Locarno, Lyon, Hong Kong, Würzburg, Göttingen), 1379 patients were selected according to following criteria: a) presence of a previously untreated lymphoma, b) onset of disease between 1st January 1988 and 31st December 1990, c) geographical rappresentativeness of the lymphomas, d) availability of enough material for the diagnosis and classification of the tumour, e) presence of immunophenotypic studies, f) complete clinical data (age, sex, race, date, and site of biopsy, nodal/extranodal involvements, stage, presence of bulky disease, laboratory findings, immunological status, performance status, first therapy, response to treatment, follow up). All cases were diagnosed by 5 experienced haematopathologists (Jacques Diebold, Kenneth A. Mac Lennan, Hans-Konrad Müller-Hermelink, Bharat Natwani, Dennis Weinsenburger), only one being member of the ILSG (HKMH), according to the REAL,[17] UKC[11,12] and the WF.[13] The statistical analysis was performed by James R. Anderson and Pascal Roy.

The study was organized with a first review of the histological slides (bone marrow biopsies included) with minimal clinical information (age, sex, site of biopsy, and stage), followed by the addition of the immunophenotypic and cytogenetic/molecular findings, and subsequently the clinical information; in the end, 20% of all cases were re-reviwed in the light of all the above mentioned parameters and consensus was established when 4/5 panel pathologists made the same diagnosis. The results of this project[50] showed that the REAL[17] provided the highest correspondence-index between the diagnoses of each panelist and the consensus ones (ranging from 86% and 95%), being 20 and 40 points higher than that obtained when applying the UKC[11,12] and the WF,[13] respectively. While morphology alone turned out to be *per se* diagnostic for few categories (B-cell chronic lymphocytic leukaemia, and extranodal marginal zone lymphomas), for the remaining types, the exact diagnosis could not be confidentially made without the employment of ancillary techniques (immunohistochemistry and molecular biology). The knowledge of clinical data was essential for diagnostic purposes only in the case of primary mediastinal large B-cell lymphoma.

Regarding the epidemiological point of view, the study established that the most common type of lymphoma worldwide was the diffuse large B-cell one (30%), followed by the follicle centre cell type (22%), the marginal zone one (8%), the small B-cell and the unspecified peripheral T-cell (7%), the mantle cell type (5%), the Burkitt's lymphoma (3%), and the types anaplastic large cell, primary mediastinal large B-cell, and precursor T-cell lymphoblastic lymphomas (2%), being the other categories rarer (12% of all lymphoid neoplasms).

The application of the REAL[17] was also found to have an important impact on survival. The group of the peripheral small B-cell lymphomas, classified as low grade in the WF[13] and the UKC,[11,12] actually showed a wide range of 8-year-survivals, corresponding to 80% for the marginal zone lymphoma, to 50% for the B-cell chronic lymphocytic leukaemia and to 18% for the mantle cell lymphoma. Analogously, the large cell lymphomas, classified as high-grade neoplasms in the UKC[11,12] and the WF,[13] showed the following values: anaplastic large cell type 80%, diffuse large B cell type 50%, peripheral T-cell types <20%. Moreover, statistically significant differences in terms of survival emerged when subdiving patients affected by the same lymphoma type and alive after 5 years, according to the International Prognostic Index (0,1 vs 4,5) (e.g.: mantle cell lymphoma: 59% vs 12%, large B-cell diffuse lymphoma 72% vs 22%).

As further confirmation of such a variability, if the histotypes were lumped according to their 5-year-survival, three prognostic groups could be identified: the first provided with a survival >50%, comprehensive of follicle centre cell lymphoma, small B-cell/B chronic lymphocytic leukaemia, marginal zone lymphoma, and anaplastic large cell lymphoma, the second with intermediate prognosis (survival between 30% and 50%), inclusive of diffuse large B-cell lymphomas, Burkitt's lymphoma, and precursor lymphoblastic lymphoma, and the third one with a survival <30% that include mantle cell lymphoma and peripheral T-cell lymphomas.

These data indeed contrasts with the definition of grades of malignacies present in the UKC[11,12] and the WF,[13] confirms other groups' results[51–54] and further support the choice of the ILSG to avoid it.

2. WHO CLASSIFICATION

In 1994 the WHO arranged the writing of an update classification of all the haematopoietic tumours and nominated boards of experienced heamatopathologists, in part members of the ILSG. As far as lymphomas are concerned, a classification was elaborated, the final and official version of which was presented at the Consensus Conference in Airlie (Virginia, USA) last november (Table 3). The classification is conceptually based on the principles proposed in the REAL and few changes were made.

Regarding the B-cell lymphomas, the main differences can be listed as follows:

1) the immunocytoma was re-termed lymphoplasmacytic lymphoma, thus keeping the definition present in the UKC[11,12] for those lymphomas composed of small CD5 negative lymphocytes and plasma cells and often associated with a Waldenstrom syndrome;
2) the nodal and splenic marginal zone lymphomas, indicated as provisional entities in the REAL, were switched to accepted on the basis of the data meanwhile published;[55,56]
3) the prolymphocytic B-cell leukaemia was kept separate from the chronic B-cell lymphocytic leukaemia due to the different and more aggressive clinical course, the higher lymphocyte count and the slightly different phenotype;[12]
4) the Burkitt-like type was eliminated as a separate entity and not distinguished from diffuse large B-cell lymphomas, due to the same response to therapy;[50]
5) the grading classes of the follicle centre cell lymphomas were decreased from 3 to 2, by lumping grades I and II, although the use of the REAL grading system is still accepted;
6) it was stated that the morphological and clinical variants of the diffuse large B-cell lymphomas, some of which already reported in the REAL, do not necessarily have to be specified. It has to be reminded however, that the following histological subtypes are mentioned: centroblastic, immunoblastic (so called if immunoblasts represent >90% of the neoplastic cells), anaplastic, rich in reactive lymphocytes and/or histiocytes,[57–59] while the clinical variants are the primary mediastinal <$F not quoted in the REAL >,[60–62] intravascular <$F not quoted in the REAL >,[63,64] and pyothorax associated <$F not quoted in the REAL >.[65] The latter variant can also be classified among the lymphoproliferative disorders of immunodepressed subjects according to the WHO.[66]

As far as the T-cell lymphomas are concerned, the below listed differences were recorded:

1) the T-cell chronic lymphocytic leukaemia was redenominated T-prolymphocytic leukaemia in order to underline its aggressive course, although indolent forms and/or small cell variants can rarely be encountered;
2) the term nasal/nasal-type lymphoma was introduced in the place of the angiocentric T-cell lymphoma:[67] the nasal form refers to those tumours arising in

Table 3. WHO classification of lymphoid tumours

B-cell neoplasms
Precursor B-cell lymphoblastic leukemia/lymphoma
Peripheral B-cell Neoplasms
B-cell chronic lymphocytic leukemia/small lymphocytic lymphoma
B-cell prolymphocytic leukemia
Lymphoplasmacytic lymphoma
Mantle cell lymphoma
Follicular lymphoma
Marginal zone B-cell lymphoma of mucosa associated lymphoid tissue (MALT) type (± monocytoid B-cells)
Nodal marginal zone B-cell lymphoma (± monocytoid B-cells)
Splenic marginal zone B-cell lymphoma (± villous lymphocytes)
Hairy cell leukemia
Diffuse large B-cell lymphoma^
Sutypes: mediastinal (thymic), intravascular
Burkitt lymphoma
Plasmacytoma
Plasma cell myeloma
T-cell Neoplasms
Precursor-cell lymphoblastic leukemia/lymphoma
Peripheral T-cell and NK-cell neoplasms
T-cell prolymphocytic leukemia
T-cell large granular lymphocytic leukemia
NK cell leukemia
NK/T cell lymphoma, nasal, and nasal-type
Mycosis fungoides
Sézary syndrome
Angioimmunoblastic T-cell lymphoma
Peripheral T-cell lymphoma (unspecified)^
Adult T-cell Leukemia/lymphoma (HTLV-1+)^
Anaplastic large cell lymphoma (T and null cell types)
Primary cutaneous CD30-positive T-cell lymprolipherative disorders^
Subcutaneous panniculitis-like T-cell lymphoma
Enteropathy-type intestinal T-cell lymphoma
Hepatosplenic γ/δ T-cell lymphoma
Hodgkin Lymphoma (Hodgkin Disease)
Nodular lymphocyte predominance Hodgkin's lymphoma
Classical Hodgkin lymphoma
Hodgkin lymphoma, nodular sclerosis (Grades I and II)
Classical Hodgkin lymphoma, lymphocyte-rich
Hodgkin lymphoma, mixed cellularity
Hodgkin lymphoma, lymphocytic depletion

^Morphologic and/or clinical variants of these diseases are not listed, for the purpose of clarity and ease of presentation.

the nose and more frequently in oriental populations, while the nasal type one is observed in Western countries at other extranodal sites;[68,69]

3) the entities hepato-splenic γ/δ T-cell lymphoma and the subcutaneous panniculitis-like T-cell lymphoma,[17] previously included in the "provisional" categories of the REAL, were moved into the accepted types;
4) the NK- and skin lymphomas were classified apart in the WHO (while the REAL list them as subvarieties);
5) an autonomous position was reserved to CD30+ lymphoproliferative disorders of the skin that comprise patterns with varying clinical and pathological aggressiveness (ranging from lymphomatoid papulosis to anaplastic large cell lymphoma);[70]

6) the recognition of subclasses among the peripheral T-cell lymphomas, unspecified was considered optional.

ACKNOWLEDGMENTS

The present study was supported by grants from A.I.R.C. (Milan) and M.U.R.S.T. (Rome)

REFERENCES

1. Banks P et al. Mantle cell lymphoma: a proposal for unification of morphologic, immunologic, and molecular data. Am. J. Surg. Pathol. 1992; 16:637–640.
2. Mason DY et al. Nodular lymphocyte predominance Hodgkin's disease. A distinct clinicopathological entity. Am. J. Surg. Pathol. 1994; 18:526–530.
3. Rappaport H. Tumors of the Hematopoietic System. In: Atlas of Tumor Pathology. Sect. 3, Fasc. 8. Washington, D.C.: Armed Forces Institute of Pathology 1966.
4. Lukes RJ, Collins RD. Immunologic characterization of human malignant lymphomas. Cancer 1974; 34:1488–1503.
5. Dorfman RF. Classification of non-Hodgkin's lymphomas. Lancet 1974, i:1295–1296.
6. Bennett HD et al. Classification of non-Hodgkin's lymphomas. Lancet 1974: ii:405–406.
7. Gérard-Marchant R et al. Classification of non-Hodgkin's lymphomas. Lancet 1974; ii:406–408.
8. Lennert K. Malignant lymphomas other than Hodgkin's disease. New York, Heidelberg, Berlin, Springer-Verlag; 1978.
9. Mathé G et al. Histological and cytological typing of neoplastic diseases of haematopoietic and lymphoid tissues. In: International Histological Classification of Tumors, No. 14. Geneva: World Health Organization, 1976.
10. Kay HEM. Classification of non-Hodgkin's lymphomas. Lancet 1974; ii:586.
11. Stansfeld A et al. Updated Kiel classification for lymphomas. Lancet 1988; i:292–293.
12. Lennert K, Feller AC. Histopathology of non-Hodgkin's lymphomas (based on the Updated Kiel Classification). Berlin, Heidelberg, New York, London, Paris, Tokyo, Hong Kong, Barcellona, Budapest, Spinger-Verlag; 1992.
13. Non-Hodgkin's lymphoma pathologic classification project. National Cancer Institute sponsored study of classifications of non-Hodgkin's lymphomas: Summary and description of a Working Formulation for clinical usage. Cancer 1982; 49:2112–2135.
14. Isaacson PG, Norton AJ. Extranodal lymphomas. Edinburgh, London, Madrid, Melbourne, New York, Tokyo, Churchill Livingstone; 1994:1–4.
15. Nathwani B et al. What should be the morphologic criteria for the subdivision of follicular lymphomas? Blood 1986; 68:837–845.
16. Hastrup N et al. Peripheral T-cell lymphomas: an evaluation of reproducibility of the updated Kiel classification. Histopathol. 1991; 18:99–105.
17. Harris N et al. A Revised European-American Classification of lymphoid neoplasms: a proposal from the International Lymphoma Study Group. Blood 1994; 84:1361–1392.
18. Lippman SM et al. The prognostic significance of the immunophenotype in diffuse large-cell lymphoma: A comparative study of the T-cell and B-cell phenotype. Blood 1988; 72:436–441.
19. Coiffier B et al. Peripheral T-cell lymphomas have a worse prognosis than B-cell lymphomas: A retrospective study of 361 immunophenotyped patients treated with the LNH-84 regimen. Ann. Oncol. 1990; 1:45–50.
20. Gerdes J et al. Prognostic relevance of tumour-cell growth fraction in malignant non-Hodgkin's lymphomas. Lancet 1987; ii:448–449.
21. Gerdes J et al. Immunohistochemical determination of the proliferative capacity of malignant lymphomas with Ki-67 and other monoclonal antibodies. In: Cell proliferation in lymphomas. Crocker, J, Editor. Blackwell Scientific Publications, Oxford. 1993:131–144.
22. Leoncini L et al. Correlations between apoptotic and proliferative indices in malignant non-Hodgkin's lymphomas. Am. J. Pathol. 1993; 142:755–763.

23. Pileri S et al. Apoptosis as programmed cell death (PCD): cupio dissolvi in cell life. Curr. Diagn. Pathol. 1994; 1:48–55.
24. Lo Coco F et al. Rearrangements of bcl-6 gene in diffuse large cell non-Hodgkin's lymphoma. Blood 1994; 83:757–759.
25. Pileri SA et al. Immunohistochemical detection of the multidrug transport protein p170 in human normal tissue and malignant lymphomas. Histopathol. 1991; 19:131–140.
26. Mason DY et al. CD30-positive large cell lymphomas ("Ki-1 lymphoma") are associated with a chromosomal translocation involving 5q35. Br. J. Haematol. 1990; 74:161–168.
27. Orscheschek K et al. Large-cell anaplastic lymphoma-specific translocation (t[2;5] [p23;q35]) in Hodgkin's disease: indication of a common pathogenesis? Lancet 1995; i:87–90.
28. Bullrich F et al. Nucleophosmin (NPM) gene rearrangement in Ki-1-positive lymphomas. Cancer Res. 1994; 54:2873–2877.
29. Shiota M et al. Diagnosis of t(2;5)(p23;q35)-associated Ki-1 lymphoma with immunohistochemistry. Blood 1994; 84:3648–3652.
30. Pulford K et al. Detection of anaplastic lymphoma kinase (ALK) and nucleolar protein nucleophosmin(NPM)-ALK proteins in normal and neoplastic cells with the monoclonal antibody ALK1. Blood 1997; 89:1394–1404.
31. Pileri SA et al. Frequent Expression of the NPM-ALK Chimeric Fusion Protein in Anaplastic Large Cell Lymphoma, Lympho-Histiocytic Type. Am. J. Pathol. 1997; 150:1207–1211.
32. Nakamura S et al. Anaplastic large cell lymphoma: a distinct molecular pathologic entity. A reappraisal with special reference to $p80^{NPM/ALK}$ expression. Am. J. Surg. Pathol. 1997; 21:1420–1434.
33. Wotherspoon AC et al. Regression of primary low-grade B-cell gastric lymphoma of mucosa-associated lymphoid tissue type after eradication of Helicobacter pylori. Lancet 1993; ii:575–577.
34. Isaacson PG. The MALT lymphoma concept updated. Ann. Oncol. 1995; 6:319–329.
35. Zucca E et al. B-cell lymphoma of MALT type: a review with special emphasis on diagnostic and management problems of low-grade gastric tumours. Br. J. Haematol. 1998; 100:3–14.
36. Bonadonna G. Modern treatment of malignant lymphomas: a multidisclipinary approach? Ann. Oncol. 1994; 5 (suppl 2):5–16.
37. Bonadonna G. Will new treatment strategies improve the control of non-Hodgkin's lymphomas? Curr. Opinion. Oncol. 1994; 6:453–455.
38. Pileri S, Falini B. Revised European-American Lymphoma Classification: a new approach to the study and diagnosis of malignant lymphomas. Editorial, Current Science; 1995; 7:401–407.
39. Rosenberg SA. Classification of lymphoid neoplasms. Blood 1994; 84:1359–1360.
40. Harris N et al. Lymphoma classification proposal: clarification. Blood 1995; 85:857–860.
41. Stein H. Critique to the critique: Response to the editorial entitled "Classification of lymphoid neoplasms" by Dr. Saul Rosenberg which accompanied the publication of the "Revised European-American Lymphoma Classification proposal" in the September 1994 issue of Blood. Ann. Oncol. 1995; 6:109–111.
42. Lennert K. The proposal for a Revised European American classification—a new start of a transatlantic discussion. Histopathol. 1995; 26:481–483.
43. Meijer CJLM et al. The Revised European-American Lymphoma (REAL) Classification of non-Hodgkin's lymphoma: a missed opportunity? Blood 1995; 85:1971–1972.
44. Demeter J et al. Observations regarding hairy cell leukemia and chronic lymphocytic leukemia within the proposed new classification of lymphoid neoplasms. Blood 1995; 85:1972–1973.
45. Harris NL et al. Response. Blood 1985; 85:1973–1974.
46. Al-Saleem T. Classification of lymphoid neoplasms between the hemato-pathologists and the "common person". Blood 1995; 85:1974–1975.
47. Lister TA et al. The proposed revised European-American Classification of lymphoma. Blood 1995; 85:1975.
48. O'Connor N. New classification for lymphomas. Lancet 1995; i:1521–1522.
49. Mason DY, Gatter K. Not another lymphoma classification! Br. J. Haematol. 1995; 90:493–497.
50. The non-Hodgkin's lymphoma classification project. A clinical evaluation of the International Lymphoma Study Group Classification of non-Hodgkin's lymphoma. Blood 1997; 89:3909–3918.
51. Falini B et al. Mediastinal large B-cell lymphoma: clinical and immunohistochemical findings of 18 patients treated with two different third generation regimens. Br. J. Haematol. 1995; 89:780–789.
52. Pileri S et al. Anaplastic large cell lymphoma (CD30+/KI-1+): results of the prospective clinico-pathologic study of 69 cases. Br. J. Haematol. 1994; 86:513–523.
53. Zinzani PL et al. Anaplastic large cell lymphoma (Ki-1/CD30+): clinical and prognostic evaluation of 90 adult patients. Ann. Oncol. 1996; 14:955–962.

54. Ascani S et al. Peripheral T-cell lymphomas: Clinico-pathologic study of 168 cases diagnosed according to the R.E.A.L. Classification. Annals of Oncology 1997; 8:583–592.
55. Wu CD et al. Splenic marginal zone cell lymphoma. An immunophenotypic and molecular study of five cases. Am. J. Clin. Pathol. 1996; 105:277–285.
56. Solè F et al. Frequent involvement of chromosomes 1, 3, 7, and 8 in splenic marginal zone-Bcell lymphoma. Br. J. Haematol. 1997; 98:446–449.
57. Ramsey AD et al. T-cell rich B-cell lymphoma. Am. J. Surg. Pathol. 1988; 12:433–443.
58. Chittal SM et al. Large B-cell lymphoma rich in T-cells and simulating Hodgkin's disease. Histopathol. 1991; 19:211–220.
59. Delabie J et al. Histiocyte-rich B-cell lymphoma: a distinct clinicopathologic entity possibly related to lymphocyte predominant Hodgkin's disease, paragranuloma subtype. Am. J. Surg. Pathol. 1992; 16:37–48.
60. Addis BJ, Isaacson PG. Large cell lymphoma of the mediastinum: a B cell tumour of probable thymic origin. Histopathol. 1986; 10:379–390.
61. Perrone T et al. Mediastinal diffuse large-cell lymphoma with sclerosis: a clinicopathologic study of 60 cases. Am. J. Surg. Pathol. 1986; 10:176–191.
62. Menestrina F et al. Mediastinal large-cell lymphoma with sclerosis: histopathological and immunohistochemical study of eight cases. Histopathol. 1986; 10:589–600.
63. Fukuki M et al. An autopsy case of intravascular malignant lymphoma presenting with intracranial B-cell type malignant lymphoma. Noshuyo Byori 1996; 13:119–125.
64. Sanna P et al. Angiotropic (intravascular) large cell l ymphoma: case report and short discussion of the literature. Tumori 1997; 83:772–775.
65. Ascani S et al. Pyothorax-associated lymphoma: description of the first two cases detected in Italy. Ann. Oncol. 1998; in press.
66. Harris NL, Swerdlow SH. Posttransplant lymphoproliferative disorders: summary of Society for Hematopathology Workshop. Semin. Diagn. Pathol. 1997; 14:8–14.
67. Guinee D et al. Pulmonary lesions with features of lymphomatoid granulomatosis/angiocentric immunoproliferative lesion (LYG/AIL): evidence for Epstein-Barr virus within B lymphocytes. Mod. Pathol. 1994; 7:151A.
68. Jaffe ES et al. Report of the Workshop on Nasal and Related Extranodal Angiocentric T/Natural Killer Cell Lymphomas. Definitions, differential diagnosis, and epidemiology. Am. J. Surg. Pathol. 1996; 20:103–111.
69. Chan JK et al. Nonnasal lymphoma expressing the natural killer cell marker CD56: a clinicopathologic study of 49 cases of an uncommon aggressive neoplasm. Blood 1997: 89: 4501–4513.
70. Willemze R et al. EORTC Classification for primary cutaneous lymphomas: a proposal from the Cutaneous Lymphoma Study Group of the European Organization for Research and Treatment of Cancer. Blood 1997; 90:354–371.

25

THE PATHOLOGIC SPECTRUM OF AIDS-RELATED NON-HODGKIN'S LYMPHOMAS

Antonino Carbone,[1] Umberto Tirelli,[2] Annunziata Gloghini,[1]
Brunangelo Falini,[3] Riccardo Dalla-Favera,[4] and Gianluca Gaidano[5]

[1] Division of Pathology
Centro di Riferimento Oncologico
Istituto Nazionale Tumori
IRCCS, Aviano, Italy
[2] Division of Medical Oncology and AIDS Program
Centro di Riferimento Oncologico
Istituto Nazionale Tumori
IRCCS, Aviano, Italy
[3] Institute of Hematology
University of Perugia
Perugia, Italy
[4] Division of Oncology
Department of Pathology
College of Physicians and Surgeons
Columbia University
New York, NY 10032 USA
[5] Division of Internal Medicine
Department of Medical Sciences
Amedeo Avogadro
University of Eastern Piedmont
Novara, Italy

Patients with acquired immunodeficiency syndrome (AIDS) are at increased risk of developing non Hodgkin's lymphomas (NHL). Current estimates indicate that 5–10% of HIV-infected individuals develop AIDS-related NHL (AIDS-NHL), a life-threatening disease.

---

* *Corresponding author*: Antonino Carbone, MD, Division of Pathology, Centro di Riferimento Oncologico, IRCCS, Istituto Nazionale Tumori, Via Pedemontana Occidentale, Aviano I-33081, Italy. Tel (39) (0434) 659.292; Fax: (39) (0434) 659.370; e-mail: acarbone@ets.it

Molecular Biology of Hematopoiesis 6, edited by Abraham *et al.*
Kluwer Academic/Plenum Publishers, New York, 1999.

Table 1. Clinico-pathologic spectrum of AIDS-related non-Hodgkin's lymphomas

→ Systemic non-Hodgkin's lymphoma (NHL) (75–80%)
→ Primary central nervous system lymphoma (PCNSL) (15–20%)
→ Primary effusion lymphoma (PEL) (2–3%)
→ Plasmablastic lymphoma (PBL) of the oral cavity (2–3%)

AIDS-NHL share several clinical features, including frequent extranodal presentation, aggressive clinical course, and poor outcome. From a pathologic point of view, AIDS-NHL are characterized by a diffuse growth pattern, cellular pleomorphism, high-grade morphology, and B-cell derivation. Low-grade B-cell and T-cell NHL, which may occasionally develop in the HIV-infected population, are not considered AIDS-related tumors because their incidence has not significantly increased since the start of the AIDS-epidemic.

AIDS-NHL traditionally included systemic and primary brain lymphomas, but nowadays their updated clinicopathologic spectrum also comprises two novel entities, namely primary effusion lymphoma and plasmablastic lymphoma of the oral cavity. Table 1 shows the relative frequency of these clinicopathologic categories.

The present paper deals with recent data concerning clinicopathologic, pathogenetic, and histogenetic aspects of AIDS-NHL.

1. SYSTEMIC NHL

Systemic AIDS-NHL are a heterogeneous group of malignancies displaying a B-cell phenotype. The overwhelming majority of systemic AIDS-NHL fall within three Working Formulation histologic categories[1]: small noncleaved cell lymphoma (SNCCL) which includes classic Burkitt's lymphoma and Burkitt-like lymphoma; large noncleaved cell lymphoma (LNCCL); and large cell immunoblastic lymphoma plasmacytoid (IBL-P) (Fig. 1). It has been proposed to classify LNCCL and IBL-P as a single group under the term 'diffuse large cell lymphoma (DLCL)' (Fig. 1), because these subtypes display a certain degree of overlap when classified on pure histologic grounds.[2] The category termed DLCL has been further expanded to include also B-cell lymphomas exhibiting anaplastic large cell morphology.[2] Notably, CD30 positive anaplastic large cell lymphoma (ALCL), a heterogeneous group of high grade lymphomas at the borderline between Hodgkin's disease and NHL, have recently been described in association with AIDS.[3–5]

An intriguing pathologic characteristic common to systemic lymphomas in HIV patients is the occurrence of cases having some overlap between established histologic

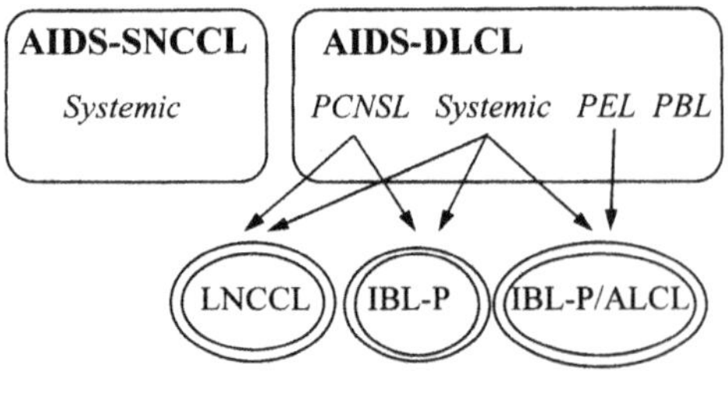

Figure 1. Main histologic categories of AIDS-related non-Hodgkin's lymphomas.

subtypes.[6] In this regard, an atypical variant mainly made up of blastic cells exhibiting features 'intermediate' between SNCCL with plasma cell differentiation and IBL-P has also been described in HIV patients.[7] This 'intermediate' variant may be part of the Burkitt-like lymphomas.[2] Deceptively, its atypical morphologic features may hamper a correct discrimination between SNCCL and IBL-P. A correct differential diagnosis between SNCCL and IBL-P is clinically relevant, because of the different behaviour and prognosis of these lymphomas.

2. PRIMARY CENTRAL NERVOUS SYSTEM LYMPHOMA (PCNSL)

PCNSL is a major cause of morbidity and mortality among individuals infected by HIV. The relative risk of PCNSL among HIV-infected hosts approximates 1000 when compared to the HIV-negative population. Virtually all AIDS-related PCNSL derive from B-cells and are clinically aggressive tumors.

Unlike the heterogeneous systemic AIDS-NHL, lymphomas arising in the central nervous system represent a more uniform group. Histologically, the overwhelming majority of PCNSL is represented by DLCL, which usually display histologic features consistent with IBL-P.[8,9] However, cases with LNCCL morphology can also be observed (Fig. 1).[8,9]

3. BODY CAVITY-BASED LYMPHOMA/PRIMARY EFFUSION LYMPHOMA (BCBL/PEL)

The broad category of lymphomas involving the serous body cavities includes all NHL associated with a malignant effusion (Fig. 2). Cases of NHL presenting in the serous body cavities in the absence of clinically identifiable tumor masses are defined as primary lymphomatous effusions, whereas secondary lymphomatous effusions are

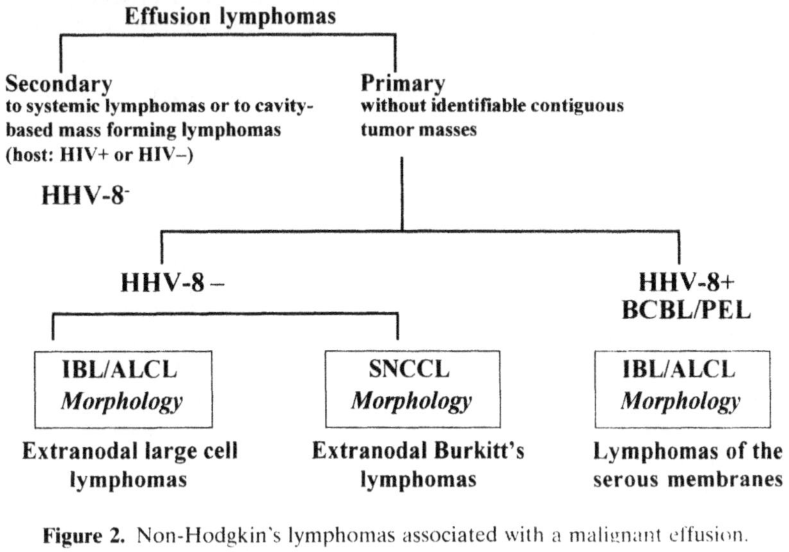

Figure 2. Non-Hodgkin's lymphomas associated with a malignant effusion.

malignant effusions complicating a clinically evident solid NHL through contiguous spread of tumor cells into the body cavities.[10]

Recently, both in the USA and Europe, a peculiar type of primary lymphomatous effusion has been recognized as an individual entity based on its distinctive biologic features and its consistent infection by a novel herpesvirus, termed human herpesvirus-8 (HHV-8).[11–15] Because of its peculiar and predominant tropism for the serous cavities of the body, this type of B-cell NHL has been provisionally designated body cavity-based lymphoma (BCBL).[10,11] In the USA, BCBL has also been termed as primary effusion lymphoma (PEL).[14]

Like most systemic lymphomas, the clinical outcome of patients with BCBL/PEL is poor, with a survival ranging from 1 to 6 months. Most patients have advanced AIDS and are poor candidates for aggressive chemotheraphy.[11,12,14] The epidemiology of BCBL/PEL is peculiar, since this lymphoma is found preferentially, though not exclusively, in association with HIV infection.[12–15] In AIDS setting, BCBL/PEL represents about 3% of all AIDS-NHL.[12] Among immunocompetent hosts, BCBL/PEL is very rare, since it accounts for less than 1% of cases of high-grade NHL.[12] Remarkably, BCBL/PEL appears to develop at a substantially older age in immunocompetent hosts than in HIV-infected individuals.[12,14,15]

Clinically, BCBL/PEL is characterized by growth in liquid phase in the absence of detectable tumor masses (Fig. 2).[10–12,14–21] Pathologically, however, autopsy and computed tomography scan studies have consistently revealed that BCBL/PEL presents as multiple small tumor foci involving the serous membranes, which appear irregularly thickened.[10,21–23] Thus, the basic feature of BCBL/PEL is a striking predilection for diffuse spreading along the serous membranes without infiltrative or destructive growth patterns.[10,21,22] These clinical and gross pathological features would indicate a primary serous membrane neoplasm. Notably, BCBL/PEL cells derived from lymphomatous effusions and those derived from solid tissue localizations display similar morphologic, immunophenotypic, and molecular features.[10,23] Overall, these findings suggest that BCBL/PEL is a lymphoma of the serous membranes typically giving rise to lymphomatous effusions without mass formation (Fig. 2).

Cytomorphologic features of BCBL/PEL cells bridge those of large cell immunoblastic and anaplastic large cell lymphomas (Fig. 1).[12,14,15,17,20] The most common appearance is that of a pleomorphic mixture of immunoblast-like cells and other cell types including anaplastic, multilobated or multinucleated large cells resembling the Reed-Sternberg cells of Hodgkin's disease.

BCBL/PEL exhibits an indeterminate immunophenotype.[11,12,14,15,17] Tumor cells lack surface immunoglobulin (Ig) and B- and T-cell-associated antigens, but express the leucocyte common antigen (LCA) and various markers associated with activation, including CD30, CD38, CD71, and the epithelial membrane antigen (EMA). Notably, BCBL/PEL cells express markers associated with the late stages of B-cell differentiation, including the plasma cell specific antigen CD138 recognized by the B-B4 monoclonal antibody.[24] Despite the fact that BCBL/PEL frequently fails to express the most common B-cell associated antigens, it is well demonstrated that BCBL/PEL is consistently represented by a monoclonal B-cell population. This notion is documented by the observation that BCBL/PEL cases display a unique band of Ig gene rearrangement in immunogenotypic studies and show restriction in the expression pattern of immunoglobulin light chains.[11,12,14,15,17]

Infection of the tumor clone by HHV-8 constitutes the genetic hallmark of BCBL/PEL. HHV-8 infection occurs in all cases of BCBL/PEL, whereas it is absent in

Table 2. Histopathologic classification of AIDS-related non-Hodgkin's lymphomas—Adapted from IARC/WHO Working Group on AIDS and Tumours

- Systemic lymphomas
  - a) "Blastic" cell lymphomas
    - large non-cleaved-cell (G-WF)
    - immunoblastic (H-WF) with or without plasma cell differentiation
    - small non-cleaved-cell (J-WF) with or without plasma cell differentiation
    - blastic cell with "intermediate" features
  - b) "Anaplastic" cell lymphomas
    - anaplastic large-cell (CD30/Ki-1+)
- Primary brain lymphomas
- Others (including PEL and PBL*)

*PBL has not been formally recognized yet.

all other B-cell and T-cell lymphomas.[11,12,14,19,25–27] Apart from infection by HHV-8 and frequently by EBV, no other known genetic lesions have been demonstrated to associate with BCBL/PEL. Rather, BCBL/PEL is generally devoid of alterations of *BCL*-2, *BCL*-6, c-*MYC*, and *p53* which, conversely, are detected at sustained frequencies in other types of aggressive B-cell NHL.[11,12,14] Available karyotypes of BCBL/PEL have failed to reveal a recurrent cytogenetic abnormality unique to this lymphoma.[10,17] However, in a recent study, cytogenetic and fluorescent in situ hybridization analysis of 7 BCBL/PEL cell lines showed frequent occurrence of complete or partial trisomy 12 (7/7 cases), trisomy 7 (4/7 cases), and abnormalities of bands 1q21–25 (5/7 cases).[28]

As shown in Table 2, the classification of AIDS-related lymphoproliferative disorders recently adopted by the IARC/WHO Working Group on AIDS and Tumours[29] includes BCBL/PEL in the spectrum of AIDS-NHL.

4. PLASMABLASTIC LYMPHOMA OF THE ORAL CAVITY (PBL)

AIDS-NHL may occur anywhere in the head and neck area. The tumor has been observed in a variety of sites including the gingiva, palate, tongue, tonsil, nasopharynx, orbit, parotid, maxillary sinus, soft tissue, and hypopharynx.[30–37] Most AIDS-NHL of the head and neck are morphologically similar to other lymph nodal or extranodal systemic AIDS-NHL and express B-cell markers.

A particular type of lymphoma has been recently identified as a specific entity among AIDS-DLCL involving the oral cavity.[38] This lymphoma has been designated as "plasmablastic lymphoma" (PBL) in accordance with its plasmablastic morphology and immunohistologic features.[38] The real incidence of this new entity is still unknown. In a consecutive series of 191 AIDS-NHL observed at one institution between 1984 and 1997, AIDS-PBL accounted for 2.6% of cases.[39]

From a clinical standpoint, AIDS-PBL is associated with a distinctive clinical feature that greatly contributes to its diagnosis, i.e. origin in the oral cavity. In all reported AIDS-PBL, the tumors were localized in the mucosa of the oral cavity and often involved the gingiva. In several instances, the bulk of the tumor infiltrated the adjacent bone, as assessed by x-ray analysis[38] (and our unpublished observation). In most patients, the tumor was limited to the oral cavity at the time of diagnosis, but the extension of AIDS-PBL to distant sites (abdomen, retroperitoneum) and other organs (bones and ovary) could be identified shortly after the diagnosis. All AIDS-PBL proved clinically very aggressive and the prognosis was poor (survival: 1 to 16 months) despite chemotherapy[38] (and our unpublished observation), with or without combined radiotherapy.[38] Only 3 patients with AIDS-PBL limited to the oral cavity responded well to radiotherapy alone.[38]

The histologic appearance of AIDS-PBL is relatively monomorphic. AIDS-PBL displays a diffuse and cohesive growth pattern with large neoplastic cells exhibiting a squared-off appearance. These cells have a centrally- or eccentrically-placed nucleus with a single prominent nucleolus or several nucleoli.[38,39] According to their morphologic features, AIDS-PBL tumor cells display a blastic cell morphology with plasma cell differentiation. This is consistent with their unusual immunophenotype, which is characterized by a weak or absent expression of the most commonly used lymphoid markers. A monotypic light chain expression can be demonstrated in about one third of cases. Conversely, AIDS-PBL displays strong immunostaining with the new plasma cell reacting antibodies VS38c and CD138/B-B4.[38,39] Overall, the AIDS-PBL immunophenotype ($CD20^-$, $CD45^{-/+}$, $CD79a^{-/+}$, EMA^+, $VS38c^+$, $CD138/B\text{-}B4^+$) is consistent with a marked plasma cell differentiation of the tumor clone.

At present, AIDS-PBL is not formally recognized as an individual entity in AIDS-NHL classification schemes (Table 2). Yet, the particular clinical, morphologic, and immunophenotypic features of AIDS-PBL mandate that it be singled out as an individual clinicopathologic category.

Pathogenetic and Histogenetic Heterogeneity of AIDS-NHL

From a molecular point of view, AIDS-NHL are a heterogeneous group of malignancies. In the last few years, several studies have shown that the pathologic heterogeneity of AIDS-related lymphomas correlates with the heterogeneity of the molecular lesions associated with these lymphomas. As summarized in Fig. 3, SNCCL displaying features of classic Burkitt's lymphoma selectively associates with activation of c-*MYC*, whereas rearrangements of *BCL*-6 are restricted to a subset of DLCL. Infection by HHV-8 clusters with BCBL/PEL; conversely, infection by EBV occurs, though at different rates, in different lymphoma types. However, only IBL-P express the EBV-encoded latent membrane protein-1 (LMP-1) (see[40,41] for a general review).

The precise stage of B-cell differentiation reflected by the different categories of AIDS-NHL has not been clarified in detail. In order to get further information on the

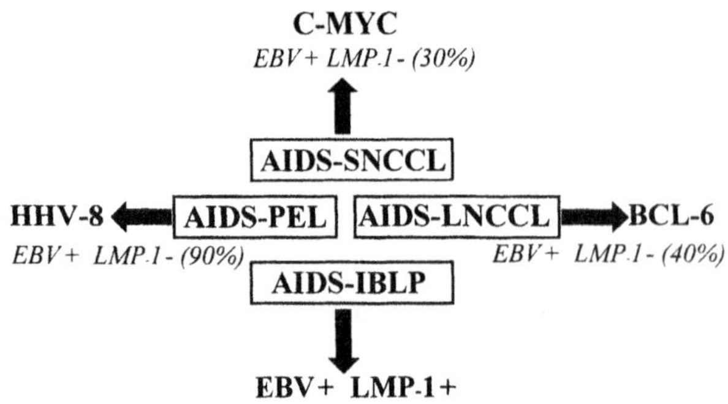

Figure 3. Pattern of association of histologic subtypes of AIDS-related non-Hodgkin's lymphomas with molecular lesions.

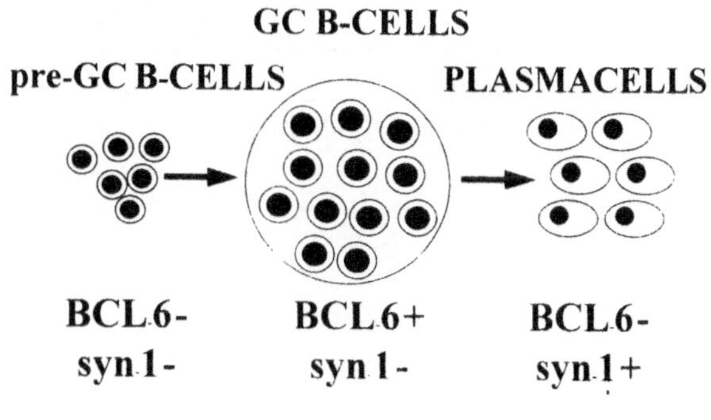

Figure 4. Combined expression pattern of BCL-6 and syndecan-1 (syn-1) throughout the stages of physiologic B-cell differentiation.

histogenesis of AIDS-NHL, we have recently carried out an original study[42] using biologic markers identifying distinct subsets of mature B-cells. Two such markers are protein expression of BCL-6 and CD138 or syndecan-1, which may discriminate between germinal center and post-germinal center B-cells.[42] Within the mature B-cell compartment, expression of BCL-6, which is a zinc finger transcriptional repressor, clusters with germinal center (GC) B-cells.[40] Conversely, syndecan-1, a proteoglycan belonging to the syndecan family, is not expressed in GC B-cells, but it is expressed in post-GC B-cells, including immunoblasts and plasmacells.[24] As shown in Figs. 4 and 5, B-cells within the germinal center display the BCL-6$^+$/syndecan-1$^-$ phenotype, whereas

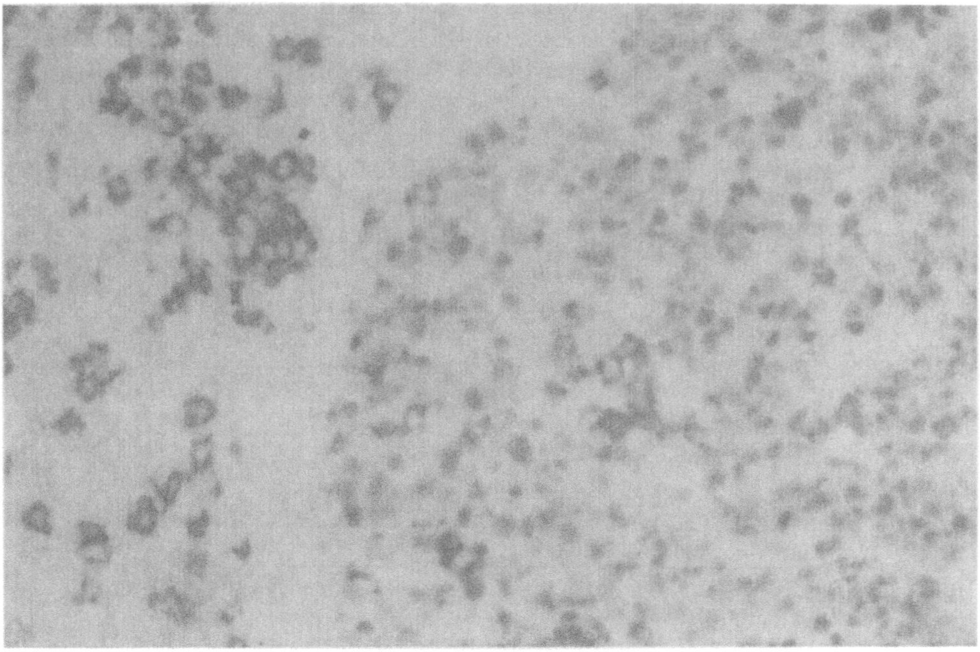

Figure 5. Two-color staining demonstration in a reactive lymph node from an HIV-infected person with PGL. B-cells within the GC are BCL-6$^+$ (the staining pattern is nuclear, and the color is blue), whereas plasmacells, which predominate in the interfollicular area, are syndecan-1$^+$ (the staining pattern is cytoplasmic and membranous, and the color is reddish).

Table 3. Case material

Samples		Morphology	
Systemic AIDS-NHL (27)		SNCCL/Burkitt's	(11)
		LNCCL	(7)
		IBL-P	(9)
AIDS-PCNSL	(8)	LNCCL	(4)
		IBL-P	(4)
AIDS-PEL	(5)	IBL-P/ALCL	(5)
AIDS-PEL Cell lines	(6)	IBL-P/ALCL	(6)
AIDS-PBL	(7)	Plasmablastic	(7)

B-cells which have exited the GC and are reaching the plasmacell stage exhibit the BCL-6$^-$/syndecan-1$^+$ phenotype.[42]

On these bases, we analyzed the expression pattern of BCL-6 and syndecan-1 in a large series of AIDS-NHL, in order to obtain new insights into their histogenesis. Table 3 shows the clinicopathologic groups and the histopathologic categories investigated. The study included neoplastic samples from patients with systemic lymphomas (n = 27), PCNSL (n = 8), PEL (n = 5), and PBL (n = 7). Six PEL cell lines were also studied (HBL-6, BC-1, BC-2, BC-3, BCBL-1, and CRO-AP/2).[42] Among systemic lymphomas, SNCCL cases included only classic Burkitt's lymphomas. DLCL cases were further subdivided into LNCCL and IBL-P according to the Working Formulation for non-Hodgkin's lymphomas.[1]

Table 4 summarizes the data of BCL-6 and syndecan-1 expression throughout the spectrum of AIDS-NHL. LNCCL and IBL-P from both systemic lymphomas and PCNSL have been grouped together.

To sum up, our investigation led to the following findings: first, expression of BCL-6 and of syndecan-1 are mutually exclusive; second, the combined expression pattern of BCL-6 and syndecan-1 separates two major phenotypic subsets, namely BCL-6$^+$/syndecan-1$^-$ and BCL-6$^-$/syndecan-1$^+$. The BCL-6$^+$/syndecan-1$^-$ pattern is associated with SNCCL and LNCCL, whereas the BCL-6$^-$/syndecan-1$^+$ pattern is associated with IBL-P, BCBL/PEL, and PBL. Lastly, the connection between phenotype and histology appears to be independent of the primary site of the disease. Only two cases classified as IBL-P were associated with BCL-6$^+$/syn-1$^-$ phenotype. These two cases predominantly consisted of tumor cells with immunoblastic features without plasmacytoid differentiation.

Table 4. Expression of BCL-6 and Syndecan-1 in AIDS-NHL

Samples	BCL-6+syn-1–	BCL-6–syn-1+	BCL-6–syn-1–
AIDS-SNCCL (11)	11	0	0
AIDS-LNCCL (11)	11	0	0
AIDS-IBL-P (13)	2	8*	3**
AIDS-PEL (11)	0	11	0
AIDS-PBL (7)	0	7	0

The expression of BCL-6 and syn-1 is mutually exclusive.
*Two of 8 were PCNSL. **Two of 3 were PCNSL.

Table 5. BCL-6 and Syndecan-1 among LMP-1 + AIDS-NHL

(LMP-1+ cases/total)		BCL-6–syn-1+	BCL-6–syn-1–
Systemic	SNCCL (0/11)		
	LNCCL (0/7)		
	IBLP (5/9)	5	
PCNSL	LNCCL (0/4)		
	IBLP (3/4)	1	2
AIDS-PEL	(3/11)	3	

Expression of LMP-1 is mutually exclusive with expression of BCL-6.

An additional target of our investigation was the relationship of BCL-6 and syndecan-1 expression with LMP-1 expression (Table 5). Among lymphomas infected by EBV, expression of LMP-1 is mainly associated the with BCL-6$^-$/syndecan-1$^+$ phenotype and, as expected, with immunoblastic subtype. Furthermore, LMP-1 and BCL-6 expression are mutually exclusive throughout the entire spectrum of AIDS lymphomas.

Figure 6 proposes a histogenetic model for the development of AIDS-NHL. The model is based on the combined expression pattern of BCL-6 and syndecan-1 throughout the stages of physiologic B-cell differentiation. Thus, AIDS-NHL displaying the BCL-6$^+$/syndecan-1$^-$ phenotype, namely SNCCL and LNCCL, are postulated to originate from GC B-cells. Conversely, AIDS-NHL displaying the BCL-6$^-$/syndecan-1$^+$ phenotype, namely IBL-P, BCBL/PEL, and PBL are believed to derive from B-cells at the pre-terminal stage of differentiation. In other words, the immunophenotypic characteristics of IBL-P, BCBL/PEL, and PBL suggest that these lymphomas represent the malignant counterpart of a B-cell that has reached a mature stage of development and is conceivably shifting toward terminal plasma cell differentiation.

These findings may also have clinical implications. As distinct pathologic categories associate with different phenotypic patterns, the combined expression of BCL-6 and syndecan-1 may be of help in diagnosing these lymphomas.

Finally, since expression of LMP-1 is associated with the syndecan-1$^+$ phenotype and is mutually exclusive with expression of BCL-6, post GC maturation seems to be an essential requirement for LMP-1 expression in B-cell tumors. Further investigations are needed to understand whether LMP-1 expression may play a role in the regulation of the level of BCL-6 protein.

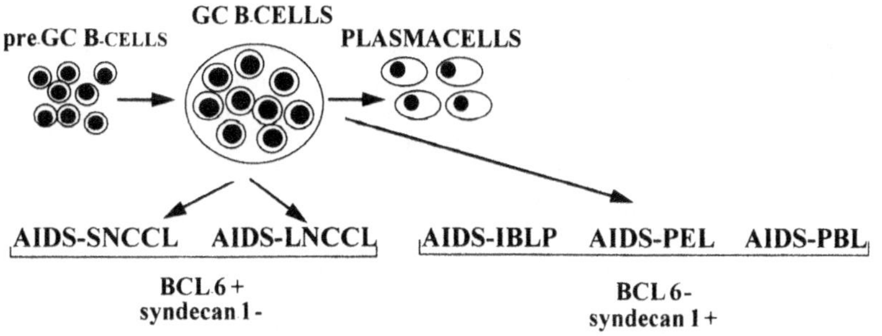

Figure 6. Histogenetic model for AIDS-lymphoma development.

ACKNOWLEDGMENTS

Work by the authors described in this review has been supported by ISS, Programma nazionale di ricerca sull'AIDS—Progetto Patologia clinica e terapia dell'AIDS, Rome, Italy; by "Fondazione CRT", Torino, Italy; by "Fondazione Piera, Pietro e Giovanni Ferrero", Alba, Italy; and by National Institutes of Health grant CA-37295, USA.

REFERENCES

1. The Non-Hodgkin's Lymphoma Pathologic Classification Project. National Cancer Institute sponsored study of classifications of non-Hodgkin's lymphomas. Summary and description of a working formulation for clinical usage. Cancer 49:2112, 1982.
2. Harris NL, Jaffe ES, Stein H, Banks PM, Chan JKC, Cleary ML, Delsol G, De Wolf-Peeters C, Falini B, Gatter KC, Grogan TM, Isaacson PG, Knowles DM, Mason DY, Muller-Hermelink H-K, Pileri S, Piris MA, Ralfkiaer E, Warnke RA: A revised European-American classification of lymphoid neoplasms: a proposal from the International Lymphoma Study Group. Blood 84:1361, 1994.
3. Carbone A, Tirelli U, Vaccher E, Volpe R, Gloghini A, Bertola G. De Re V, Rossi C, Boiocchi M, Monfardini S: A clinicopathologic study of lymphoid neoplasias associated with human immunodeficiency virus infection in Italy. Cancer 68:842, 1991.
4. Chadburn A, Cesarman E, Jagirdar J, Subar M, Mir RN, Knowles DM: CD30 (Ki-1) positive anaplastic large cell lymphomas in individuals infected with the human immunodeficiency virus. Cancer 72:3078, 1993.
5. Tirelli U, Vaccher E, Zagonel V, Talamini R, Bernardi D, Tavio M, Gloghini A, Merola MC, Monfardini S, Carbone A: CD30 (Ki-1)—positive anaplastic large-cell lymphomas in 13 patients with and 27 patients without human immmunodeficiency virus infection: The first comparative clinicopathologic study from a single institution that also includes 80 patients with other human immunodeficiency virus-related systemic lymphomas. J Clin Oncol: 13:373, 1995.
6. Raphael M, Gentilhomme O, Tulliez M, Byron P-A, Diebold J, for the French Study Group of Pathology for Human Immunodeficiency Virus-Associated Tumors: Histopathologic features of high-grade non-Hodgkin's lymphomas in acquired immunodeficiency syndrome. Arch Pathol Lab Med 115:15, 1991.
7. Carbone A, Gloghini A, Gaidano G, Cilia AM, Bassi P, Polito P, Vaccher E, Saglio G, Tirelli U: AIDS-related Burkitt's lymphoma. Morphologic and immunophenotypic study of biopsy specimens. Am J Clin Pathol 103:561, 1995.
8. Remick SC, Diamond C, Migliozzi JA, Solis O, Wagner H, Haase RF, Ruckdeschel JC: Primary central nervous system lymphoma in patients with and without the acquired immune deficiency syndrome. A retrospective analysis and review of the literature. Medicine 69:345, 1990.
9. Camilleri-Broët S, Davi F, Feuillard J, Bourgeois C, Seilhean D, Hauw J-J, Raphaël M: High expression of latent membrane protein 1 of Epstein-Barr virus and BCL-2 oncoprotein in acquired immunodeficiency syndrome-related primary brain lymphomas. Blood 86:432, 1995.
10. Carbone A, Gaidano G: HHV-8 positive body cavity-based lymphoma: A novel lymphoma entity. Br J Haematol 97:515, 1997.
11. Cesarman E, Chang Y, Moore PS, Said JW, Knowles DM. Kaposi's sarcoma-associated herpesvirus-like DNA sequences in AIDS-related body cavity-based lymphomas. N Engl J Med 332:1186, 1995.
12. Carbone A, Gloghini A, Vaccher E, Zagonel V, Pastore C, Dalla Palma P, Branz F, Saglio G, Volpe R, Tirelli U, Gaidano G: Kaposi's sarcoma-associated herpesvirus DNA sequences in AIDS-related and AIDS-unrelated lymphomatous effusions. Br J Haematol 94:533, 1996.
13. Nador RG, Cesarman E, Knowles DM, Said JW: Herpes-like DNA sequences in a body-cavity-based lymphoma in an HIV-negative patient. N Engl J Med 333:943, 1995.
14. Nador RG, Cesarman E, Chadburn A, et al. Primary effusion lymphoma: a distinct clinicopathologic entity associated with the Kaposi's sarcoma-associated herpes virus. Blood 88:645–656, 1996.
15. Said JW, Tasaka T, Takeuchi S, Asou H, de Vos S, Cesarman E, Knowles DM, Koeffler HP: Primary effusion lymphoma in women: report of two cases of Kaposi's sarcoma herpes virus-associated effusion-based lymphoma in human immunodeficiency virus-negative women. Blood 88:3124, 1996.

16. Said JW, Chien K, Tasaka T, Takeuchi S, Asou H, Cho SK, Cesarman E, Knowles DM, Koeffler HP: Kaposi's sarcoma associated herpesvirus (KSHV or HHV8) in primary effusion lymphomas: ultrastructural demonstration of herpesvirus in lymphoma cell lines. Blood 87:4937, 1996.
17. Ansari MQ, Dawson DB, Nador R, Rutherford C, Schneider NR, Latimer MJ, Picker L, Knowles DM, Mckenna RW: Primary body cavity-based AIDS-related lymphomas. Am J Pathol 105:221, 1996.
18. Cesarman E, Nador RG, Aozasa K, Delsol G, Said JW, Knowles DM: Kaposi's sarcoma-associated herpesvirus in non-AIDS-related lymphomas occurring in the body cavities. Am J Pathol 149:53, 1996.
19. Gaidano G, Pastore C, Gloghini A, Cusini M, Nomdedeu J, Volpe G, Capello D, Vaccher E, Bordes R, Tirelli U, Saglio G, Carbone A: Distribution of human herpesvirus-8 sequences throughout the spectrum of AIDS-related neoplasia. AIDS 10:941, 1996.
20. Jaffe ES. Primary body-cavity-based AIDS-related lymphomas. Evolution of a new disease entity. Am J Clin Pathol 105:141, 1996.
21. Morassut S, Vaccher E, Balestreri L, Gloghini A, Gaidano G, Volpe R, Tirelli U, Carbone A: HIV-associated HHV-8 positive primary lymphomatous effusions: Radiological findings in six patients. Radiology 205:459, 1997.
22. Komanduri KV, Luce JA, McGrath MS, Herndier BG, Ng VL: The natural history and molecular heterogeneity of HIV-associated primary malignant lymphomatous effusions. J Acquir Immune Defic Syndr 13:215, 1996.
23. Hsi ED, Foreman KE, Duggan J, Alkan S, Kauffman CA, Aronow HD, Nickoloff BJ: Molecular and pathologic characterization of an AIDS-related body cavity-based lymphoma, including ultrastructural demonstration of human herpesvirus-8. Am J Surg Pathol 22:493, 1998.
24. Gaidano G, Gloghini A, Gattei V, Rossi MF, Cilia AM, Godeas C, Degan M, Perin T, Canzonieri V, Aldinucci D, Saglio G, Carbone A, Pinto A: Association of KSHV positive primary effusion lymphoma with expression of the CD138/syndecan-1 antigen. Blood 90:4894, 1997.
25. Pastore C, Gloghini A, Volpe G, Nomdedeu J, Leonardo E, Mazza U, Saglio G, Carbone A, Gaidano G: Distribution of Kaposi's sarcoma herpesvirus sequences among lymphoid malignancies in Italy and Spain. Br J Haematol 91:918, 1995.
26. Carbone A, Tirelli U, Gloghini A, Vaccher E, Gaidano G: Herpesvirus-like DNA sequences selectively cluster with body-cavity-based lymphomas throughout the spectrum of AIDS-related lymphomatous effusions. Eur J Cancer 32A:555, 1996.
27. Otsuki T, Kumar S, Ensoli B, Kingma DW, Yano T, Stetler-Stevenson M, Jaffe ES, Raffeld M: Detection of HHV-8/KSHV DNA sequences in AIDS-associated extranodal lymphoid malignancies. Leukemia 10:1358, 1996.
28. Gaidano G, Daniela Capello, Anna Maria Cilia, Annunziata Gloghini, Tiziana Perin, Silvia Quattrone, Anna Migliazza, Francesco Lo Coco, Giuseppe Saglio, Valeria Ascoli, Antonino Carbone: Genetic characterization of HHV-8/KSHV positive primary effusion lymphoma reveals frequent mutations of *BCL*-6: implications for disease pathogenesis and histogenesis. Genes Chromosom Cancer 24:16, 1999.
29. IARC Monographs on the Evaluation of Carcinogenic Risks to Humans, Vol. 67, Human Immunodeficiency Viruses and Human T-Cell Lymphotropic Viruses. International Agency for Research on Cancer / World Health Organization, Lyon, 1996.
30. Carbone A, Vaccher E, Barzan L, Gloghini A, Volpe R, De Re V, Boiocchi M, Monfardini S, Tirelli U: Head and neck lymphomas associated with human immunodeficiency virus infection. Arch Otolaryngol Head Neck Surg 121:210, 1995.
31. Brahim JS, Katz RW, Roberts MW: Non-Hodgkin's lymphoma of the hard palate mucosa and buccal gingiva associated with AIDS. J Oral Maxillofac Surg 46:328, 1988.
32. Green TL, Eversole LR: Oral lymphomas in HIV-infected patients: Association with Epstein-Barr virus DNA. Oral Surg Oral Med Oral Pathol 67:437, 1989.
33. Kaugars GE, Burns JC: Non-Hodgkin's lymphoma of the oral cavity associated with AIDS. Oral Surg Oral Med Oral Pathol 67:433, 1989.
34. Leess FR, Kessler DJ, Mickel RA: Non-Hodgkin's lymphoma of the head and neck in patients with AIDS. Arch Otolaryngol Head Neck Surg 113:1104, 1987.
35. Shapiro AL, Shechtman FG, Guida RA, Kimmelman CP: Head and neck lymphoma in patients with the acquired immune deficiency syndrome. Otolaryngol Head Neck Surg 106:258, 1992.
36. Groot RH, Van Merkesteyn JP, Bras J: Oral manifestations of non-Hodgkin's lymphoma in HIV-infected patients. Int J Oral Maxillofac Surg 19:194, 1990.
37. Epstein JB, Silverman S Jr: Head and neck malignancies associated with HIV infection. Oral Surg Oral Med Oral Pathol 73:193, 1992.

38. Delecluse HJ, Anagnostopoulos I, Dallenbach F, Hummel M, Marafioti T, Schneider U, Huhn D, Schmidt-Westhausen A, Reichart PA, Gross U, Stein H: Plasmablastic lymphomas of the oral cavity: a new entity associated with the human immunodeficiency virus infection. Blood 89:1413, 1997.
39. Carbone A, Gloghini G, Canzonieri V, Tirelli U, Gaidano G: AIDS-related extranodal non-Hodgkin's lymphomas with plasma cell differentiation. Blood 90:1337, 1997.
40. Gaidano G, Carbone A, Dalla-Favera R: Pathogenesis of AIDS-related lymphomas: molecular and histogenetic heterogeneity. Am J Pathol 152:623, 1998.
41. Gaidano G, Carbone A, Dalla-Favera R: Genetic basis of acquired immunodeficiency syndrome-related lymphomagenesis. J Natl Cancer Inst Monographs 23:95, 1998.
42. Carbone A, Gaidano G, Gloghini A, Larocca LM, Capello D, Canzonieri V, Antinori A, Tirelli U, Falini B, Dalla-Favera R: Differential expression of BCL-6, CD138/syndecan-1, and EBV-encoded latent membrane protein-1 identifies distinct histogenetic subsets of acquired immunodeficiency syndrome-related non-Hodgkin's lymphomas. Blood 91:747, 1998.

26

GENESIS AND PROGRESSION OF GASTRIC MUCOSA-ASSOCIATED LYMPHOID TISSUE (MALT) LYMPHOMA

Ming-Qing Du and Peter G. Isaacson

Department of Histopathology
University College London Medical School
London, UK

Keywords: gastric MALT lymphoma, biology, pathogenesis.

ABSTRACT

MALT lymphomas typically arise in extranodal tissues, such as the stomach, which are normally devoid of lymphoid tissue but which have acquired MALT as the result of a chronic inflammatory disorder. In the stomach this process is exemplified by chronic gastritis caused by infection with *H. pylori*. The lymphoma is at first confined to the mucosa and later progresses with deeper invasion of the gastric wall, local and systemic dissemination and finally, transformation to high grade lymphoma. In its early phase the growth of gastric MALT lymphoma is stimulated by *H. pylori* specific T-cells and to a lesser extent, by an autoantigen. It is in this phase that eradication of H pylori may result in clinical regression of the lymphoma. Molecular genetic studies have suggested that early events in the evolution of gastric MALT lymphoma from "acquired" MALT include "genetic instability" as indicated by replication error phenotype, t(11;14), trisomy 3, partial p53 inactivation, and c-myc mutation. Later molecular events include, probably t(1;14) and other uncharacterised, which cause complete malignant transformation and result in escape of T-cell dependency. Finally, further genetic events such as complete inactivation of the tumour suppressor genes *p53* and p16, and possi-

---

Corresponding author: Dr Ming-Qing Du, Department of Histopathology, UCL Medical School, Rockefeller Building, University Street, London WC1E 6JJ, United Kingdom, Tel: +44 171 2096021; Fax: +44 171 3873674; E-mail: m.du@ucl.ac.uk

Molecular Biology of Hematopoiesis 6, edited by Abraham *et al.*
Kluwer Academic/Plenum Publishers, New York, 1999.

ble activation of *c-myc* oncogene by translocation or other undetermined abnormalities can result in high grade transformation.

1. INTRODUCTION

The concept that mucosa-associated lymphoid tissue (MALT) lymphomas constitute a distinct clinicopathological entity was proposed by Isaacson and Wright based on studies of immunoproliferative small intestinal disease (IPSID) and low grade primary B-cell lymphoma of the stomach in 1983.[1] Since then, the MALT lymphoma concept has been extended to include lymphomas arising from other extranodal sites such as the salivary gland, lung, thyroid, ocular adnexa, liver etc.[2–7] Of these various extranodal sites, the stomach is the most common. Gastric MALT lymphoma is therefore the best characterised and serves as the paradigm for MALT lymphoma group as whole.

Gastric MALT lymphomas arise from acquired MALT as a consequence of *H. pylori* infection[8–10] and run an unusually indolent clinical course.[11] Although these tumours are characterised by their tendency to remain localised for long periods, progression is well documented. Wotherspoon et al. systematically examined gastrectomy specimens of 5 overt low grade gastric MALT lymphomas[12] and showed that numerous small tumour foci were distributed throughout the gastric mucosa including macroscopically normal regions. Subsequent investigations by microdissection and molecular genetic analysis of the rearranged Ig gene confirmed the clonal identity of these multiple tumour foci [13] and further demonstrated that tumour cells were frequently present in reactive lymphoid tissue that showed no histological evidence of lymphoma.[14] In a recent study by Hoshida et al., intramucosal dissemination without involvement of the muscularis propria was demonstrated in the majority of low grade gastric MALT lymphoma at stages I or II1E.[15] Hence, it appears that low grade gastric MALT lymphoma is a mutifocal disease of the gastric mucosa.

Regional lymph nodes are also frequently involved by gastric MALT lymphoma,[11,16] although dissemination beyond the gastric lymph nodes is rarely seen at the time of diagnosis. Du et al. have studied 5 cases of concurrent gastric and intestinal MALT lymphoma and showed that the intestinal lesion was secondary to the gastric lymphoma.[17] Gastric MALT lymphoma has recently been shown preferentially to disseminate to the splenic marginal zone.[18]

Low grade MALT lymphoma is also capable of transforming into a high grade tumour. Initial suggestion for this comes from the observation of coexistence of low grade MALT and high grade components in the same primary gastric lymphoma.[19] This has now been confirmed by the finding of identical Ig gene sequences in both low and high grade components of such cases.[20]

2. THE NORMAL CELL COUNTERPART OF MALT LYMPHOMA

Several lines of evidence indicate that the normal cell counterpart of MALT lymphoma is the IgM$^+$ memory B-cell of the marginal zone. Histologically, MALT lymphoma cells predominantly infiltrate in the region around the reactive lymphoid follicles. Immunophenotypically, MALT lymphoma cells show a memory B-cell phenotype, which is almost identical to that of marginal zone B-cells of the spleen and

Peyer's patch, being $CD20^+$, $CD5^-$, $CD10^-$, $CD11c^-$, $CD23^-$, $CD43^{-/+}$, IgM^+, IgD^-.[11,21] Genotypically, both MALT lymphoma cells[17,22,23] and marginal zone B-cells of the spleen[24] and Peyer's patch[25] display a similar degree of frequent somatic mutations in their rearranged Ig heavy chain variable (VH) genes. The mutation pattern in the majority of MALT lymphomas shows features of either positive or negative selection, suggesting they have gone through the germinal centre reaction.

The property of the precursor cells of gastric MALT lymphoma appears to be autoreactive, despite the intimate association of *H. pylori* infections with the tumour development. Autoimmue phenomena are a well recognised component of *H. pylori* associated gastritis.[26] Hussell et al. have studied the tumour Ig specificity in 3 cases of gastric MALT lymphoma using immunohistochemistry and ELISA and demonstrated distinct specific immuno-reactivity to normal tissue components.[27] More recent studies on VH germline usage show that gastric MALT lymphomas frequently use the VH germline segments commonly involved in autoimmune diseases, such as DP10 and DP54.[22,23,28,29]

3. THE IMMUNOLOGICAL AND MOLECULAR BASIS OF GASTRIC MALT LYMPHOMA

MALT lymphomas do not share the chromosome translocations which characterise nodal low grade B-cell lymphomas such as t(14;18)(q32;q21) in follicular lymphoma.[30] The genesis of MALT lymphoma involves the acquisition of a number of other distinctive genetic abnormalities and also fundamentally dependent on immune stimulation.

3.1. Immunological Stimulation

An increasing body of evidence indicates that persistent antigen stimulation plays an important role in the evolution of monoclonal low grade gastric MALT lymphoma. Antigen stimulation generates an actively proliferating B-cell population and may also directly trigger and sustain, at least in part, the expansion of the tumour clone. Recent studies indicate that both direct and indirect antigen stimulation mechanisms are involved.

3.1.1. Direct Antigen Stimulation. Several histological features of low grade MALT lymphoma, such as the intimate interaction of tumour cells with reactive follicles (known as follicular colonisation) and its association with active tumour cell proliferation and blast transformation,[31,32] suggest that the tumour cells may receive antigen stimulation via antigen receptors. This notion is reinforced by more recent studies of somatic mutations of tumour derived Ig genes. Of the low grade gastric MALT lymphomas examined, the majority show evidence of ongoing mutations.[17,23,33] As somatic mutations occur at the Ig locus specifically in B-cells engaging in the germinal centre reaction, an antigen driven immune response,[34] the finding of ongoing mutations strongly suggests a role of antigen stimulation in the tumour clonal expansion. This is also supported by genetic evidence that evolution of the tumour clone is constantly under a positive or negative selection pressure.[17,23,33] Moreover, crosslinking surface Ig on tumour cells in some low grade MALT lymphomas has been shown to stimulate the tumour cells to proliferate.[35] There are some preliminary data suggesting that the rate of ongoing mutations gradually declines during the evolution from an early to a late stage of low grade tumour and that finally the activity disappears in high grade lesions.[17,23] Thus, it is likely that the role of direct antigen stimulation in the pathogenesis of this tumour progressively decreases during tumour evolution.

3.1.2. Indirect Antigen Stimulation. By co-culture of tumour cells with different strains of heat-killed *H. pylori*, Hussell et al. demonstrated that *H. pylori* induced active proliferation of low grade tumour cells and that the effect was strain specific but T-cell mediated.[36] Subsequent study by these authors confirmed that *H. pylori* did not directly stimulate the lymphoma cells but specifically activated tumour infiltrating T-cells,[37] which then helped the malignant B-cells to proliferate through direct T-B-cell interaction via CD40-mediated signalling.[37,38] Despite this indirect role, the stimulatory effect of *H. pylori* is clearly of major importance since it has been shown in clinical studies that antibiotic eradication of the organism often leads to regression of low grade MALT lymphoma.[39–41]

Unlike low grade tumour cells, high grade tumour cells do not show any growth response to *H. pylori* mediated T-cell stimulation *in vitro*.[36] Similarly, high grade lymphomas do not respond to *H. pylori* eradication therapy.[41,42] These findings suggest that the growth of high grade tumour is no longer dependent on an *H. pylori* mediated immune response.

3.2. Genetic Abnormalities

In addition to driving tumour cell proliferation, *H. pylori* infection also provides a setting that promotes induction of genetic abnormalities. On the one hand, *H. pylori* infection induces and sustains an actively proliferating B-cell population at risk of developing genetic abnormalities, on the other neutrophillic response to the infection constantly generates oxygen reactive species,[43,44] which are potentially genotoxic and can cause a wide range of genetic abnormalities.[45] The $CagA^+$ strain of *H. pylori* is known to be more virulent and to induce epithelial cells to secrete IL-8, which activates neutrophils[44,46] and may ultimately leads to a higher rate of genetic damage in susceptible cells. Recently, the $CagA^+$ strain has been found to be frequently associated with high grade but not low grade gastric MALT lymphomas.[47–49]

3.2.1. Replication Error (RER) Phenotype. The mutator phenotype is caused by a reduced efficacy of DNA mismatch repair machinery[50,51] and cells with the RER^+ phenotype are highly prone to accumulation of somatic mutations and thus at high risk of malignant transformation.[52] Peng et al. examined 40 MALT lymphomas and demonstrated the RER+ phenotype in 52.5% cases, with a slightly higher frequency in high grade (56%) than in low grade (50%) tumours,[53] recently confirmed by Chong et al.[54] These findings are in sharp contrast to those observed in nodal lymphomas in nonimmunocompromised patients, where the RER^+ phenotype was only reported in 0–20% of cases. High grade MALT lymphomas always showed more heterogeneous microsatellite alterations than low grade tumours.[53,54] These findings strongly suggest that microsatellite alterations are an ongoing event in gastric MALT lymphoma, a true tumourigenesis associated phenomenon rather than a reflection of the heterogeneous nature of the normal cell populations.[55] This genetic instability may promote the accumulation of genetic abnormalities during the evolution of gastric MALT lymphoma.

3.2.2. Chromosomal Abnormalities. The limited availability of fresh tissue and poor in vitro growth of the MALT lymphoma cells has hampered cytogenetic study of the tumour. Using both metaphase and interphase cytogenetic techniques, Wotherspoon et al. reported a high frequency of trisomy 3 (60%) in a large series of low grade MALT lymphoma.[56,57] Similar results were observed in subsequent studies and further extended to monocytoid B-cell lymphoma, the nodal equivalent of MALT lymphoma.[58,59] Two recently studies, in which a relatively large series (9 and 44 cases) of MALT lymphomas were examined, revealed the t(11;18)(q21;q21) in 35% of low grade MALT lymphomas,[60,61] confirming several previous case reports of this translocation.[62-65] Interestingly, in all these cases, the t(11;18) was the sole aberration and therefore was thought to represent an early event in the genesis of MALT lymphoma. Curiously, this translocation was not observed in a single case of the 24 high grade MALT lymphomas examined.[60]

Another recurrent translocation observed in low grade MALT lymphoma is the t(1;14)(p22;q32), which has been found in 3 cases.[56,66] Interestingly, tumour cells bearing the t(1;14)(p22;q32) grow spontaneously in vitro without addition of mitogens.

3.2.3. p53 Abnormalities. The transcription factor p53 plays an important role in maintaining the integrity of the genome. p53 triggers G1 arrest or apoptosis in response to DNA damage[67,68] and involves in mitotic spindle checkpoint[69] and in the regulation of centrosome function.[70] Inactivation of p53 function could result in inappropriate replication of damaged DNA, inappropriate cell survival after cellular stress or abnormal segregation of chromosomes during mitosis, therefore contributing to malignant transformation.[71] Loss of p53 function is commonly caused by mutation in one allele and loss of the other and is the most common genetic defect found in human cancer, contributing to both early tumour development and subsequent disease progression.[72,73]

In low grade MALT lymphoma, p53 allele loss and mutation are found in 6.8% and 18.8% cases respectively, with concomitant p53 mutation and allelic loss being rare (1/11 affected cases).[74] This indicates that p53 function is only partly inactivated. High grade MALT lymphomas show more frequent p53 allele loss (28.6%) and mutation (33.3%) than low grade lesions.[74] More importantly, the majority of high grade cases affected (6/9) show both p53 allele loss and mutation, indicating complete p53 inactivation.

3.2.4. c-myc Abnormalities. c-myc acts as a transcription factor and modulates different sets of genes regulating cellular proliferation, differentiation, and apoptosis.[75] In lymphoid malignancies, translocations involving the c-myc locus such as t(8;14)(q24;q32), t(2;8)(p11;q24) and (8;22)(q24;q11) are the major cause of c-myc aberrant expression. In addition, mutations in the regulatory regions of the c-myc gene, which potentially abolish the binding of factors that negatively regulate its expression, are also thought to be responsible for c-myc over-expression.[76-80] These c-myc abnormalities are believed to be the critical genetic event underlying the development of Burkitt's lymphoma[81] and high grade transformation of a proportion of follicular lymphomas.[82]

In low grade MALT lymphoma, both cytogenetic and Southern blot studies have failed to show c-myc translocation or rearrangement.[56,58,60,83] The involvement of c-myc translocation in high grade MALT lymphoma is controversial. Of the limited cytogenetic studies, high grade MALT lymphoma was examined only in one, in which t(8;14)(q24;q32) was reported in 3 of 24 cases analysed.[60] Using Southern blot analysis, van Krieken et al. showed c-myc rearrangement in 6 of 12 high grade gastric B-cell lymphomas.[84] However, this high frequency of c-myc rearrangement was not seen in a recent study in which 11 cases of histologically well defined high grade MALT lymphoma were studied using the same approach.[83] Sequence analysis of the regulatory regions of the c-myc gene has also been carried out in a single study and somatic mutation was found in 16% of low grade and 18% of high grade MALT lymphomas.[83] The significance of these mutations in the MALT lymphomagenesis remains to be determined.

3.2.5. p16. The p16 gene, also known as CDKN2, MTS1, and INK4A, is one of the major negative regulators at the early G1 phase of the cell cycle. It specifically inhibits the cyclin-dependent kinases Cdk4 and Cdk6.[85] Inactivation of the p16 gene, mainly by homozygous deletion, has been found in a wide range of human tumours. In lymphoid malignancies, p16 deletion has been almost exclusively found in high grade tumours.[85,86] Neumeister et al examined 28 low grade and 14 high grade gastric MALT lymphomas and showed homozygous deletion of p16 in 2 high grade cases, suggesting a role for p16 deletion in high grade transformation in some cases.[87]

4. CONCLUSIONS

The evolution of gastric MALT lymphoma comprises sequential development of chronic *H. pylori* associated gastritis, low grade and high grade tumour. On the basis of the observations summarised above, a tentative scheme for the pathogenesis of this tumour is illustrated in Fig. 1. As a result of *H. pylori* infection, B and T cells together with neutrophils, the primary effector cells in the host defence, are recruited to the gastric mucosa and form MALT. The host immune response to *H. pylori* induces and sustains an actively proliferating B-cell population, which is constantly subject to geno-

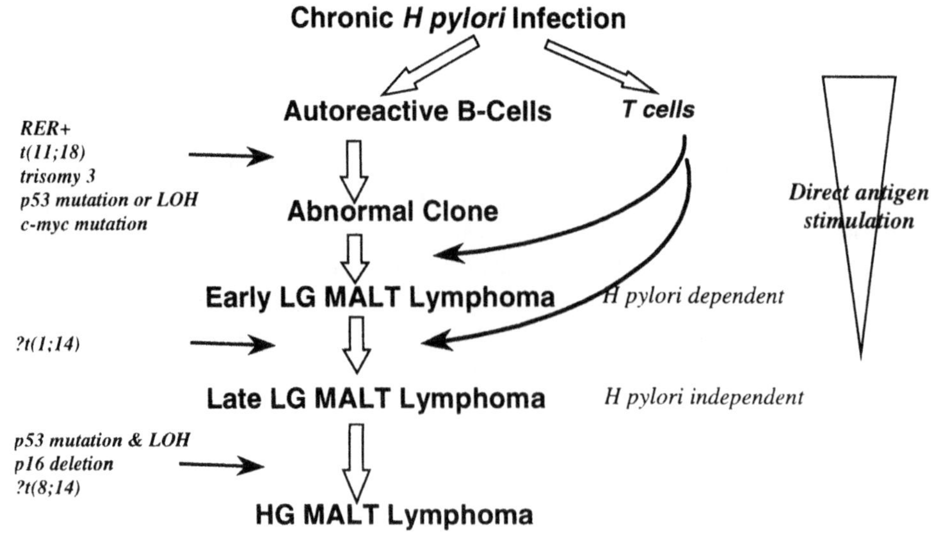

Figure 1. Proposed scheme for the pathogenesis of gastric MALT lymphoma. LOH, loss of heterozygosity. (from Du & Isaacson, Forum 8.2, 1998).

toxic insult by activated neutrophils. During the course of this chronic inflammatory disease, the RER+ phenotype and genetic abnormalities such as t(11;18), trisomy three, partial p53 inactivation may occur in B-cells, particularly in those which recognise autoantigen and already possess a growth advantage, and consequently lead to a partially transformed B-cell clone in rare cases. In the presence of growth help from *H. pylori* specific T-cells, this abnormal B-cell clone may undergo clonal expansion and gradually form a frank low grade MALT lymphoma. At this stage, the tumour is confined to the stomach and will regress following eradication of *H. pylori*. Additional genetic abnormalities, such as the t(1;14) translocation and others uncharacterised, could completely transform this abnormal B-cell clone and result in escape from T-cell dependency. Tumour may now present beyond the stomach and is no longer responsive to *H. pylori* eradication therapy. Finally, further genetic events such as complete inactivation of the tumour suppressor genes p53 and p16, possibly activation of c-myc oncogene by translocation and other undetermined abnormalities can result in high grade transformation.

REFERENCES

1. Isaacson P, Wright DH: Malignant lymphoma of mucosa-associated lymphoid tissue. A distinctive type of B-cell lymphoma. Cancer 52: 1410, 1983.
2. Isaacson P, Wright DH: Extranodal malignant lymphoma arising from mucosa-associated lymphoid tissue. Cancer 53: 2515, 1984.
3. Hyjek E, Smith WJ, Isaacson PG: Primary B-cell lymphoma of salivary glands and its relationship to myoepithelial sialadenitis. Hum Pathol 19: 766, 1988.
4. Addis BJ, Hyjek E, Isaacson PG: Primary pulmonary lymphoma: a re-appraisal of its histogenesis and its relationship to pseudolymphoma and lymphoid interstitial pneumonia. Histopathology 13: 1, 1988.
5. Hyjek E, Isaacson PG: Primary B cell lymphoma of the thyroid and its relationship to Hashimoto's thyroiditis. Hum Pathol 19: 1315, 1988.
6. Wotherspoon AC, Diss TC, Pan LX, Schmid C, Kerr Muir MG, Lea SH, Isaacson PG: Primary low-grade B-cell lymphoma of the conjunctiva: a mucosa-associated lymphoid tissue type lymphoma. Histopathology 23: 417, 1993.
7. Isaacson PG, Banks PM, Best PV, McLure SP, Muller Hermelink HK, Wyatt JI: Primary low-grade hepatic B-cell lymphoma of mucosa-associated lymphoid tissue (MALT)-type. Am J Surg Pathol 19: 571, 1995.
8. Wotherspoon AC, Ortiz Hidalgo C, Falzon MR, Isaacson PG: Helicobacter pylori-associated gastritis and primary B-cell gastric lymphoma [see comments]. Lancet 338: 1175, 1991.
9. Isaacson PG, Spencer J: Gastric lymphoma and Helicobacter pylori. Important Adv Oncol: 111, 1996.
10. Isaacson PG: Recent developments in our understanding of gastric lymphomas. Am J Surg Pathol 20 Suppl 1: S1, 1996.
11. Isaacson PG, Norton AJ: Extranodal lymphomas. Edinburgh, London, Madrid, Melbourne, New York, and Tokyo: Churchill Livergstone, 1994.
12. Wotherspoon AC, Doglioni C, Isaacson PG: Low-grade gastric B-cell lymphoma of mucosa-associated lymphoid tissue (MALT): a multifocal disease. Histopathology 20: 29, 1992.
13. Du MQ, Diss TC, Xu CF, Wotherspoon AC, Pan LX, Isaacson PG: Clonal origin of micro-lymphomas in low grade B-cell gastric MALT lymphoma. J Pathol 181S: A57, 1997 (abstr).
14. Du MQ, Diss TC, Dogan A, Wotherspoon AC, Doglioni C, Pan LX, Isaacson PG: Clone specific PCR reveals wide dissemination of gastric MALT lymphoma to the gastrointestinal mucosa. J Pathol 182S: A12, 1997 (abstr).
15. Hoshida Y, Kusakabe H, Furukawa H, Kasugai T, Miwa H, Ishiguro S, Aozasa K: Reassessment of gastric lymphoma in light of the concept of mucosa-associated lymphoid tissue lymphoma: analysis of 53 patients. Cancer 80: 1151, 1997.
16. Diss TC, Wotherspoon AC, Pan LX, Isaacson PG: PCR detection of tumour clones in local lymph nodes in low grade B-cell gastric MALT lymphoma. J Pathol 176: A33, 1995 (abstr).

17. Du MQ, Xu CF, Diss TC, Peng HZ, Wotherspoon AC, Isaacson PG, Pan LX: Intestinal dissemination of gastric mucosa-associated lymphoid tissue lymphoma. Blood 88: 4445, 1996.
18. Du MQ, Peng HZ, Dogan A, Diss TC, Liu H, Pan LX, Mosley RP, Briskin MJ, Chan JK, Isaacson PG: Preferential dissemination of B-cell gastric mucosa-associated lymphoid tissue (MALT) lymphoma to the splenic marginal zone. Blood 90: 4071, 1997.
19. Chan JK, Ng CS, Isaacson PG: Relationship between high-grade lymphoma and low-grade B-cell mucosa-associated lymphoid tissue lymphoma (MALToma) of the stomach. Am J Pathol 136: 1153, 1990.
20. Peng H, Du M, Diss TC, Isaacson PG, Pan L: Genetic evidence for a clonal link between low and high-grade components in gastric MALT B-cell lymphoma. Histopathology 30: 425, 1997.
21. Isaacson PG, Spencer J: Malignant lymphoma of mucosa-associated lymphoid tissue. Histopathology 11: 445, 1987.
22. Qin Y, Greiner A, Trunk MJ, Schmausser B, Ott MM, Muller Hermelink HK: Somatic hypermutation in low-grade mucosa-associated lymphoid tissue-type B-cell lymphoma. Blood 86: 3528, 1995.
23. Du MQ, Diss TC, Xu CF, Peng HZ, Isaacson PG, Pan LX: Ongoing mutation in MALT lymphoma immunoglobulin VH gene suggests that antigen stimulation plays a role in the clonal expansion. Leukemia 10: 1190, 1996.
24. Dunn Walters DK, Isaacson PG, Spencer J: Analysis of mutations in immunoglobulin heavy chain variable region genes of microdissected marginal zone (MGZ) B cells suggests that the MGZ of human spleen is a reservoir of memory B cells. J Exp Med 182: 559, 1995.
25. Dunn Walters DK, Isaacson PG, Spencer J: Sequence analysis of rearranged IgVH genes from microdissected human Peyer's patch marginal zone B cells. Immunology 88: 618, 1996.
26. Negrini R, Lisato L, Zanella I, Cavazzini L, Gullini S, Villanacci V, Poiesi C, Albertini A, Ghielmi S: Helicobacter pylori infection induces antibodies cross-reacting with human gastric mucosa. Gastroenterology 101: 437, 1991.
27. Hussell T, Isaacson PG, Crabtree JE, Dogan A, Spencer J: Immunoglobulin specificity of low grade B cell gastrointestinal lymphoma of mucosa-associated lymphoid tissue (MALT) type. Am J Pathol 142: 285, 1993.
28. Chapman CJ, Dunn Walters DK, Stevenson FK, Hussell T, Isaacson PG, Spencer J: sequence analysis of immunoglobulin variable region genes which encode autoantigenes expressed by lymphomas of mucosa-asociated lymphoid tissue. J Clin Pathol -Clin Mol Pathol 49: M29, 1996.
29. Thiede C, Alpen B, Morgner A, Ritter M, Wundisch T, Ehninger G, Bayerdorffer E, Stolte M, Neubauer A: Biased VH-gene usage and ongoing somatic mutation after Helicobacter pylori eradication in gastric low grade B-cell lymphoma of MALT. Blood 90S: 414a, 1997 (abstr).
30. Pan L, Diss TC, Cunningham D, Isaacson PG: The bcl-2 gene in primary B cell lymphoma of mucosa-associated lymphoid tissue (MALT). Am J Pathol 135: 7, 1989.
31. Isaacson PG, Wotherspoon AC, Diss T, Pan LX: Follicular colonization in B-cell lymphoma of mucosa-associated lymphoid tissue. Am J Surg Pathol 15: 819, 1991.
32. Isaacson PG, Androulakis Papachristou A, Diss TC, Pan L, Wright DH: Follicular colonization in thyroid lymphoma. Am J Pathol 141: 43, 1992.
33. Qin Y, Greiner A, Hallas C, Haedicke W, Muller Hermelink HK: Intraclonal offspring expansion of gastric low-grade MALT-type lymphoma: evidence for the role of antigen-driven high-affinity mutation in lymphomagenesis. Lab Invest 76: 477, 1997.
34. Jacob J, Kelsoe G, Rajewsky K, Weiss U: Intraclonal generation of antibody mutants in germinal centres [see comments]. Nature 354: 389, 1991.
35. Hussell T, Isaacson PG, Spencer J: Proliferation and differentiation of tumour cells from B-cell lymphoma of mucosa-associated lymphoid tissue in vitro. J Pathol 169: 221, 1993.
36. Hussell T, Isaacson PG, Crabtree JE, Spencer J: The response of cells from low-grade B-cell gastric lymphomas of mucosa-associated lymphoid tissue to Helicobacter pylori. Lancet 342: 571, 1993.
37. Hussell T, Isaacson PG, Crabtree JE, Spencer J: Helicobacter pylori-specific tumour-infiltrating T cells provide contact dependent help for the growth of malignant B cells in low-grade gastric lymphoma of mucosa-associated lymphoid tissue [see comments]. J Pathol 178: 122, 1996.
38. Greiner A, Knorr C, Qin Y, Sebald W, Schimpl A, Banchereau J, Muller Hermelink HK: Low-grade B cell lymphomas of mucosa-associated lymphoid tissue (MALT-type) require CD40-mediated signaling and Th2-type cytokines for in vitro growth and differentiation. Am J Pathol 150: 1583, 1997.
39. Wotherspoon AC, Doglioni C, Diss TC, Pan L, Moschini A, de Boni M, Isaacson PG: Regression of primary low-grade B-cell gastric lymphoma of mucosa-associated lymphoid tissue type after eradication of Helicobacter pylori. Lancet 342: 575, 1993.

40. Wotherspoon AC: Gastric MALT lymphoma and Helicobacter pylori. Yale J Biol Med 69: 61, 1997.
41. Thiede C, Morgner A, Alpen B, Wundisch T, Herrmann J, Ritter M, Ehninger G, Stolte M, Bayerdorffer E, Neubauer A: What role does Helicobacter pylori eradication play in gastric MALT and gastric MALT lymphoma? gastroenterology 113: s61, 1997.
42. Bayerdorffer E, Neubauer A, Rudolph B, Thiede C, Lehn N, Eidt S, Stolte M: Regression of primary gastric lymphoma of mucosa-associated lymphoid tissue type after cure of Helicobacter pylori infection. MALT Lymphoma Study Group. Lancet 345: 1591, 1995.
43. Neilsen H, Andersen LP: Activation of human phagocyte oxidative metabolism by Helicobacter pylori. Gastroenterology 103: 1747, 1992.
44. Crabtree JE, Spencer J: Immunologic aspects of Helicobacter pylori infection and malignant transformation of B cells. Semin Gastrointest Dis 7: 30, 1996.
45. Cerutti PA, Trump BF: Inflammation and oxidative stress in carcinogenesis. Cancer Cells 3: 1, 1991.
46. Rushton PJ, Hooley R, Lazarus CM: Aleurone nuclear proteins bind to similar elements in the promoter regions of two gibberellin-regulated alpha-amylase genes. Plant Mol Biol 19: 891, 1992.
47. Peng HZ, Ranaldi R, Diss TC, Issacson PG, Bearzi I, Pan LX: High grequency of CagA+ Helicobacter pylori infection in high grade gastric MALT B-cell lymphomas. J Pathol 1997 185: 409, 1998.
48. de Jong D, van der Hulst RW, Pals G, van Dijk WC, van der Ende A, Tytgat GN, Taal BG, Boot H: Gastric non-Hodgkin lymphomas of mucosa-associated lymphoid tissue are not associated with more aggressive Helicobacter pylori strains as identified by CagA [see comments]. Am J Clin Pathol 106: 670, 1996.
49. Eck M, Schmausser B, Haas R, Greiner A, Czub S, Muller Hermelink HK: MALT-type lymphoma of the stomach is associated with Helicobacter pylori strains expressing the CagA protein. Gastroenterology 112: 1482, 1997.
50. Jiricny J: Colon cancer and DNA repair: have mismatches met their match? Trends Genet 10: 164, 1994.
51. Leach FS, Nicolaides NC, Papadopoulos N, Liu B, Jen J, Parsons R, Peltomaki P, Sistonen P, Aaltonen LA, Nystrom Lahti M, et al: Mutations of a mutS homolog in hereditary nonpolyposis colorectal cancer. Cell 75: 1215, 1993.
52. Parsons R, Li GM, Longley MJ, Fang WH, Papadopoulos N, Jen J, de la Chapelle A, Kinzler KW, Vogelstein B, Modrich P: Hypermutability and mismatch repair deficiency in RER+ tumor cells. Cell 75: 1227, 1993.
53. Peng H, Chen G, Du M, Singh N, Isaacson PG, Pan L: Replication error phenotype and p53 gene mutation in lymphomas of mucosa-associated lymphoid tissue. Am J Pathol 148: 643, 1996.
54. Chong JM, Fukayama M, Hayashi Y, Hishima T, Funta N, Koike M, Matsuya S, Konishi M, Miyaki M: Microsatellite instability and loss of heterozygosity in gastric lymphoma. Lab Invest 77: 639, 1997.
55. Simpson AJG: The natural somatic mutation frequency and human carcinogenesis. Advances in cancer research 71: 209, 1997.
56. Wotherspoon AC, Pan LX, Diss TC, Isaacson PG: Cytogenetic study of B-cell lymphoma of mucosa-associated lymphoid tissue. Cancer Genet Cytogenet 58: 35, 1992.
57. Wotherspoon AC, Finn TM, Isaacson PG: Trisomy 3 in low-grade B-cell lymphomas of mucosa-associated lymphoid tissue. Blood 85: 2000, 1995.
58. Dierlamm J, Pittaluga S, Wlodarska I, Stul M, Thomas J, Boogaerts M, Michaux L, Driessen A, Mecucci C, Cassiman JJ, et al: Marginal zone B-cell lymphomas of different sites share similar cytogenetic and morphologic features [see comments]. Blood 87: 299, 1996.
59. Brynes RK, Almaguer PD, Leathery KE, McCourty A, Arber DA, Medeiros LJ, Nathwani BN: Numerical cytogenetic abnormalities of chromasomes 3, 7, and 12 in marginal zone B-cell lymphomas. Hum Pathol 9: 995, 1996.
60. Ott G, Katzenberger T, Greiner A, Kalla J, Rosenwald A, Heinrich U, Ott MM, Muller Hermelink HK: The t(11;18)(q21;q21) chromosome translocation is a frequent and specific aberration in low-grade but not high-grade malignant non-Hodgkin's lymphomas of the mucosa-associated lymphoid tissue (MALT-) type. Cancer Res 57: 3944, 1997.
61. Auer IA, Gascoyne RD, Connors JM, Cotter FE, Greiner TC, Sanger WG, Horsman DE: t(11;18)(q21;q21) is the most common translocation in MALT lymphomas. Ann Oncol 8: 979, 1997.
62. Levine EG, Arthur DC, Machnicki J, Frizzera G, Hurd D, Peterson B, Gajl Peczalska KJ, Bloomfield CD: Four new recurring translocations in non-Hodgkin lymphoma. Blood 74: 1796, 1989.
63. Horsman D, Gascoyne R, Klasa R, Coupland R: t(11;18)(q21;q21.1): a recurring translocation in lymphomas of mucosa-associated lymphoid tissue (MALT)? Genes Chromosom Cancer 4: 183, 1992.

64. Griffin CA, Zehnbauer BA, Beschorner WE, Ambinder R, Mann R: t(11;18)(q21;q21) is a recurrent chromosome abnormality in small lymphocytic lymphoma. Genes Chromosomes Cancer 4: 153, 1992.
65. Leroux D, Seite P, Hillion J, Le Marc'Hadour F, Pegourie Bandelier B, Jacob MC, Larsen CJ, Sotto JJ: t(11;18)(q21;q21) may delineate a spectrum of diffuse small B-cell lymphoma with extranodal involvement. Genes Chromosom Cancer 7: 54, 1993.
66. Kristoffersson U, Heim S, Olsson H, Akerman M, Mitelman F: Cytogenetic studies in non-Hodgkin lymphomas–results from surgical biopsies. Hereditas 104: 1, 1986.
67. Kastan MB, Onyekwere O, Sidransky D, Vogelstein B, Craig RW: Participation of p53 protein in the cellular response to DNA damage. Cancer Res 51: 6304, 1991.
68. Lane DP: Cancer. p53, guardian of the genome [news; comment]. Nature 358: 15, 1992.
69. Cross SM, Sanchez CA, Morgan CA, Shimke MK, Ramel S, Idzerda RL, Raskind WH, Reid BJ: A p53-dependent mouse spindle checkpoint. Science 267: 1353, 1998.
70. Ishizaki K, Nishizawa K, Mimaki S, Aizawa S: UV-induced mutations of supF gene on a shuttle vector plasmid in p53-deficient mouse cells are qualitatively different from those in wild-type cells. Mutat Res 364: 43, 1996.
71. Morgan SE, Kastan MB: p53 and ATM: cell cycle, cell death, and cancer. Adv Cancer Res 71: 1, 1997.
72. Imamura J, Miyoshi I, Koeffler HP: p53 in hematologic malignancies. Blood 84: 2412, 1994.
73. Hall PA, Lane DP: Tumor suppressors: a developing role for p53? Curr Biol 7: R144, 1997.
74. Du M, Peng H, Singh N, Isaacson PG, Pan L: The accumulation of p53 abnormalities is associated with progression of mucosa-associated lymphoid tissue lymphoma. Blood 86: 4587, 1995.
75. Evan GI, Littlewood TD: The role of c-myc in cell growth. Curr Opin Genet Dev 3: 44, 1993.
76. Eick D, Berger R, Polack A, Bornkamm GW: Transcription of c-myc in human mononuclear cells is regulated by an elongation block. Oncogene 2: 61, 1987.
77. Zajac Kaye M, Levens D: Phosphorylation-dependent binding of a 138-kDa myc intron factor to a regulatory element in the first intron of the c-myc gene. J Biol Chem 265: 4547, 1990.
78. Johnston JM, Carroll WL: c-myc hypermutation in Burkitt's lymphoma. Leuk Lymphoma 8: 431, 1992.
79. Yu BW, Ichinose I, Bonham MA, Zajac Kaye M: Somatic mutations in c-myc intron I cluster in discrete domains that define protein binding sequences. J Biol Chem 268: 19586, 1993.
80. Hann SR: Regulation and function of non-AUG-initiated proto-oncogenes. Biochimie 76: 880, 1994.
81. Korsmeyer SJ: Chromosomal translocations in lymphoid malignancies reveal novel proto-oncogenes. Annu Rev Immunol 10: 785, 1992.
82. Yano T, Jaffe ES, Longo DL, Raffeld M: MYC rearrangements in histologically progressed follicular lymphomas. Blood 80: 758, 1992.
83. Peng HZ, Diss TC, Isaacson PG, Pan LX: C-myc gene abnormalities in mucosa associated lymphoid tissue (MALT) lymphomas. J Pathol 181: 381–6, 1997.
84. van Krieken JH, Raffeld M, Raghoebier S, Jaffe ES, van Ommen GJ, Kluin PM: Molecular genetics of gastrointestinal non-Hodgkin's lymphomas: unusual prevalence and pattern of c-myc rearrangements in aggressive lymphomas. Blood 76: 797, 1990.
85. Hunter T, Pines J: Cyclins and cancer. II: Cyclin D and CDK inhibitors come of age [see comments]. Cell 79: 573, 1994.
86. Hall M, Peters G: Genetic alterations of cyclins, cyclin-dependent kinases, and cdk inhibitors in human cancer. Adv Cancer Res 68: 67, 1996.
87. Neumeister P, Hoefler G, Beham Schmid C, Schmidt H, Apfelbeck U, Schaider H, Linkesch W, Sill H: Deletion analysis of the p16 tumor suppressor gene in gastrointestinal mucosa-associated lymphoid tissue lymphomas. Gastroenterology 112: 1871, 1997.

27

CYTOKINES, THROMBOPOIETIN, AND SELECTINS IN IMMUNE THROMBOCYTOPENIC PURPURA

İbrahim C. Haznedaroğlu, Yahya Büyükaşık, Ali Koşar, Osman İ. Özcebe, Şerafettin Kirazlı, and Semra V. Dündar

Hacettepe University Medical School
Department of Hematology
Ankara/TURKEY

Megakaryocytopoiesis is a perplexing enigmatic process dependent on both early and late-acting hematopoietic growth factors. The proliferation and maturation steps of this process are regulated by lineage non-specific megakaryocytopoietic cytokines and thrombopoietin (TPO).[1,2] TPO is the most critical cytokine regulator of platelet production and maturation. Likewise, Interleukin-6 (IL-6) has the most potent thrombopoietic activity of any agent except thrombopoietin and likely to have a physiological role in this process.[3]

Autoimmune (idiopathic) thrombocytopenic purpura (ITP) is a heterogeneous disease characterized by immune platelet destruction and reactive/compensatory megakaryocytopoiesis.[4] Ineffective/impaired platelet production have also been identified in most ITP patients, suggesting that the mechanism of thrombocytopenia in ITP may be due to both impaired platelet production and increased platelet clearance.[5] Evaluation of circulating cytokine levels in various thrombocytopenias, particularly in ITP, is a current debate.[6–8] Determination of circulating "de novo" megakaryocyte-related cytokine levels and after therapeutic intervention affects their levels may represent an important clue for the complex regulation of thrombopoiesis in physiologic and pathologic conditions and also for understanding the pathobiology of thrombocytopenia. In this review, we will summarize our recently published three studies[9–11] which were designed to investigate circulating cytokine, TPO, and selectin concentrations in ITP.

The mechanisms that regulate serum TPO concentration and platelet count in circulation have been a subject of debate. In patients with aplastic anemia (AA), decrements in both platelet and megakaryocyte mass may contribute to the marked elevation

Molecular Biology of Hematopoiesis 6, edited by Abraham *et al.*
Kluwer Academic/Plenum Publishers, New York, 1999.

of circulating TPO levels. Although circulating TPO is inversely proportional to platelet number in thrombocytopenias, serum TPO concentrations are well below the elevation observed in other thrombocytopenias, low or undetectable in ITP. This observation has been tried to be explained with the increased megakaryocyte mass capable of absorbing TPO or release of sufficient number of platelets to adsorb serum TPO prior to their rapid destruction by immune mechanisms.[1,5–7] We have proposed an alternative novel hypothesis different from this prevailing view of TPO in ITP.[12–14] We suggested that the leading reason of thrombocytopenia in ITP may be faulty platelet formation from megakaryocytes inducing rapid platelet clearance. Despite the increased megakaryocytopoietic cytokines in the proliferation phase of megakaryocytopoiesis in ITP.[12–14] TPO which stimulate the final maturation of megakaryocytes is relatively inadequate.[6,7] Accordingly, endogenous TPO deficiency in respect to the elevated megakaryocytopoietic cytokines may provoke an excess of immature megakaryocytes invading the bone marrow of ITP patients. Immature megakaryocytes might blossom bizarre platelets which is "foreign" to the immune system and the clearance of autoreactive antibody-coated "non-self" platelets by activated tissue macrophages is facilitated. The findings of undetectable/deficient circulating TPO levels and increased concentrations of megakaryocytopoietic cytokines in patients with ITP may be considered as important initial clues for the precision of this hypothesis. In our studies,[9–11] TPO level was below the detection limit in most of the patients with ITP, while IL-6 concentrations were significantly elevated. These results gave support to our previously published hypothesis. One of the most well known functions of the IL-6 is in the regulation of the inflammatory and acute phase responses. It plays role in stimulation and differentiation T and B lymphocytes and monocytes. So, it may be assumed that IL-6 concentration may be elevated as a reflection of the immunologic response in ITP. L-selectin is expressed on the surface of most leukocytes. After leukocyte activation, L-selectin is rapidly shed from the cell surface. Accordingly, upon leukocyte inducement plasma sL-selectin increases, when L-selectin expression decreases on the leukocyte. Plasma sE-selectin concentration is also known to be increased by the inflammatory cytokines, a sign of endothelial activation. Normal levels of sL-selectin and sE-selectin (i.e. normal leukocyte and endothelial activity) in ITP suggest that increment of IL-6 may not be a sign of acute phase response.[11] Instead of that, it may be related to the compensatory megakaryocytopoiesis and platelet hyperactivity.[9–11]

However, there are some serious flaws in this developing scenario of immune platelet destruction. Labeling studies suggest that the rate of platelet production in ITP is not reduced.[6] Although, McCarty et al. reported that TPO mRNA levels in bone marrow are modulated by platelet count after thrombocytopenia.[15] The current weight of evidence suggests that TPO production does not fluctuate at the transcriptional level and direct evidence that production is faulty in ITP patients may be required to support the alternative hypothesis. McCarty et al. reported that TPO mRNA levels in bone marrow are increased by thrombocytopenia after irradiation/carboplatinum therapy and anti-platelet sera administration.[15] They also described that TPO transcripts decreased in bone marrow and spleen in TPO-treated mice. In contrast to McCarty and coworkers, Stoffel et al. suggested that TPO mRNA level is not regulated during thrombocytopenia, but platelets have a direct regulatory role on TPO activity at a post-transcriptional level.[16] Fielder et al., in consistent with Stoffel et al., have shown that TPO bound to the circulating platelets was degraded into fragments, and that platelets regulate plasma TPO concentrations by binding to the Mpl on platelets.[17] Finally, Cohen-Solal et al. suggested constitutive expression of Mpl ligand transcripts during thrombocytopenia or thrombocytosis.[18] Thus, literature data, excluding the opposite study by McCarty et al., agree with the conclusion that TPO mRNA levels do not change in response to platelet counts.[15-19] Abnormalities in TPO-knockout mice have shown that TPO is not purely a maturation factor and that TPO deficiency significantly reduces normal megakaryocyte progenitor development as well as differentiation.[20] If ineffective TPO concentrations operate in ITP, a reduced number of megakaryocyte progenitors as well as megakaryocytes of all differentiation stages might be expected but is not observed in the disease. Moreover, TPO-knockout mice provide a model where TPO deficiency leads to thrombocytopenia and relative immaturity of those megakaryocytes which do develop. Nevertheless, the residual platelets that are produced appear structurally and functionally normal[21] and there is no evidence of immune destruction.

If the rate of platelet destruction in ITP can be overcome by pharmacological doses of TPO, the cytokine may indeed prove useful in managing the disease. However, there is little current evidence to support the hypothesized role for TPO in disease development. Higher uptake and degradation of TPO by platelets in ITP or increased megakaryocyte mass seems to be the most appropriate explanation.

In summary, it seems that low TPO concentration of ITP in our studies,[9,10] 1) may be a result of disease process and also 2) it might play a role in the pathobiology of the disease. 1) TPO is produced in a constant rate and catabolised by the end-cells, i.e. platelets and megakaryocytes. Platelet and megakaryocyte masses are well known factors determining the rate of TPO catabolism. So, different results of TPO in ITP also can be attributed to the heterogenity of platelet production rate (increased, normal or decreased) and megakaryocyte mass (increased, normal or even occasionally low) in this disease. We can suggest that increased megakaryocyte mass and consequently TPO clearance were responsible for low baseline TPO concentrations in our patients. Enhanced c-mpl expression and consequently TPO-sensitivity of megakaryocytes could also be responsible for decreased plasma TPO concentrations in active ITP. 2) It has been proposed that deficient values of TPO could lead to insufficient maturation of the megakaryocytes and impaired thrombopoiesis. Immature megakaryocytes, which are proliferated by the act of compensatorily increased non-lineage-specific cytokines consume more TPO leading to persisting deficiency of this molecule and a vicious circle goes on. In the study,[9] increase of TPO after steroid-induced recovery of ITP was possibly due to decrease of megakaryocyte mass and/or TPO-sensitivity and accordingly decrease of megakaryocyte-related TPO catabolism. The positive relationship between TPO level and platelet count, which is the other end-cell for this molecule suggests that platelets clear TPO less efficiently from circulation compared to megakaryocytes.

Compensatory megakaryocytopoiesis, a hallmark of ITP, is usually associated with the appearance of a high percentage of young platelets, i.e. compound platelets in the peripheral blood which are hemostatically more active. So, despite what may be severe thrombocytopenia patients with classic ITP generally do not suffer from severe hemorrhage.[22] In agreement with this clinical observation, we found the platelet activation marker sP-selectin to be significantly elevated in patients with ITP at diagnosis and it decreased during steroid-induced recovery.[9,11] Previously other α-granule proteins, which are also known to reflect platelet activity, have been reported to be elevated in ITP in consistence with our observation.[23] Platelet microparticles, which are known to arise with platelet activation and commonly rich for P-selectin, are significantly increased in ITP and they are suggested to decrease bleeding. Although our results may be taken as clues for platelet activation in classical ITP, microparticles and sP-selectin should also be evaluated as possible markers of platelet destruction.

REFERENCES

1. Kaushansky K. Thrombopoietin: the primary regulator of platelet production. Blood 1995;86:419–431.
2. Haznedaroğlu IC, Benekli M, Dündar SV. Thrombopoietin, megakaryocytopoietic cytokines, and reactive versus clonal megakaryocytopoiesis: an enigma complex. Blood 1996;87:1654.
3. Williams N, Bertoncello I, Jackson H, Kavnoudias H. The role of interleukin 6 in megakaryocyte formation, megakaryocyte development and platelet production. Ciba Found Symp 1992;167:160–170.
4. Karpatkin S. Autoimmune (idiopathic) thrombocytopenic purpura. Lancet 1997;349:1531–1536.
5. Ballem PJ, Segal GM, Stratton JR, Gernsheimer T, Adamson JW, Slichter SJ. Mechanisms of thrombocytopenia in chronic autoimmune thrombocytopenic purpura. Evidence of both impaired platelet production and increased platelet clearence. J Clin Invest 1987;80:33.
6. Kuter DJ. The physiology of platelet production. Stem Cells 1996;14(suppl 1):88–101.
7. Emmons RVB, Reid DM, Cohen RL, Meng G, Young NS, Dunbar CE, Shulman NR. Human thrombopoietin levels are high when thrombocytopenia is due to megakaryocyte deficiency and low when due to increased platelet destruction. Blood 1996;87:4068–4071.
8. Kosugi S, Kurata Y, Tamiyama Y, Tahara T, Kato T, Tadokoro S, Shiraga M, Honda S, Kanakura Y, Matsuzawa Y. Circulating thrombopoietin level in chronic immune thrombocytopenic purpura. Br J Haematol 1996;93:704–706.
9. Haznedaroğlu İC, Büyükaşık Y, Koşar A, Kirazlı Ş, Dündar SV. Thrombopoietin, interleukin-6, and P-selectin at diagnosis and during post-steroid recovery period of patients with autoimmune thrombocytopenic purpura. Ann Hematol 1998;77:165–170.
10. Koşar A, Haznedaroğlu İC, Büyükaşık Y, Özcebe OI, Kirazlı Ş, Dündar SV. Circulating thrombopoietin and interleukin-6 in newly diagnosed autoimmune versus aplastic thrombocytopenia. Haematologica 1998;83:1055–1056.
11. Haznedaroğlu İC, Büyükaşık Y, Koşar A, Özcebe OI, Kirazlı Ş, Dündar SV. Selectins and IL-6 during the clinical course of idiopathic thrombocytopenic purpura. Acta Haematol 1999;101:16–20.
12. Haznedaroğlu İC, Savaş MC, Benekli M, Güllü İH, Dündar SV. The significance of megakaryocytopoietic cytokines and thrombopoietin in autoimmune thrombocytopenic purpura: a hypothesis. NZ Med J 1996;109:389.
13. Haznedaroğlu İC, Güllü İH, Dündar SV, Kirazlı Ş. The significance and distinct interactions of various growth factors in physiologic and pathologic megakaryocytopoiesis/ thrombocytopoiesis. Aust NZ J Med 1997;27:191–192.
14. Haznedaroğlu IC, Sayınalp NM, Özcebe OI, Özdemir O, Dündar SV, Kirazlı Ş. Megakaryocytopoietic cytokines in autoimmune thrombocytopenic purpura. Am J Hematol 1995;49:265.
15. McCarty JM, Sprugel KH, Fox NE, Sabath DE, Kaushansky K. Murine thrombopoietin mRNA levels are modulated by platelet count. Blood 1995;86:3668–75.
16. Stoffel R, Wiestner A, Skoda RC: Thrombopoietin in thrombocytopenic mice: Evidence against regulation at the mRNA level and for a direct regulatory role of platelets. Blood 1996;87:567.
17. Fielder PJ, Gurney AL, Stefanich E, Marian M, Moore MW, Carver-Moore K, de Sauvage FJ: Regulation of thrombopoietin levels by c-mpl-mediated binding to platelets. Blood 1996;87:2154–2161.
18. Cohen-Solal K, Villeval JL, Titeux M, Lok S, Vainchenker W, Wendling F: Constitutive expression of Mpl ligand transcripts during thrombocytopenia or thrombocytosis. Blood 1996;88:2578–2584.
19. Nagata Y, Shozaki Y, Nagahisa H, Nagasawa T, Abe T, Todokoro K: Serum thrombopoietin level is not regulated by transcription but by the total counts of both megakaryocytes and platelets during thrombocytopenia or thrombocytosis. Thromb Haemostas 1997;77:808.
20. de Sauvage FJ, Carver-Moore K, Luoh SM, Ryan A, Dowd M, Eaton DL, Moore MW. Physiological regulation of early and late stages of megakaryocytopoiesis by thrombopoietin. J Exp Med 1996;183:651–5.
21. Eaton DL, de Sauvage FJ. Thrombopoietin: the primary regulator of megakaryocytopoiesis and thrombocytopoiesis. Exp Hematol 1997;25:1–7.
22. Schattner E, Bussel J. Mortality in immune thrombocytopenic purpura: report of seven cases and consideration of prognostic indicators. Am J Hematol 1994;46:120–126.
23. Wang ZY, Shen ZX. Megakaryocytes and platelets in immune thrombocytopenic purpura. Bailliéres Clin Haematol 1997;10:89–107.

28

LYMPHOCYTE FRACTIONATION IN A MAGNETIC FLOW CELL SORTER

Lee R. Moore,[1] Maciej Zborowski,[1] Liping Sun,[1] and Jeffrey J. Chalmers[2]

[1]Department of Biomedical Engineering
The Cleveland Clinic Foundation
9500 Euclid Avenue
Cleveland, OH 44195-5254
[2]Department of Chemical Engineering
The Ohio State University
140 West 19th Street
Columbus, OH 43210

ABSTRACT

We have developed a novel cell sorter, the dipole magnet flow sorter (DMFS) which separates cells into multiple fractions according to theirsurface marker expression, as determined by flow cytometry. A narrow stream of cells is introduced into a sheath of carrier fluid in a rectangular channel while subjected to a perpendicular magnetic force. The special design of the pole pieces ensures a constant magnetic force acting on the cells in the separation area. Cells are spread across the flow in relation to their magnetophoretic mobilities. Separation is achieved by control of the positions of the stream boundaries dividing outlet fractions. CD8 T-lymphocytes labeled with primary antibody-FITC conjugate and anti-FITC-magnetic colloid is the chosen cell system. Flow cytometry analysis shows that a four-fold increase in total marker number per cell is observed over the range of recovered fractions. We believe that this type of separation is applicable to any cells in suspension for which a suitable antibody exists, and—due to the comparatively gentle nature of the process—is particularly suitable for the sorting of fragile cells.

Corresponding author: Lee R. Moore, Department of Biomedical Engineering/ND-20, The Cleveland Clinic Foundation, 9500 Euclid Avenue, Cleveland, OH 44195-5254, tel. 216-445-3230, Fax. 216-444-9198, e-mail: moore@bme.ri.ccf.org

Molecular Biology of Hematopoiesis 6, edited by Abraham *et al.*
Kluwer Academic/Plenum Publishers, New York, 1999.

INTRODUCTION

Cell fractionation by magnetic *flow* sorting is unique among the existing magnetic separation methods. The fast growth and proliferation of magnetic cell separation methods in the past ten years is the result of their meeting the demands for fast, high volume, relatively inexpensive cell immunoseparation.[1–3] Current methods rely on cell trapping on solid supports, and require a separate step for elution of the magnetic cell fraction. The current "magnetic affinity" columns separate cells into two fractions, magnetic and non-magnetic, and are particularly useful for sample depletion of unwanted cell populations. Such binary separations are particularly easy to achieve using relatively large, micrometer-size particulate magnetic labels (Dynabeads). According to previous reports, magnetic flow sorting utilizing large, particulate magnetic labels is limited to such binary separations.

The emergence of nanometer-size magnetic labels bound to antibodies opens the possibilities of separating cells based on surface composition and cell function.[4–8] Small colloidal- or molecular-size magnetic labels may provide sufficiently-high spatial resolution of the label binding, necessary for gradation of cell magnetization according to cell surface-antigen expression levels.[7–9] The nominal diameter of the commercially-available Magnetic Activated Cell Sorting (MACS) microbead, 50 nm, is comparable to the size of important lymphocyte antigenic determinants.[2,10] Thus, ideally, the quantity of bound immunomagnetic colloid becomes a measure of cell surface antigen expression.

Magnetic colloid bound to the cell surface behaves as superparamagnetic material.[11–13] The orientation of single-domain colloidal particles is random during binding to the cell surface; the resulting total magnetic dipole moment of the cell increases with the increasing strength of the magnetic field due to domain realignment. For a freely suspended cell labeled with immunomagnetic colloid, the magnetic force becomes proportional to the magnetic field energy gradient, grad $B^2/2\mu_o$, where B is field intensity, and μ_0 is a constant (magnetic permeability of vacum).[14,15] For a population of cells distributed in receptor density and in magnetic label content, one may expect a well-defined spatial separation when placed in a constant magnetic field energy gradient.

Our objectives include: 1) building a dipole magnet flow sorter which provides a constant magnetic energy gradient across a significant volume of the flow channel, 2) achieving a cell distribution across the flow channel following cell labeling with immunomagnetic colloid, and 3) obtaining fractions of the feed cell distribution by control of fluid stream boundaries. In the DMFS cells are separated in a flowing medium due to unequal body forces acting in the direction perpendicular to the flow.[16] Cell displacements in the magnetic field are correlated with the CD8 antigen expression by flow cytometry analysis of their immunofluorescein intensity.

MATERIALS AND METHODS

Cell model. Nucleated cells are isolated from peripheral blood of normal healthy volunteers using a Ficoll sodium diatrizoate density gradient (Pharmacia, Uppsala, Sweden). The cells are then rendered paramagnetic through a double-antibody labeling method. The primary antibody, reactive to the CD8 markers of lymphocytes is conjugated to fluorescein isothiocyanate (FITC) (B-D Immunocytometry Systems, San Jose, CA). The secondary antibody is an anti-FITC$^{(\text{Isomer 1})}$ magnetic bead conjugate

(Miltenyi Biotec, Inc., Auburn, CA). The colloidal beads of polymeric latex doped with Fe_3O_4, have a mean diameter of 50 nm. The labeled cells are suspended in carrier fluid buffer: calcium- and magnesium-free PBS containing 1.1% bovine albumin and 5 mM EDTA.

Apparatus. The Dipole Magnet Flow Sorter (DMFS) consists of a magnet assembly, a flow channel with connecting tubing, and two multi-syringe pumps, Fig. 1. A magnetic field is generated with a permanent magnet assembly consisting of four pairs of neodymium-iron-boron magnets. Two magnetic circuits in parallel convey the flux through the center pole pieces and air gap. The tips of the pole pieces are specially shaped in the *x-y* plane. The magnetic field in the interpolar gap is measured by a gauss meter and Hall-effect probe (Model 9200 Gaussmeter and transverse probe STG920404, F.W. Bell, Orlando, FL). Measured data are used to calibrate output from the magnetic field modeling software, MAGNETO (Integrated Engineering Software, Winnipeg, Manitoba). MAGNETO output is then used for field-related calculations.

The flow channel consists of a borosilicate rectangular glass conduit of internal dimensions (depth, 2a; width, 2b; length, L) of 1 mm × 15 mm × 152 mm (Vitrocom, Inc., Mountain lakes, NJ). Its position along the *y* axis inside the interpolar gap is controlled with the use of a precision translation stage (Newport Corp., Irvine, CA). Dissimilar-sized syringes, driven by two precision syringe pumps ("44" model, Harvard

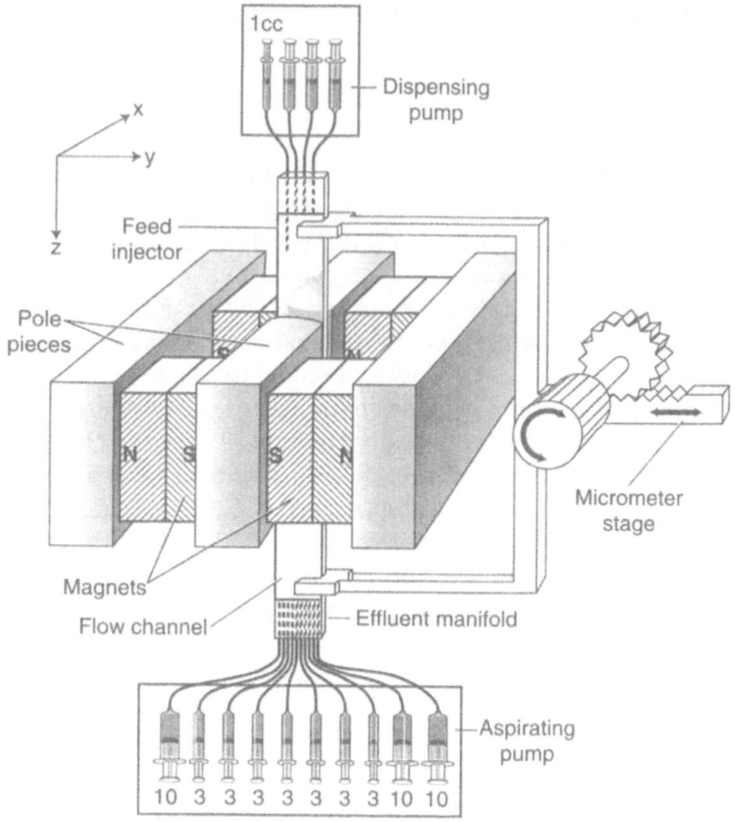

Figure 1. Components of the Dipole Magnet Flow Sorter.

Apparatus, Inc., Holliston, MA) dispensing into and withdrawing from the flow channel, are used to focus the feed stream and control the split between exit streams, as described below. The syringes are coupled to the flow cell through Teflon tubes, stainless steel injectors and fitted ten-port manifolds; see Fig. 1.

Operation. The flow channel is primed with de-gassed carrier buffer. The feed syringe is filled with cell suspension and the remaining infusion syringes are filled with carrier buffer; these are attached to the dispensing syringe pump. Preweighed empty syringes are attached to the aspirating pump. The rectangular channel is carefully moved into the desired position in the gap with the horizontal translation stage. After the run, the eluate tube lines are purged of adhering cells by aspiration of additional buffer into the syringes. The cell concentration in each syringe is determined by a Z-1 Coulter Counter (Coulter Corp., Hialeah, Fla.); the collected volumes are determined gravimetrically. Finally, samples of the feed and eluate fractions are fixed with 0.5% para-formaldehyde and submitted to the flow cytometry facility for analysis.

In a typical run 60–70% of the positive cells fed are recovered in the outlets, the remainder being retained in the channel. This is due to the relatively high ratio of the channel hold-up volume to the volume of dispensed fluid (2.28 ml to 24.4 ml, typically). Further, cell trajectories have a weak component in x (see Fig. 1), and some cells are deflected toward and adhere on the channel walls. Upon completion of the sorting process, the retained cells can be flushed from the system for analysis.

Magnetic and drag forces. Continuous flow magnetic separation depends on a balance of magnetic, viscous drag, and sedimentation forces.[17] The equation defining the x-y-plane velocity, $\boldsymbol{v}$, due to the interaction of magnetic and drag forces is[18]

$$\boldsymbol{v} = \frac{2R^2\Delta\chi}{9\eta}\, grad\left(\frac{B^2}{2\mu_0}\right) \tag{1}$$

V is the cell volume, $\Delta\chi$ is the net cell volumetric susceptibility relative to the medium, μ_0 is the permeability of free space, $B^2/2\mu_0$ is the magnetostatic field energy density, η is the carrier fluid viscosity and R is the cell radius. From equation 1, the magnetic migration velocity follows lines of the magnetic field energy gradient, grad $B^2/2\mu_0$. In these experiments the calculated cell settling velocity, compared with the bulk axial fluid velocity, is negligible. Also, there is no axial component of the magnetic force. Therefore, along the z-axis, the cell velocity is equal to the local fluid velocity, $\boldsymbol{w}$(x,y).

The pole piece tips in the *x-y* plane are designed to ensure that the magnitude of the magnetic field energy gradient is nearly constant. Taking the channel center as the origin, across the channel depth (from $x = -0.5$ to 0.5 mm or from $\phi = -1$ to 1) and over a significant portion of the channel width (from $y = -5.5$ to 1.6 mm, or from $\theta = -0.733$ to 0.213), grad $B^2 = 0.290 \pm 0.011\ T^2/mm$ and grad $B^2/2\mu_0 = 1.15 \pm 0.04 \times 10^8\ kg/m^2\text{-}s^2$; see Fig. 2. Therefore, cells entering the magnetic field in a narrow stream focused around $x = 0$, experience the same magnetic field energy gradient, Eq. 1, and the same residence time in the field. In such conditions the magnetic displacement depends on cell properties only, R and $\Delta\chi$, and cells will distribute across the stream boundaries and exit ports based solely on their sizes and magnetic susceptibilities.

Flow field. The theoretical determination of the velocity profile in a duct of rectangular section, as well as split stream boundary calculations for the QMFS channel, have been described elsewhere.[18] For the syringe complement of Fig. 1, the theoretical split stream positions which divide the channel flow between the outlets are shown in Fig. 2. Also shown is the dimensionless axial fluid velocity, $\boldsymbol{w}'$, defined as $\boldsymbol{w}/\boldsymbol{w}_{mean}$, at the

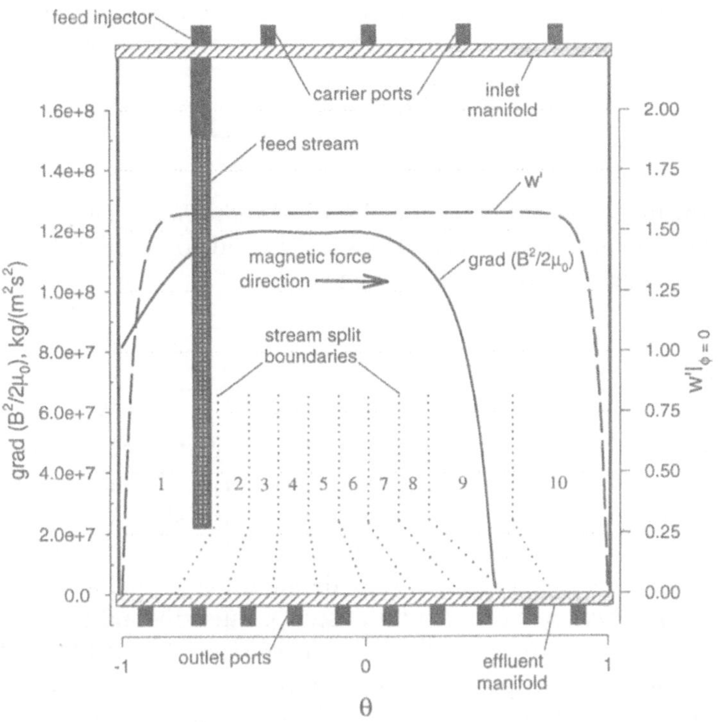

Figure 2. Elevation view of the flow channel comprising the superposition of magnetic energy gradient, grad $(B^2/2\mu_o)$ (solid line); dimensionless axial fluid velocity, w', (broken line) at midsection ($\phi = 0$); and boundary flow streamlines (dotted lines). Channel dimensions are not to scale.

mid-plane of the channel, where x = ϕ = 0. The dimensionless channel coordinates, ϕ and θ, are defined by ϕ = x/a and ϕ = y/b; *a* and *b* are the channel half-depth and half-width, and the origin is at the channel center.

The syringe complement is chosen to maximize the number of effluent fractions (eight) in the region where the magnetic force and axial mid-plane fluid velocity, are constant. Also, the feed injector is positioned to introduce cells into this constant region. See Fig. 2. Unlabeled cells follow the axial flow streamlines and are contained within the first stream split boundary, and thus exit through port 1. This is verified by separate control runs with unlabeled cells where at least 90% of the injected cells are collected from port 1, with virtually all of the remaining cells collected from port 2 (results not shown).

Magnetophoretic mobility. We propose to define a magnetophoretic mobility, μ_{mp}, as the ratio of the magnetic migration velocity to magnetic field energy gradient, $v/|\text{grad } B^2/2\mu_0|$. Our definition of magnetophoretic mobility differs from that used by others ($v/|\text{grad } B|$) because we deal with induced, not constant, magnetic dipoles in our paramagnetic labels.[19] From Eq.1 it can be shown that the dimensions of magnetophoretic mobility are volume*time/mass. For cells in the region of constant field energy gradient, μ_{mp} is a linear function of the magnetic migration velocity, $\boldsymbol{v}$. If we assume that the positive cells are introduced and migrate near the mid-plane of the channel, where $\boldsymbol{w} = \boldsymbol{w}_{max} = 1.572\boldsymbol{w}_{mean}$ = constant, then $\boldsymbol{v}$ is computed directly from the total flow rate,

Q, channel cross section, $S = 2a \times 2b$, axial length of the channel in the magnetic field, z_m, and distance between the feed stream and exit stream midpoints, d_i:[18]

$$v_i = \frac{1.572Q}{z_m S} d_i; \quad i = 1 \ldots 10 \tag{2}$$

Experimental values are given as: $Q = 15.27$ ml/hr, $S = 15\,\text{mm}^2$, $z_m = 76.2$ mm. The resulting magnetophoretic mobility values, calculated for each effluent fraction, are shown in Fig. 5, which will be discussed later.

RESULTS

The efficacy of cell sorting with the DMFS is investigated by targeting the cytotoxic lymphocyte subset, CD8 T cells, of the human peripheral mononuclear cell (PMC) fraction, with immunofluorescent and immunomagnetic labels. Our objective is to separate positive (labeled) from negative (unlabeled) cells, and to fractionate the positive cells according to their magnetization. With the fluorescent and magnetic double label, we expect a correlation between cell fluorescence intensity and cell magnetization in sorted fractions. The feed concentration is approximately 3×10^6 PMC/ml and the feed flow rate is 0.5 ml/hr in all experiments. A total of 3 experiments were performed. The best cell sorting results are illustrated in the histogram of Figs. 3 and 4 and summarized in Table 1, while the three experiments are summarized in Fig. 5.

The histograms of Fig. 3 plot the number of events per fluorescence intensity (FI) interval. The CD8-negative cells appear as a distinct low-emitting peak arising from intrinsic cellular fluorescence. The position of the gate discriminating between labeled and unlabeled cells (indicated by a bar on the histograms) is selected from a histogram of IgG_1 isotype-fluorescein conjugate control (results not shown). The purity in CD8 antigen is indicated on each histogram as the percentage of sampled cells lying above the gate. The feed sample histogram, Fig. 3A, has a positive cell distribution (cells to the right of the gate) which is considerably more disperse than other T cell subsets we've sorted (not shown). This is due to the fact that the CD8 marker is found on at least three subsets of lymphocytes, each of varying expression.[20,21]

Flow cytometry histograms of the cells collected at the outlets are given in Fig. 3B. The overall trend, from left to right, is toward increasing purity in CD8-positive cells. Likewise, we encounter a gradual increase in the proportion of high emitters to low emitters within each positive fraction, with increasing fraction number. This is consistent with the expectation that the most magnetic—and correspondingly, most fluorescent—cells should migrate farthest. Numerically, this is manifested by an increasing geometric mean fluorescence intensity, as shown in Table 1. No significant number of cells are collected in fractions 9 and 10, as expected from the rapid decrease in field energy gradient beyond port 8, indicated in Fig. 2.

The paradoxical result of a decrease in purity beyond outlet 6 can be conjectured by cell clumping. Purely geometrical considerations show that an aggregate of labeled and unlabeled cells can have an equal or higher migration velocity than a single positive cell, because magnetic force increases with the volume of the magnetic cells in the cluster, whereas the opposing viscous drag increases only with the linear dimension of the cluster. A cluster of cells which migrates to a high-numbered outlet, such as the eighth, will be broken apart during the flow cytometry analysis, into cells of various florescence intensities lowering the purity of that outlet.

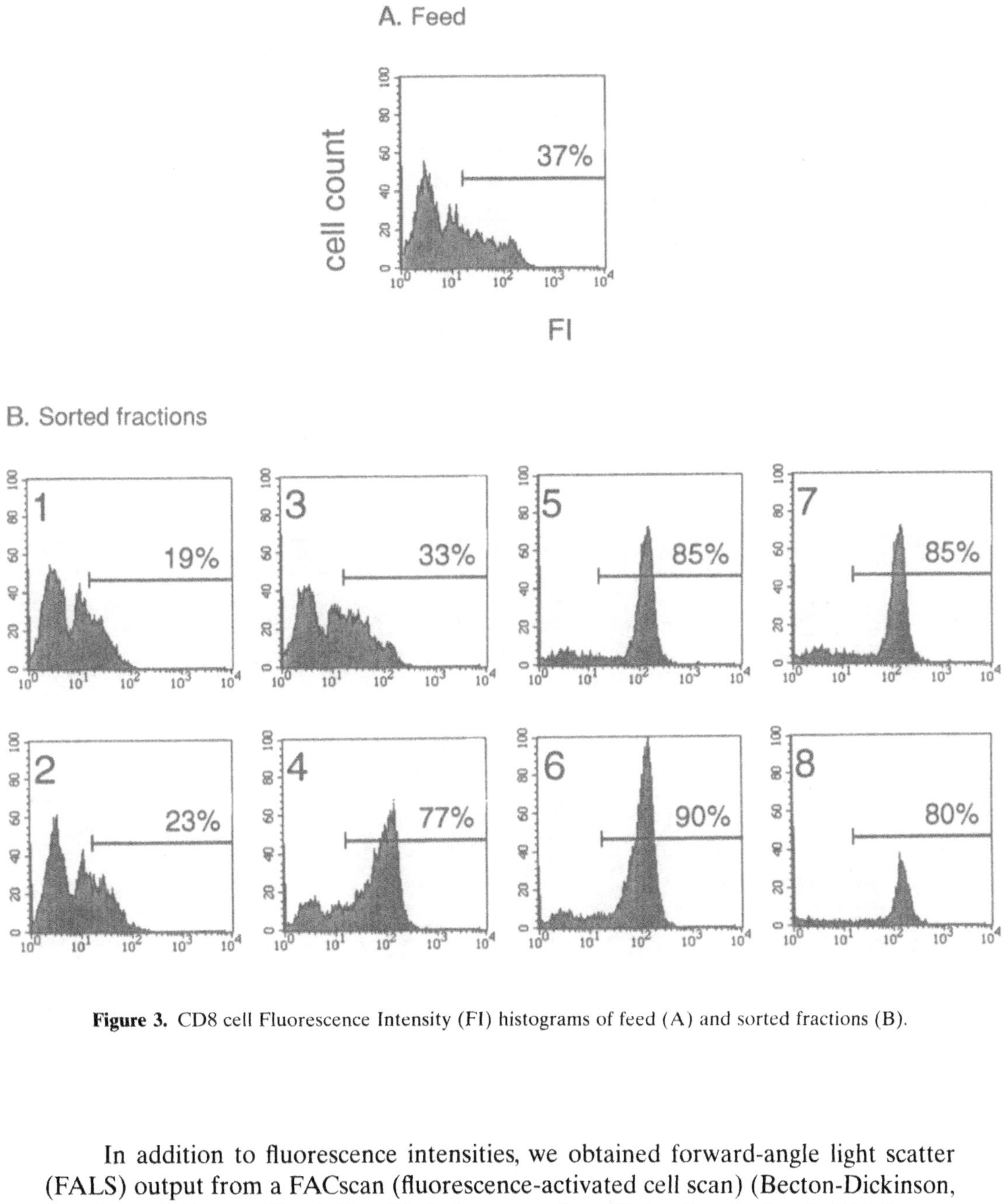

Figure 3. CD8 cell Fluorescence Intensity (FI) histograms of feed (A) and sorted fractions (B).

In addition to fluorescence intensities, we obtained forward-angle light scatter (FALS) output from a FACscan (fluorescence-activated cell scan) (Becton-Dickinson, San Jose, CA); see Fig. 4. Provided the range is sufficiently restricted, FALS is an approximately linear function of cell diameter. Our PMC samples are comprised of two fractions, "small cells" and "large cells" which correspond to lymphocytes and a few contaminating granulocytes, and monocytes. Fig. 4 shows that the purity in small cells, identified by FALS analysis, increases with higher outlet number. This illustrates that monocytes, which are CD8 negative, linger near the parent feed stream and are eluted in the lower fractions. The fact that monocytes are apparently recovered at higher port numbers suggests that some adhere to the CD8 T cells, consistent with the above-stated clumping conjecture.

Table 1 summarizes results for a single experiment. It includes the flow cytome-

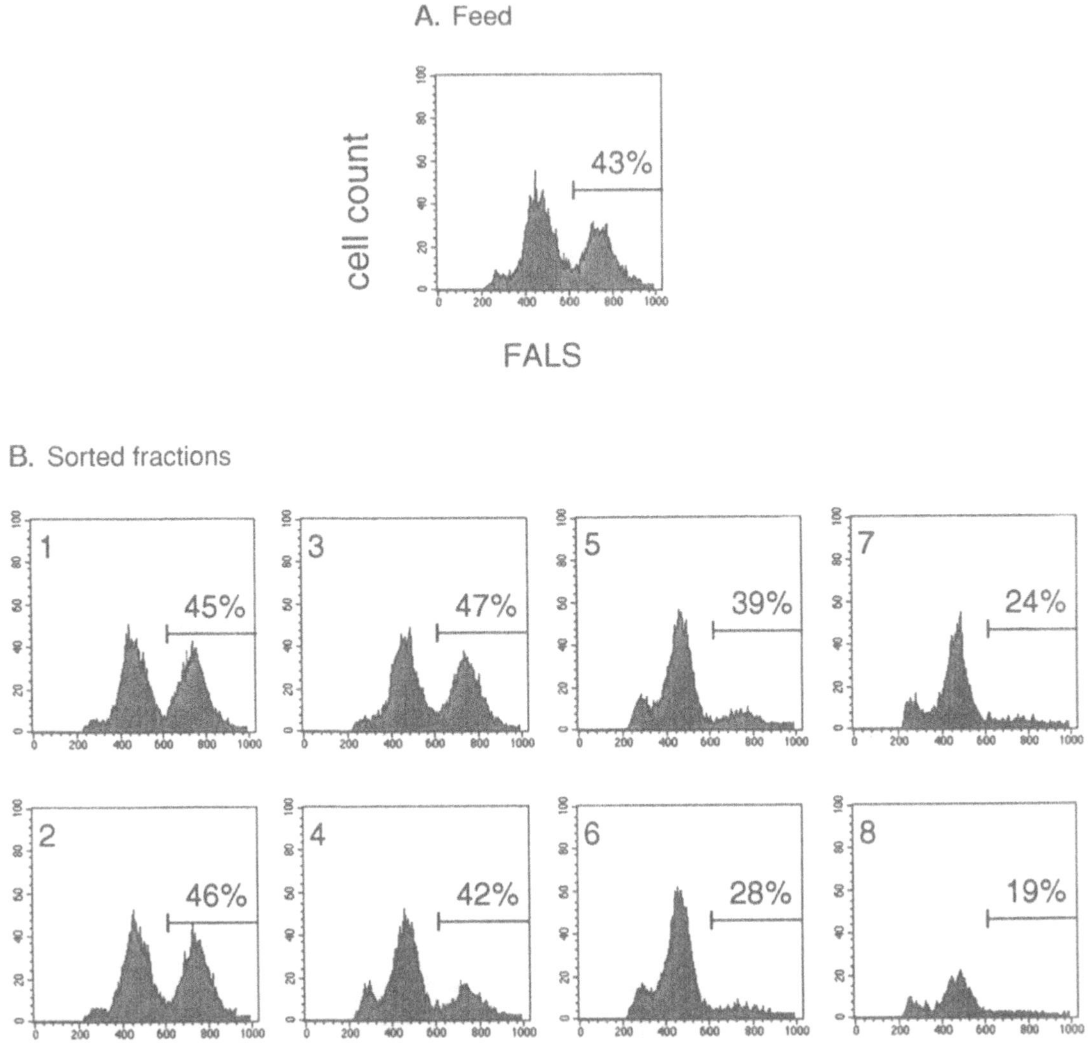

Figure 4. CD8 cell Forward Angle Light Scatter (FALS) histograms of feed (A) and sorted fractions (B).

try data as well as concentrations, flow rates, purities, and recoveries. Recovery is defines as the percentage of the labeled cells in the feed that were recovered in the sorted fraction. In Fig. 5, cell fractions with higher magnetophoretic mobility have higher relative mean fluorescence: the expected result. Large variances arise from differences between blood donors.

DISCUSSION

The unique design of the DMFS includes a constant energy gradient across the cell flow path resulting in a constant magnetic force on the cell tagged with the superparamagnetic colloid. The relative orientation of fluid flow and body force fields is

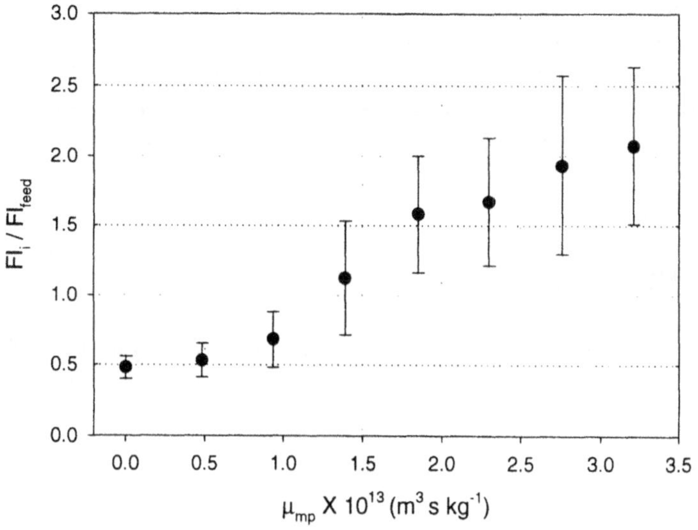

Figure 5. Relative cell fluorescence intensity as a function of magnetophoretic mobility. The data shown are for three experiments of CD8 cells collected at the eight effluent ports.

similar to that employed in free-flow (or curtain flow) cell electrophoresis[22] and in earlier attempts at continuous magnetic cell separation, including modifications of dry separation.[23–25] Our design resulted from optimizing the features of cell sorting using magnetic colloid in viscous (aqueous) media. CD8 cells magnetized with immunomagnetic colloid and fractionated with the DMFS had markedly different fluorescence intensities: the ratio of cell mean geometric fluorescence intensities between the highest and the lowest magnetic mobility fractions is 4-to-1.

Cell fluorescence intensity of a given cell type is a measure of cell surface marker expression.[26] The correlation between cell surface antigen expression and the associated cell fluorescence is the subject of quantitative fluorescence-activated cell scan (FACS) analysis.[27,28] When used on a well-defined cell population, such as the mature, peripheral T lymphocytes of this study, the quantitative FACS analysis provides information about the number of antigenic determinants on the cell surface. Without attempting to quantify the number of fluorescence-associated antigenic sites, we

Table 1. Results summary of CD8+ lymphocyte separation using the Dipole Magnet Separator (experiment #100197)

		Outlets, *I*							
Parameter	Feed	1	2	3	4	5	6	7	8
Conc. $\times 10^6$ (cells/ml)	3.0	0.21	0.44	0.12	0.25	0.019	0.14	0.008	0.005
Flow rate (ml/hr)	0.5	2.78	0.99	0.99	0.99	0.99	0.99	0.99	0.99
Purity, % CD8+ cells	27.4	18.6	22.9	33.0	76.6	89.5	89.2	84.7	80.5
Recovery, % CD8+ cells	—	26.0	23.6	9.5	4.9	3.9	3.3	1.7	0.9
Geometric mean	64	34	38	51	96	122	135	152	167
Florescence Intensity $(FI)_I$ $(FI)_I/(FI)_{feed}$	1.00	0.53	0.59	0.80	1.50	1.91	2.11	2.38	2.61
small cell purity, %, from FALS	57.2	53.7	53.9	53.5	58.3	61.1	71.6	75.9	81.1

followed the usual assumption that the geometrical mean T cell fluorescence is directly proportional to the geometrical mean number of antigen sites on the cell. The results indicate that the cells collected at eight different outlet ports differ significantly in their ability to bind the target antibody. Therefore, our results demonstrate the possibility of using the DMFS to fractionate lymphocytes according to cell surface marker expression.

ACKNOWLEDGMENT

Supported by National Institutes of Health (CA 62349) and the Whitaker Foundation (20010308).

REFERENCES

1. Ugelstad J, Stenstad P, Kilaas L, Prestvik WS, Herje R, Berge A, Hornes E: Monodisperse magnetic polymer particles. Blood Purif. 11:347, 1993.
2. Radbruch A, Mechtold B, Thiel A, Miltenyi S, Pfluger E: High-gradient magnetic cell sorting. Methods in Cell Biology 42:387, 1994.
3. DePalma A: Developments in biomagnetic separations focus on new affinity mechanisms. Genetic Engineering News July:11, 1997.
4. Miltenyi S, Müller W, Weichel W, Radbruch A: High gradient magnetic cell separation with MACS. Cytometry 11:231, 1990.
5. Zborowski M, Fuh CB, Green R, Sun L, Chalmers JJ: Analytical magnetapheresis of ferritin-labeled lymphocytes. Anal. Chem. 67:3702, 1995.
6. Thomas TE, Lansdorp PM: Selective separation of cells using magnetic colloids. Proceedings of the Fourth International Symposium on Bone Marrow Purging and Processing. New York, 1994, p 65.
7. Liberti PA, Feeley BP: Analytical- and process-scale cell separation with bioreceptor ferrofluids and high gradient magnetic separation, in Compala DS, Todd P (eds): Cell Separation Science and Technology, vol. 464. Washington, ACS Symposium Series, 1991, p 268.
8. Chalmers JJ, Zborowski M, Moore L, Mandal S, Fang B, Sun L: Theoretical analysis of cell separation based on cell surface marker density. Biotechnol Bioeng 58:(in print), 1998.
9. Thomas TE, Abraham SJR, Otter AJ, Blackmore EW, Lansdorp PM: High gradient magnetic separation of cells on the basis of expression levels of cell surface antigens. J. Immunol. Meth. 154:245, 1992.
10. Barclay AN, Birkeland ML, Brown MH, Beyers AD, Davis SJ, Somoza C, Williams AF: The Leukocyte Antigen FactsBook. London, Academic Press, Harcourt Brace Jovanovich, Publishers, 1993.
11. Bean CP, Livingstone JD: Superparamagnetism. J. App. Phys. 30:120S, 1959.
12. Zborowski M: Physics of the magnetic cell sorting, in Häfeli U, Schütt W, Teller J, Zborowski M (eds): Scientific and Clinical Applications of Magnetic Microcarriers: An Overview. New York, Plenum Press, 1997, p 205.
13. Chikazumi S, Charap SH: Physics of Magnetism. Malabar, Florida, Robert E. Krieger Publishing Co., 1986.
14. Gill SJ, Malone CP, Downing M: Magnetic susceptibility measurements of single small particles. The Review of Scientific Instruments 31:1299, 1960.
15. Reddy S, Moore LR, Sun L, Zborowski M, Chalmers JJ: Determination of the magnetic susceptibility of labeled particles by video imaging. Chemical Engineering Science 51:947, 1996.
16. Zborowski M, Moore LR, Reddy S, Chen G-H, Sun L, Chalmers JJ: Magnetic flow sorting using a model system of human lymphocytes and a colloidal magnetic label. ASAIO Journal 42:M666, 1996.
17. Zborowski M, Moore LR, Sun L, Chalmers JJ: Continuous-flow magnetic cell sorting using soluble immunomagnetic label, in Häfeli U, Schütt W, Teller J, Zborowski M (eds): Scientific and Clinical Applications of Magnetic Microcarriers: An Overview. New York, Plenum Press, 1997, p 247.
18. Moore LR, Zborowski M, Sun L, Chalmers, JJ: Lymphocyte fractionation using immunomagnetic colloid and a dipole magnet flow cell sorter. Journal of Biochemical and Biophysical Methods: in press, 1998.

19. Schramm LL, Clark BW: On the measurement of magnetophoretic mobilities. Colloids and Surfaces 7:135, 1983.
20. Terry LA, DiSanto JP, Small TN, Flomenberg N: Differential expression and regulation of the human CD8-alpha and CD8-beta chains. Tissue Antigens 35:82, 1990.
21. Moebius U, Kober G, Girscelli AL, Hercend T, Meuer SC: Expression of different CD8 isoforms on distinct lymphocyte subpopulations. Eur. J. Immunol. 21:1793, 1991.
22. Vanderhoff JW, Micale FJ, Krumrine PH: Continuous flow electrophoresis, in Righetti PG, van Oss CJ, Vanderhoff JW (eds): Electrokinetic Separation Methods. New York, Elsevier/North Holland Biomedical Press, 1979, p 121.
23. Takayasu M, Kelland DR: Selective continuous magnetic separation of two-component particulate suspensions. IEEE Trans. MAG-22:1125, 1986.
24. Hartig R, Hausmann M, Schmitt J, Herrmann DBJ, Riedmiller M, Cremer C: Preparative continuous separation of biological particles by means of free-flow magnetophoresis in a free-flow electrophoresis chamber. Electrophoresis 13:674, 1992.
25. Hartig R, Hausmann M, Cremer C: Continuous focusing of biological particles by continuous immuno magnetic sorter: Technique and applications. Electrophoresis 16:789, 1995.
26. Shapiro HM: Flow sorting, in Shapiro HM (ed): Practical Cytometry. New York, Wiley-Liss, 1995.
27. Poncelet P, Carayon P: Cytofluorometric quantification of cell-surface antigens by indirect immunofluorescence using monoclonal antibodies. J. Immunol. Meth. 85:65, 1985.
28. Dawson CD, Scheppler JA, Nicholson JKA, Holman RC: Enumeration of antigen sites on cells by flow cytometry. Pathobiology 59:57, 1991.

THE BASIS FOR A RATIONAL NUTRITIONAL APPROACH TO PATIENTS WITH CANCER

Filippo Rossi Fanelli, Maurizio Muscaritoli, Carlo Cangiano, Antonia Cascino, Alessandro Laviano, and Alessandra Fava

Department of Clinical Medicine
University 'La Sapienza' Rome
Italy

Keywords: Cancer, Cachexia, Anorexia, Nutrition, Amino Acids, Tryptophan, Cytokines, Serotonin

ABSTRACT

Malnutrition negatively affects outcome of cancer patients. Nutritional support is ineffective in correcting malnutrition and its clinical consequences. The lack of efficacy is secondary to: a) the changes in host's metabolism induced by the tumor; b) the presence of anorexia, and c) nutrient stealing by tumor tissues. A) *Metabolic Alterations*: the most relevant changes in carbohydrate metabolism are glucose intolerance and increased gluconeogenesis. Lipid mobilization and oxidation are increased in host's tissues, while exogenous fat hydrolysis by lipoprotein lipase is reduced. Host's protein metabolism is characterized by an increase in hepatic synthesis and muscle degradation of protein. Changes in amino acid metabolism involving glucogenic, branched-chain amino acids, free tryptophan, glutamic acid, and taurine have also been described. These metabolic abnormalities are likely the consequence of a number of cytokines produced by the tumor or by the host in response to the tumor. B) *Cancer Anorexia*: it is likely the consequence of both an increase in brain serotonergic activity and of a direct effect of cytokines on the feeding centers. C) *Tumor metabolism*: tumor protein synthesis is very active, but attempts to inhibit it by depriving single

---

Corresponding author: Filippo Rossi Fanelli, MD, Professor of Medicine, Department of Clinical Medicine, University "La Sapienza", Viale dell'Università 37, 00185 Rome, Italy; Tel. ++39 6 4452929; Fax ++39 6 4461341; E-mail medint11@uniroma1.it

Molecular Biology of Hematopoiesis 6, edited by Abraham *et al.*
Kluwer Academic/Plenum Publishers, New York, 1999.

amino acids did not yield convincing results. Glucose is the tumor preferred energy fuel, while fatty acids are poorly oxidized. Thus, inhibition of tumor growth by lipid-enriched TPN mixtures has been proposed. Results are however controversial. Nutrient administration may promote cancer growth. However, stimulation of tumor cells to enter a proliferative phase could improve the efficacy of cycle-specific antineoplastic treatments.

INTRODUCTION

Although the final outcome of cancer patients is essentially determined by the prognosis of the underlying disease, the presence of malnutrition has *per se* a negative impact on morbidity, mortality, and quality of life in cancer states. The presence and severity of malnutrition in fact shorten patients' overall survival, heavily contribute to the onset of post-operative complications, increase radio-chemotherapy toxicity while decreasing tumor sensitivity to antineoplastic regimens, dampen the immune response and finally is of considerable psychological discomfort to patients and their families.[1] The rationale behind the use of a nutritional support to a neoplastic patient is therefore based on such premises.

Artificial nutrition has been in use in the supportive care of malignancies for more than two decades, Surprisingly, the real efficacy of this approach in correcting malnutrition and its clinical consequences in the presence of cancer is still debated. There is indeed clear evidence that, even after a prolonged course of nutritional support in malnourished cancer patients lean body mass is only moderately increased, despite a significant increment of body weight which is actually accounted for by an increase in body fat and water.[2]

The profound changes in host metabolism induced by the tumor leading to wasting of energy and impaired substrate utilization[3] may well be responsible for the rather disappointing effects of artificial nutrition in cancer. Reduced food intake, mainly as a consequence of tumor-induced anorexia, and nutrients stealing by growing tumor tissues also participate in determining host's nutritional deprivation and unsatisfactory effects of artificial nutrition in cancer. In the attempt to improve the efficacy of the nutritional approach to cancer patients, all the major factors participating in determining malnutrition in these patients (i.e., changes in host's metabolism, reduced food intake and specific tumor nutritional needs) should be considered jointly.

METABOLIC ALTERATIONS

Solid tumors may share the ability to induce a number of metabolic alterations in the host involving all aspects of intermediary metabolism including energy, carbohydrate, fat, and protein metabolism, irrespective of their site of origin.

Carbohydrate metabolism is profoundly impaired during tumor growth. The most relevant alterations are glucose intolerance, and increased gluconeogenesis. Glucose intolerance in cancer patients is known since 1920s, and has been subsequently confirmed by several clinical studies.[5,6] The incidence of glucose intolerance in cancer patients is quite high. In a recent study we have carried out in 42 cancer patients undergoing an oral glucose tolerance test, we found that approximately 60% of patients bearing a neoplastic disease were either intolerant or overtly diabetic upon diagnosis.[7]

Hepatic gluconeogenesis from lactate and gluconeogenic amino acids is markedly increased in cancer patients. It does not respond to physiological inhibitory stimuli, such as glucose administration, and is one of the main determinants of energy wasting observed in these patients.[8]

Also, lipid metabolism is deranged during tumor growth. Body fat stores are avidly consumed in cancer patients as a consequence of an increased lipid mobilization and oxidation in the host's tissues.[3] Unlike it is observed in uncomplicated starvation, fatty acid oxidation in cancer patients is not glucose suppressible and also participate in the waste of energy in these patients.[9] Another aspect of fat metabolism with nutritional relevance is the significant reduction of exogenous fat hydrolysis by lipoprotein lipase. We first described this abnormality in cancer patients by studying the plasma clearance of exogenously administered triglycerides, as expressed by the two constants K_1 and K_2, which are calculated at two different infusion rates.[10] Both were found significantly reduced in cancer patients.

Host's protein metabolism in cancer is also impaired, with increased whole body protein turnover, secondary to increased hepatic synthesis and muscle degradation of protein.[11] Changes in amino acid metabolism resulting in abnormal concentrations of circulating amino acids also characterize impaired protein metabolism in cancer.[12] Several studies have shown a reduction of glucogenic amino acids in cancer patients,[13,14] accounted by the high rate of hepatic glucose synthesis. Surprisingly, branched-chain amino acids, which are known to be avidly consumed during simple malnutrition, are constantly found normal in cancer patients, further confirming the profound differences existing between cancer cachexia and non-cancer malnutrition. In a neoplastic rat model, we demonstrated that the reduction of plasma glucogenic amino acids is a very early phenomenon during tumor growth.[15] It became evident as early as six days after tumor cells inoculation, when tumor was not even palpable and animals were normally behaving.

Other changes of circulating amino acids have been observed in cancer. In a study we carried out in untreated lung and breast cancer patients, we confirmed the reduction of glucogenic amino acids in lung cancer.[16] Free tryptophan, glutamic acid, and ornitine were markedly increased with respect to control subjects. Glucogenic amino acids were normal in breast cancer, while free tryptophan, glutamic acid, and ornitine were increased. This study demonstrates that different tumors may induce different changes in plasma amino acids.[17]

Plasma free tryptophan deserves particular consideration since it is consistently increased in cancer patients. By reviewing more than 200 patients bearing tumors of the breast, lung, colon, stomach and of other origins and included in different studies from our group in the last 10 years, we found that free tryptophan concentrations were consistently and significantly higher when compared to controls. This prompted us to propose free tryptophan as a novel tumor marker, characterized by low specificity but very high sensitivity.[17] As most of the metabolic changes occurring in cancer, also the rise in free tryptophan is strictly related to the presence of the tumor. In fact, after radical surgical removal of the tumor, free tryptophan concentrations returned within normal or near normal values both in lung and breast cancer patients.[18]

During the last decade, data from clinical and experimental studies have accumulated suggesting that cancer-associated metabolic abnormalities are primarily the consequence of a number of cytokines (TNF, IL-1, IL-6, IFN-γ, and others) produced by the host in response to the tumor. Cytokines have been shown to induce similar metabolic changes in other wasting diseases such as sepsis, trauma, or when they are administered to normal animals.[19] Cytokines then act synergistically on specific receptors on target organs, probably via an autocrine or paracrine mechanism, to affect metabolic homeostasis and thus participating in the multi-event process leading to cancer cachexia. Based on this conceptual scheme, different therapeutic strategies may be proposed acting at different levels.[20] One may inhibit cytokines production or their binding to specific receptors on target organs, thus counteracting the metabolic abnormalities induced by cytokines.

REDUCED FOOD INTAKE

Also anorexia, the other major determinant of the onset of malnutrition in cancer, is primarily the consequence of the changes induced by the tumor on host's metabolism. There is in fact a large body of evidence linking the increased plasma concentrations of free tryptophan to anorexia in cancer bearing animals and patients.[21] Plasma and consequent brain accumulation of free tryptophan results in an increased serotonin synthesis and activity, particularly in the hypothalamic feeding centers, leading to a marked reduction of food intake.[21]

In a neoplastic rat model, a significant reduction of food intake was observed five days following tumor inoculation and it was paralleled by a marked increase in brain tryptophan and serotonin concentrations.[22] In the clinical setting, we studied 42 patients with cancer of different sites and origin, and we found that plasma free tryptophan was significantly increased only in cancer patients with anorexia, remaining unchanged in non-anorectic cancer patients.[23] Further supporting the role of free tryptophan in the pathogenesis of cancer anorexia, we found in anorectic cancer patients that the reduction of tryptophan brain availability, obtained by the oral administration of branched-chain amino acids, competing with free tryptophan for brain entry, was followed by a significant reduction of anorexia and a steady increase in daily caloric intake.[24]

Cytokines, and IL-1 in particular, are also likely to participate in the pathogenesis of cancer anorexia. Interleukin-1 receptors have been identified in the hypothalamic areas involved in eating behavior.[25] Moreover, in a recent study we found that blocking IL-1 receptors with a specific IL-1 receptor antagonist in neoplastic rats results in a significant improvement of food intake when compared to the control group.[26] On the other hand, peripheral administration of IL-1 as well as TNF have been shown to increase tryptophan concentrations in the brain.[27] Interleukin-1 was also shown to increase the firing rate of hypothalamic serotonergic neurons.[28] Thus, it is conceivable that IL-1 and tryptophan may act sinergistically in stimulating hypothalamic serotonin activity and consequently producing anorexia in cancer.[29]

METABOLISM OF THE TUMOR CELL

The development of an optimal nutritional approach to a cancer patient should also take into account the needs for tumor metabolism. Although tumors differ from each other in many aspects, they do share a number of metabolic peculiarities (for review, see ref. 3). Protein synthesis is very active to keep up with the high replication rate of cancer cells. Several attempts to inhibit protein synthesis in tumor tissues by selectively depriving single amino acids thought to be essential to the tumor have produced controversial results.

Glucose is the preferred energy fuel used mainly through the anaerobic pathway by neoplastic tissue.[30] Fatty acids are poorly oxidized in cancer cells, which indeed lack most of the key enzymes for fatty acid oxidation.[31] Based on these metabolic peculiarities, a number of attempts have been made to reduce tumor growth in experimental animals by feeding them with lipid-enriched TPN mixtures.[32–34] Results are however controversial. Using the same approach for the first time in a group of patients with cancer of the gastrointestinal tract we found that a lipid-based total parenteral nutrition was rather effective in reducing neoplastic cell replication, which was in contrast stimulated by a glucose-based total parenteral nutrition.[35] Such enhancement of tumor growth could be potentially exploited. In fact by stimulating more tumor cells to enter a proliferative phase, cycle-specific chemiotherapic agents might be more effective.[36]

CONCLUSIONS

Cancer cachexia is the result of a multi-event process involving host metabolism and food intake as well as the nutritional needs of tumor tissues. These major determinants of malnutrition in cancer should be considered together in order to improve the quality of the nutritional approach to cancer patients. The growing understanding of the mechanisms underlying tumor-induced impairment of host metabolism and anorexia may help improving food intake and nutrient utilization, while decreasing energy wasting in neoplastic patients. Specific nutrients might play a role in reducing tumor cell replication, enhancing cancer sensitivity to antineoplastic regimens or in ameliorating tolerance to aggressive chemotherapy.

It is therefore conceivable that, based on the better understanding of the integrated mechanisms leading nutritional wasting during tumor growth, the nutritional management of cancer patients will undergo profound changes in the very next future, thus offering all those involved in cancer care a new important tool in the fight against cancer.

REFERENCES

1. De Wys WD, Begg C, Lavin PT, et al. Prognostic effect of weight loss prior to chemotherapy in cancer patients. Am J Med 1980; 69:491.
2. Lipman TO. Clinical trials of nutritional support in cancer. Parenteral and enteral nutrition. Hematol Oncol Clin North Am 1991; 5:91.
3. Rossi Fanelli F, Cascino A, Muscaritoli M. Abnormal substrate metabolism and nutritional strategies in cancer management. J Parenter Enteral Nutr 1991; 15:680.
4. Campos ACL, Waitzberg DL, Meguid MM. A comparison of the changes in carbohydrate, fat, and protein metabolism occurring with malignant and benign tumors and the impact of nutritional support. In: Alfin-Slater RB, Kritchewsky D (eds.), "Cancer and Nutrition". Philadelphia, PA; Plenum Publishing, 1991:69.
5. Marks PA, Bishop JS. The glucose metabolism of patients with malignant disease and of normal subjects as studied by means of an intravenous glucose tolerance test. J Clin Invest 1957; 36:254.
6. Lundholm K, Holm G, Schersten T. Insulin resistance in patients with cancer. Cancer Res 1978; 38:4665.
7. Cascino A, Cangiano C, Ceci F, et al. Plasma amino acids in human cancer. Individual role of tumor, malnutrition, and glucose tolerance. Clin Nutr 1988; 7:213.

8. Shaw JH, Wolfe RR. Glucose and urea kinetics in patients with early and advanced gastrointestinal cancer: the response to glucose infusion, parenteral feeding, and surgical resection. Surgery 1987; 101:181.
9. Shaw JH, Wolfe RR. Fatty acid and glycerol kinetics in septic patients and in patients with gastrointestinal cancer. The response to glucose infusion and parenteral feeding. Ann Surg 1987; 205:368.
10. Muscaritoli M, Cangiano C, Cascino A, et al. Plasma clearance of exogenous lipids in patients with malignant disease. Nutrition 1990; 6:147.
11. Norton JA, Shamberger R, Stern TP, et al. The influence of tumor-bearing on protein metabolism in the rat. J Surg Res 1981; 30:456.
12. Kurzer M, Janiszewski J, Meguid MM. Amino acid profiles in tumor-bearing and pair-fed nontumor-bearing malnourished rats. Cancer 1988; 62:1492.
13. Norton JA, Gorschboth CM, Wesley RA, et al. Fasting plasma amino acid levels in cancer patients. Cancer 1985; 56:1181.
14. Naini AB, Dickerson JW, Brown MM. Preoperative and postoperative levels of plasma protein and amino acid in esophageal and lung cancer patients. Cancer 1988; 62:355.
15. Muscaritoli M, Meguid MM, Beverly JL, et al. The mechanism of early tumor anorexia. J Surg Res 1996; 60:389.
16. Cascino A, Muscaritoli M, Cangiano C, et al. Plasma amino acid imbalance in patients with lung and breast cancer. Anticancer Res 1995; 15:507.
17. Meguid MM, Muscaritoli M, Cascino A, et al. Plasma amino acid in cancer patients: a novel diagnostic tool comes of age. In: Latifi R (ed.), "Amino Acids in Critical Care and Cancer". Austin, TX: CRC Press; 1994:159.
18. Cascino A, Cangiano C, Ceci F, et al. Increased plasma free tryptophan levels in human cancer: a tumor related effect? Anticancer Res 1991; 11:1313.
19. Hardin TC. Cytokine mediator of malnutrition: clinical implications. Nutr Clin Prac 1993; 8:55.
20. Laviano A, Renvyle T, Yang Z-J. From laboratory to bedside: new strategies in the treatment of malnutrition in cancer patients. Nutrition 1996; 12:112.
21. Rossi Fanelli F, Cangiano C. Increased availability of tryptophan in brain as a common pathogenic mechanism for anorexia associated with different diseases. Nutrition 1991; 7:364.
22. Krause R, James JH, Ziparo V, Fischer JE. Brain tryptophan and neoplastic anorexia-cachexia syndrome. Cancer 1979; 44:1003.
23. Rossi Fanelli F, Cangiano C, Ceci F, et al. Plasma tryptophan and anorexia in human cancer. Eur J Cancer Clin Oncol 1986; 22:89.
24. Cangiano C, Laviano A, Meguid MM, et al. Effects of administration of oral branched-chain amino acids on anorexia and caloric intake in cancer patients. J Natl Cancer Inst 1996; 88:550.
25. Yabuuchi K, Minami M, Katsumata S, et al. Localization of type I interleukin-1 receptor mRNA in the rat brain. Mol Brain Res 1994; 27:27.
26. Laviano A, Meguid MM, Renvyle T, et al. IL-1a acts directly on hypothalamic VMH to mediate cancer anorexia. Surg Forum 1995; 46:492.
27. Dunn AJ. Endotoxin-induced activation of cerebral catecholamine and serotonin metabolism: comparison with interleukin-1. J Pharmacol Exp Therap 1992; 261:96.
28. Mohankumar PS, Thyagarajan S, Quadri SK. Interleukin-1β increases 5-hydroxy-indoleacetic acid release in the hypothalamus in vivo. Brain Res Bull 1993; 31:745.
29. Laviano A, Meguid MM, Yang Z-J, et al. Cracking the riddle of cancer anorexia. Nutrition 1996; 12:706.
30. Warburg O. The metabolism of carcinoma cells. J Cancer Res 1925; 9:148.
31. Tisdale MJ, Brennan RA. Loss of acetoacetate coenzyme A transferase activity in tumors of peripheral tissues. Br J Cancer 1983; 47:293.
32. Buzby GP, Mullen JL, Stein TP, et al. Host-tumor interaction and nutrient supply. Cancer 1980; 45:2940.
33. Dills WL, Kivong E, Covey TR, et al. Effects of diet deficient in glucose and glucose precursors on the growth of the Walker carcinosarcoma 256 in rats. J Nutr 1984; 114:2097.
34. Mares-Perlman JA, Francis AM, Shrago E. Host and tumor growth and energy substrates in blood of hepatoma-beraing rats receiving high-fat parenteral infusions. Am J Clin Nutr 1988; 48:50.
35. Rossi Fanelli F, Franchi F, Mulieri M, et al. Effect of energy substrate manipulation on tumor cell proliferation in parenterally fed neoplastic patients. Clin Nutr 1991; 10:228.
36. Torosian MM, Mullen JL, Miller EE, et al. Enhanced tumor response to phase-specific chemotherapy by parenteral amino acid administration. J Parenter Enteral Nutr 1983; 7:337.

30

DIFFERENTIAL EXPRESSION OF IL-3 AND GM-CSF RECEPTOR COMMON SIGNAL TRANSDUCING SUBUNIT (βC) DURING NORMAL HEMATOPOIETIC DIFFERENTIATION

Ugo Testa, Stefania Militi, Roberta Riccioni, Nadia Maria Sposi, Isabella Parolini, Paola Samoggia, Elvira Pelosi, and Cesare Peschle

Department of Hematology and Oncology
Istituto Superiore di Sanità
Viale Regina Elena 299, 00161 ROME
Italy

INTRODUCTION

The hematopoietic system is a complex differentiation mechanism sustained by a pool of hematopoietic stem cells (HSCs) that are able to both self-renew and differentiate into multipotent progenitor cells.[1] Hematopoietic progenitor cells (HPCs) are committed to a specific lineage(s) and are functionally defined as colony or burst-forming units (CFUs, BFUs), i.e., early and late HPCs of the erythroid series (BFU-E and CFU-E, respectively), the megakaryocytic lineage (BFU-Mk and CFU-Mk), the granulo-monocytic series (CFU-GM, CFU-G, and CFU-M), as well as multipotent CFUs for all four lineages (CFU-GEMM).[1] HPCs in turn differentiate into morphologically recognizable precursors that mature to terminal elements circulating in peripheral blood (PB).

The survival, growth and differentiation of hematopoietic cells is regulated by a complex network of hematopoietic growth factors (HGFs) termed colony stimulating factors (CSFs) or interleukins (Ils). Among these cytokines, granulocyte-macrophage CSF (GM-CSF) and IL-3 play an important role in regulating both proliferation and differentiation of early HPCs (CFU-GEMM), early erythroid (BFU-E), CFU-GM, and BFU-Mk progenitors, whereas GM-CSF exerts comparable effects but acts on more differentiated progenitors than IL-3.[2] In addition to these effects on HPCs, IL-3 and GM-CSF sustain also the proliferation and survival of differentiated precursors, particularly of the granulocytic and monocytic lineages.[2]

Molecular Biology of Hematopoiesis 6, edited by Abraham *et al.*
Kluwer Academic/Plenum Publishers, New York, 1999.

IL-3, GM-CSF, and IL-5 exert their biological activities through interaction with cell surface receptors that consist of two subunits, α subunit specific to each and the common β subunit (βc).[3] The α subunits of the IL-3R and GM-CSFR belong together with other cytokine receptors, to the type I cytokine receptor superfamily.[4]

The α-chain (IL-3Rα, GM-CSFRα or IL-5Rα) binds IL-3, GM-CSF, and IL-5 respectively, with high specificity but with low affinity.[5] The interaction of an α chain with a β chain leads to the formation of a high-affinity receptor complex which is able to bind the respective ligand in the range of its physiological concentrations and to transduce proliferative, anti-apoptotic and differentiative signals.[5–7] The βc expressed alone, in the absence of a specific α chain, confers little binding affinity for either IL-3, GM-CSF or IL-5.[8–9]

The βc is not only required for high-affinity binding of ligands, but it also plays a key role in mediating the origin of the signaling pathways through its intracytoplasmic tail.[10] No known protein kinase consensus sequences are observed in the cytoplasmic tail of the βc; however, functional analysis of truncated mutants of this receptor chain provided evidence for the existence of several functional domains: (i) the membrane proximal region binds JAK kinases,[11] is necessary for the induction of c-myc and pim-1[12,13] and is essential and sufficient for control of cell proliferation;[10] (ii) the membrane distal domains are responsible for control of cell differentiation,[14] induction of tyrosine phosphorylation of proteins, induction of c-fos and c-jun, and prevention of apoptosis.[10,12,15] Finally, recent studies indicate that a point mutation occurring at the level of the βc chain DNA is responsible for human pulmonary alveolar proteinosis, a rare genetic disease characterized by severe respiratory distress.[16]

A recent study[17] showed that normal as well as leukemic cells possess two different isoforms of the common βc: a full-length isoform and a truncated isoform caused by a deletion of a 104-bp exon in the membrane-proximal region of the chain. As a consequence of this deletion, the β intracytoplasmic truncated chain (β_{IT}) has a cytoplasmic tail of 46 amino acids, instead of 432 for βc.[17] The β_{IT} maintains the capacity to bind JAK2, but is unable to transduce the mitogenic signal.

Few studies have been carried out to characterize the expression of IL-3R and GM-CSFR on human hematopoietic progenitors. In this context, double labeling experiments of total human[18,19] or Rhesus monkey bone marrow cells[20,21] with anti-CD34 and either anti-IL-3R or GM-CSFR α chain mAbs have shown that $CD34^+$ cell subsets express IL-3Rs and/or GM-CSFRs. These observations were confirmed using preparations of highly purified human HPCs.[22,23] On the other hand, the few data available on βc chain expression on HPCs indicate that these cells only scarcely express this receptor chain on their surface.[19,24]

In the present study we have investigated the pattern of βc expression during the process of normal hematopoietic differentiation, starting from HPC to the late stages of maturation.

MATERIALS AND METHODS

Adult Peripheral Blood Purification

Adult PB was obtained from male donors after informed consent. HPCs were purified as reported[25] and modified.[26,27] Briefly, (IA) PB samples were separated over a Ficoll-Hypaque density gradient (Pharmacia, Pîscataway, NJ), and (IB) PB mononuclear cells resuspended in IMDM containing 20% heat-inactivated fetal calf serum

(FCS, GIBCO) for three cycles of plastic adherence. Thereafter, (II) cells were separated by centrifugation on a discontinuous Percoll gradient (Biochrom KG, Berlin, Germany). (III) Step III purification was potentiated (Step IIIP) as indicated.[26,27]

Unilineage HPC Liquid Suspension Culture

Step IIIP HPCs were seeded (5×10^4 cells/ml) and grown in liquid suspension culture supplemented with the following HGFs: (1) in FCS^- erythropoietic cultures, very low doses of IL-3 (0.01 U/ml) and GM-CSF (0.001 ng/ml) and a saturating level of Epo (3 U/ml);[28] (2) in FCS- granulopoietic culture, low amounts of IL-3 (1 U/ml) and GM-CSF (0.1 ng/ml) and saturating amounts of G-CSF (500 U/ml);[29] (3) in FCS^- megakaryopoietic culture, saturating doses of Tpo;[27] and (4) in FCS+ monocytopoietic culture, saturating amounts of both flt3 ligand (FL, 100 ng/ml) and M-CSF (500 U/ml).[30] Cells were periodically counted and analyzed for morphology, membrane phenotype and HGFR expression at both mRNA and protein levels.

Control Cells

Normal human granulocytes and monocytes were purified according to standard procedures. Human TF-1 and UT-7 erythroleukemic cell lines were grown under standard conditions (i.e., RPMI-1640 medium supplemented with 10% heat-inactivated FCS).

HPC Immunofluorescence Analysis

Monoclonal Antibodies. Anti-CD34 (Clone 8G12), -HLA-DR, -CD38, -CD61, -CD11b (Becton Dickinson, Mountain View, CA, USA), -c-kit, -GlycophorinA (Immunotech, Marseille, France) directly conjugated with fluorochrome (fluorescein isothiocyanate (FITC) or phycoerythrin (PE)) were used to characterize the membrane phenotype of Step IIIP HPCs.

Labeling Procedure. Cells were washed twice in phosphate-buffered saline (PBS) and then incubated for 60 min at 4°C in the presence of an appropriate amount of mAb. After three washes with cold PBS, cells were resuspended in PBS containing 2.5% formaldehyde and then analyzed by FACS (FACScan, Becton Dickinson) for fluorescence intensity with a program suitable for double fluorescence analysis. At least 4000 cells were analyzed for each determination. The gating for analysis of the living cell population was determined by propidium iodide staining.

Fluorescence Analysis of HGFR Expression. Two types of reagents were used: (i) anti-HGFR mAb either simply purified or fluorochrome-conjugated; and (ii) recombinant HGFs either biotin or fluorochrome conjugated.

Anti-IL-3R and -GM-CSFRα chain PE-labeled anti-IL-3Rα chain mAb, clone 9G5 was purchased from Pharmingen; PE-labeled anti-GM-CSFRα mAb was obtained from Immunotech. The antibodies were incubated as indicated above for fluorochrome-conjugated antibodies.

Anti-βc mAb. Purified unconjugated anti-βc mAb was purchased from Santa Cruz Biotechnology. Cells were first incubated for 60 min at 4°C either in the presence of 5 mg/ml of this antibody or in the presence of an appropriate negative control (mouse isotype IgG), washed with cold PBS, and then incubated for 30 min at 4°C with PE-conjugated sheep anti-mouse IgGs (Dakopatts, Denmark). After three washings with cold PBS, cells were analyzed by fluorescence activated cell sorter (FACS).

Ligands. Biotin-conjugated IL-3 and PE-labeled GM-CSF were both obtained from R&D Systems. For detection of IL-3 binding, the cells were incubated for 60 min at 4°C in the presence of biotin-IL-3, incubated for an additional 30 min at 4°C with FITC-conjugated streptavidin (R&D Systems), and then analyzed by FACS. For GM-CSF binding the cells were incubated for 60 min at 4°C with PE-labeled GM-CSF, washed, and then analyzed by FACS.

The specificity of biotinylated IL-3 and PE-conjugated GM-CSF binding was assessed by using TF-1 and UT-7 erythroleukemic cells incubated with ligand alone or combined with an appropriate amount of blocking anti-ligand mAb: the difference in fluorescence intensity between these two conditions reflects the level of specific binding and the number of receptors.

Reverse Transcriptase-Polymerase Chain Reaction (RT-PCR) Analysis of βc mRNA

Total RNA was extracted by the guanidinium isothiocyanate/Cl method[31] and normalized by dot hybridization with a human rRNA probe.[28] The normalized RNA was reverse transcribed according to the manufacturer's instructions (GIBCO-BRL, Gaithersburg, MD). The RT-PCR was normalized for β2-microglobulin (amplification within the linear range was achieved by 20 PCR cycles: denaturation at 95°C for 30 sec, annealing at 54°C for 30 sec, and extension at 72°C for 45 sec). To evaluate the expression of βc and β_{IT} gene, an aliquot of RT-RNA (corresponding to 2 ng RNA) was amplified within the linear range by 35 PCR cycles: denaturation at 95°C for 30 sec, annealing at 64°C for 30 sec and extension at 72°C for 1 minute were performed.[17] Each sample was electrophoresed in 2% agarose, transferred to nylon membrane, and hybridized with a specific probe. The following primers and probes were used for βc and β_{IT} detection:

Primer 5′: GCACCGGCTACAACGGGATCT (1281–1301)
Primer 3′: TCCCCGAATCCTACAGGGAAC (1600–1620)
Probe: CAGAAAGTGGGAGGAGAAGAT (1441–1461)

RT-PCR of RNA using these primers ielded two products, one of 339 bp, as expected from the published sequence,[6] and another of 240 bp corresponding to the truncated isoform β_{IT} chain.

Positive controls were normalized for β2-microglobulin and included in all PCR assays. For negative controls, an aliquot of RNA (2 ng) from each sample was amplified to exclude the presence of contaminant DNA; a mock reaction was also included.

RESULTS

Expression of IL-3Rα, GM-CSFRα, and βc Chains in Leukemic Cell Lines

Anti-IL-3Rα antibody, in combination with anti-GM-CSFRα, anti-βc, and anti-c-kit antibodies, was used to examine the expression of these cytokine receptors on human leukemic cell lines TF-1 and UT-7 grown in the absence of exogenous growth factors or in the presence of either IL-3 (100 U/ml) or GM-CSF (10 ng/ml) or both cytokines. Both these cell lines express all the four receptor subunits (Fig. 1 and data not shown). Addition of IL-3 to TF-1 and UT-7 cells elicited a significant downmodulation of IL-3Rα chain, associated with a significant downmodulation of GM-CSFRα in UT-7, but not in TF-1 cells and no modification of the c-kit expression (Fig. 1 and data not shown). Similarly, GM-CSF induced a marked decline of GM-CSFRα, a moderate (TF-1) or pronounced (UT-7) decline of βc, without significant modification of IL-3Rα and c-kit expression (Fig. 1 and data not shown). Finally, treatment with both IL-3 and GM-CSF elicited a downmodulation of both IL-3Rα and GM-CSFRα, a moderate decline of βc and no change in c-kit expression (Fig. 1 and data not shown).

These observations indicate that IL-3 and GM-CSF elicited a marked downmodulation of their respective α chain, associated with only a moderate decline of βc expression.

RT-PCR analysis of βc and β_{IT} mRNAs in these two cell lines showed that the mRNA encoding the full-length β chain is more highly abundant than that encoding for truncated β chain (data not shown).

Expression of βc Chain in Quiescent HPCs and during E, G, Mk, and Mo Differentiation

In a first set of experiments we evaluated βc chain expression in quiescent Step IIIP HPCs by immunofluorescence (Fig. 2) and RT-PCR (Fig. 3). Immunofluorescence experiments showed that quiescent purified HPCs displayed only a slight reactivity with anti-βc antibodies (Fig. 2). RT-PCR experiments showed that these cells possess low, but significant levels of β chain mRNA, with predominant expression of full-length βc with respect to truncated β_{IT} mRNA.

βc chain expression increased during the initial stages of HPC differentiation along the E, G, Mo or Mk lineages. Starting from day 5, βc chain expression increased in both the G and Mo lineages, remained moderately expressed in the Mk lineage and markedly declined to undetectable level in the E lineage. Analysis of relative fluorescence intensity showed that the highest level of βc chain expression was observed on cells pertaining to the G lineage.

Comparable observations were made by RT-PCR analysis of βc chain mRNA levels. In fact, during the initial stages of HPC differentiation along either the E, G, Mo or Mk lineages an increase in β chain mRNA expression was observed, with prevalent expression of full-length βc mRNA as compared to βIT mRNA (data not shown). Starting from day 5, a sustained level of βc chain mRNA expression was observed only in cells differentiating along the G and Mo lineages, with predominant expression of βc mRNA as compared to βIT mRNA at all stages of differentiation/maturation. In contrast, cells differentiating along the E and Mk lineages exhibited a marked decline of βc chain mRNA, this phenomenon being more evident for βc than for βIT mRNA (Fig. 3 and data not shown).

IL-3Rα and GM-CSFRα Chain Expression during HPC E, G, Mk, and Mo Differentiation/Maturation

In a previous study we reported the evaluation of IL-3Rα and GM-CSFRα chain expression in purified HPCs.[23] In the present study we have evaluated IL-3R and

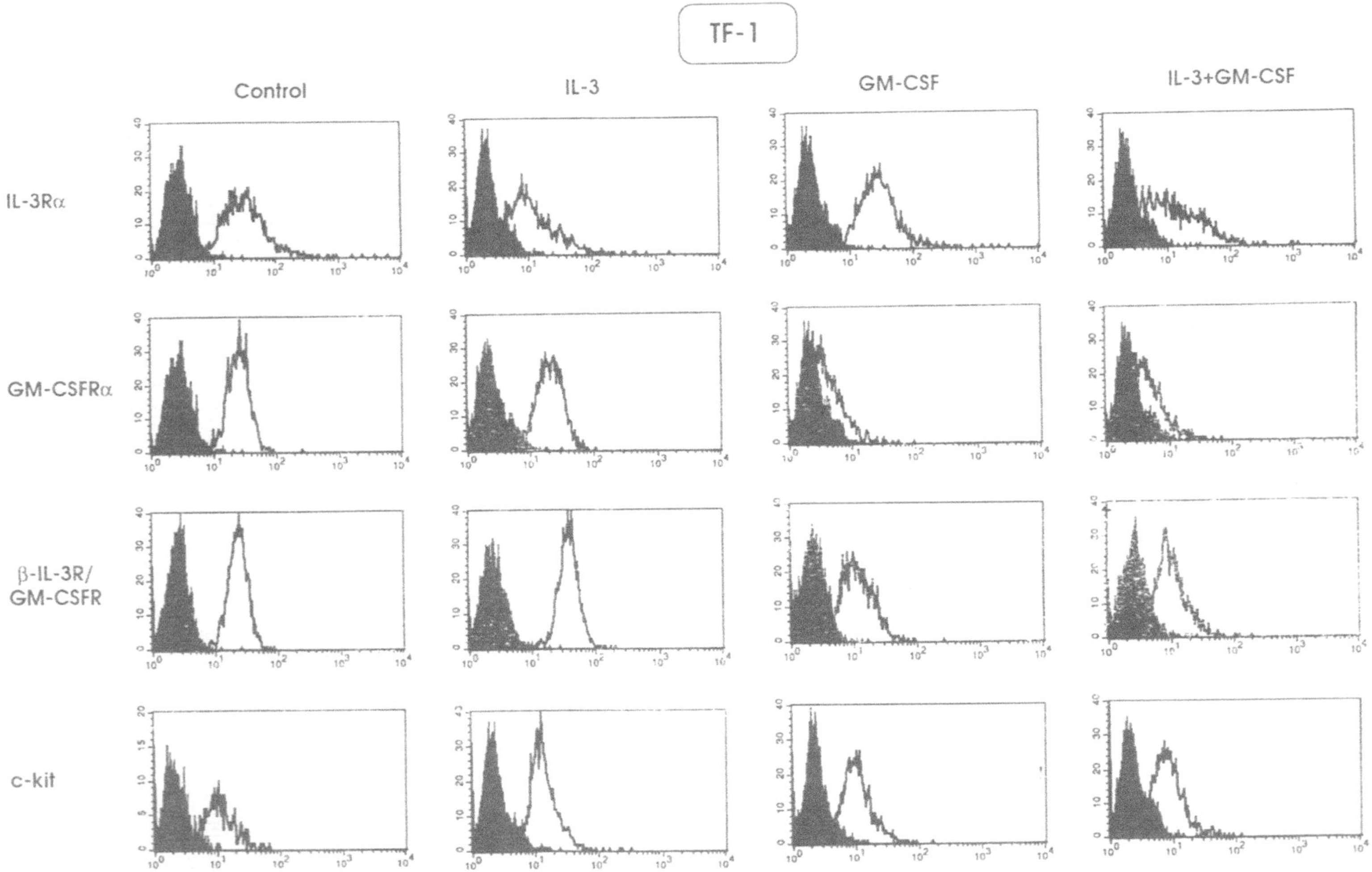

Figure 1. Flow cytometric analysis of IL-3Rα, GM-CSFRα, βc, and c-kit in erythroleukemic TF-1 cells grown for 24h either in the absence or in the presence of IL-3 (100 U/ml), GM-CSF (10 ng/ml) or both ligands.

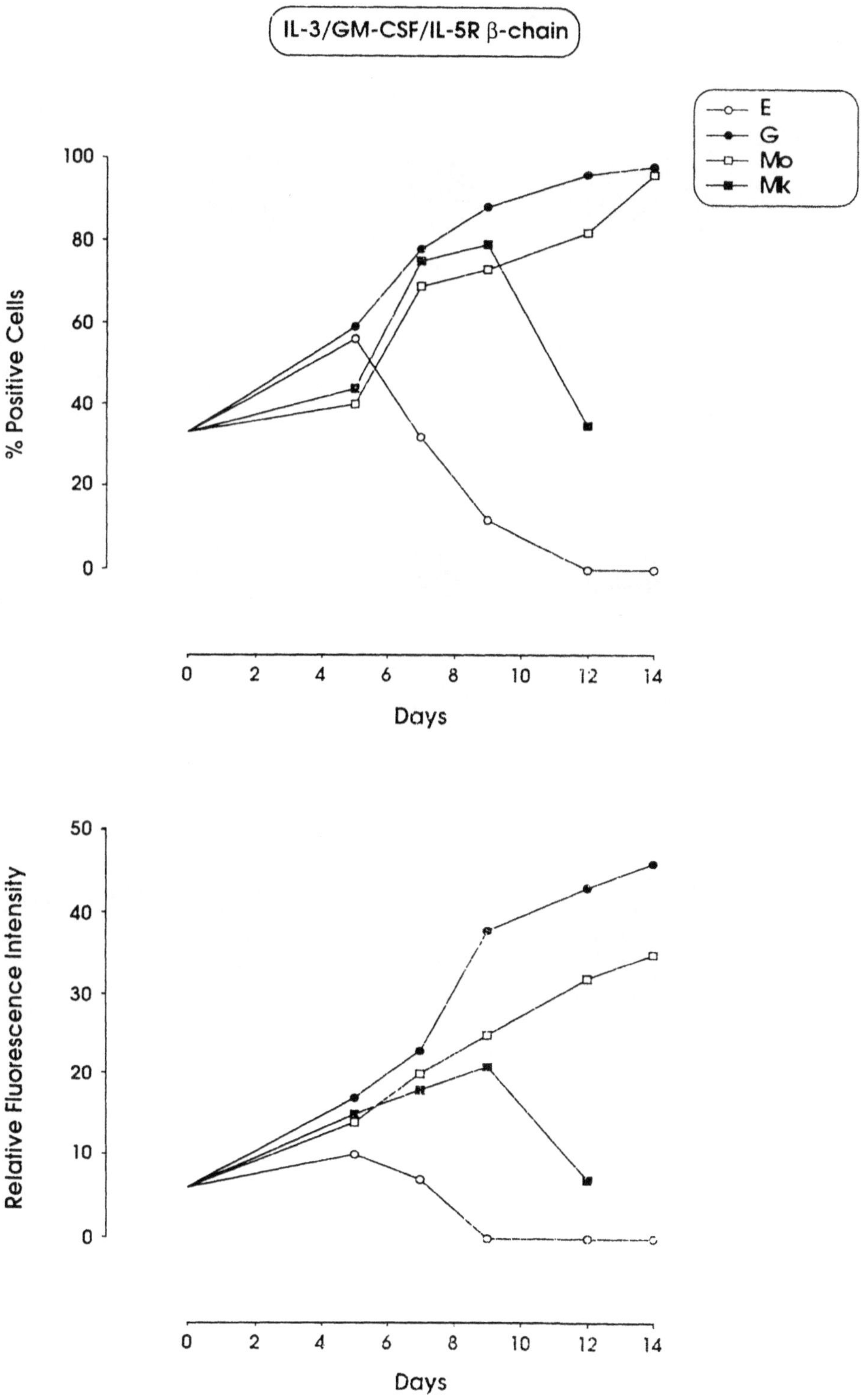

Figure 2. Flow cytometric analysis of βc expression in unilineage cultures of purified human HPCs. HPCs were induced to selective erythroid (E), granulocytic (G), megakaryocytic (Mk) or monocytic (Mo) differentiation and the expression of βc was investigated at different days of culture by flow cytometry using a specific mAb. The results are expressed as both percentage of positive cells and of relative fluorescence intensity. Results represent the mean value observed in three separate experiments.

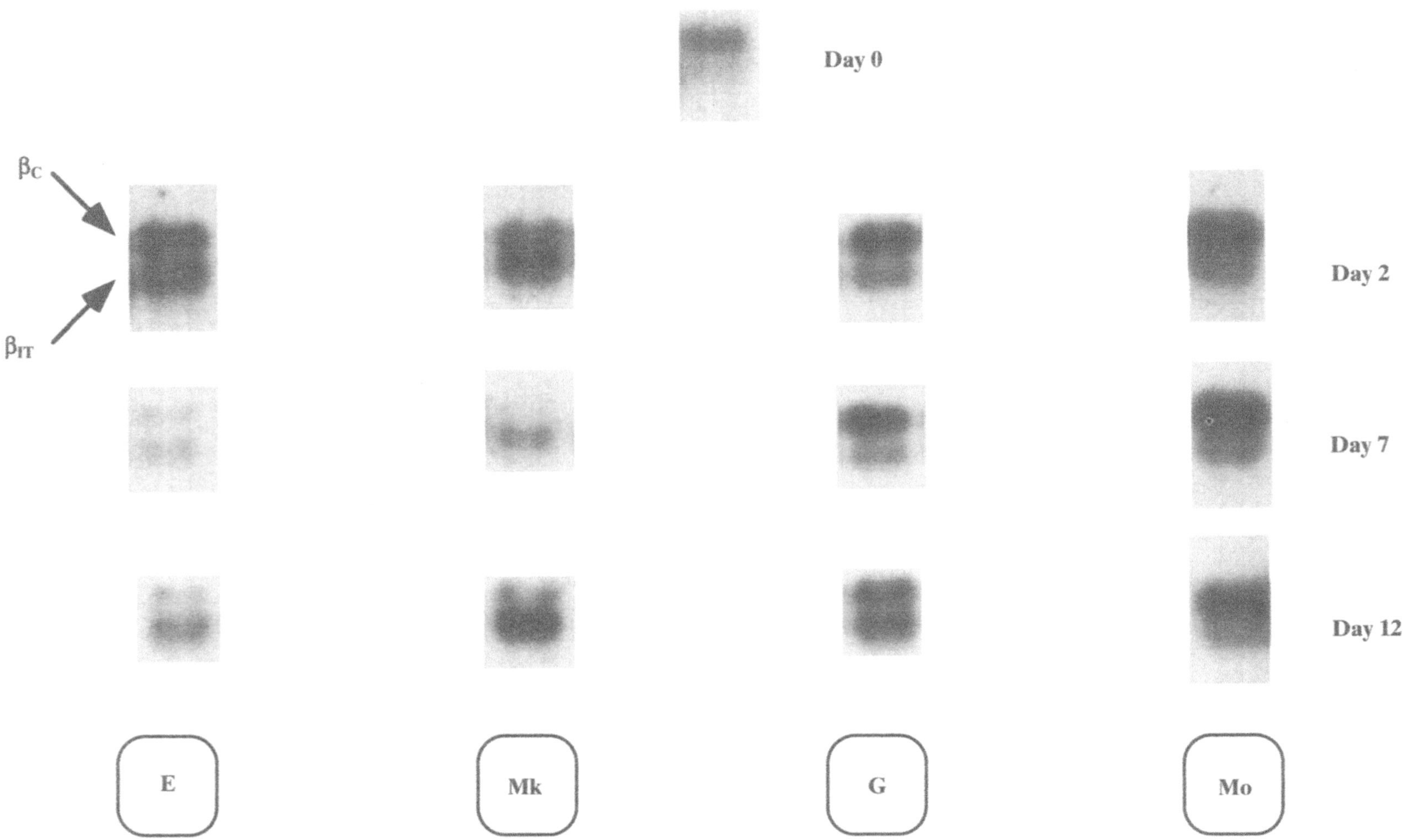

Figure 3. RT-PCR analysis of βc and β_{IT} mRNA in purified HPCs grown in unilineage conditions to induce selective E, G, Mk, or Mo differentiation.

GM-CSFR expression by using either anti-IL-3Rα or anti-GM-CSFRα mAbs or labeled ligands (either biotin IL-3 or PE-labeled GM-CSF). The study of IL-3R and GM-CSFR showed a substantially similar pattern of expression: IL-3Rα and GM-CSFRα chains are already expressed in quiescent HPCs, and their expression is maintained only in the G and Mo lineages, while it is progressively lost during E and Mk differentaition (Fig. 4); IL-3R and GM-CSFR expression as evaluated by analysis of their respective labeled ligands showed a pattern comparable to that observed by analysis of the expression of their α chains, with the exception that the initial reactivity of quiescent purified HPCs was lower for the labeled ligands than for anti-α chains mAbs (data not shown).

The Initial Induction of βc Is Related to a Differentiation Event and Not to the Cycling Rate of HPCs

Since βc expression is induced and maintained at high levels during the induction of granulocytic differentiation, additional experiments were focused on a possible link between the cycling of HPCs and expression of βc. The results showed that: (i) IL-3 doses as low as 0.1 U/ml are sufficient for stimulation of βc expression, while only high doses (100 U/ml) are able to induce a significant stimulation of HPC cycling (data not shown); (ii) on the contrary, GM-CSF doses as low as 0.1 ng/ml induce a significant cycling of HPCs, but are not sufficient to induce a significant upmodulation of βc expression.

These observations suggest that upmodulation observed during the initial phases of HPC differentiation is related to the differentiation of HPCs rather than to their cycling. This concept is further reinforced by the observation that the proportion of cycling cells particularly at late days of culture, is higher in erythroid cells as compared with granulocytic or monocytic cells, in which βc chain expression is maximal (data not shown).

DISCUSSION

In this study we have analyzed the expression pattern of IL-3, GM-CSF, and IL-5Rs common βc chain during normal hematopoietic differentiation starting from immature hematopoietic progenitors to terminal differentiated cells.

Our results indicate that βc is expressed at low, but significant levels, in quiescent HPCs. In the initial stages of HPC differentiation the expression of βc is enhanced in all lineages (E, G, Mk, and Mo) and this phenomenon is related to a differentiation event and not to the rate of HPC cycling. In the later stages of HPC differentiation (i.e., from day 5 of culture) βc expression was preferentially observed in cells differentiating along the G or Mo lineage, while its expression is progressively lost in cells pertaining to the E and Mk lineages.

This pattern of βc expression corresponds to, and may partially explain the sensitivity of the different types of hematopoietic cells to IL-3 and GM-CSF: in fact, cells of the E and Mk lineages are responsive to IL-3/GM-CSF only during their initial stages of differentiation (i.e., at the level of the progenitor compartment and during the initial phases of progenitor differentiation), whereas cells pertaining to the G and Mo lineages are responsive to IL-3/GM-CSF during both initial and late stages of differentiation/maturation.

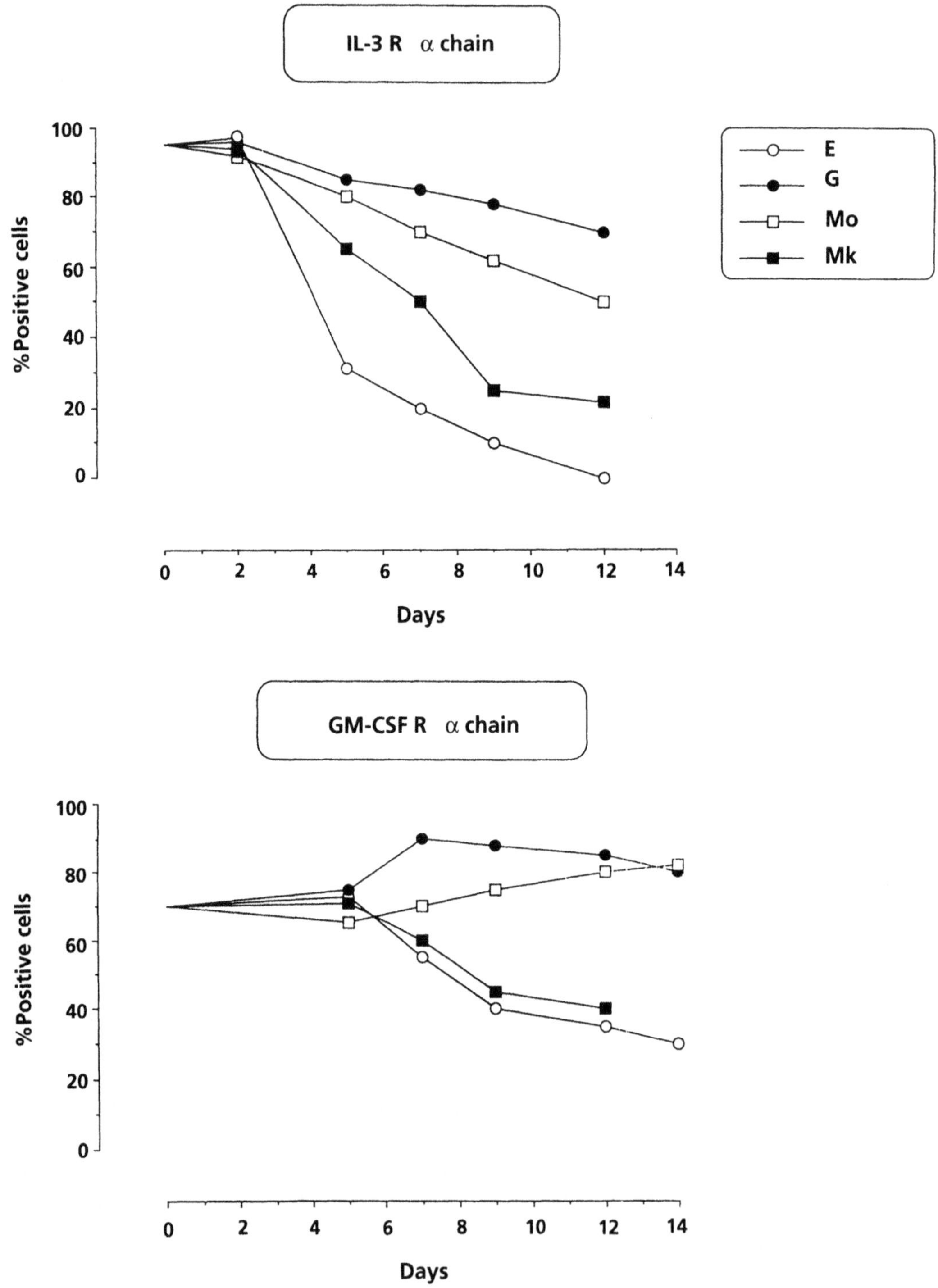

Figure 4. Flow cytometric analysis of IL-3Rα and GM-CSFRα chain expression in HPCs triggered to E, G, Mk, or Mo differentiation. Results (mean values of three separate experiments) are expressed as percentage of positive cells.

At the molecular level this pattern of βc expression probably implies the existence of different regulation mechanisms operating during the early and late stages of hematopoietic differentiation.

Recent studies have shown the existence of two different β chain isoforms: βc and β_{IT}. The full-length βc is involved in mediating the mitogenic/differentiative signal to the cells originated following interaction of the ligands with the IL-3/GM-CSF receptors. The intracytoplasmic truncated β_{IT} isoform was unable to convey the receptor-originated signal to the cell in that it lacks the majority of the receptor cytoplasmic tail.[17] β_{IT} retains the capacity to interact with and to form heterodimers with both the GM-CSFRα and IL-3Rα chains, thus competing with the full-length βc, it was proposed that these mechanisms may represent a system to buffer the response of the cell to IL-3/GM-CSF.[17] The analysis of βc and β_{IT} mRNAs during the different stages of hematopoietic differentiation indicates that the expression of βc is always dominant over that of β_{IT}, with the exception of the late stages of E and Mk differentiation. Preliminary observations show a similar situation at protein level with βc protein expression higher the β_{IT} protein expression.

We have also investigated here and in a previous study[23] the expression of IL-3Rα, GM-CSFRα chain expression during hematopoietic differentiation. Generally, the pattern of IL-3Rα and GM-CSFRα chains is comparable to that observed for the βc, the only exception being that in quiescent HPCs βc is expressed at low levels, while IL-3Rα and GM-CSFRα chains are already expressed. This last observation explains the relatively low IL-3 and GM-CSF binding capacity of quiescent HPCs which markedly increases during the early phases of progenitor differentiation, in parallel to the increase of βc protein expression.

Finally, we have investigated in leukemic cell lines the regulation of surface α and βc chains of the IL-3R/GM-CSFR following incubation with their respective ligands added alone or in combination. As shown in previous studies, both IL-3R and GM-CSFRα chains are rapidly and markedly downmodulated following incubation with their respective ligands; in contrast, βc expression is only slightly modified by ligand exposure. This differential modulation of surface α and βc chains of the IL-3R/GM-CSFR may have important physiological implications: in fact, the downmodulation of IL-3Ra chains observed following incubation with exogenous IL-3 may represent a mechanism to avoid over-response of the cell to this growth factor; in contrast, the absent/low downmodulation of βc expression following IL-3 stimulation may allow the cell to retain the capacity of responding to GM-CSF or IL-5 stimulation.

REFERENCES

1. Ogawa M. Differentiation and proliferation of hematopoietic stem cells. Blood 81:2844,1993.
2. Metcalf D. Hematopoietic regulators: redundancy or subtlety? Blood 82:3515,1993.
3. Miyajima A, Mui LF, Orogochi T, Sakamaki K. Recptors for granulocyte-macrophage colony-stimulating factor, interleukin-3, and interleukin-5. Blood 82:1960,1993.
4. Bazan JF. Structural design and molecular evolution of a cytokine receptor superfamily. Proc. Natl. Acad. Sci. USA 87:6934,1990.
5. Gearing DP, King JA, Gough NM, Nicola NA. Expression and cloning of a receptor for human granulocyte-macrophage colony-stimulating factor. EMBO J 8:3667,1989.
6. Hayashida K, Kitamura T, Gorman DM, Arai K, Yokota T, Miyajima A. Molecular cloning of a second subunit of the receptor for human granulocyte-macrophage colony-stimulating factor (GM-CSF): reconstitution of a high-affinity GM-CSF receptor. Proc. Natl. Acad. Sci. USA 87:9655,1990.

7. Kitamura T, Hayashida K, Sakamaki K, Yokota T, Arai K, Miyajima A. Reconstitution of functional receptors for human granulocyte/macrophage colony-stimulating factor (GM-CSF): evidence that the protein encoded by the AIC2B cDNA is a subunit of the murine GM-CSF receptor. Proc. Natl. Acad. Sci. USA 88:5082,1991.
8. Tavernier J, Devos R, Cornelis S, Tuypens T, Van Der Heyden J, Fiers W, Plaetnick G. A human high-affinity interleukin-5 receptor (IL-5R) is composed of an IL-5 specific alpha chain and a beta chain shared with the receptor for GM-CSF. Cell 66:1175,1991.
9. Hara T, Miyajima A. Two distinct functional high-affinity receptors for mouse interleukin-3 (IL-3). EMBO J. 11:1875,1992.
10. Sakamaki K, Miyajima I, Kitamura T, Miyajima A. Critical cytoplasmic domains of the common β subunit of the human GM-CSF, IL-3, and IL-5 receptors for growth signal transduction and tyrosine phosphorylation. EMBO J 11:3541,1992.
11. Quelle F, Sato WN, Witthuhn BA, Inborn RC, Eder M, Miyajima A, Griffin JD, Ihle JN. JAK2 associates with the βc chain of the receptor for granulocyte-macrophage colony-stimulating factor, and its activation requires the membrane-proximal region. Mol Cell Biol 14:4335,1994.
12. Sato N, Sakamaki K, Terada N, Arai K, Miyajima A. Signal transduction by the high-affinity GM-CSF receptor: two distinct cytoplasmic regions of the common β subunit responsible for different signaling. EMBO J 12:4181,1993.
13. Yoshimma A, Ohkubo T, Kiguchi T, Jenkins NA, Gilbert DJ, Copeland NG, Hara T, Miyajima A. A novel cytokine inducible gene CIS encodes a SH2-containing protein that binds to tyrosine-phosphorylated interleukin 3 and erythropoietin receptors. EMBO J 14:2816,1995.
14. Smith A, Metcalf D, Nicola NA. Cytoplasmic domains of the common beta-chain of the GM-CSF/IL-3/IL-5 receptors that are required for inducing differentiation or clonal suppression in myeloid leukemic cell lines. EMBO J 16:451,1997.
15. Kinoshita T, Yokota T, Arai K, Miyajima A. Suppression of apoptotic death in hematopoietic cells by signalling through the IL-3/GM-CSF receptors. EMBO J 14:266,1995.
16. Dirksen U, Nishinakamura R, GraneK P, Hattenhorst U, Nagee L, Murray R, Burdach S. Human pulmonary alveolar proteinosis associated with a defect in GM-CSF/IL-3/IL-5 receptor common beta chain expression. J Clin Invest 100:2211,1997.
17. Gale RE, Freeburn RW, Khwaya A, Chopra R, Linch DC. A truncated isoform of the human β chain common to the receptors for granulocyte-macrophage colony-stimulating factor, interleukin-3 (IL-3), and IL-5 with increased mRNA expression in some patients with acute leukemia. Blood 91:54,1998.
18. Jubinsky PT, Laurie AS, Nathan DG, Yetz-Adepe J, Sieff CA. Expression and function of the human granulocyte-macrophage colony-stimulating factor receptor α subunit. Blood 84:4174,1994.
19. Kurata H, Arai T, Yokota T, Arai K. Differential expression of granulocyte-macrophage colony-stimulating factor and IL-3 receptor subunits on human CD34+ cells and leukemic cell lines. J. Allergy Clin Immunol 96:1083,1995.
20. Wognum AW, Westerman Y, Visser TP, Wagemaker G. Distribution of receptors for granulocyte-macrophage colony-stimulating factor on immature CD34+ bone marrow cells, differentiating mono-myeloid progenitors, and mature blood cell subsets. Blood 84:764,1994.
21. Wognum AW, Visser TP, De Jong MO, Egeland T, Wagemaker G. Differential expression of receptors for interleukin-3 on subsets of CD34- expressing hematopoietic cells of Rhesus monkey. Blood 86:581,1995.
22. Testa U, Pelosi E, Gabbianelli M, Fossati C, Campisi S, Isacchi G, Peschle C. Cascade transactivation of growth factor receptors in early human hematopoiesis. Blood 81:1442,1993.
23. Testa U, Fossati C, Samoggia P, Masciulli R, Mariani G, Hassan JH, Sposi NM, Guerriero R, Rosato V, Gabbianelli M, Pelosi E, Valtieri M, Peschle C. Expression of growth factor receptors in unilineage differentiation culture of purified hematopoietic progenitors. Blood 88:3391–3406, 1996.
24. Sato N, Caux C, Kiatamura T, Watanabe Y, Banchereau J, Mayajima A. Expression and factor-dependent modulation of the interleukin-3 receptor subunits on human hematopoietic cells. Blood 82:752,1993.
25. Gabbianelli M, Sargiacomo M, Pelosi E, Testa U, Isacchi G, Peschle C. "Pure" human hematopoietic progenitors: permissive action of fibroblast growth factor Science 249:1561,1990.
26. Labbaye C, Valtieri M, Testa U, Giampaolo A, Meccia E, Sterpetti P, Parolini I, Pelosi E, Bulgarini D, Cayre Y, Peschle C. Retinoic acid downmodulates erythroid differentiation and GATA-1 expression in purified adult progenitor culture. Blood 83:651,1994.

27. Guerriero R, Testa U, Gabbianelli M, Mattia G, Montesoro E, Macioce G, Pace A, Ziegler B, Hassan HJ, Peschle C. Unilineage megakaryocytic proliferation and differentiation of purified hematopoietic progenitors in serum-free liquid culture. Blood 86:3725,1995.
28. Sposi NM, Zon LI, Carè A, Valtieri M, Testa U, Gabbianelli M, Mariani G, Bottero L, Mather C, Orkin SH, Peschle C. Cell cycle-dependent initiation and lineage-dependent abrogation of GATA-1 expression in pure differentiating hematopoietic progenitors. Proc Natl Acad Sci USA 89:6353,1992.
29. Condorelli GL, Testa U, Valtieri M, Vitelli L, De Luca A, Barberi T, Montesoro E, Campisi S, Giordano A, Peschle C. Modulation of retinoblastoma gene in normal adult hematopoiesis: peak expression and functional role in advanced erythroid differentiation. Proc. Natl Acad Sci USA 92:4808,1995.
30. Gabbianelli M, Pelosi E, Montesoro E, Valtieri M, Luchetti L, Samoggia P, Vitelli L, Barberi T, Testa U, Lyman S, Peschle C. Multi-level action of flt3 ligand on human hematopoiesis: expansion of putative stem cells and proliferation of granulomonocytic progenitors/monocytic precursors. Blood 86:1661,1995.
31. Chirwing JM, Przybyla AE, MacDonald RJ, Rutter WJ. Isolation and biologically active ribonucleic acid from sources enriched in ribonuclease. Biochemistry 18:5294,1979.

31

EXPRESSION OF DISTAL *Gata1* TRANSCRIPTS IN ERYTHROID CELLS

Alessandro M. Vannucchi, Silvia Linari, and Anna Rita Migliaccio[1]

Division of Hematology
University of Florence and Azienda Ospedale Careggi
Florence and Laboratory of Cell Biology
Istituto Superiore di Sanita', Rome
Italy

ABSTRACT

Gata1 is expressed from either one of two alternative promoters, the erythroid (or proximal with respect to the AUG) and the testis (or distal with respect to the AUG) promoter, both of which are utilized by erythroid cells. We have determined by specific RT-PCRs the relative levels of the two *Gata1* transcripts during the differentiation of erythroid precursors purified from the spleen of phenylhydrazine (PHZ)-induced anemic mice. Day 1 PHZ-spleen cells are mainly composed by erythroid precursors that synchronously progress *in vivo*, and *in vitro* in the presence of erythropoietin (EPO), to mature erythroblasts. The total levels of *Gata1* expression increased by 5–8-fold during *in vivo* and *in vitro* differentiation of PHZ-cells. Such increase was correlated with increased levels of the distal, but not of the proximal *Gata1* transcripts. The day 1 PHZ-cells which survived 18 hours in culture in the absence of EPO, were benzidine-negative and expressed detectable levels of the distal, but not of the proximal, *Gata1* transcripts. When re-exposed to EPO, these cells activated the expression of the globin genes within 2 hrs. The existence of an erythroid precursor which expresses only distal *Gata1* transcripts, and the fact that the increase of *Gata1* expression observed during erythroid differentiation is due to increased levels of transcripts from the distal promoter, suggest that regulation of the expression of *Gata1* from its distal promoter is an important step in the erythroid differentiation program.

[1] *Address Correspondence to*: Anna Rita Migliaccio, Laboratory of Cell Biology, Istituto Superiore di Sanita', Ciale Regina Elena 299, 00161 Rome, Italy. Phone no.: 0039-6-49902576. Fax no.: 0039-6-49387143. E-mail: migliar@iss.it.

Molecular Biology of Hematopoiesis 6, edited by Abraham *et al.*
Kluwer Academic/Plenum Publishers, New York, 1999.

INTRODUCTION

Gata1[1] is a member of the highly conserved family of zinc finger proteins which bind to the conserved motif, GATA and represents one of the most studied erythroid transcription factors.[2] *Gata1* is also expressed in the Sertoli cells of prepuberal mouse testis.[3] In this case, *Gata1* transcription utilizes a testis-specific promoter (the distal promoter with respect to the AUG) located 8 Kb upstream of the erythroid (or proximal to the AUG) promoter. Recently we have described that transcripts from the distal promoter represent 1% of the total *Gata1* mRNA of adult bone marrow.[4,5] Low levels of distal *Gata1* transcripts are also expressed by cultured hematopoietic cells but only when induced toward the erythroid differentiation pathway by erythropoietin (EPO). However, also in this case, the low levels of expression observed obscured the significance of the result for the erythroid differentiation program.

The level of *Gata1* expression increases during the erythroid differentiation[1] and such an increase appears to be the cause of the establishment of the erythroid phenotype itself. In fact, only the avian myelo-monocytic cell clones which establish the highest levels of ectopic *Gata1* expression, acquire an erythroid phenotype.[6] Furthermore, mice reconstituted with stem cells engineered to costitutively express *Gata1* have lower white blood cell counts and absolute neutrophil counts and higher red blood cell counts than animals reconstituted with normal stem cells.[7] These animals also require more extensive phlebotomy to become anemic and recover from anemia sooner than the controls.

These data[6,7] and our previous results[4,5] suggested to us that the increased level of expression of *Gata1* observed during erythroid differentiation could be the consequence of increased levels of the distal transcripts. To test this hypothesis, we have analyzed by semi-quantitative reverse transcriptase polymerase chain reaction (RT-PCR) the total level of *Gata1* expression, as well as the levels of the proximal and distal transcripts, during *in vivo* and *in vitro* differentiation of purified erythroid precursors isolated from the spleen of mice 1–3 days after treatment with phenylhydrazine (PHZ).[8] PHZ induces a profound hemolytic anemia from which the animals recover by undergoing an acute expansion of erythroid cells in the spleen that becomes a transient "erythroid" organ. Mononuclear spleen cells collected on day 1 and 3 after phenylhydrazine treatment are mainly (>95%) composed of pro-erythroblasts and of mature erythroblasts, respectively.[9] The pro-erythroblasts contained in the spleen at day 1 proliferate and differentiate simultaneously also *in vitro* if stimulated with erythropoietin (Epo).[9] These cells represent, therefore a unique model to investigate variations of gene expression in normal cells synchronized at specific stages of erythroid differentiation. The results obtained confirmed that the expression of *Gata1* increases during the progression of erythroid differentiation and indicated that such an increase is due to increased levels of the distal, but not of the proximal, *Gata1* transcripts.

MATERIALS AND METHODS

Phenylhydrazine-induced Anemia. CBA (Charles River, Calco, Italy) mice were sacrified 1 and 3 days after i.v. injection with phenylhydrazine (PHZ, 60 mg/kg b.w., Sigma, St. Louis, Mo) as described.[8] All of the animal manipulations were approved by the Institutional Animal Care Committee.

Cell Purification and Culture. Mononuclear cells were isolated from single cell suspensions of normal or PHZ-treated spleens by 20 min centrifugation at 1650 g on a discontinuous Ficoll-Paque (1.077 g/ml, Pharmacia Biotec, Upsala, Sweden) gradient. Day 1 PHZ-cells were cultured at 37°C for up to 32–72 hours in Iscove's modified Dulbecco's medium (IMDM, Sigma) supplemented with 30% fetal calf serum (FCS, Boehringer Mannheim, Mannheim, Germany), 1% deionized bovine serum albumin, 0.1 mM α-thioglycerol, and 2 U/ml recombinant human EPO (Boehringer Mannheim). Alternatively, day 1 PHZ-cells were growth factor starved for 18 hrs in IMDM supplemented with 10% Nutridoma (Boehringer Mannheim) and then cultured either in the absence or in the presence of EPO (5 U/ml) for additional 48 hours.

RNA Isolation. Total RNA was prepared according to the guanidine thiocyanate/phenol protocol[10] as described by the manifacture (Trizol, Gibco BRL, Paisley, UK).

Reverse-Transcription and PCR Amplification. Total RNA (1 μg/reaction) was reverse transcribed at 42°C for 30 min in 20 μl of 10 mM Tris-HCl (pH 8.3) containing 5 mM $MgCl_2$, 1 U RNAse inhibitor, dNTPs (200 μM each), 2.5 U Moloney Murine Leukemia Virus reverse-transcriptase, and 2.5 μM random hexamers (all from Perkin-Elmer, New Jersey, USA). An aliquot (1:10) of the obtained cDNA was then dissolved in 100 μL of 10 mM Tris-HCl (pH 8.3) containing 2 mM $MgCl_2$, dNTPs (200 μM each), 0.1 μCi of [α^{32}P]dCTP (Amersham International, Buckinghamshire, UK, specific activity 3000 Ci/mmol), 2 U AmpliTaq DNA polymerase and specific *Gata1* sense and antisense primers (100 nM each) as described.[4,5] The amplification reaction (30 cycles) was performed in a Perkin Elmer thermocycler (GeneAmp PCR System 2400) with the following parameters: denaturing 60 sec at 95°C, annealing 60 sec at 66°C, and primer extention 60 sec at 72°C. As control, primers specific for β-actin (50 nM each) were added to each tube after the first 20 cycles of amplification.[11] Aliquots (10 μL) of PCR mixture were electrophoresed through a 4% polyacrylamide gel according to standard protocols.[12] Gels were dried using a Biorad apparatus, and exposed to Hyperfilm-MP (Amersham) for 2 hours at –70°C. The intensity of the X-ray bands was quantified using the Seprascan software (Integrated Separation Systems, Natick, MA) and expressed in arbitrary optical density units.

RESULTS

Expression of Distal *Gata1* Transcripts in Normal Tissues and in the Spleen of PHZ-Treated Animals

Figure 1 shows the RT-PCR amplifications of *Gata1* transcripts in normal murine tissues, and in spleens from PHZ-treated animals.

The levels of total *Gata1* mRNA and of the proximal *Gata1* transcripts were only slightly (by less than 2-fold) different among the normal tissues analyzed (the range of total and of proximal *Gata1* transcripts were 0.77–1.1 and 0.8–1.4 of the level of actin, respectively). In contrast, the levels of the distal *Gata1* transcripts ranged widely from barely detectable in the bone marrow, to highly detectable (0.8–1.5 of the actin level) levels in the spleen. Interestingly, day 3 PHZ-treated cells (mature erythroblasts) expressed levels of total and of distal *Gata1* transcripts 3–4-fold higher than the levels expressed by day 1 PHZ-treated cells (pro-erythroblasts) while the levels of proximal *Gata1* transcripts expressed by the two cell types were similar.

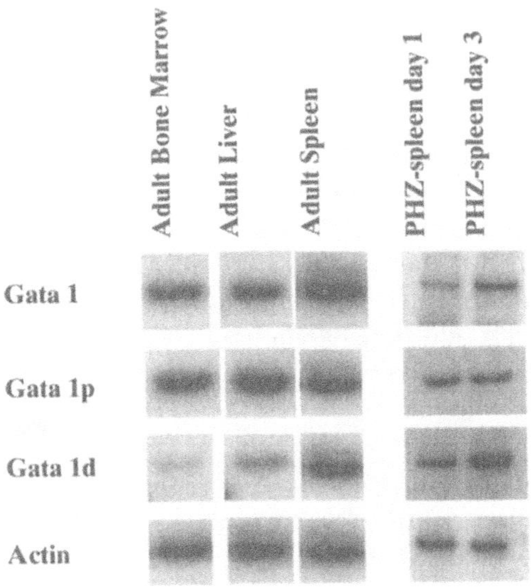

Figure 1. RT-PCR analysis of the level of total Gata1 mRNA and of the proximal (Gata1p) and distal (Gata1d) transcripts in normal (adult bone marrow, spleen and liver) and anemic (spleen cells from day 1–3 PHZ-treated mice) tissues. Actin was co-amplified as control of the amount of cDNA used in each reaction.

Expression of Distal *Gata1* Transcripts during *in vitro* Differentiation of Day 1 PHZ-Cells

To further characterize the levels of the two *Gata1* transcripts during erythroid differentiation, Gata1 transcripts were amplified from RNA extracted from day 1 PHZ-cells induced to differentiate in culture with EPO as described[8,9] (Fig. 2).The total level of *Gata1* increased up to 5–8-fold after 24 hrs of culture with Epo. The levels of proximal transcripts remained constant during the entire period of incubation while the levels of the distal transcripts increased from 1/3 of the levels of proximal transcripts (time 0) to levels equivalent or even higher than those of the proximal transcripts in cells cultured for 24–32 hrs (Fig. 2).

In the experimental models used in this study, Epo is required not only for cell survival, but also for proliferation and differentiation.[8,9] In order to discriminate whether the increase of the distal transcripts observed in PHZ-cells cultured with EPO is a direct consequence of Epo stimulation or was a property intrinsic to the cellular modifications induced by Epo, PHZ-cells were cultured for 18 hours without growth factors under defined culture conditions (Fig. 3). At that point (arrow in panel A), half

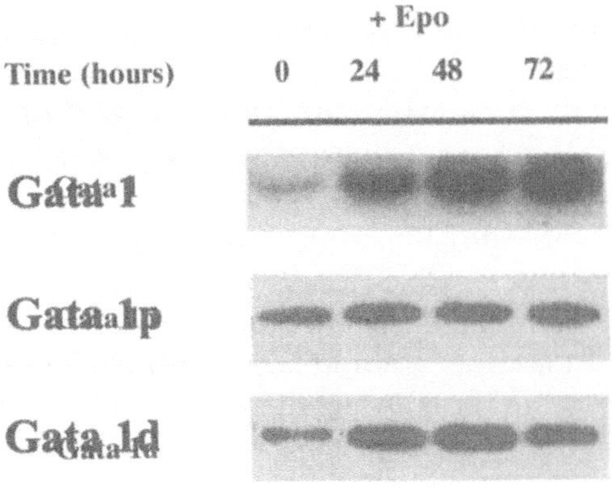

Figure 2. Time-course analysis of the levels of total, proximal and distal Gata1 transcripts in day 1 PHZ cells cultured with EPO for up to 72 hrs.

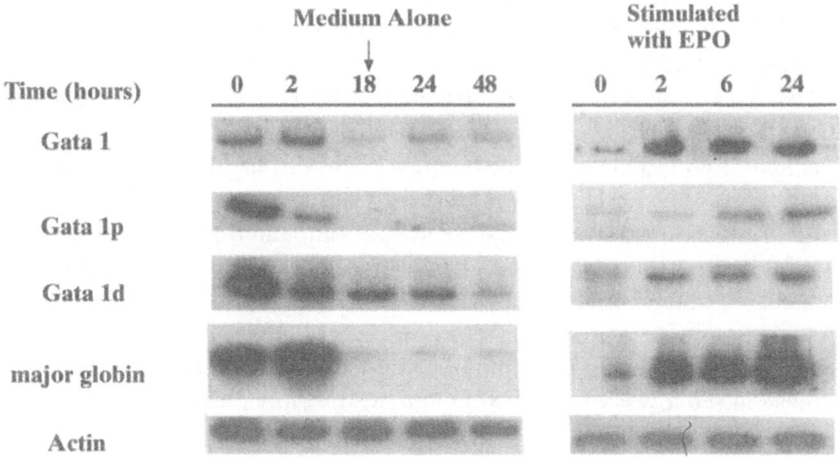

Figure 3. Effects of EPO on the levels of total, proximal, and distal Gata1 transcripts expressed by day 1 PHZ-cells. The levels of β-globin and actin are also presented as a control of the erythroid differentiation and of the levels of cDNA used in the amplification, respectively.

of the cells were removed, washed and incubated with EPO (5 U/ml) up to 24 hours (panel B, the 0 time point in this panel is a duplicate of the 18 hrs point in panel A) while the other half of the cells remained in culture not supplemented with EPO up to 48 hours (panel A).

Since day 1 PHZ-cells contain not only immature erythroid precursors, but also more primitive, less EPO-dependent erythroid progenitor cells such as BFU-E (0.5% of the cells), 25–45% of the cells survive a brief growth factor starvation. The surviving cells proliferate and double in number by 24 hrs when EPO is added to the culture (from $0.33 \pm 0.06 \times 10^6$ to $0.68 \pm 0.14 \times 10^6$ cells/ml). The PHZ-cells which survive 18 hours of EPO deprivation do not express β-globin (Fig. 3) suggesting that these cells have not activated the differentiation pathway as yet. Interestingly, there was a progressive decrease in the levels of proximal *Gata1* transcripts expressed by the cells surviving Epo-starvation down to undetectable levels by 18 hrs. In contrast, the levels of distal transcripts in these cells decreased only by 3-fold over the same period and were still detectable after 48 hours of EPO deprivation. When re-exposed to EPO, the cells started to differentiate, as shown by the detection of globin transcripts (Fig. 3). The levels of *Gata1* mRNA originating from the distal promoter did not change significantly and accounted for almost the entire levels of *Gata1* (amplification primers within the coding region) mRNA until 6 hrs of culture in the presence of EPO when also the proximal transcripts became detectable.

DISCUSSION

The tissue distribution of the expression of the two *Gata1* transcripts in normal and anemic mice presented in this paper supports the hypothesis that expression of *Gata1* from its distal promoter is a molecular event associated with erythroid differentiation. In fact, not only distal *Gata1* transcripts were expressed in normal erythroid tissues, as previously shown,[4,5] but their relative levels of expression in a tissue was found to be directly correlated with the frequency and stage of maturation of erythroid cells in that tissue (day 3 PHZ-spleen > day 1 PHZ-spleen > normal marrow). In particular, in PHZ-treated spleens, the mature erythroid cells on day 3 expressed levels of total and distal *Gata1* transcripts that were higher than the less mature erythroid cells of day 1, while the levels of proximal transcripts expressed by the two cell populations were the same (Fig. 1).

The increase in the levels of distal *Gata1* transcripts during erythroid maturation, was confirmed in day 1 PHZ-cells induced to differentiate *in vitro* with EPO.[8,9] Erythroid maturation of PHZ-cells *in vitro* was accompanied by parallel increases in the levels of total and distal *Gata1* transcripts while the levels of proximal transcripts did not change over time (Fig. 2).

PHZ-cells proliferate and differentiate in response to EPO. The expression of *Gata1* has been reported to be stimulated by EPO[13] and to be cell cycle dependent.[14] The increase in the levels of the distal *Gata1* transcripts observed in the present study could be due either directly to EPO stimulation or could be mediated by EPO through the induction of proliferation and/or of differentiation. In order to discriminate among these possibilities, we have analyzed the effect of EPO starvation on the expression of *Gata1* in PHZ-cells. PHZ-cells are extremely sensitive to EPO starvation and more than 55–75% of the cells died when they were starved for up to 18 hrs. The cells which survive, do not express detectable levels of globin genes but express *Gata1*. Interestingly, the levels of proximal *Gata1* transcripts expressed by the starved cells were below the levels detectable by RT-PCR while the distal *Gata1* transcripts were easely detected (Fig. 3). This result indicates that the majority of *Gata1* expressed by these cells comes from the distal promoter whose activation appears to be EPO-independent. When the EPO-starved PHZ-cells were re-exposed to EPO, they activate the expression of the globin genes within 2 hrs (Fig. 3) and double in cell number within 24 hrs. This fact indicates that the cells which express only distal *Gata1* transcripts are globin-negative erythroid precursors which have all the machinery to activate the expression of the globin genes.

It is not clear yet which molecular mechanism might mediate the differential activation of the distal and proximal *Gata1* promoter in erythroid cells described here. It is well established that the proximal *Gata1* promoter is under the control of GATA binding proteins.[15,16] Although, the regulation of transcription from the distal promoter has been much less investigated, the rat and mouse distal promoter sequences are known and display a structure similar to the proximal promoter with putative GATA binding sites.[17,18] Since, in transgenic mice models, these sequences are sufficient to direct specifically the expression of a reporter gene in erythroid-tissues,[18] sequences which confer erythroid specificity to the distal *Gata1* promoter, altough not identified as yet, must exist.

In conclusion, the data presented here demonstrate that the increase in the expression of *Gata1* observed during erythroid differentiation is mainly due to increased expression of distal *Gata1* transcripts.

REFERENCES

1. Tsai SF, Martin DI, Zon LI, D'Andrea AD, Wong GG, Orkin SH: Cloning of cDNA for the major DNA-binding protein of the erythroid lineage through expression in mammalian cells. Nature 339: 446, 1989.
2. Orkin SH: GATA-binding transcription factors in hematopoietic cells [review]. Blood 80: 575, 1992.

3. Ito E, Toki T, Ishihara H, Ohtani H, Gu L, Yokoyama M, Engel JD, Yamamoto M. Erythroid transcription factor GATA-1 is abundantly transcribed in mouse testis. Nature 362: 466, 1993.
4. Migliaccio AR, Migliaccio G, Ashihara E, Moroni E, Giglioni B, Ottolenghi S: Erythroid-specific activation of the distal (testis) promoter of *GATA1* during differentiation of purified normal murine hematopoietic stem cells. Acta Haematol 95: 229, 1996.
5. Moroni E, Cairns L, Ottolenghi S, Giglioni B, Ashihara E, Migliaccio G, Migliaccio AR: Expression in haematopoietic cells of GATA-1 transcripts from the alternative "testis" promoter during development and cell differentiation. Bioch Biophys Res Comm 231: 299, 1997.
6. Kulessa H, Frampton J, Graf T: GATA-1 reprograms avian myelomonocytic cell lines into eosinophils, thromboblasts, and erythroblasts. Genes & Dev 9: 1250, 1995.
7. Farina SF, Girard LJ, Vanin EF, Ninhuis AW, Bodine DM. Dysregulated expression of Gata-1 following retrovirus-mediated gene transfer into murine murine hematopoietic stem cells increases erythropoiesis. Blood 86: 4124, 1995.
8. Krystal G: A simple microassay for erythropoietin based on ^{3}H-thymidine incorporation into spleen cells from phenylhydrazine treated mice. Exp Hematol 7: 649, 1983.
9. Vannucchi AM, Grossi A, Rafanelli D, Vannucchi L, Rossi Ferrini P: Binding of recombinant human ^{125}I-erythropoietin to CFU-E from the spleen of anemic mice. Haematologica 75: 21, 1990.
10. Chomczynski P, Sacchi N: Single-step method of RNA isolation by acid guanidinium thiocyanate-phenol-chloroform extraction. Anal Biochem 162: 156, 1987.
11. Weiss MJ, Orkin SH. Transcription factor GATA-1 permits survival and maturation of erythroid precursors by preventing apoptosis. Proc Natl Acad Sci USA 92: 9623, 1995.
12. Sambrook J, Fritsch EF, Maniatis T. (eds.) Molecular Cloning: A Laboratory Manual. Cold Spring Harbor Laboratory Press, Cold Spring Harbor, 1989.
13. Chiba T, Ikawa Y, Todokoro K: GATA-1 transactivates erythropoietin receptor gene, and erythropoietin receptor-mediated signals enhance GATA-1 gene expresson. Nucl Acids Res 19: 3843, 1991.
14. Sposi NM, Zon LI, Care A, Valtieri M, Testa U, Gabbianelli M, Mariani G, Bottero L, Mather C, Orkin SH: Cell cycle-dependent initiation and lineage-dependent abrogation of GATA-1 expression in pure differentiating hematopoietic progenitors. Proc Natl Acad Sci USA 89: 6353, 1992.
15. Tsai SF, Strauss E, Orkin SH: Functional analysis and in vivo footprinting implicate the erythroid transcription factor GATA-1 as a positive regulator of its own promoter. Genes Dev 5: 919, 1991.
16. Baron MH, Farrington SM: Positive regulators of the lineage-specific transcription factor GATA-1 in differentiating erythroid cells. Mol Cell Biol 14: 3108, 1994.
17. Onodera K, Yomogida K, Suwabe N, Takahashi S, Muraosa Y, Hayashi N, Ito E, Gu L, Rassoulzadegan M, Engel JD, Yamamoto M: Conserved structure, regulatory elements and transcriptional regulation from the GATA-1 gene testis promoter. J Biochem 121: 251, 1997.
18. Onodera K, Takahashi S, Nishimura S, Ohta J, Motohashi H, Yomogida K, Hayashi N, Engel JD, Yamamoto M: GATA-1 transcription is controlled by distinct regulatory mechanisms during primitive and definitive erythropoiesis. Proc Natl Acad Sci 94: 4487, 1997.

32

EXPRESSIONS OF FAS (CD95) AND FAS LIGAND IN HIV INFECTED INDIVIDUALS

Naoki Hosaka,[1,2] Naoki Oyaizu,[1] Susumu Ikehara,[2] and Savita Pahwa[1]

[1] Department of Pediatrics, Division of Allergy/Immunology
North Shore University Hospital
New York University School of Medicine
350 Community Drive, Manhasset, NY, 11030.
[2] First Department of Pathology
Kansai Medical University
10–15 Fumizono-Cho, Moriguchi City
Osaka 570–8506, Japan

ABSTRACT

To gain insights into mechanisms of accelerated lymphocyte apoptosis in HIV disease, we evaluated the expression of membrane bound Fas ligand (FasL) and Fas in PBMC and soluble forms of FasL (sFasL) and Fas (sFas) in plasma from HIV$^+$ individuals and HIV$^-$ healthy volunteers. Surface FasL was detectable on monocytes, but poorly so on lymphocytes. Unexpectedly, monocytes expressed less FasL in HIV$^+$ patients than in HIV$^-$ volunteers. Levels of sFasL in plasma of HIV$^+$ individuals were elevated and correlated with plasma HIV RNA burden. Levels of sFas in plasma of HIV$^+$ subjects were also elevated and correlated with Fas expression in apoptotic lymphocytes. Finally, culture-induced lymphocyte apoptosis of HIV$^+$ subjects was enhanced by anti-Fas agonistic Ab, but was not inhibited by anti-FasL blocking Abs. These results suggest that significant dysregulation of both Fas and FasL occurs in HIV infection, and that this contributes to increased sensitivity of lymphocytes to undergo apoptosis.

INTRODUCTION

Fas (CD95), a member of the TNF receptor superfamily, and its ligand (FasL), a member of TNF family, have been cloned, and signaling through Fas has been shown to induce apoptosis in activated T cells and in mammalian cells expressing this antigen.[1] Soluble forms of both Fas and FasL have also been reported.[2,3] In HIV-infected patients,

Molecular Biology of Hematopoiesis 6, edited by Abraham *et al.*
Kluwer Academic/Plenum Publishers, New York, 1999.

lymphocytes showed accelerated spontaneous apoptosis in culture.[4] These cells expressed high membrane bound Fas and showed high susceptibility to apoptosis with α-Fas stimulation.[5] These results suggest that Fas-mediated accelerated lymphocyte apoptosis may play an important role in the pathogenesis of HIV disease. However, the nature and the role of the Fas/FasL interaction resulting in lymphocyte apoptosis remains controversial in patients with HIV infection.[6] Some reports underscore the important role of accessory cells (monocytes, macrophages, and dendritic cells) as producers of FasL and inducers of apoptosis for themselves and for T cells.[7,8]

To gain further insights into the role of Fas/FasL interaction in accelerated lymphocyte apoptosis in HIV disease, we investigated the expression of FasL and Fas in monocytes and lymphocytes and simultaneously determined levels of sFasL and sFas in plasma of normal volunteers and in HIV^+ individuals.

MATERIALS AND METHODS

Study Subjects

The study utilized peripheral blood samples from 41 HIV-infected patients with a mean CD4 count of 201.7 ± 154.4 cells/mL. Twenty healthy HIV seronegative donors were evaluated as well.

Reagents

Mouse anti-human FasL Abs, NOK-1 (IgG_1, κ), and NOK-3 (IgM, κ) were kindly provided by Dr. H Yagita.[2] Mouse anti-human Fas Ab, CH11 (IgM) was purchased from KAMIYA Biomedical Co. (Tukwila, WA).

Flow Cytometry

Surface staining was performed using appropriate concentrations of mAbs or isotype matched control Abs at 4C° for 40 minutes in staining buffer.

ELISA for sFasL and sFas

Levels of sFasL in the plasma of patients and healthy volunteers were evaluated by sandwich ELISA using NOK-1 and NOK-3 mAbs.[2] Levels of sFas in the plasma were quantified by ELISA using a Soluble Fas ELISA kit (KAMIYA) in accordance with the manufacturer's instructions.

Measurement of Apoptosis

We used PI (propidium iodide) staining for study of apoptosis. Apoptotic cells were analyzed 48–72 hours after PI staining by flow cytometry.

RESULTS

We evaluated membrane bound FasL and Fas expression in PBMC and plasma levels of sFasL and sFas in HIV infected and non-infected individuals (Table 1). FasL

Table 1. Membrane and soluble forms of Fas and FasL in PBMC and in plasma from HIV infected and non-infected individuals

Souse	HIV^+	HIV^-
FasL in monocytes	+	++
FasL in lymphocytes	–	–
sFasL in plasma	++	±
Fas in lymphocytes	++	+
sFas in plasma	++	±

in $CD14^+$ monocytes increased rapidly in 24 hour-PBMC cultures and returned to baseline by day 2.[9] However, the levels in HIV^+ subjects were much lower than those in HIV^- subjects in freshly isolated and 1 day-culture PBMC. In contrast to monocytes, the levels of FasL expression in $CD45^+$ lymphocytes and in the subsets ($CD4^+$, $CD8^+$ T cells, and $CD19^+$ B cells, data not shown) of both HIV^+ and HIV^- subjects were very low or non-detectable on day 0 to day 2. We also examined FasL expression in these cells in the presence of KB8301, a matrix metalloprotease inhibitor, which inhibits the cleavage of membrane FasL into its soluble form.[2] Unexpectedly, with this treatment, FasL expression on monocytes did not change but decreased slightly and the percent of monocytes in PBMC decreased in both HIV^- and HIV^+ subjects (data not shown); it may be that accumulated FasL on monocytes resulted in accelerated autocrine FasL/Fas mediated culture-induced apoptosis, as previously reported.[7] Lymphocytes did not show any significant increase of FasL expression with KB8301 treatment in either HIV^- and HIV^+ subjects. Membrane bound Fas in lymphocytes and plasma levels of sFasL or sFas in HIV^+ patients were elevated compared to HIV^- controls.

Next, we examined the effect of α-Fas agonistic Ab (CH11) and α-FasL blocking Ab (NOK-1) on spontaneous apoptosis in HIV infected patients (Table 2). CH11 enhanced spontaneous lymphocytes apoptosis in HIV^+ subjects, but not in HIV^- subjects. However, NOK-1 did not block the spontaneous apoptosis in HIV^+ subjects and showed no effect in HIV^- subjects.

DISCUSSION

Although FasL expression in monocytes was lower in HIV-infected subjects than healthy volunteers, the levels of sFasL in plasma were found to be increased and correlated with plasma HIV RNA burden.[9] Additionally, we observed the same levels of FasL mRNA expression in monocyte-enriched populations by RT-PCR from HIV^- and HIV^+ subjects.[9] Considering the fact that monocyte-macrophage tropic HIV predomi-

Table 2. Effect of Fas agonistic Ab (CH11) and FasL blocking Ab (NOK-1) on spontaneous lymphocytes apoptosis from HIV infected and non-infected individuals

Treatment	HIV^+	HIV^-
None	+	–
CH11	++	–
NOK-1	+	–

nates until the development of progressive disease and AIDS,[10] it is possible that HIV infection of monocytes in vivo may stimulate monocytes to release sFasL from the cell surface.[11] Such mechanisms may contribute to the decreased membrane FasL and increased levels of sFasL in plasma of HIV⁺ patients. In this respect, Sieg et al. also reported a lower expression of functional FasL on monocytes in PBMC from HIV⁺ patients.[12] Membrane bound Fas was increased in lymphocytes of HIV⁺ patients, as previously reported.[6] Plasma level of soluble Fas was also increased in HIV⁺ subjects and correlated with Fas expression in the apoptotic cell fraction,[9] suggesting that sFas may be released during the process of lymphocyte apoptosis.

The patients' lymphocytes were highly sensitive to apoptosis by treatment with anti-Fas agonistic Ab, but addition of anti-FasL antagonistic Ab failed to block the spontaneous apoptosis. We contend that cells undergoing apoptosis in vitro have been primed for apoptosis in vivo, having already received the death signal via FasL from activated monocytes/macrophages. Thus, attempts to block this interaction in vitro are ineffective in preventing apoptosis.

Recent reports have shown that macrophages infected with HIV in vitro mediate not only $CD4^+$, but also $CD8^+$ T cell apoptosis by FasL, TNF-α, and chemokines.[13,14] Our data also suggest that FasL in monocytes/macrophages may play an important role in the induction of lymphocyte apoptosis and immnodeficiency in HIV disease in vivo.

ACKNOWLEDGMENTS

The authors gratefully acknowledge for providing materials: Dr Hideo Yagita at Juntendo University, Tokyo, Japan, the assistance of colleagues at North Shore University Hospital: Drs. Surendra Chavan, Thomas McCloskey, Surya Tetali, and Soe Than for helpful discussions, Maria Marecki and Regina Kowalski for technical assistance.

REFERENCES

1. Nagata S: Apoptosis by death factor. Cell 88:355, 1997.
2. Kayagaki N, Kawasaki A, Ebata T, Ohmoto H, Ikeda S, Inoue S, Yoshino K, Okumura K, Yagita, H: Metalloptoteinase-mediated release of human Fas ligand. J Exp Med 182:1777, 1996.
3. Cheng J, Zhou T, Liu C, Shapiro JP, Brauner MJ, Kiefer MC, Barr PJ, Mountz JD: Protection from Fas-mediated apoptosis by a soluble form of the Fas molecule. Science 263:1759, 1994.
4. Oyaizu N, McCloskey TW, Coronesi M, Chirmule N, Kalyanaraman VS, Pahwa S: Accelerated apoptosis in peripheral blood mononuclear cells (PBMCs) from human immunodeficiency virus type-1 infected patients and in CD4 cross-linked PBMCs from normal individuals. Blood 82:3392, 1993.
5. Katsikis PD, Wunderlich ES, Smith CA, Herzenberg LA, Herzenberg LA: Fas antigen stimulation induces marked apoptosis of T lymphocytes in human immunodeficiency virus-infected individuals. J Exp Med 181:2029, 1995.
6. Kaplan D, Sieg S: Role of the Fas/FasL apoptotic pathway in human immunodeficiency virus type 1 disease. J Virol 726:279, 1998.
7. Kiener PA, Davis PM, Starling GC, Mehlin C, Klebanoff AJ, Ledbetter JF, Liles WC: Differential induction of apoptosis by Fas-Fas ligand interactions in human monocytes and macrophages. J Exp Med 185:1511, 1997.
8. Wu MX, Ao Z, Hengen M, Morimoto C, Schlossman SF: Requirement of Fas (CD95), CD45, and CD11a/CD18 in monocytes-dependent apoptosis of human T cell. J Immunol 157:707, 1996.
9. Hosaka N, Oyaizun, Kaplan MH, Yagita H, Pahwa S: Membrane and soluble forms of Fas (CD95) and Fas ligand in peripheral blood mononuclear cells and in plasma from HIV infected individuals. J Infect Dis 178:1030, 1998.
10. Schuitemaker H, Koot M, Kootstra NA, Dercksen MW, de Goede RE, van Steenwijk RP, Lange JM,

Schattenkerk JK, Miedema F, Tersmette M: Biological phenotype of human immunodeficiency virus type 1 clones at different stages of infection: progression of disease is associated with a shift from monocytotropic to T-cell-tropic virus populations. J Virol 3:1354, 1992.

11. Kiener PA, Davis PM, Rankin BM, Klebanoff SJ, Ledbetter JA, Starling GC, Liles WC: Human monocytic cells contain high levels of intracellular Fas ligand. Rapid release following cellular activation. J Immunol 159:1594, 1997.
12. Sieg S, Smith D, Yildirim M, Kaplan M: Fas ligand deficiency in HIV disease. Proc Natl Acad Sci USA 94:4860, 1997.
13. Herbein G, Mahlknecht U, Batliwalla F, Gregersen P, Pappas T, Butler J, O'Brien WA, Verdin E: Apoptosis of $CD8^+$ T cells is mediated by macrophages through interaction of HIV gp120 with chemokine receptor CXCR4. Nature 395:189, 1998.
14. Badley AD, Dockrell D, Simpson M, Schut R, Lynch DH, Leibson P, Paya CV: Macrophage-dependent apoptosis of $CD4^+$ T lymphocytes from HIV-infected individuals is mediated by FasL and Tumor necrosis factor. J Exp Med 185:55, 1997.

33

PATTERN OF PU.1 BINDING TO M-CSF RECEPTOR PROMOTER DURING MONOCYTE TO MACROPHAGE MATURATION

Nadia Maria Sposi, Giovanna Marziali, Elvira Pelosi, Angela Battistini, Ugo Testa, and Cesare Peschle

Department of Hematology and Oncology and
Department of Virology
Istituto Superiore di Sanità
Viale Regina Elena 299, 00161 Rome
Italy

ABSTRACT

PU.1 has been reported to control the expression of an increasing number of myeolid-specific genes, including the macrophage CSF receptor (M-CSFR). Macrophage colony stimulating factor (M-CSF) and its receptor are required for proliferation, differentiation, and maturation of monocytes/macrophages.

To investigate the mechanism of M-CSFR expression, we studied the binding of PU.1 to M-CSFR promoter during monocyte to macrophage differentiation. We show that induction of a highly phosphorylated form of PU.1 positively correlates with monocyte to macrophage maturation.

INTRODUCTION

Hematopoiesis is sustained by a pool of hematopoietic stem cells that can extensively self-renew and undergo a commitment to multipotent progenitor cells, which in turn give rise to mature blood cells. Transcription factors play a major role in differentiation in a number of cell types, including the various hematopoietic lineages.[1-4] Different transcription factors have been identified which orchestrate, at molecular level, hematopoiesis inducing a cascade of events responsible for differentiation/

Molecular Biology of Hematopoiesis 6, edited by Abraham *et al.*
Kluwer Academic/Plenum Publishers, New York, 1999.

maturation of hematopoietic progenitor cells into the different blood lineages.[1–4] Myeloid cells arise from a common precursor that differentiates into monocytes or granulocytes. Monocytes isolated from peripheral blood are committed cells, not yet fully differentiated, which undergo terminal differentiation upon migration in extravascular tissues.[4] These cells can also spontaneously mature in vitro in a time-dependent process that highly mimics their in vivo maturation.

Many studies have demonstrated that macrophage colony-stimulating factor (M-CSF) is required for the proliferation, differentiation, and maturation of monocytes/macrophages.[5,6] The macrophage colony-stimulating factor receptor (M-CSFR) is encoded by the c-fms proto-oncogene, which is a member of a family of growth factor receptors with an intrinsic tyrosine kinase activity[7] and exhibits a pattern of expression largely restricted to monocytes/macrophages.[8] During the process of normal hematopoietic differentiation c-fms expression is virtually absent in quiescent hematopoietic progenitors and is then induced by hematopoietic growth factors when these cells are triggered to proliferate; at later stages of hematopoietic differentiation c-fms expression is restricted to cells pertaining to the monocytic lineage, while in other lineages, including the granulocytic one, c-fms expression is very low or absent.[9]

Many studies have focused on PU.1 as a master regulator of myeloid development through the control of the expression of an increasing number of myeloid-specific genes, including the macrophage CSF-R.[3,10,11]

PU.1 is a tissue-specific ets-family member that is expressed in various hematopoietic lineages.[3] It is encoded by the proto-oncogene Spi-1, whose deregulated expression caused by insertion of the spleen focus-forming provirus leads to erythroleukemias in the mouse.[12] It is expressed at the highest levels in myeloid and B cells, but not in T cells and is specifically upregulated with myeloid differentiation.[3] Gene targeting studies have demonstrated that PU.1 is required for the development of both myeloid and lympoid lineages.[3] The targeted PU.1 allele results in embryonic lethality and fetal liver cells derived from the PU.$1^{-/-}$ embryos fail to generate macrophage or granulocyte colonies or differentiate into B cells in vitro.[13] Recent studies on PU.$1^{-/-}$ embryonic stem cells demonstrate that PU.1 controls both cytokine-dependent proliferation and differentiation of granulocyte/macrophage progenitors.[14]

In the present study we show that induction of a peculiar PU.1 binding activity positively correlates with macrophagic differentiation. Preliminary observations indicate that this band corresponds to a highly phosphorylated form of PU.1 and suggest a link between PU.1 phosphorylation and cell differentiation.

MATERIALS AND METHODS

Mature resting monocytes were obtained from peripheral blood buffy coat of healthy donors by enrichment on a Ficoll-Hypaque cushion and subsequent purification (>90%) by centrifugation on a Percoll density gradient.[15] Cells were then cultured in medium containing FCS and, at seven days of culture, were recovered and analyzed. In vitro cultivation of blood monocytes results in their adherence to the plastic surface and in the initiation of a series of changes that resemble their in vivo maturation to macrophages.

Whole cell extracts were prepared from cells by standard techniques. Briefly, the cells were washed twice in cold phosphate-buffered saline and then collected by centrifugation. The pellet was resuspended in lysis buffer containing 20mM HEPES,

50mM NaCl, 10mM EDTA, 2mM EGTA, and 0.5% (vol/vol) Nonidet P-40(pH8) supplemented with 0.5mM dithiothreitol, 10mM sodium molybdate, 10mM sodium orthovanadate, 100mM NaF, 10µg of leupeptin per ml, and 0.5mM phenylmethylsulfonylfluoride (PMSF). After incubation for 30min. on ice, the suspension was centrifuged at 10,000 × g for 10min. The supernatants were aliquoted and stored at—80°C. The protein concentration in the soluble supernatant was determined using the Bio-Rad protein assay (Bio-Rad, Richmond, USA).

Identification and characterization of PU.1 binding to the M-CSFR promoter was assessed by gel mobility shift using a double-stranded M-CSFR promoter oligonucleotide extending from bp -61 to -22 and corresponding to the sequence 5′-TGCCTAGCTAAAAGGGGAAGAAGAGGATCAGCCCAAGGAG-3′. The binding-reaction mixture (20-µl final volume) contained labeled oligonucleotide probes (10,000cpm) in binding buffer (75mM KCl, 20mM Tris-HCl pH 7.5, 1mM dithiothreitol) containing 5µg of bovine serum albumin per ml, 14% (vol/vol) glycerol, and 3µg of poly (dI-dC). Total cell lysates (10µg) were added, and the reaction mixture was incubated for 20min. at room temperature. Samples were electrophoresed in a 5% polyacrylamide gel in 0.5 × Tris-borate-EDTA (TBE) buffer for 2h at 200V at 18°C.

The gels were then dried and autoradiographed. Competition studies were performed by adding unlabeled double-stranded oligonucleotides at a 100-fold molar excess over the labeled probe. The DNA sequences of the oligonucleotides used are the following: (i) wt PU.1 box: 5′-TGCCTAGCTAAAAGGGGAAGAAGAGGATCAGCCCAAGGAG-3′; mutant PU.1 box: 5′-TGCCTAGCTAAAAGGGATCGTAGCGGATCAGCCCAAGGAG-3′.

RESULTS AND DISCUSSION

Previous results indicated that PU.1 was expressed at low level in quiescent human hematopoietic progenitor cells and upmodulated during monocytic differentiation before the appearance of the M-CSF R and monocytic membrane markets, i.e., CD11b and CD14 (data not shown).

To investigate the mechanism of M-CSFR expression, we studied the binding of PU.1 to M-CSFR promoter during the late stage of monocyte to macrophage maturation. In freshly isolated monocytes one high-mobility PU.1 binding activity, designated band A, was detected. In 7-day macrophages one slower migrating PU.1 binding activity (band C) was detected (Fig. 1). Competition with a 100 fold molar excess of homologous oligonucleotide abolished the formation of all complexes (band A and C), while competition with the oligonucleotide mutated in the PU.1 site did not affect complex formation (Fig. 1). Preincubation of whole cell extracts with monoclonal antibody anti-human PU.1 resulted in a supershift of both A and C mobility binding activities (Fig. 1).

Modification by phosphorylation may influence either DNA binding or the functional activity of a number of transcription factors. Several observations indicated that the capacity of PU.1 to modulate transcription might be regulated by post-translational phosphorylation-dependent events.[10,14,16]

To directly examine whether the slower migrating PU.1 binding activity represented a phosphorylated form of this transcription factor, macrophage cell extracts were treated with phosphatase for 30 minutes at 30°C before the EMSA.[16] Phosphatase treatment of the samples either reduced or abolished band C, while band A was not

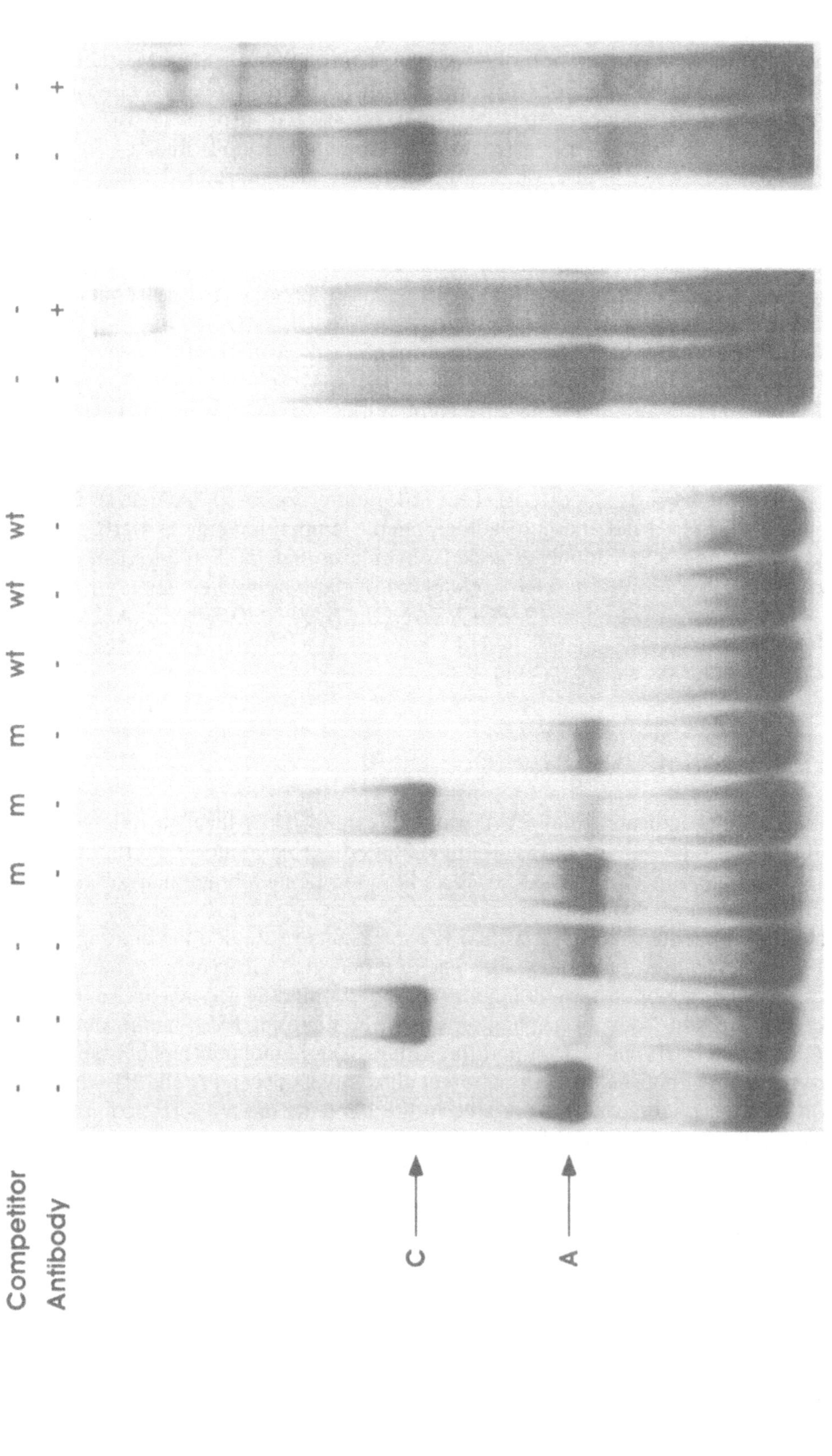

Figure 1. EMSA (gel shift) analysis of PU.1 DNA binding as described in Material and Methods. The legend above the autoradiograph indicates the competitor and antibody used. Competitors are the wild-type M-CSFR promoter oligonucleotide (wt) and the M-CSFR promoter mutant (m). Antibody added is a monoclonal anti-human PU.1 (Santa Cruz). Lanes 1,4,7 and 10 and 11: whole cell extracts from freshly isolated monocytes. Lanes 2,5,8 and 12 and 13: whole cells extracts from 7-day macrophages. Land 3,6 and 9: whole cell extracts from freshly isolated granulocytes.

significantly altered by the treatment (data not shown). To confirm that this phenomenon was due to dephosphorylation rather than to a non specific effect by the phosphatase preparation, heat-denatured phosphatase was used. Unlike native phosphatase, the denatured phosphatase preparation did not reduce the intensity of the band C (data not shown).

RNA analysis of samples derived during monocyte to macrophage differentiation indicated a down-modulation of M-CSFR, a finding confirmed also at protein level by Western blot and flow cytometry analyses (data not shown).

In conclusion we show that induction of a specific PU.1 binding activity (band C) positively correlates with macrophagic differentiation. We suggest that the capacity of PU.1 to modulate transcription can be regulated by a phosphorylation mechanism.

ACKNOWLEDGMENTS

The authors thank Giampiero Macioce for his expert preparation of oligonucleotides, Domenico Falcone for technical assistance and Mauro Teragnoli for graphics.

REFERENCES

1. Shivdasani RA, Orkin SH: The transcriptional control of hematopoiesis. Blood 87:4025, 1996.
2. Orkin SH: Transcription factors and hematopoietic development. J.Biol.Chem. 270:4955, 1995.
3. Tenen DG, Hromas R, Licht JD Zhang DE: Transcription factors, normal myeloid development, and leukemia. Blood 90:489, 1997.
4. Kehrl JH: Hematopoietic lineage commitment: role of transcription factors. Stem Cells 13:223, 1995.
5. Clark SC, Kamen R: The human hematopoietic colony-stimulating factors. Science 236:1229, 1987.
6. Roth P, Stanley E: The biology of CSF-1 and its receptor. Current Topics Microbiol Immunol 181:141, 1992.
7. Sherr CJ, Retteumier CW, Sacca R, Roussel MF, Look AT, Stanley ER: The c-fms proto-oncogene product is related to the receptor for the mononuclear phagocyte growth factor, CSF-1. Cell 41:665, 1985.
8. Sherr CJ: Colony-stimulating factor-1 receptor. Blood 75:1, 1990.
9. Testa U, Fossati C, Samoggia P, Masciulli R, Mariani G, Hassan HJ, Sposi NM, Guerriero R, Rosato V, Gabbianelli M, Pelosi E, Valtieri M, Peschle C: Expression of growth factor receptors in unilineage differentiation culture of purified hematopoietic progenitors. Blood 88:3391, 1996.
10. Fisher RC, Scott EW: Role of PU.1 in Hematopoiesis. Stem Cells 16:25, 1998.
11. Hume DA, Yue X, Ross IL, Favot P, Lichanska A, Ostrowski MC: Regulation of CSF-1 R expression. Molecular Reproduction and Development 46:46, 1997.
12. Moreau-Gachelin F, Tavitian A, Tambourin P: Spi-1 is a putative oncogene in virally induced murine erythroleukemias. Nature 331:277, 1988.
13. Scott EW, Fisher RC, Olson MC, Kahrli EW, Simon MC, Singh H: PU.1 functions in a cell-autonomous manner to control the differentiation of mulipotential lymphoid-myeloid progenitors. Immunity 6:437, 1997.
14. Dekoter RP, Walsh JC, Singh H: PU.1 regulates both cytokine-dependent proliferation and differentiation of granulocyte/macrophage progenitors. The EMBO Journal 17:4456, 1998.
15. Testa U, Conti L, Sposi NM, Varano B, Tritarelli E, Malorni W, Samoggia P, Rinaldi G, Peschle C, Belardelli F, Gessani S: IFN- selectively down regulates transferrin receptor expression in human peripheral blood macrophages by a post-translation mechanism. J.Jmmunol. 155:427, 1995.
16. Carey JO, Posekany KJ, E deVente J, Pettit GR, Ways DK: Phorbol Ester-Stimulated Phosphorylation of PU.1: Association with Leukemic Cell Growth Inhibition. Blood 87:4316, 1996.

34

MOLECULAR BIOLOGY OF HYPOXIA-INDUCIBLE FACTOR-1

Roland H. Wenger and Max Gassmann

Institute of Physiology
University of Zürich-Irchel
CH-8057 Zürich, Switzerland

SUMMARY

The hypoxia-inducible factor-1 (HIF-1) is a basic-helix-loop-helix-PAS heterodimeric transcription factor that confers oxygen-regulated expression to a number of genes involved in oxygen homeostasis including erythropoietin (Epo), transferrin, glycolytic enzymes, and vascular endothelial growth factor (VEGF). Hypoxic exposure stabilizes the HIF-1α protein by a mechanism involving redox processes. Following heterodimerization with HIF-1β, better known as the aryl hydrocarbon receptor nuclear translocator (ARNT), HIF-1 binds to the DNA consensus sequence CGTG, known as a potential target of CpG methylation in mammalian cells. We showed that CpG methylation blocks HIF-1 DNA-binding as well as transactivation of reporter gene expression. The hypoxia-responsive 3′ enhancer of the Epo gene was found to be methylation-free in Epo-producing cells despite its location outside of a CpG island. Intriguingly, this site was also methylation-free in cells that do not express Epo, indicating a general selective pressure to prevent CpG methylation, even in the absence of HIF-1 under normoxic conditions. We previously identified the constitutively expressed ATF-1/CREB-1 family members as candidate factors capable of binding the HIF-1 site. We cloned the mouse HIF-1α gene (designated *Hif1a*) and found that it consists of 15 exons dispersed over 45 kb. Interestingly, mouse *Hif1a* contains two alternative first exons whose expression is driven by a tissue-specific promoter (exon I.1) or a housekeeping-type promoter located within a methylation-free CpG island (exon I.2). The exon I.1-containing mRNA isoform encodes a predicted polypeptide that is 12 amino acids shorter than the exon I.2-derived mRNA isoform. So far, however, we did not find any functional differences between the two isoforms. The genomic *Hif1a* clone was

---

Address reprint requests to: Roland H. Wenger, PhD, Physiologisches Institut der Universität Zürich-Irchel, Winterthurerstrasse 190, CH-8057 Zürich, Switzerland.

Molecular Biology of Hematopoiesis 6, edited by Abraham *et al.*
Kluwer Academic/Plenum Publishers, New York, 1999.

used to introduce a null mutation into the mouse *Hif1a* locus by gene targeting in embryonic stem cells. HIF-1α deficiency is embryonic lethal, suggesting that HIF-1 serves as a non-redundant master regulator of oxygen homeostasis.

INTRODUCTION

Stimulation of erythropoiesis during periods of limited oxygen supply improves the oxygen transport capacity of the blood. This physiological adaptation to hypoxia is regulated by the glycoprotein hormone Epo that is mainly produced in the fetal liver and the adult kidney. Identification of a hypoxia-responsive element in the 3′ flanking region of the gene encoding Epo led to the discovery of the hypoxia-inducible factor-1α (HIF-1α).[1] HIF-1α is a basic-helix-loop-helix (bHLH) transcription factor that is activated by exposure of cells to physiologically relevant reductions in oxygen partial pressure (reviewed in Wenger and Gassmann).[2] This activation involves hypoxic stabilization of the protein that is otherwise ubiquitinated and rapidly degraded in proteasomes under normoxic conditions.[3–6] Moreover, the activity of the transactivation domain(s) can also be hypoxically induced.[4,7,8] So far, the mechanisms leading to these hypoxic effects are not clearly understood, but there is evidence that redox processes[3,6,9] as well as phosphorylation[10,11] might be involved. Following hypoxic exposure, HIF-1α forms a heterodimeric complex with the aryl hydrocarbon receptor nuclear translocator (ARNT), also termed HIF-1β, which activates expression of oxygen-dependent genes.[2] These "oxy-genes"[2] include Epo, transferrin,[12] VEGF, glycolytic enzymes, inducible nitric oxide synthase, heme oxygenase 1, α_{1B}-adrenergic receptor as well as other genes involved in the adaptation of an organism to reduced oxygenation at the cellular, local, and systemic level.[2]

CpG METHYLATION IMPAIRS HIF-1 DNA BINDING AND GENE ACTIVATION

CpG methylation plays an important regulatory role in mammalian gene expression, contributing to X-chromosome inactivation, genomic imprinting as well as tissue- and developmental stage-specific transcriptional regulation. CpG dinucleotides are underrepresented in the mammalian genome and are often methylated if located outside of the so-called CpG islands. The human genome contains around 45,000 CpG islands of approximately 1 kb length. In contrast to the bulk genome, these regions are G+C rich and entirely methylation-free. CpG islands are associated with the promoters of all housekeeping genes and of 40% of the tissue-specific genes. Methylated CpG (containing 5-methylcytosine) interferes with transcription factor binding to DNA through both direct steric hindrance and the binding of repressor proteins. Since the conserved core HIF-1 binding site (HBS) CGTG contains a CpG dinucleotide,[2] we analyzed the effect of CpG methylation on HIF-1 DNA binding. Therefore, we first assessed whether HIF-1 is capable of binding a CpG methylated oligonucleotide derived from the human Epo HBS. Oligonucleotides with or without a methylated HBS were used to probe nuclear extracts prepared from human Hep3B and HeLaS3 cells, as well as mouse L929 and Hepa1 cells exposed to either normoxic (20% O_2) or hypoxic (1% O_2) conditions for four hours. Cytosine methylation within the CGTG core HBS abolished binding of HIF-1 in electrophoretic mobility shift assays.[13] Binding of ATF-

1/CREB-1 was also abolished, confirming our previous conclusion that ATF-1/CREB-1 bind to the same position as HIF-1.[14] Hemimethylated oligonucleotide probes also did not bind to HIF-1, irrespective of whether the sense or antisense strand contained 5-methylcytosine.

Impaired HIF-1 binding to a 5-methylcytosine-containing HBS *in vitro* implies that HIF-1-mediated transactivation of *in vivo* gene transcription would also be inhibited. To test this notion directly, a hypoxia-responsive reporter gene construct, containing the firefly luciferase gene under the control of the simian virus 40 promoter and three concatamerized copies of the Epo 3′ HBS,[14–16] was methylated *in vitro* with the CpG dinucleotide-specific bacterial enzyme *Sss*I methylase. Unmethylated and methylated constructs were transiently transfected into Hep3B and HeLaS3 cells, which were then split and exposed to either normoxic (20% O_2) or hypoxic (1% O_2) conditions. Reporter gene expression derived from methylation-free plasmids increased under hypoxic conditions 7.0- and 5.3-fold in Hep3B and HeLaS3 cells, respectively, whereas the parental vector lacking the three HBSs was only weakly hypoxia-responsive (1.3- and 1.7-fold in Hep3B and HeLaS3, respectively). In contrast, hypoxic inducibility of CpG methylated reporter gene constructs was drastically reduced, which is about equal with the empty parental vector.[13] Basal luciferase expression was slightly reduced by CpG methylation (approximately 5- and 2-fold in Hep3B and HeLaS3 cells, respectively), but this effect most probably did not account for the loss of hypoxic inducibility since luciferase expression was still clearly above background. However, to rule out that CpG methylation of the promoter region or the luciferase gene itself affected hypoxic inducibility, a second series of experiments was performed where annealed, Epo 3′ HBS-containing oligonucleotides were quantitatively inserted into the parental vector pGL3Promoter. Transient transfection of ligation products containing wild-type oligonucleotides resulted in 3.4- and 2.7-fold hypoxic stimulation of luciferase expression in Hep3B and HeLaS3 cells, respectively. In contrast, transfection with ligation products containing mutant HBSs or CpG methylated HBSs abolished hypoxic induction of reporter gene expression.[13]

Epo EXPRESSION IN HepG2 CELLS IS RELATED TO THE DEGREE OF CpG METHYLATION OF THE Epo 3′ HBS

Based on the results presented above, one would predict that the HBS located in the hypoxia-responsive Epo 3′ enhancer should be CpG methylation-free in Epo-producing cells, despite the fact that this site is not located within a CpG island. However, the question still remains whether this site is methylated in non-Epo-producing cells, thereby inhibiting hypoxia-dependent Epo gene expression. To answer this, we took advantage of the restriction enzyme *Tai*I that recognizes the palindromic sequence 5′-ACGT-3′ present in the Epo 3′ HBS. Since this enzyme does not cleave cytosine methylated target DNA, it is ideally suited to assess the *in vivo* methylation state of the HBS in the Epo-expressing human hepatoma cell lines Hep3B and HepG2 by Southern blotting. As expected, the Epo 3′ HBS in Hep3B cells was almost completely methylation-free. Intriguingly, this site was repeatedly found to be partially methylated (45%) in the second Epo-expressing cell line HepG2. Conversely, the HBS was only partially methylated (25 to 50%) in non-Epo-producing control cell lines, suggesting that the mechanism keeping this site methylation-free is independent of Epo expression. To determine if the methylation status of the Epo 3′ HBS affects endogenous Epo expression, we established several sublines of the HepG2 cell line. Three sublines were chosen which secreted 4.2, 7.7, and 14.2 U/l Epo, respectively, following plating of an equal number of cells and incubation at 1% O_2 for 48 hours. Interestingly, methylation of the Epo 3′ HBS was 52%, 26%, and 7%, respectively, indicating that the degree of methylation was inversely related to hypoxic Epo production.[13] This clonally inherited methylatin pattern can be explained by the tendency of cultured cell lines to inactivate genes that are not required for the cell's metabolism. This inactivation often goes along with CpG methylation of these genes.[17] This might explain why HepG2 cells lose their capability of efficiently expressing Epo after multiple rounds of passaging, an effect that forced us to sporadically re-clone this cell line in order to maintain high-level hypoxic inducibility of the Epo gene

WHAT KEEPS THE Epo 3′ HBS METHYLATION-FREE?

In contrast to most HIF-1 dependent genes where the HBSs are located in methylation-free CpG islands, the Epo 3′ HBS is present in a locus with average G+C content and suppressed CpG dinucleotide frequency. Thus, there must be a selective pressure to keep this site methylation-free. It is tempting to speculate that one such mechanism could be the constitutive binding of a transcription factor to the HBS. Since HIF-1 activity can only be detected under hypoxic conditions, the specific constitutive binding of ATF-1/CREB-1 to the HBS[14] might prevent methylation of the HBS. In support of this model, Hu and collaborators[18] showed by *in vivo* footprinting that the HBS is also occupied during normoxic growth (i.e. in the absence of HIF-1). Consequently, since ATF-1/CREB-1 expression is ubiquitous and not restricted to Epo-producing tissues, such a model would imply that non-Epo-expressing cell lines and tissues are also CpG methylation-free at the Epo 3′ HBS. Indeed, we found that non-Epo-producing cell lines and mouse tissues contain a significant proportion of methylation-free cells.[13]

CLONING OF THE MOUSE *Hif1a* GENE

Cloning of the human,[1] mouse[7,19] and rat HIF-1α[20] cDNAs revealed a conserved predicted protein (overall similarity is around 90%) of 826 amino acids (for human HIF-1α) and a relatively long 3′ untranslated region (UTR) containing several mRNA destabilisation elements. While the mouse HIF-1α cDNA reported by Li et al.[7] corresponds to the human cDNA, the mouse mRNA isoform that we reported differed at its 5′ end by an alternative UTR and the lack of the first twelve codons. Cloning and characterisation of the mouse gene encoding HIF-1α (termed *Hif1a*) provided the explanation for this discrepancy. Mouse *Hif1a* is a single-copy gene[19] containing 15 exons spread over approximately 45 kb.[21] As indicated in Fig. 1, *Hif1a* contains two alternative first exons (I.1 and I.2) giving rise to two different mRNA isoforms. We demonstrated that expression of exons I.1 and I.2 is driven by their own promoters.[21,22] Whereas the exon I.1 promoter exhibits tissue-specific features, exon I.2 is associated with a CpG island as typically found in housekeeping-type promoters.[22] Consequently, expression of the exon I.1-derived mRNA isoform was detected by RNase protection in kidney, tongue, stomach, and testis, while the exon 1.2-derived isoform was ubiquitously expressed. However, the exon I.1-derived mRNA isoform has not yet been found

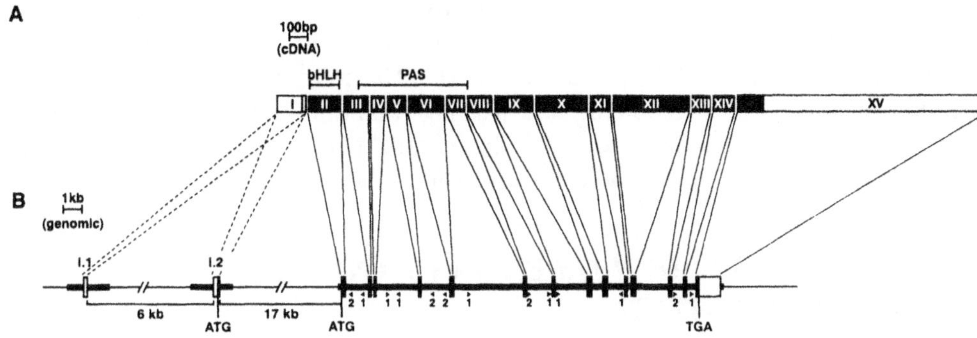

Figure 1. Structure of the gene encoding mouse HIF-1α. (A) Structure of the mouse HIF-1α cDNA. Exons are numbered with roman numerals. The translated region is indicated by black boxes and the 5′ and 3′ UTRs by open boxes. The bHLH and PAS domains of HIF-1α, are indicated. (B) Structure of the mouse *Hif1a* gene. Sequenced regions are indicated by the thick line and non-sequenced regions by the thin line. Filled and open boxes represent coding and non-coding regions, respectively. B1 and B2 mouse repetitive elements are indicated by arrowheads.

in humans and the function of the predicted 12 amino acid truncated translation product remains to be determined. Recently, also Luo et al.[23] reported the structure of the *Hif1a* locus. Closer inspection of their sequence revealed a third mouse HIF-1α mRNA isoform that contained a 14 amino acids insertion due to alternative splicing at the 5′ end of exon XI (Fig. 2). Again, the functional significance of this mRNA isoform needs to be defined.

The human *HIF1A* gene was assigned to chromosome 14q21–q24; and the mouse *Hif1a* gene was assigned to a region on chromosome 12 that shows conserved synteny with the corresponding region bearing *HIF1A* on the human chromosome.[19,24] Neither the human *HIF2A* gene (human chromosome 2), nor the human *ARNT* (human chromosome 1) and mouse *Arnt* genes (mouse chromosome 3) are syntenic with the *HIF1A/Hif1a* genes.[25,26] A comparison of the exon-intron structures of the genes encoding mouse HIF-1α,[21] human HIF-2α,[25] and mouse AhR[27] is shown in Fig. 2 (the genomic structure of other mammalian bHLH-PAS family members, including ARNT, has not

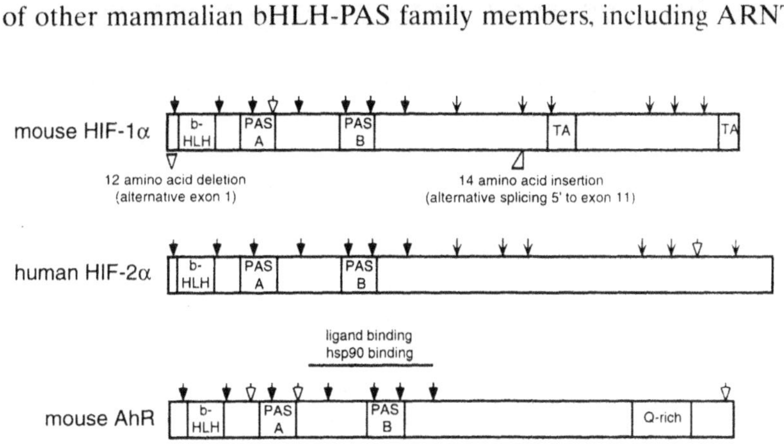

Figure 2. Comparison of the genes encoding mouse HIF-1α, human HIF-2α, and mouse AhR. Conserved intron positions are indicated by filled arrowheads and non-conserved positions by open arrowheads. Introns that are conserved only between HIF-1α and HIF-2α are depicted by small arrowheads. The basic-helix-loop-helix (bHLH) domains, PAS A and B repeats, the transactivation domains (TA), binding sites for aryl hydrocarbon ligands and hsp90, and a glutamine (Q) rich region are indicated.

yet been completely resolved). Whereas the positions of seven out of eight mouse *Hif1a* introns in the 5′ part are conserved in *HIF2A* and *Ahr*, the six introns of the *Hif1a* 3′ part can be found only in *HIF2A* but not in *Ahr*. These results suggest more general functions for the bHLH-PAS domains (such as DNA binding and protein-protein interactions) and more specialized roles of the C-terminal domain (such as hypoxia-responsiveness and transactivation/transrepression of specific genes).

COMPARISON OF THE *Hif1a* EXON I.1 AND I.2 PROMOTER ACTIVITIES

To directly compare the promoter activities of exon I.1 and exon I.2 upstream sequences, reporter genes under the transcriptional control of the respective promoters were transiently transfected into L929 fibroblasts and Hepa1 hepatoma cells. Following normalization to a co-transfected β-galactosidase expression vector, the exon I.2 promoter construct was 10 and 5-fold more active than the exon I.1 promoter construct in L929 and Hepa1 cells, respectively.[22] Interestingly, while in these experiments the exon I.1 promoter stimulated basal luciferase expression 9-fold in L929 fibroblasts, only very low promoter activity could be detected in Hepa1 hepatoma cells, suggesting that the exon I.1 promoter was not active in hepatoma cells. This finding is in line with the lack of detectable exon I.1 mRNA isoform expression in mouse liver.[22] Interestingly, exposure of the cells to 1% oxygen reduced luciferase expression about 5-fold in Hepa1 cells, but only marginally in L929 cells. Conversely, parallel transfections with an SV40 promoter-driven luciferase vector containing three concatamerized Epo-derived HIF-1 binding sites[14–16] showed about 2-fold hypoxic induction. These results imply that the previously observed[15,16] time-dependent hypoxic decrease in HIF-1α steady-state mRNA levels in Hepa1 cells might be due to a concomitant reduction in exon I.2 promoter activity.

GENE TARGETING EXPERIMENTS REVEAL HIF-1 AS A MASTER REGULATOR OF OXYGEN HOMEOSTASIS

One of the most impressive demonstrations for the importance of HIF-1-mediated oxygen homeostasis is provided by the phenotype of mice bearing null mutations in the HIF-1 subunits. Homozygote HIF-1α "knock-out" mice die at around embryonic day 10.5 to 11.0 mainly due to neural tube defects, cardiovascular malformations, cell death within the cephalic mesenchyme and impaired vascularization. On the single cell level, expression of "(oxy)genes" and proliferation is impaired.[28] Also ARNT "knock-out" mouse embryos die at midgestation due to blood vessel malformations. It is a matter of debate whether defective angiogenesis of the yolk sac[29] or failure of the embryonic compartment of the placenta to vascularize and to form labyrinthine spongiotrophoblasts[30] is the primary cause for lethality. However, even if HIF-1 has additional functions during embryonic development,[28] failure of hypoxically induced VEGF expression might be the cause of the embryonic lethality in HIF-1α and ARNT "knock-out" mice since their phenotypes are reminiscent of those observed in heterozygote VEGF "knock-out" mouse embryos.[31,32] Obviously, HIF-1α and ARNT are non-redundant genes and HIF-2α, ARNT2, and ARNT3/BMAL1 cannot functionally replace them.

ACKNOWLEDGMENT

Supported by the Swiss National Science Foundation (grant 31-47111.96). R.H.W. is a recipient of the Sondermassnahmen des Bundes zur Förderung des akademischen Nachwuchses.

REFERENCES

1. Wang GL, Jiang BH, Rue EA, Semenza GL: Hypoxia-inducible factor 1 is a basic-helix-loop-helix-PAS heterodimer regulated by cellular O_2 tension. Proc Natl Acad Sci USA 92:5510, 1995.
2. Wenger RH, Gassmann M: Oxygen(es) and the hypoxia-inducible factor-1. Biological Chemistry 378:609, 1997.
3. Huang LE, Arany Z, Livingston DM, Bunn HF: Activation of hypoxia-inducible transcription factor depends primarily upon redox-sensitive stabilization of its α subunit. J Biol Chem 271:32253, 1996.
4. Pugh CW, JF OR, Nagao M, Gleadle JM, Ratcliffe PJ: Activation of hypoxia-inducible factor-1; definition of regulatory domains within the α subunit. J Biol Chem 272:11205, 1997.
5. Kallio PJ, Pongratz I, Gradin K, McGuire J, Poellinger L: Activation of hypoxia-inducible factor 1α: posttranscriptional regulation and conformational change by recruitment of the Arnt transcription factor. Proc Natl Acad Sci USA 94:5667, 1997.
6. Salceda S, Caro J: Hypoxia-inducible factor 1α (HIF-1α) protein is rapidly degraded by the ubiquitin-proteasome system under normoxic conditions. Its stabilization by hypoxia depends on redox-induced changes. J Biol Chem 272:22642, 1997.
7. Li H, Ko HP, Whitlock JP: Induction of phosphoglycerate kinase 1 gene expression by hypoxia. Roles of ARNT and HIF1α. J Biol Chem 271:21262, 1996.
8. Jiang BH, Zheng JZ, Leung SW, Roe R, Semenza GL: Transactivation and inhibitory domains of hypoxia-inducible factor 1α. Modulation of transcriptional activity by oxygen tension. J Biol Chem 272:19253, 1997.
9. Wang GL, Jiang BH, Semenza GL: Effect of altered redox states on expression and DNA-binding activity of hypoxia-inducible factor 1. Biochem Biophys Res Commun 212:550, 1995.
10. Wang GL, Jiang BH, Semenza GL: Effect of protein kinase and phosphatase inhibitors on expression of hypoxia-inducible factor 1. Biochem Biophys Res Commun 216:669, 1995.
11. Salceda S, Beck I, Srinivas V, Caro J: Complex role of protein phosphorylation in gene activation by hypoxia. Kidney Int 51:556, 1997.
12. Rolfs A, Kvietikova I, Gassmann M, Wenger RH: Oxygen-regulated transferrin expression is mediated by hypoxia-inducible factor-1. J Biol Chem 272:20055, 1997.
13. Wenger RH, Kvietikova I, Rolfs A, Camenisch G, Gassmann M: Oxygen-regulated erythropoietin gene expression is dependent on a CpG methylation-free hypoxia-inducible factor-1 DNA-binding site. Eur J Biochem 253:771, 1998.
14. Kvietikova I, Wenger RH, Marti HH, Gassmann M: The transcription factors ATF-1 and CREB-1 bind constitutively to the hypoxia-inducible factor-1 (HIF-1) DNA recognition site. Nucleic Acids Res 23:4542, 1995.
15. Gradin K, McGuire J, Wenger RH, Kvietikova I, Whitelaw ML, Toftgard R, Tora L, Gassmann M, Poellinger L: Functional interference between hypoxia and dioxin signal transduction pathways: competition for recruitment of the Arnt transcription factor. Mol Cell Biol 16:5221, 1996.
16. Gassmann M, Kvietikova I, Rolfs A, Wenger RH: Oxygen- and dioxin-regulated gene expression in mouse hepatoma cells. Kidney Int 51:567, 1997.
17. Antequera F, Boyes J, Bird A: High levels of de novo methylation and altered chromatin structure at CpG islands in cell lines. Cell 62:503, 1990.
18. Hu B, Wright E, Campbell L, Blanchard KL: In vivo analysis of DNA-protein interactions on the human erythropoietin enhancer. Mol Cell Biol 17:851, 1997.
19. Wenger RH, Rolfs A, Marti HH, Guénet JL, Gassmann M: Nucleotide sequence, chromosomal assignment and mRNA expression of mouse hypoxia-inducible factor-1α. Biochem Biophys Res Commun 223:54, 1996.
20. Ladoux A, Frelin C: Cardiac expressions of HIF-1α and HLF/EPAS, two basic loop helix/PAS domain transcription factors involved in adaptative responses to hypoxic stresses. Biochem Biophys Res Commun 240:552, 1997.

21. Wenger RH, Rolfs A, Kvietikova I, Spielmann P, Zimmermann DR, Gassmann M: The mouse gene for hypoxia-inducible factor-1α. Genomic organization, expression, and characterization of an alternative first exon and 5′ flanking sequence. Eur J Biochem 246:155, 1997.
22. Wenger RH, Rolfs A, Spielmann P, Zimmermann DR, Gassmann M: Mouse hypoxia-inducible factor-1α is encoded by two different mRNA isoforms—expression from a tissue-specific and a housekeeping-type promoter. Blood 91:3471, 1998.
23. Luo G, Gu Y-Z, Jain S, Chan WK, Carr KM, Hogenesch JB, Bradfield CA: Molecular characterization of the murine HIF-1α locus. Gene Expression 6:287, 1997.
24. Semenza GL, Rue EA, Iyer NV, Pang MG, Kearns WG: Assignment of the hypoxia-inducible factor 1α gene to a region of conserved synteny on mouse chromosome 12 and human chromosome 14q. Genomics 34:437, 1996.
25. Tian H, McKnight SL, Russell DW: Endothelial PAS domain protein 1 (EPAS1), a transcription factor selectively expressed in endothelial cells. Genes Dev 11:72, 1997.
26. Johnson B, Brooks BA, Heinzmann C, Diep A, Mohandas T, Sparkes RS, Reyes H, Hoffman E, Lange E, Gatti RA, Xia Y-R, Lusis AJ, Hankinson O: The Ah receptor nuclear translocator gene (ARNT) is located on q21 of human chromosome 1 and on mouse chromosome 3 near Cf-3. Genomics 17:592, 1993.
27. Schmidt JV, Carver LA, Bradfield CA: Molecular characterization of the murine Ahr gene. Organization, promoter analysis, and chromosomal assignment. J Biol Chem 268:22203, 1993.
28. Iyer NV, Kotch LE, Agani F, Leung SW, Laughner E, Wenger RH, Gassmann M, Gearhart JD, Lawler AM, Yu AY, Semenza GL: Cellular and developmental control of O_2 homeostasis by hypoxia-inducible factor 1α. Genes Dev 12:149, 1998.
29. Maltepe E, Schmidt JV, Baunoch D, Bradfield CA, Simon MC: Abnormal angiogenesis and responses to glucose and oxygen deprivation in mice lacking the protein ARNT. Nature 386:403, 1997.
30. Kozak KR, Abbott B, Hankinson O: Arnt-deficient mice and placental differentiation. Dev Biol 191:297, 1997.
31. Ferrara N, Carver-Moore K, Chen H, Dowd M, Lu L, KS OS, Powell-Braxton L, Hillan KJ, Moore MW: Heterozygous embryonic lethality induced by targeted inactivation of the VEGF gene. Nature 380:439, 1996.
32. Carmeliet P, Ferreira V, Breier G, Pollefeyt S, Kieckens L, Gertsenstein M, Fahrig M, Vandenhoeck A, Harpal K, Eberhardt C, Declercq C, Pawling J, Moons L, Collen D, Risau W, Nagy A: Abnormal blood vessel development and lethality in embryos lacking a single VEGF allele. Nature 380:435, 1996.

35

IDENTIFICATION AND CHARACTERIZATION OF CONSTITUTIVELY ACTIVE STAT5

Tetsuya Nosaka,[1] Mayumi Onishi,[2] Toshiyuki Kawashima,[1] Koji Yamada,[1] Kazuhide Misawa,[1] Kouichi Ariyoshi,[1] Masayuki Towatari,[3] Hidehiko Saito,[3] Kenzaburo Tani,[4] Shigetaka Asano,[4] Atsushi Miyajima,[5] and Toshio Kitamura[1]

[1]Department of Hematopoietic Factors
Institute of Medical Science
University of Tokyo
[2]Third Department of Internal Medicine
University of Tokyo
[3]First Department of Internal Medicine
University of Nagoya
[4]Department of Hematology-Oncology
Institute of Medical Science
University of Tokyo, and
[5]Institute of Molecular and Biosciences
University of Tokyo

SUMMARY

We have recently identified several constitutively active STAT5s using polymerase chain reaction (PCR)-driven random mutagenesis followed by retrovirus-mediated expression screening. One such mutant STAT5A1*6 harbors two point mutations; one in the effector domain and the other in the DNA binding domain, and induces factor-independent growth of several IL-3-dependent cell lines. The mutant STAT5A1*6 shows constitutive phosphorylation on its tyrosine residues, translocates into the nucleus, binds the target DNA sequence, and stimulates transcription in the absence of IL-3 stimulation. Further biochemical analyses suggested that a molecular basis for the constitutive activity of the mutant STAT5A is the increased stability

Correspondence: Toshio Kitamura, M.D., Ph.D., Department of Hematopoietic Factors, Institute of Medical Science, University of Tokyo, 4-6-1 Shirokanedai, Minato-ku, Tokyo 108-8639, Japan, TEL: +81-3-5449-5758, FAX: +81-3-5449-5453, e-mail: kitamura@ims.u-tokyo.ac.jp

Molecular Biology of Hematopoiesis 6, edited by Abraham *et al.*
Kluwer Academic/Plenum Publishers, New York, 1999.

of the phosphorylated form of STAT5. In addition to the ability of inducing factor-independent growth of IL-3-dependent cell lines, the mutant STAT5A induces apoptosis or differentiation under some conditions or in a different cell line, implicating STAT5 as a molecular switch for pleiotropic functions of cytokine receptors. The relationship between active STAT5 and leukemogenesis will also be discussed.

INTRODUCTION

Cytokine binding to cognate receptors induces activation of multiple signal transduction pathways including Ras/Raf/MEK/MAPK and Janus kinase (JAK)-signal transducers and activators of transcription (STAT) pathways.[1–5] The former signaling pathway is well characterized and plays important roles in various cellular responses including proliferation, differentiation, survival, cell mobility, and transformation. The JAK-STAT pathway, originally identified downstream of the interferon receptors, is now recognized as a common pathway downstream of many cytokine receptors, and is thought to play important roles to exert common functions as well as specific functions of cytokines. Following cytokine-induced dimerization of the receptors, receptor-associated JAKs are cross-activated and phosphorylate the receptor on its tyrosine residues, which recruits SH2 domains of STAT transcription factors. Then, the JAKs phosphorylate the STAT proteins on their tyrosine residues, and the phosphorylated STAT molecules form homo- or heterodimers through SH2-mediated interactions and translocate into nuclei to activate transcription of various target genes. Seven members of the STAT gene family (STAT1-4, 5A, 5B, and 6) have been identified, and STAT5A and STAT5B are closely related. Except for STAT4 and STAT6 which are specifically activated by one or two cytokines, IL-12 and IL-4/IL-13 respectively,[6–8] most STAT proteins are activated by multiple cytokines. In particular, STAT5A and STAT5B are activated by a number of cytokines including IL-2, IL-3, IL-5, IL-7, GM-CSF, G-CSF, M-CSF, epidermal growth factor (EGF), erythropoietin, thrombopoietin, growth hormone (GH), and prolactin.

We previously suggested that STAT5 was involved in cell proliferation by a set of experiments using deletion mutants of the human GM-CSF receptor[9,10] as well as by an experiment using a dominant negative STAT5 molecule which partially inhibited IL-3-induced cell proliferation of IL-3-dependent Ba/F3 cells.[11] Similarly, some researchers demonstrated that the mutant IL-2 and Epo receptors defective in activating STAT5[12–14] were also deficient in transmitting a proliferative signal, implicating a role of STAT5 in cell proliferation. In addition, two groups have reported that STAT5 is involved in erythroid differentiation using dominant negative forms of STAT5.[15,16] On the contrary, some concluded that STAT5 was not involved in cell proliferation because a particular mutant IL-2 receptor[17] or a mutant EPO receptor[18] which could not activate STAT5 was still capable of transmitting a proliferative signal. Recently mice nullizygous for STAT5A[19] or STAT5B[20] as well as for both STAT5A and STAT5B[21] have been isolated. Since these mice did not exhibit major abnormality in normal hematopoiesis, the contribution of the STAT5 pathway to proliferation or differentiation of hematopoietic cells in normal mice remains unclear.

We have recently identified constitutively active forms of STAT5 by random mutagenesis (ref. 22 and Yamada et al. submitted). The mutant STAT5 protein is phosphorylated on tyrosine residues and is transcriptionally active in the absence of IL-3 stimulation, and the expression of the mutant STAT5 in IL-3-dependent cell lines induced cytokine-independent growth, thus implicating a role of STAT5 in cell proliferation. Interestingly, constitutive activation of STAT5 proteins was demonstrated in some leukemic patients, implicating a role of STAT5 in cellular transformation. In this article, we will describe the isolation and characterization of the constitutively active STAT5s and discuss our preliminary results concerning STAT5 activation in patients' leukemic cells as well as the pleiotropic cellular responses caused by the expression of active STAT5 mutants.

RESULTS AND DISCUSSION

Identification of the constitutively active STAT5s. To identify constitutively active forms of STAT5A, we introduced random mutations into the STAT5A sequence by PCR, and the mutated STAT5As were transduced in a murine IL-3-dependent cell line Ba/F3 via retrovirus infection.[22,23] The infected Ba/F3 cells were then screened for autonomous growth in the absence of IL-3, and several factor-independent (FI) clones were isolated. Interestingly, STAT5 proteins are constitutively phosphorylated on the tyrosine residues in all these clones (Fig. 1 and data not shown). Ability of the mutant STAT5A to induce factor-independent growth of Ba/F3 cells was confirmed by reintroduction of the integrated STAT5A into Ba/F3 cells via retrovirus infection. The integrated STAT5A sequences, which were recovered from five FI clones (FI-5, FI-10, FI-11, FI-13, and FI-15) by genomic PCR and reproduced the factor-independent phenotype in Ba/F3 cells, are shown in Fig. 2; some mutations are rather common and some are sporadic. In order to determine which mutations are responsible for the phenotype, we constructed STAT5 mutants each of which harbors one of the rather common mutations (#1–#9), and tested them for the ability to induce factor-independent growth of Ba/F3 cells. Among these clones, STAT5A#2 harboring a point mutation in the SH2 domain did give rise to factor-independent growth of Ba/F3 cells as well as several IL-3-dependent cell lines (Ariyoshi et al., manuscript in preparation), which explains the phenotype of FI-10 and FI-11. However, expression of other single mutants

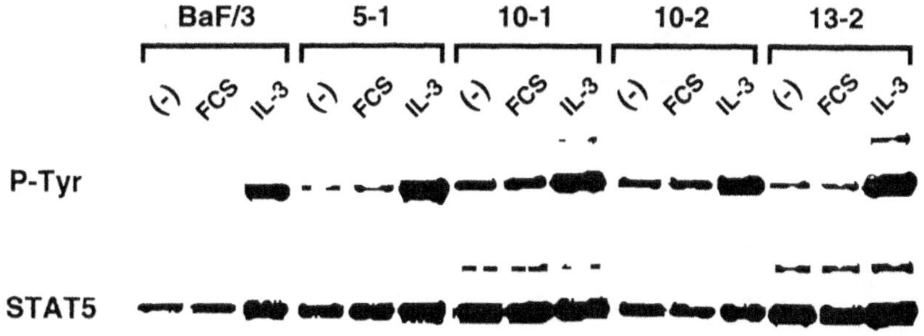

Figure 1. Constitutive phosphorylation of STAT5 in IL-3-independent Ba/F3 clones expressing the mutated STAT5As. IL-3-independent Ba/F3 (FI) clones which had been selected in the absence of IL-3 after the mutated STAT5A library transduction were maintained in the absence of IL-3. These cells were then depleted of serum for 6 hours, stimulated with serum or mIL-3 for 10 min, and then the cell lysates prepared from these cells were subjected to the immunoprecipitation and Western blot analysis. STAT5A was immunoprecipitated by the anti-STAT5A antibody (R&D Systems) and blotted with the antiphosphotyrosine (Upstate Biotechnology Inc.) or with the anti-STAT5A antibody. Stimulation with FCS had no effect on phosphorylation of STAT5A.

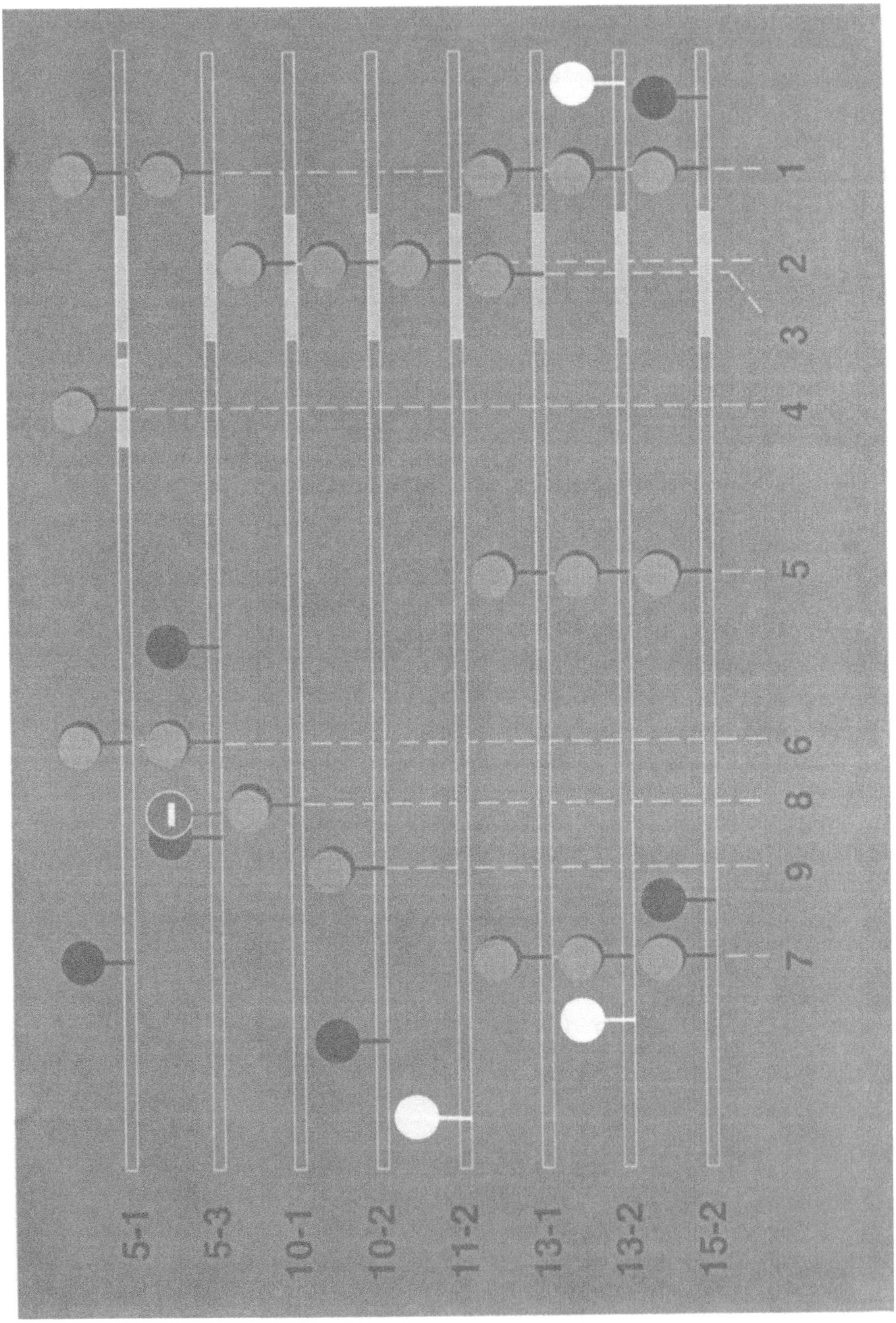

Figure 2. Mutations found in the STAT5A sequences recovered from FI clones by PCR. Each set of clones (e.g. #5–1, #5–2, and #5–3) was derived from the corresponding FI clone (e.g. #5), and may include mutations incorporated during the recovery from the FI clones. Among the mutations, some are sporadic (shown in black circles), some are sense mutations (shown in white circles), and some are stop codons (shown in stop marks). To identify mutations which are responsible for the phenotype, single point mutants harboring one of the common mutations (#1–#9) are generated (shown in grey color).

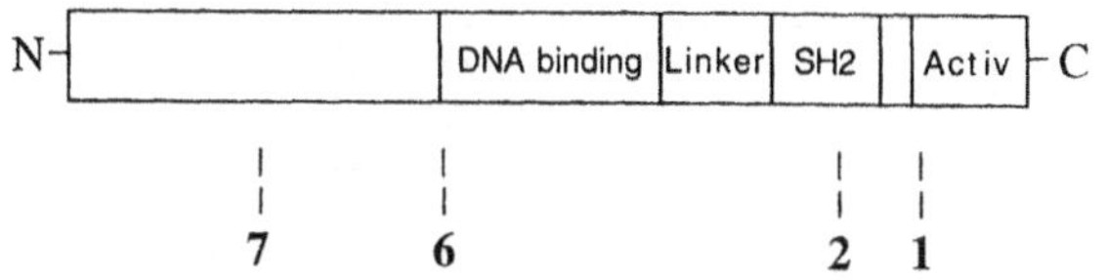

Figure 3. Mutations of the constitutively active STAT5A. A schematic structure and the positions of the mutations are shown. DNA binding, DNA binding domain; SH2, *src* homology (SH) 2 domain; Activ, effector (transcriptional activation) domain.

(STAT5A#1, and #3–#9) did not induce continuous factor-independent growth of Ba/F3 cells, and some of them (STAT5A#1, #6, and #7) induced only transient and weak growth of Ba/F3 cells in the absence of IL-3. Therefore, we thought that the combination of the mutations was required to constitutively activate STAT5 in FI-5, FI-13, and FI-15, and tested the mutant STAT5A harboring #1 and #6 (STAT5A 1*6) or #1 and #7 (STAT5A 1*7). Interestingly, both of the double mutants STAT5A 1*6 and STAT5A 1*7 induced factor-independent growth of Ba/F3 cells as strong as the original mutants #5–1, #13–1, #13–2, and #15–2, indicating that these double mutations (1*6 or 1*7) were responsible for the IL-3-independent growth of FI-5, FI-13, and FI-15.

As shown in Fig. 3, point mutation #1 localizes in the effector (activation) domain and caused an amino acid substitution from serine to phenylalanine (S710F). Point mutations #6 and #7, on the other hand, localize in and upstream of the DNA binding domain and resulted in an amino acid substitution from histidine to arginine (H298R) and glutamic acid to glycine (E150G), respectively (Fig. 3). Interestingly, the two double mutants STAT5A (1*6 and 1*7) gave rise to nearly identical phenotypes (Nosaka et al., unpublished results), and therefore we did further experiments using the mutant STAT5A 1*6. Because the amino acid sequence surrounding the mutation #1 and #6 are conserved between STAT5A and STAT5B, two closely related STAT5 proteins, we introduced the same mutations into STAT5B and found that the mutant STAT5B1*6 could also induce IL-3-independent growth of IL-3-dependent cell lines, which again confirmed the significance of the mutations (Fig. 4).[22]

Characterization of the active STAT5s. First, we expressed the mutant STAT5A and 5B into Ba/F3 cells using an efficient retrovirus vector pMX[24] (infection efficiency is 40–50%), and examined proliferation in the absence of IL-3. As shown in Fig. 4, both the mutant STAT5A and the mutant STAT5B induced IL-3-independent growth of Ba/F3 cells while expression of the wild type STAT5A or 5B was not able to induce IL-3-independence of Ba/F3 cells.[22] The mutant STAT5-induced growth of Ba/F3 cells was slower than IL-3-driven growth of parental Ba/F3 cells.

Next, we established stable transfectants of Ba/F3 cells expressing the wild type STAT5A, STAT5A #1, STAT5A #6, and STAT5A1*6, and examined the tyrosine phosphorylation of STAT5 proteins by immunoprecipitation followed by the Western blot analysis.[22] In this experiment, we found that #1 (S710F) mutation was required for constitutive tyrosine phosphorylation of the mutant STAT5A; both STAT5A1*6 and STAT5A#1 but not the wild type STAT5A and STAT5A#6 showed constitutive tyrosine phosphorylation in the absence of IL-3. Consistent with this result, both STAT5A1*6 and STAT5A#1 showed DNA binding capability in the gel mobility shift experiment.[22]

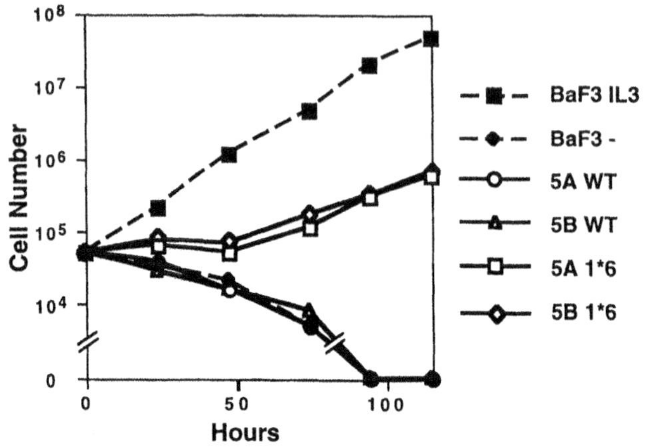

Figure 4. The constitutively active STAT5A1*6 and the mutant STAT5B1*6 harboring the same mutations induced IL-3-independent growth of Ba/F3 cells. Proliferation was examined with bulk populations of Ba/F3 cells after retrovirus-mediated transduction of the wild-type STAT5A (5A WT), the wild-type STAT5B (5B WT), the mutant STAT5A (5A 1*6), and the mutant STAT5B (5B 1*6). The estimated infection efficiency in this experiment was 30 to 50%. Proliferation of parental Ba/F3 cells with and without mIL-3 is also shown.

Nuclear localization of the mutant STAT5A1*6 (Fig. 5A) and STAT5B1*6 (data not shown) has been demonstrated in the experiments in which GFP-fused wild type and the mutant STAT5s were expressed in NIH3T3 cells. However, the wild type STAT5A (Fig. 5B) as well as the single mutants STAT5A#1 and STAT5A#6 did not show predominant nuclear localization.[22]

The transcriptional activity was tested in transient transfection experiments in Ba/F3 cells using a luciferase construct driven by the β-casein promoter known to be a target sequence of STAT5 (Table I). In this assay, the double mutant STAT5A1*6 as well as the single mutants STAT5A#1 and STAT5A#6 transactivated β-casein promoter-driven luciferase expression even in the absence of IL-3-stimulation. Inconsistency of this result with the results of tyrosine phosphorylation, localization and DNA binding capability of the mutant STAT5 molecules (summarized in Table II) is probably caused by the difference in the sensitivity of each assay.

Mechanism of constitutive activation of STAT5s. Activation of JAK and MAPK was not detected in Ba/F3 transfectants in which the mutant STAT5 was expressed (ref. 22 and data not shown). Therefore, we were interested in how the constitutively active

Table I. Transcriptional activation of a β-casein promoter-driven luciferase construct by the wild type and mutant STAT5s

	IL-3 stimulation (–)	IL-3 stimulation (+)
Vector	1.0	9.7
wild type	4.0	29.5
1	41.9	195.6
6	22.5	111.3
1*6	186.6	679.1

The averages of three independent experiments are shown.

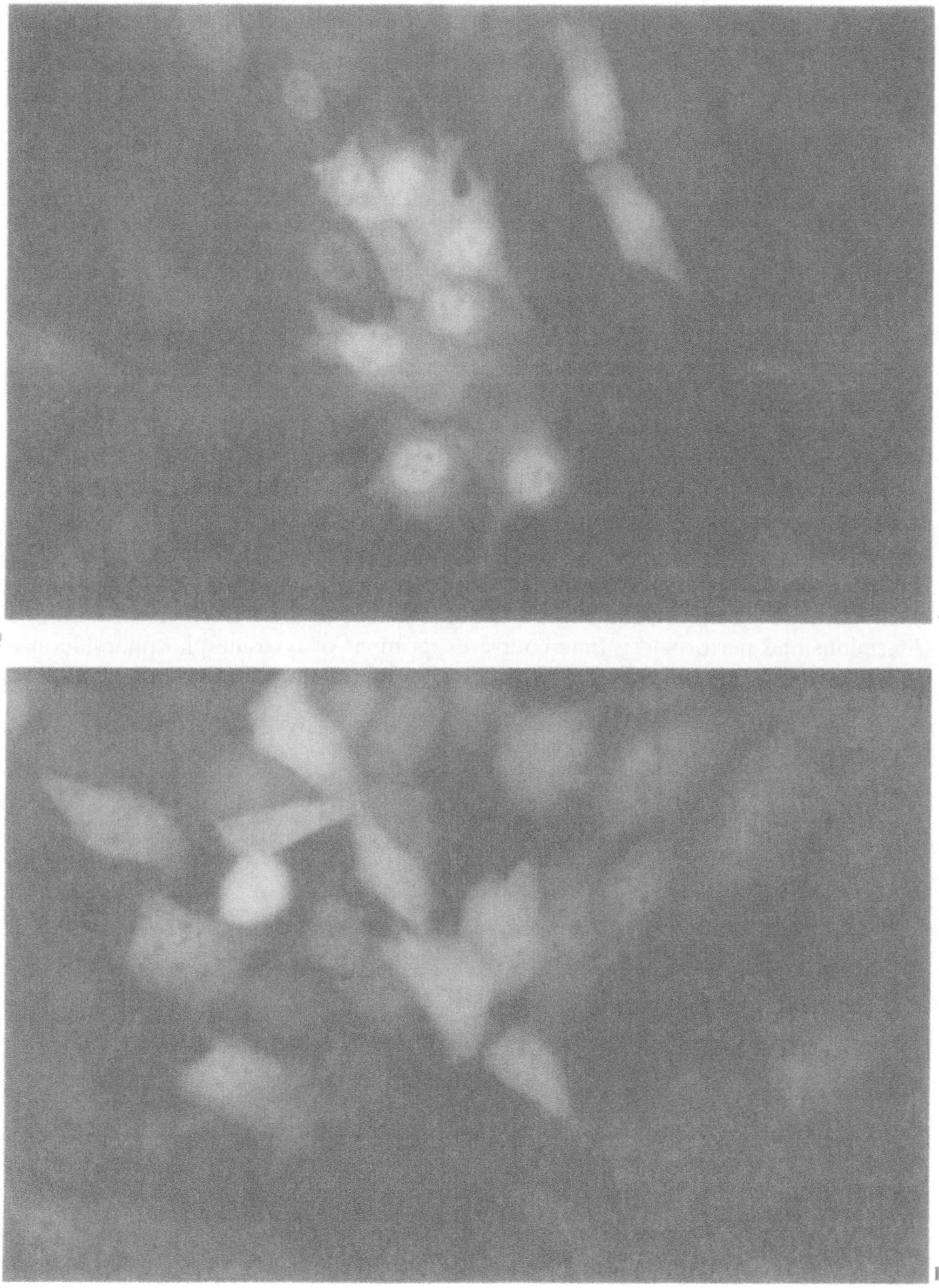

A

B

Figure 5. Nuclear localization of the mutant STAT5A. The fusion molecules STAT5A1*6-EGFP (A) and wild type STAT5A-EGFP (B) were expressed in NIH 3T3 cells and observed under a fluorescence microscope. The photographs were taken with an automatic microscopic system (PM30; Olympus).

Table II. Biochemical and biological characterization of the wild type and the mutant STAT5s

	Tyr-P/DNA binding[a]	transcriptional activation[b]	nuclear loc.[c]	FI/APO[d]
STAT5AWT	–	–	–	–
#1	+	+	–	–
#6	–	+	–	–
1*6	+	++	+	+
1*7	+	++	NT[e]	+
#2	+	++	+	+

[a]Tyrosine phosphorylation and DNA binding in the absence of IL-3 stimulation assessed by the Western blotting and EMSA, respectively.
[b]Transcriptional activation in the absence of stimulation assessed by the luciferase assay.
[c]Nuclear localization in NIH3T3 cells.
[d]Induction of proliferation or apoptosis of Ba/f3 cells in the absence or presence of IL-3, respectively.
[e]Not tested.

STAT5 was activated in the absence of IL-3-stimulation. Mutation #1 localizes closely to the C-terminal tyrosine (Tyr694 of STAT5A) which becomes tyrosine phosphorylated in response to cytokine stimulation and mediates STAT5 dimerization through its interaction with the SH2 domain, suggesting a possible role of mutation #1 in phosphorylation and dimerization. We speculated that the other mutation #6 in the DNA binding domain was required for the stability of the phosphorylated STAT5. For further biochemical analyses, we added a FLAG-tag sequence to the STAT5 constructs at the C-terminus and performed a time course experiment of tyrosine phosphorylation of the STAT5-FLAG proteins before and after IL-3 stimulation.[22] In this experiment, it is

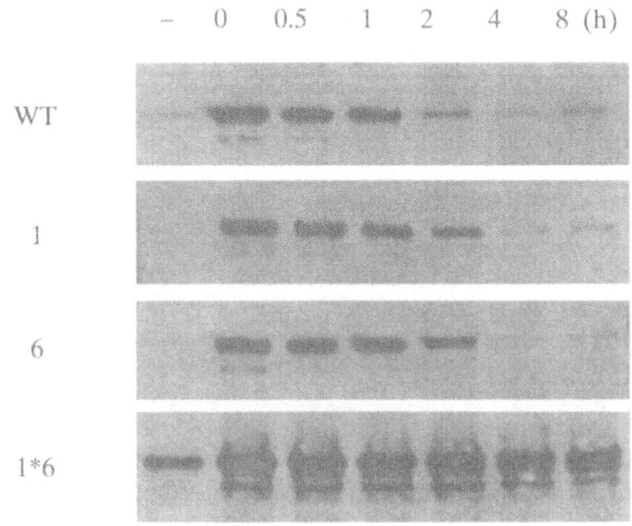

Figure 6. Stability of the phosphorylated forms of the wild-type and the mutant STAT5As. Phosphorylation of the wild type and the mutant STAT5As was examined in the transfectants (bulk population) expressing the wild type STAT5A-Flag (WT), STAT5A #1-Flag (1), STAT5A #6-Flag (6), and STAT5A 1*6-Flag (1*6). The cells were depleted of IL-3 for 12 hr (–), stimulated with 10 ng/ml mIL-3 for 30 min, cultured in the absence of mIL-3 for the indicated time periods (0 to 8 hr), and cell lysates were prepared. The cell lysates were then immunoprecipitated by the anti-Flag M2 monocloral antibody (Eastman Chemical Company), and blotted with the antiphosphotyrosine antibody.

possible to distinguish the exogeneously transduced STAT5 from the endogenous STAT5. Although the mutation #6 by itself did not increase the stability of the STAT5 protein, the double mutant STAT5A1*6 was much more stable after IL-3 stimulation (Fig. 6), suggesting that the mutation #6 was not sufficient but was required for the stabilization of the phosphorylated form of STAT5. The hyperphosphorylation of the mutant STAT5A1*6 after IL-3 stimulation extended for more than 8 hours while IL-3-induced phosphorylation of the wild type STAT5A, STAT5A#1, and STAT5A#6 returned to the basal level in 4 hours. In addition, after IL-3 stimulation, phosphorylation of the mutant STAT5A1*6 was much stronger than that of the wild type STAT5 and the single point mutant STAT5A #1 or #6. Another active mutant STAT5A#2 (the SH2 mutant) also showed hyperphosphorylation after IL-3 stimulation, suggesting a common mechanism to constitutively activate STAT5 proteins. We propose that the mechanism of the constitutive activation of the mutant STAT proteins is the increased stability of the active protein after phosphorylation which could be achieved by the basal levels of tyrosine kinase activity in the cells and would accumulate in time.

Active STAT5 in patients' leukemic cells. We are interested in whether or not constitutively active STAT5 leads to cellular transformation *in vivo*. It has been reported that STAT proteins are constitutively activated in patients' leukemic cells as well as in transformed cells such as HTLV-I-infected cells,[25] and the cells expressing *v-src*,[26] *v-abl*[27] or *bcr-abl*.[28] Although activation of STATs is usually associated with concomitant activation of Jaks or other kinases, it is also possible that STAT5 mutation is one of the primary causes of leukemogenesis. However, it is laborious to sequence the full length cDNA of STAT5A and STAT5B, and moreover, if the mutant STAT5 works as a dominant active molecule, it is required to sequence both allelles of both STAT5A and STAT5B. Even after the sequencing many cDNA clones, it is difficult to find meaningful mutations which are not PCR-derived nor simply derived from polymorphism, but are indeed invloved in leukemogenesis. Therefore, we are now taking two different approaches as described below. First, in stead of sequencing many clones from patients' leukemic cells, cDNAs for STAT5A and STAT5B cloned from patients' leukemic cells were expressed in Ba/F3 cells via retrovirus infection to see whether patient-derived STAT5 molecules show constitutive phosphorylation and even induce IL-3-independent proliferation in Ba/F3 cells. Thus, in this assay, we attempted to examine biological activities of patients' STAT proteins. In the second approach, we focused on sequencing the particular bases of the STAT5 sequence which we identified as activating mutations by random mutagenesis. Using these strategies, we are now screening for significant mutations in fresh leukemic cells from patients whose leukemic cells displayed constitutive activation of STAT5. So far we have analyzed leukemic cells from several patients, and have not yet found any significant mutations, suggesting that oncogenic mutations of STAT5A and STAT5B are rather rare if any although the constitutive activation of STAT5 is frequently observed in leukemic patients.

Active STAT5-induced pleiotropic functions other than autonomous growth of the cells. Recently, we have observed a rather unexpected result; IL-3 stimulation of Ba/F3 transfectants expressing the constitutively active STAT5 (STAT5A1*6, STAT5A1*7, and STAT5A#2) that have been selected in the absence of IL-3 leads to cell apoptosis in most clones or differentiation into macrophage-like cells in some clones (Nosaka et al. in submission). In addition, expression of the mutant STAT5 in a mouse leukemic cell line M1 leads to differentiation of the cell line into macrophages

(Kawashima et al. manuscript in preparation). We were interested in the mechanisms by which the mutant STAT5 exhibits various biological outcomes, and performed the Northern blot analysis of Ba/F3 transfectants expressing either the wild type STAT5 or the mutant STAT5 before and after IL-3 stimulation. This experiment suggested that constitutive expression of *pim-1*, *c-myc*, and *bcl-xL* was responsible for the induction of proliferation in the absence of IL-3 and that strong and prolonged expression of SOCS-1/SSI-1/JAB and $p21^{WAF1/Cip1}$ played important roles in inducing apoptosis and differentiation (Nosaka et al. in submission).

CONCLUDING REMARKS

We have identified constitutively active forms of STAT5s using PCR-driven random mutagenesis followed by retrovirus-mediated expression screening. These mutant STAT5s are constitutively phosphorylated, localize in the nucleus, bind DNA, activate transcription, and induce proliferation as well as apoptosis or differentiation of the cells. Cytokine stimulation rapidly induces the activation phase where tyrosine kinases play major roles, which is quenched by the following inactivating phase induced by phosphatases, CIS, SOCS/SSI/JAB, and $p21.^{WAF1/Cip1}$ Both phases are at least partly controlled by the genes induced by STAT5. Thus, a variety of genes can be induced by STAT5, and various biological outcomes could be manifested by various combinations of gene expression. Therefore, we believe that although our mutant STAT5 is artificial, its pleiotropic functions would represent the physiological functions of the STAT5. In that sense, STAT5 molecules are a sort of the molecular switch by which cells manifest various biological functions in response to cytokine stimulation.

ACKNOWLEDGMENT

Department of Hematopoietic Factors is supported in part by Chugai Pharmaceutical Company, Ltd.

REFERENCES

1. Taniguchi, T: Cytokine signaling through nonreceptor protein tyrosine kinases. Science 268:251, 1995.
2. Mui, AL-F, Wakao, H, O'Farrell, A, Harada, N, Miyajima, A: Interleukin-3, granulocyte-macrophage colony stimulating factor and interleukin-5 transduce signals through two STAT5 homologs. EMBO J 14:1166, 1995.
3. Schindler, C, Darnell, J, Jr.: Transcriptional responses to polypeptide ligands: the JAK-STAT pathway. Annu Rev Biochem 64:621, 1995.
4. Ihle, JN: STATs: signal transducers and activators of transcription. Cell 84:331, 1996.
5. O'Shea, JJ: Jaks, STATs, cytokine signal transduction, and immunoregulation: Are we there yet? Immunity 7:1, 1997.
6. Jacobson, NG, Szabo, SJ, Weber-Nordt, RM, Zhong, Z, Schreiber, RD, Darnell, J, Jr., Murphy, KM: Interleukin 12 signaling in T helper type 1 (Th1) cells involves tyrosine phosphorylation of signal transducer and activator of transcription (Stat)3 and Stat4. J Exp Med 181:1755, 1995.
7. Bacon, CM, Petricoin, EF, Ortaldo, JR, Rees, RC, Larner, AC, Johnston, JA, O'Shea, JJ: Interleukin 12 (IL-12) induces tyrosine phosphorylation and activation of STAT4 in human lymphocytes. Proc. Natl. Acad. Sci. USA. 92:7307, 1995.
8. Takeda, K, Kishimoto, T, Akira, S: STAT6: its role in interleukin 4-mediated biological functions. J. Mol. Med. 75:317, 1997.

9. Sato, N, Sakamaki, K, Terada, N, Arai, K, Miyajima, A: Signal transduction by the high-affinity GM-CSF receptor: two distinct cytoplasmic regions of the common beta subunit responsible for different signaling. EMBO J 12:4181, 1993.
10. Kinoshita, T, Yokota, T, Arai, K, Miyajima, A: Suppression of apoptotic death in hematopoietic cells by signalling through the IL-3/GM-CSF receptors. EMBO J 14:266, 1995.
11. Mui, ALF, Wakao, H, Kinoshita, T, Kitamura, T, Miyajima, A: Suppression of interleukin-3-induced gene expression by a C-terminal truncated Stat5: role of Stat5 in proliferation. EMBO J 15:2425, 1996.
12. Damen, JE, Wakao, H, Miyajima, A, Krosl, J, Humphries, RK, Cutler, RL, Krystal, G: Tyrosine 343 in the erythropoietin receptor positively regulates erythropoietin-induced cell proliferation and Stat5 activation. EMBO J 14:5557, 1995.
13. Friedmann, MC, Migone, T, Russell, SM, Leonard, WJ: Different interleukin 2 receptor β-chain tyrosines couple to at least two signaling pathways and synergistically mediate interleukin 2-induced proliferation. Proc. Natl. Acad. Sci. USA 93:2077, 1996.
14. Gobert, S, Chretien, S, Gouilleux, F, Muller, O, Pallard, C, Dusanter-Fourt, I, Groner, B, Lacombe, C, Gisselbrecht, S, Mayeux, P: Identification of tyrosine residues within the intracellular domain of the erythropoietin receptor crucial for STAT5 activation. EMBO J. 15:2434, 1996.
15. Iwatsuki, K, Endo, T, Misawa, H, Yokouchi, M, Matsumoto, A, Ohtsubo, M, Mori, KJ, Yoshimura, A: STAT5 activation correlates with erythropoietin-mediated erythroid differentiation of an erythroleukemia cell line. J. Biol. Chem. 272:8149, 1997.
16. Wakao, H, Chida, D, Damen, JE, Krystal, G, Miyajima, A: A possible involvement of Stat5 in erythropoietin-induced hemoglobin synthesis. Biochem. Byophys. Res. Commun. 234:198, 1997.
17. Fujii, H, Nakagawa, Y, Schindler, U, Kawahara, A, Mori, H, Goulilleux, F, Groner, B, Ihle, JN, Minami, Y, Miyazaki, T, Taniguchi, T: Activation of Stat5 by interleukin 2 requires a carboxyl-terminal region of the interleukin 2 receptor β chain but is not essential for the proliferative signal transmission. Proc. Natl. Acad. Sci. USA. 92:5482, 1995.
18. Quelle, FW, Wang, D, Nosaka, T, Thierfelder, WE, Stravopodis, D, Weinstein, Y, Ihle, JN: Erythropoietin induces activation of Stat5 through association with specific tyrosines on the receptor that are not required for a mitogenic response. Mol. Cell. Biol. 16:1622, 1996.
19. Liu, X, Robinson, GW, Wagner, K-U, Garrett, L, Wynshaw-Boris, A, Hennighausen, L: Stat5a is mandatory for adult mammary gland and lactogenesis. Genes & Dev. 11:179, 1997.
20. Udy, GB, Towers, RP, Snell, RG, Wilkins, RJ, Park, S-H, Ram, P, Waxman, DJ, Davey, HW: Requirement of Stat5b for sexual dimorphism of body growth rates and liver gene expression. Proc. Natl. Acad. Sci. U.S.A. 94:7239, 1997.
21. Teglund, S, Mckay, C, Schuetz, E, van Deursen, JM, Stravopodis, D, Wang, D, Brown, M, Bodner, S, Grosveld, G, Ihle, JN: Stat5a and Stat5b proteins have essential and nonessential, or redundant roles in cytokine responses. Cell 93:841, 1998.
22. Onishi, M, Nosaka, T, Misawa, K, Mui, AL-F, Gorman, D, McMahon, M, Miyajima, A, Kitamura, T: Identification and characterization of a constitutive active STAT5 mutant that promotes cell proliferation. Mol. Cell. Biol. 18:3871, 1998.
23. Kitamura, T, Onishi, M, Kinoshita, S, Shibuya, A, Miyajima, A, Nolan, GP: Efficient screening of retroviral cDNA expression libraries. Proc. Natl. Acad. Sci. U.S.A. 92:9146, 1995.
24. Onishi, M, Kinoshita, S, Morikawa, Y, Shibuya, A, Phillips, J, Lanier, LL, Gorman, DM, Nolan, GP, Miyajima, A, Kitamura, T: Applications of retrovirus-mediated expression cloning. Exp. Hematol. 24:324, 1996.
25. Migone, T-S, Lin, J-X, Cereseto, A, Mulloy, JC, O'Shea, JJ, Franchini, G, Leonard, WJ: Constitutive activated Jak-STAT pathway in T cells transformed with HTLV-I. Science 269:79, 1995.
26. Cao, X, Tay, A, Guy, GR, Tan, YH: Activation and association of Stat3 with src in v-src-transformed cell lines. Mol. Cell. Biol. 16:1595, 1996.
27. Danial, NN, Pernis, A, Rothman, PB: Jak-STAT signaling by the v-abl oncogene. Science 269:1875, 1995.
28. Carlesso, N, Frank, DA, Griffin, JD: Tyrosyl phosphorylation and DNA binding activity of signal transducers and activators of transcription (STAT) proteins in hematopoietic cell lines transformed by Bcr/Abl. J. Exp. Med. 183:811, 1996.

AN INCREASE IN THE LEVELS OF RETROVIRAL-MEDIATED TRANSDUCTION OF ENGRAFTING HUMAN HEMATOPOIETIC PROGENITORS CAN BE OBTAINED BY MANIPULATION OF THE HEMATOPOIETIC CELL CYCLE

Mo A. Dao and Jan A. Nolta

Division of Research Immunology/Bone Marrow Transplantation
Childrens Hospital Los Angeles, and Department of Pediatrics
University of Southern California School of Medicine
4650 Sunset Blvd., Mailstop # 62, Los Angeles, CA 90027
Phone # (323) 669-5930, FAX # (323) 660-1904
E-Mail: jnolta@chla.usc.edu

The bnx/hu xenograft system uses immune deficient beige/nude/xid (bnx) mice as recipients of transduced human hematopoietic stem and progenitor cells.[1] Long-term engraftment in the mice is supported by co-transplantation of primary human stromal cells that have been genetically engineered to secrete human interleukin 3 (IL-3). Production of IL-3, a species-specific cytokine, was determined to be both necessary and sufficient to support long-term human hematopoiesis in the mice.[1] Multilineage human hematopoiesis can be followed in bnx mice for up to 18 months. All myeloid lineages, erythroid and myeloid progenitors, αβTCR+ T lymphocytes, and B lymphocytes can be recovered from the bone marrow of long-term engrafted mice.[1-11] The bnx/hu xenograft model can be used to assess the effects of various *in vitro* transduction methods on the efficacy of gene transfer, as well as on the viability and commitment to differentiation of the target human hematopoietic stem and progenitor cells.

In order to truly understand the impact of various transduction conditions, stem cells must be individually marked and their progeny tracked into different lineages. Evidence for transduction of pluripotent HSC can be deduced from demonstration of a retroviral vector integrated into the same cellular chromosomal DNA site in myeloid

Molecular Biology of Hematopoiesis 6, edited by Abraham *et al.*
Kluwer Academic/Plenum Publishers, New York, 1999.

and lymphoid cells descended from a common precursor.[12,13] Retroviral vectors integrate randomly into different chromosomal sites in each cell they transduce, and the distinct proviral clonal integration site is carried by all of the progeny of the transduced cell. Rill and Brenner furthered the use of clonal integration analysis by reporting the technique of inverse PCR to track descendants of marked human hematopoietic cells in recipients of human gene therapy.[14] Our group modified the inverse PCR technique by developing protocols by which very small cell samples could be studied.[2] As few as 200 cells, or approximately ten picograms of DNA, can be reproducibly analyzed using the single colony inverse PCR technique. Therefore, the clonal origin of small myeloid or erythroid colonies recovered from clonogenic assays, or individual bursts of T lymphocytes stimulated to respond to IL-2 and antigen or PHA can be assessed. We have recently used these techniques to trace progeny of transduced hematopoietic progenitors in human gene therapy recipients, and are sequentially examining marked T cells and CFU-GM obtained from marrow aspirates and peripheral blood samples.[15,16] Using beige/nude/xid (*bnx*) mice as transplant recipients for cells transduced using the best conditions available in 1995–1996 (Fig. 1), and the technique of single colony inverse PCR to identify the number of marked cells, we demonstrated infrequent transduction of human hematopoietic stem cells primitive enough to form both myeloid and lymphoid progeny.[2] We next sought to define better methods for transduction of primitive, engrafting hematopoietic stem, and progenitor cells.

Using the bnx/hu xenograft system, we compared the transduction and engraft-

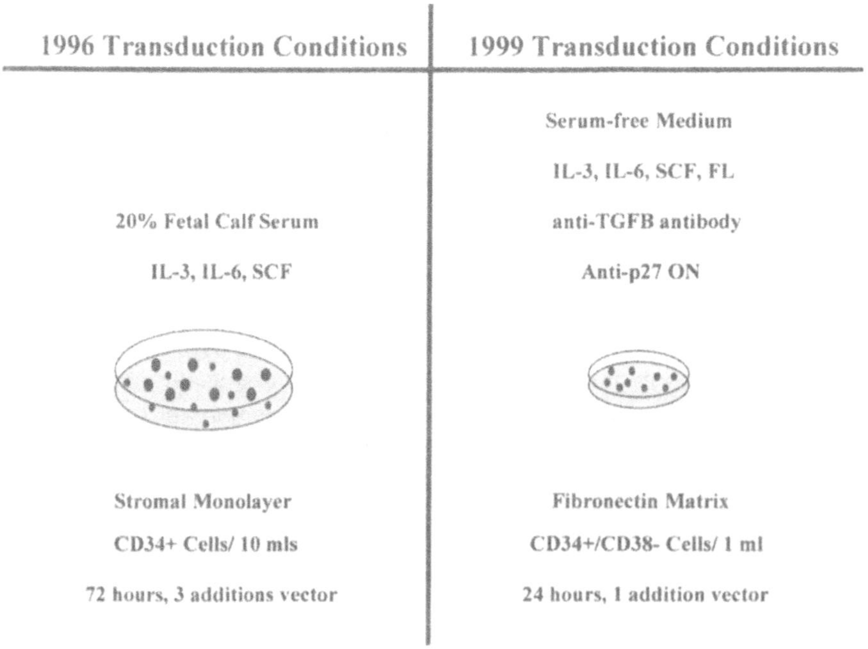

Figure 1. Updated transduction conditions. The function of each component of the transduction system is explained in Table 1. The new advances in the past several years are the use of a fibronectin matrix (the recombinant Retronectin™ molecule, available from Takara Shuzo, Otsu, Japan), cytokines including Flt3 ligand (FL), serum-free rather than serum-containing media, and additives to reduce levels of the CDK inhibitors p15 and p27, which prevent cell cycle entry.

ment of human CD34+ progenitors with a more primitive and quiescent subpopulation; CD34+CD38– cells. Comparable extents of human cell engraftment and lineage development were obtained from 5×10^5 CD34+ cells and 2000 CD34+CD38– cells.[9] Retroviral marking of long-lived progenitors from the CD34+ populations was readily accomplished, but CD34+CD38– cells capable of reconstituting bnx mice were relatively resistant to transduction. Extension of the culture and transduction period from three to seven days slightly increased the level of transduction of the CD34+CD38– cells, and the presence of FLT3 ligand was required to sustain the ability of the cells to mediate long-term engraftment.[9] The low level of marking of the primitive cells is likely due to their quiescence, since Moloney-based vectors require target cell cycling to permit integration.[17]

Although some transduction of primitive, reconstituting human hematopoietic cells could be achieved by using a seven day culture period,[9] such a long duration of culture is not acceptable for human stem cell gene therapy clinical trials. Extended culture periods can lead to differentiation of the cells and a loss of reconstituting capacity. We therefore sought to develop methods that would allow a more rapid recruitment of hematopoietic stem cells into cycle, to allow retroviral-mediated transduction without inducing differentiation. We first determined methods that would maintain the viability of the engrafting stem cells.

The potentiality of primitive human hematopoietic cells can be profoundly affected by *ex-vivo* culture. We have previously used the bnx/hu xenograft system to determine culture methods that do not diminish the capacity for multilineage engraftment. We showed that human progenitors require adhesion to retain full engraftment potential through 72 hour periods of ex-vivo culture in the presence of serum and IL-3, IL-6, and SCF.[3] Cells cultured in suspension retained the capacity to mediate short term (2–4 months) but not long-term (7–11 months) engraftment. Addition of FLT3 ligand (FL) to cells in suspension could partially restore the ability to sustain engraftment.[8] The use of the COOH terminal fragment of fibronectin (FN), Retronectin™ (Takara Shuzo, Otsu, Japan), in place of the stromal cell monolayer,[18–21] lead to full maintenance of reconstitution capacity throughout a 72 hour ex-vivo transduction period, with serum and cytokines.[10]

Next, we used culture on FN, in medium containing FL, as a starting point to further optimize retroviral—mediated transduction methods, without reducing the regenerative capacity of the human hematopoietic stem and progenitor target cells. The best scenario would be the ability to transduce significant portions of the target cells with the retroviral vectors within a 24 hour period, to minimize any potentially detrimental effects of culture. Since the external factors that recruit hematopoietic stem cells into cycle that rapidly have not yet been identified, we hypothesized that a reduction in the levels of internal cell cycle inhibitors could release HSC from quiescence to allow retroviral-mediated transduction.

Cells can be held in quiescence by intracellular cyclin-dependent kinase inhibitors (CKI). The CKI $p15^{INK4B}$ blocks association of cyclinD/CDK4, a complex mandatory for cell cycle progression (Fig. 2). In mink lung cells, p15 is upregulated by TGFβ.[22] TGFβ has been demonstrated to prevent cell cycle entry in human hematopoietic cells.[23–25] We tested human CD34+ progenitors to determine the effects of TGFβ in hematopoietic progenitors at the molecular level. Cells were incubated with cytokines ± soluble TGFβ and ± neutralizing antibody (αTGFβ) for 12–14 hours at 37°. Western blots were done and probed with αp15 antibody. In the presence of soluble TGFβ, marrow CD34+ cells had markedly elevated levels of p15, and cells incubated with

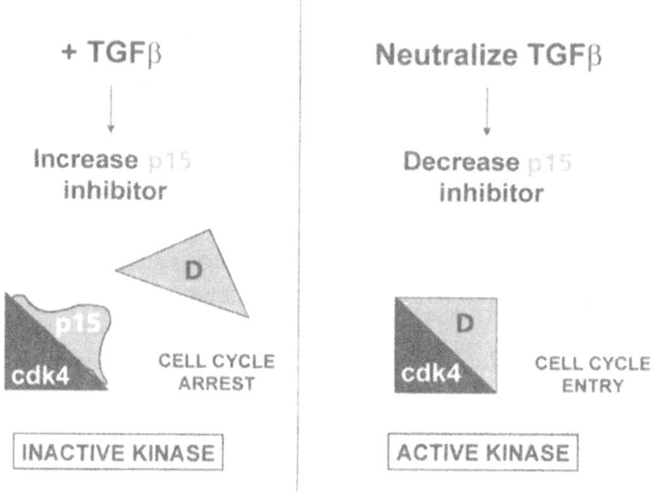

Figure 2. Mechanism of cell cycle arrest by TGFβ. TGFβ increases levels of the CDK inhibitor $p15^{INK4B}$, which prevents association of the CDK4 (and CDK6) complex with cyclin D, a required regulatory subunit. Therefore, in the presence of TGFβ, CDK4, and CDK6 are inactive. Addition of neutralizing antibody to TGFβ decreases p15 levels, allowing association of the cyclin/CDK subunits to generate active kinase complexes.

αTGFβ had undetectable p15 levels.[11] We next wanted to determine whether the reduction in p15 levels would correlate with cell cycle entry, as evidenced by an increase in transduction by the M-phase dependent retroviral vector, LN.[17]

Eight experiments were done to examine transduction with and without neutralization of TGFβ, using the LN vector. The total number of CFU obtained from 10,000 $CD34^+$ cells was significantly increased in cultures with neutralization, and the average number of transduced CFU obtained from the same cell number was increased even more dramatically (Fig. 3). We next transduced 4-HC treated CD34+ cells, a more quiescent population. 4-HC kills cycling progenitors, while sparing the reconstituting, quiescent stem cells.[26,27] Neutralization of TGFβ again increased the number of colonies obtained, but no increase in the extent of transduction was achieved (Fig. 4). This lead us to suspect that another CDK inhibitor in the more quiescent cells was preventing cell cycle entry during the period of exposure to retroviral supernatant. The CKI $p27^{kip-128}$ was a likely candidate because it is elevated in quiescent fibroblasts,[29,30] and negatively regulates entry into cell cycle, in fibroblastic cell lines.[31,32]

We targeted the p27 mRNA using antisense oligonucleotides (ON). C-5 propyne modified phosphorothioate anti-p27 16-mers[33] were introduced into human $CD34^+$ cells at a concentration of 50μM in serum-free medium + IL-3, IL-6, and SCF.[11] Mis-sense ON bearing the same modifications (mis-p27) and medium alone were used as controls. A significant increase in the number of colony-forming units was obtained from 10,000 $CD34^+$ cells incubated for 12 hours with anti-p27 ON, as compared to the same number of mis-p27-treated control cells, and those cultured in medium alone.[11] The anti-p27 ON also caused an increase in the number of colony-forming units from 4-HC purged progenitors, in comparison to those treated with mis-p27 ON, but did not permit retroviral-mediated transduction, when used alone. Both neutralization of TGFβ and antisense ON to p27 kip-1 were required to obtain transduction of a portion

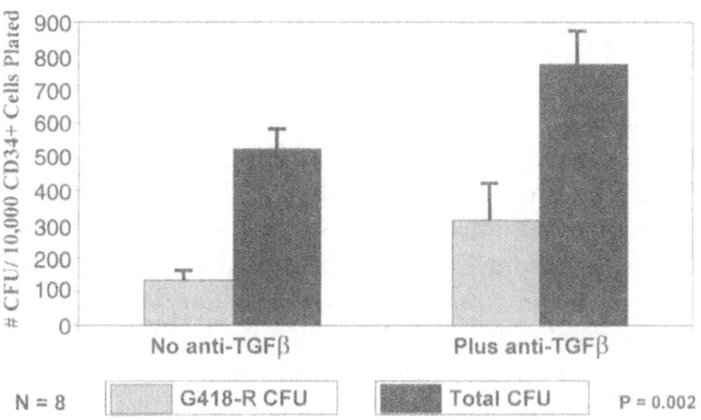

Figure 3. Transduction levels and total CFU formation are both increased from bone marrow CD34+ cells cultured with anti-TGFβ antibody in serum-free medium. CD34+ cells, isolated from human bone marrow with 95–99% purity, were cultured in serum-free medium with IL-3, IL-6, SCF, and FL, for 72 hours. Incubations were done in the presence and absence of 5 ug/ml anti-TGFβ neutralizing antibody (R&D Systems). Three additions of LN retroviral supernatant were added, at 24 hour intervals. Cells were then plated in colony-forming assay with and without the selective agent G418. Total and G418 resistant CFU were enumerated 14–21 days after plating.

of the 4-HC purged CD34+ cells.[11] We next applied the same technology to transduction of CD34+38– cells, a primitive and quiescent stem cell population.

Cells treated with anti-TGFβ antibody and antisense ON to p27$^{kip\text{-}1}$ expressed a significant increase in levels of cyclin A protein and stained positive for Ki67, characteristic of cells entering S phase.[11] As a follow-up to the *in vitro* studies, we transplanted bnx mice with the treated, transduced CD34+38– cells to address an important concern: do primitive progenitors recruited into cycle by the combination of anti-TGFβ Ab and anti-p27 ON retain their multipotentiality and capacity for long-term engraftment?

Bone marrow CD34+/38– cells were pre-incubated in serum-free medium for 12

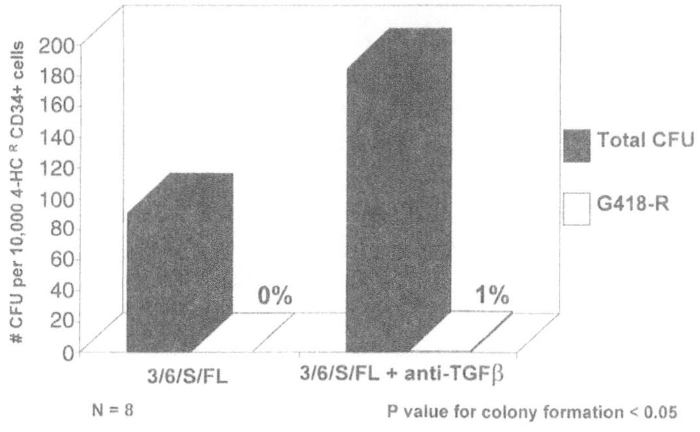

Figure 4. 4-HC resistant CD34+ cells cultured with anti-TGFβ antibody demonstrate a significant increase in total, but not transduced, CFU generation. CD34+ cells were isolated from human bone marrow that had been treated with 4-HC to kill cycling cells. Incubations, transductions, and colony-forming analyses were done as listed for Figure 3. In contrast to the results obtained from the total CD34+ cell populations (Fig. 3), there was no significant transduction of the noncycling 4-HC resistant progenitors.

hours with anti-TGFβ Ab and anti-p27 ON, vs. the scrambled mis-sense oligonucleotide. Retroviral supernatant from the PG13/LN line was then added for an additional 12 hours. Only cells that had received the dual combination of anti-TGFβ Ab and anti-p27 ON had significant CFU growth and levels of transduction (Fig. 5). Cells from each experimental condition were co-transplanted with human stroma producing hIL-3 into bnx mice. Mice were harvested and analyzed by FACS, human-specific CFU assay, neo PCR, and inverse PCR 11 months post-transplantation. Human CD45+ cell engraftment levels were equivalent between mice transplanted with cells treated with anti-TGFβ Ab and anti-p27 ON and mice transplanted with cells treated with mis-p27 ON alone.[11] There was an equal distribution of the human hematopoietic lineages that had developed in the two groups. Bone marrow samples were further analyzed by neo PCR and human-specific CFU plating for G418 resistant colonies. Only mice that had been transplanted with human progenitors transduced with anti-TGFβ Ab and anti-p27 ON had neo gene–marked human cells in their marrow.[11] Equivalent numbers of human-specific colonies developed from the bone marrow of the two groups of mice, but G418-resistant CFU only grew from mice transplanted with human progenitors transduced with anti-TGFβ Ab and anti-p27 ON.[11] The recovery, from the bnx bone marrow, of human T lymphocyte and myeloid clones with many different clonal integration sites demonstrated that multiple primitive human hematopoietic stem and progenitor cells had been transduced, using the optimized conditions.

Therefore, primitive hematopoietic progenitors induced into cell cycle entry by the combination of anti-TGFβ Ab and anti-p27 ON can be transduced by retroviral vectors in a 24 hour period, and retain the ability to reconstitute and durably engraft immune deficient mice. Figure 1 shows the current best transduction conditions for Moloney Murine Leukemia-based retroviral vectors, in comparison to the conditions that had been considered optimal in previous years. The conditions that are currently used in our

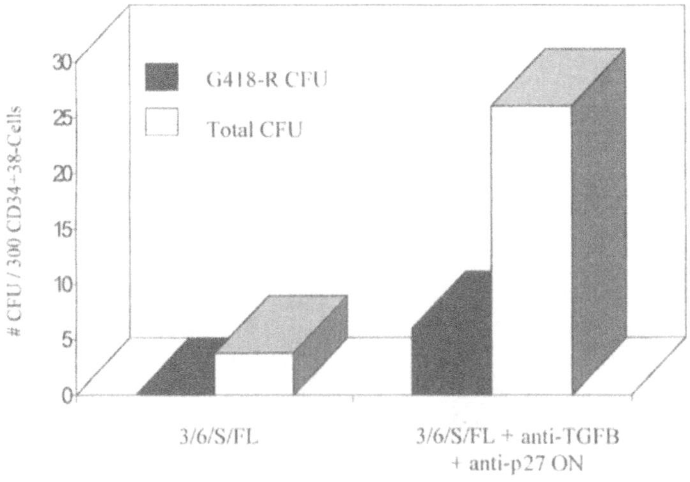

Figure 5. Antisense ON to p27 augments transduction of primitive CD34+/CD38– cells. CD34+/CD38– cells, isolated from normal human bone marrow, were subjected to an abbreviated 24 hour transduction in serum-free medium, with one addition of supernatant from the LN retroviral vector. Conditions were: cytokines alone, vs. cytokines, anti-TGFβ neutralizing antibody, and anti-p27 oligonucleotides. Significant colony growth and levels of transduction were only achieved in a 24 hour period, when anti-TGFβ neutralizing antibody and anti-p27 oligonucleotides were used.

Table 1. Components of transduction and their function

Component	Function
COOH terminal fragment of fibronectin (Retronectin™)	Co-localization of RV vector partical and target cell (18–21). Replaces the function of stromal support (10).
IL-3, IL-6, SCF, FL	Cytokines to maintain viability, and to stimulate cell division, once stem and progenitor cells exit the G_0 phase of the cell cycle.
Serum-free medium	Ex-Vivo 15 (BioWhittaker) or others (34) are defined media that contain the constituents necessary for cell survival. Serum-free media contain far lower levels of TGFβ and other inhibitory cytokines and chemokines than serum-containing media.
Anti-TGFβ antibody	Neutralizes TGFβ in the transduction medium. TGFβ is inhibitory to hematopoietic progenitor cell cycle entry (23–25). TGFβ increases levels of the CDK inhibitor p15 in human progenitors, preventing exit from quiescence (11).
Anti-p27 kip-1 oligonucleotides	Reduces levels of the CDK inhibitor p27 kip-1, which maintains primitive hematopoietic cells in a quiescent state (11).

laboratory, and that will soon be tested in human clinical gene therapy trials, are the following: Fibronectin support, serum-free medium, the cyokines IL-3, IL-6, SCF, and Flt3 ligand, anti-TGFβ antibody, and antisense ON to $p27^{kip\text{-}1}$. Retroviral supernatant will be produced with 10% fetal calf serum, and added in equal volumes to the transduction medium to generate a final serum concentration of 5%. These conditions have been demonstrated to allow retroviral-mediated transduction, in a 24 hour period, of 20–25 percent of primitive human hematopoietic cells that are capable of long-term (1 year) hematopoiesis in a xenograft model. We are currently studying the factors that prevent the remaining 75–80% of the human hematopoietic stem cells from being rapidly recruited into S phase by a reduction in the levels of the CDK inhibitors $p15^{INK4B}$ (using anti-TGFβ neutralizing antibodies and serum-free media) and $p27^{kip\text{-}1}$ (using antisense oligonucleotides). In addition, the extracellular factors that control p27 levels and activity in primitive human hematopoietic cells are currently under investigation.

ACKNOWLEDGMENT

This work was supported by a grant from the NIH NHLBI (SCOR #1-P50-HL54850), A grant from the NIH NIDDK (1RO1DK53041), Start-up funding from the John Connell Gene Therapy Foundation, a Career Development Award from the Childrens Hospital of Los Angeles Research Institute, and a James A. Shannon Directors' Award from the NIH NIDDK.

REFERENCES

1. Nolta, JA, MB Hanley, and DB Kohn. 1994. Sustained human hematopoiesis in immunodeficient mice by cotransplantation of marrow stroma expressing human interleukin-3: analysis of gene transduction of long-lived progenitors. *Blood 83:3041.*
2. Nolta, JA, MA Dao, S Wells, EM Smogorzewska, and DB Kohn. 1996. Transduction of pluripotent human hematopoietic stem cells demonstrated by clonal analysis after engraftment in immune-deficient mice. *Proc Natl Acad Sci USA 93:2414.*

3. Nolta, JA, EM Smogorzewska, and DB Kohn. 1995. Analysis of optimal conditions for retroviral-mediated transduction of primitive human hematopoietic cells. *Blood 86:101.*
4. Arakawa-Hoyt J, MA Dao, F Thiemann, QL Hao, DC Ertl, KI Weinberg, GM Crooks, and Nolta JA. The number and generative capacity of human B lymphocyte progenitors, measured in vitro and in vivo, is higher in umbilical cord blood than in adult or pediatric bone marrow. *In Press, Bone Marrow Transplantation* 1999.
5. Dao, MA, XJ Yu, and JA Nolta. 1997. Clonal diversity of primitive human hematopoietic progenitors following retroviral marking and long-term engraftment in immune-deficient mice. *Exp Hematol 25:1357.*
6. Dao, MA, KA Pepper, and JA Nolta. 1997. Long-term cytokine production from engineered primary human stromal cells influences human hematopoiesis in an in vivo xenograft model. *Stem Cells 15:443.*
7. Dao, MA, and JA Nolta. 1997. Inclusion of IL-3 during retrovirally-mediated transduction on stromal support does not increase the extent of gene transfer into long-term engrafting human hematopoietic progenitors. *Cytokines Cell Mol Ther 3:81.*
8. Dao, MA, CH Hannum, DB Kohn, and JA Nolta. 1997. FLT3 ligand preserves the ability of human CD34+ progenitors to sustain long-term hematopoiesis in immune-deficient mice after ex vivo retroviral-mediated transduction. *Blood 89:446.*
9. Dao, MA, AJ Shah, GM Crooks, and JA Nolta. 1998. Engraftment and retroviral marking of CD34+ and CD34+CD38– human hematopoietic progenitors assessed in immune-deficient mice. *Blood 91:1243.*
10. Dao, MA, K Hashino, I Kato, and JA Nolta. 1998. Adhesion to fibronectin maintains regenerative capacity during Ex vivo culture and transduction of human hematopoietic stem and progenitor cells. *Blood 92:4612.*
11. Dao, MA, N Taylor, and JA Nolta. 1998. Reduction in levels of the cyclin-dependent kinase inhibitor p27(kip-1) coupled with transforming growth factor beta neutralization induces cell-cycle entry and increases retroviral transduction of primitive human hematopoietic cells. *Proc Natl Acad Sci USA 95:13006.*
12. Keller, G, C Paige, E Gilboa, and EF Wagner. 1985. Expression of a foreign gene in myeloid and lymphoid cells derived from multipotent haematopoietic precursors. *Nature 318:149.*
13. Jordan, CT, and IR Lemischka. 1990. Clonal and systemic analysis of long-term hematopoiesis in the mouse. *Genes Dev 4:220.*
14. Rill, DR, VM Santana, WM Roberts, T Nilson, LC Bowman, RA Krance, HE Heslop, RC Moen, JN Ihle, and MK Brenner. 1994. Direct demonstration that autologous bone marrow transplantation for solid tumors can return a multiplicity of tumorigenic cells. *Blood 84:380.*
15. Kohn, DB, KI Weinberg, JA Nolta, LN Heiss, C Lenarsky, GM Crooks, ME Hanley, G Annett, JS Brooks, A el-Khoureiy, and et al. 1995. Engraftment of gene-modified umbilical cord blood cells in neonates with adenosine deaminase deficiency. *Nat Med 1:1017.*
16. Kohn, DB, MS Hershfield, D Carbonaro, A Shigeoka, J Brooks, EM Smogorzewska, LW Barsky, R Chan, F Burotto, G Annett, JA Nolta, G Crooks, N Kapoor, M Elder, D Wara, T Bowen, E Madsen, FF Snyder, J Bastian, L Muul, RM Blaese, K Weinberg, and R Parkman. 1998. T lymphocytes with a normal ADA gene accumulate after transplantation of transduced autologous umbilical cord blood CD34+ cells in ADA- deficient SCID neonates [In Process Citation]. *Nat Med 4:775.*
17. Miller, DG, MA Adam, and AD Miller. 1990. Gene transfer by retrovirus vectors occurs only in cells that are actively replicating at the time of infection [published erratum appears in Mol Cell Biol 1992 Jan;12(1):433]. *Mol Cell Biol 10:4239.*
18. Moritz, T, P Dutt, X Xiao, D Carstanjen, T Vik, H Hanenberg, and DA Williams. 1996. Fibronectin improves transduction of reconstituting hematopoietic stem cells by retroviral vectors: evidence of direct viral binding to chymotryptic carboxy-terminal fragments. *Blood 88:855.*
19. Moritz, T, VP Patel, and DA Williams. 1994. Bone marrow extracellular matrix molecules improve gene transfer into human hematopoietic cells via retroviral vectors. *J Clin Invest 93:1451.*
20. Hanenberg, H, K Hashino, H Konishi, RA Hock, I Kato, and DA Williams. 1997. Optimization of fibronectin-assisted retroviral gene transfer into human CD34+ hematopoietic cells. *Hum Gene Ther 8:2193.*
21. Hanenberg, H, XL Xiao, D Dilloo, K Hashino, I Kato, and DA Williams. 1996. Colocalization of retrovirus and target cells on specific fibronectin fragments increases genetic transduction of mammalian cells. *Nat Med 2:876.*
22. Sandhu, C, J Garbe, N Bhattacharya, J Daksis, CH Pan, P Yaswen, J Koh, JM Slingerland, and MR Stampfer. 1997. Transforming growth factor beta stabilizes p15INK4B protein, increases p15INK4B-

cdk4 complexes, and inhibits cyclin D1-cdk4 association in human mammary epithelial cells. *Mol Cell Biol 17:2458.*

23. Eaves, CJ, JD Cashman, RJ Kay, GJ Dougherty, T Otsuka, LA Gaboury, DE Hogge, PM Lansdorp, AC Eaves, and RK Humphries. 1991. Mechanisms that regulate the cell cycle status of very primitive hematopoietic cells in long-term human marrow cultures. II. Analysis of positive and negative regulators produced by stromal cells within the adherent layer. *Blood 78:110.*
24. Li, ML, AA Cardoso, P Sansilvestri, A Hatzfeld, EL Brown, H Sookdeo, JP Levesque, SC Clark, and J Hatzfeld. 1994. Additive effects of steel factor and antisense TGF-beta 1 oligodeoxynucleotide on CD34+ hematopoietic progenitor cells. *Leukemia 8:441.*
25. Cardoso, AA, ML Li, P Batard, A Hatzfeld, EL Brown, JP Levesque, H Sookdeo, B Panterne, P Sansilvestri, SC Clark, and et al. 1993. Release from quiescence of CD34+ CD38− human umbilical cord blood cells reveals their potentiality to engraft adults. *Proc Natl Acad Sci USA 90:8707.*
26. Hogge, DE, HJ Sutherland, PM Lansdorp, GL Phillips, and CJ Eaves. 1993. The elusive peripheral blood hemopoietic stem cell. *Semin Hematol 30:82.*
27. Rowley, SD, OM Colvin, and RK Stuart. 1985. Human multilineage progenitor cell sensitivity to 4-hydroperoxycyclophosphamide. *Exp Hematol 13:295.*
28. Polyak, K, JY Kato, MJ Solomon, CJ Sherr, J Massague, JM Roberts, and A Koff. 1994. p27Kip1, a cyclin-Cdk inhibitor, links transforming growth factor-beta and contact inhibition to cell cycle arrest. *Genes Dev 8:9.*
29. Kato, JY, M Matsuoka, K Polyak, J Massague, and CJ Sherr. 1994. Cyclic AMP-induced G1 phase arrest mediated by an inhibitor (p27Kip1) of cyclin-dependent kinase 4 activation. *Cell 79:487.*
30. Hengst, L, and SI Reed. 1996. Translational control of p27Kip1 accumulation during the cell cycle. *Science 271:1861.*
31. Thron, CD. 1997. Bistable biochemical switching and the control of the events of the cell cycle. *Oncogene 15:317.*
32. Reynisdottir, I, and J Massague. 1997. The subcellular locations of p15(Ink4b) and p27(Kip1) coordinate their inhibitory interactions with cdk4 and cdk2. *Genes Dev 11:492.*
33. St. Croix, B, VA Florenes, JW Rak, M Flanagan, N Bhattacharya, JM Slingerland, and RS Kerbel. 1996. Impact of the cyclin-dependent kinase inhibitor p27Kip1 on resistance of tumor cells to anticancer agents. *Nat Med 2:1204.*
34. Petzer, AL, DE Hogge, PM Landsdorp, DS Reid, and CJ Eaves. 1996. Self-renewal of primitive human hematopoietic cells (long-term-culture- initiating cells) in vitro and their expansion in defined medium. *Proc Natl Acad Sci USA 93:1470.*

GENE TRANSFER TO THE NERVOUS SYSTEM USING HIGH-TITER RETROVIRAL VECTOR

Koji Nanmoku,[1] Yasuyoshi Miyao,[2] Masakazu Tamura,[2]
Kazuyoshi Tamura,[2] Tadanori Yoshimatsu,[3] Keiji Shimizu,[2]
and Kazuhiro Ikenaka[1]

[1] Laboratory of Neural Information
National Institute for Physiological Sciences
[2] Department of Neurosurgery
Osaka University Medical School
[3] Institute for Biotechnology Research
Wakunaga Pharm. Co., Ltd.

ABSTRACT

Retroviral vector is the most commonly used vector for gene therapy. However, its use is limited because of the low titer. To overcome this problem, we improved the packaging cell line and concentrated the viral supernatant by the low-speed centrifugation and obtained retroviral titers up to 10^{12}cfu/ml. This high titer preparation enabled us to transduce foreign genes efficiently into the neural cells, which are generally resistant to the transduction. It was also possible to massively transduce RSV-glioma cells transplanted into the mouse brain *in vivo*. This study showed that high-titer retrovirus was able to overcome the limitation of infectivity to the nervous system both *in vitro* and *in vivo*.

INTRODUCTION

Retroviral vector is suitable for gene therapy because the transferred gene is stably integrated into the chromosome and is transmitted from generation to generation. In addition, there is little cytotoxicity against the target cells upon gene transfer. For these reasons, retroviral vectors are still the most frequently used vector for treatment of hereditary diseases, cancer and chronic diseases. However, the usual retroviral titers are too low to achieve sufficient therapeutic effects (usually in the range of 10^5–10^6cfu/ml). Therefore, to increase the titer of retroviral vector is crucial for them

Molecular Biology of Hematopoiesis 6, edited by Abraham *et al.*
Kluwer Academic/Plenum Publishers, New York, 1999.

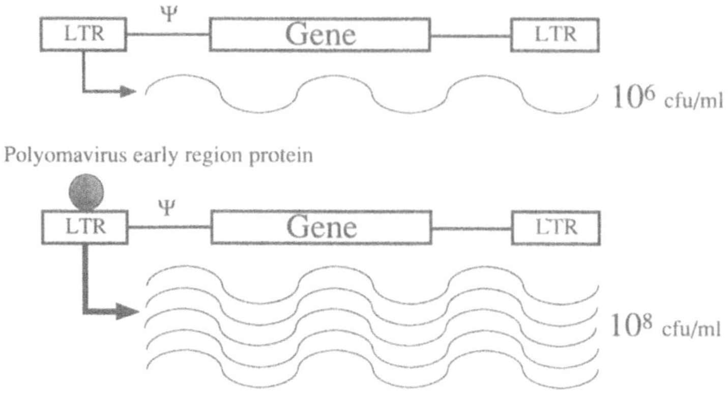

Figure 1. Mechanism underlying increased titer in ΨMP34 cells. Polyomavirus early region was introduced into the packaging cell line. The activity of the LTR promoter in the retroviral vector was up-regulated by the polyomavirus early region protein. Retroviral titer was obtained from our new packaging cell line increased 100 times than from previous one.

to be used successfully for gene therapy. Especially in the nervous system, it is very difficult to obtain even a reasonable transduction efficiency using the usual retroviral vector system. To overcome this limitation of usual retroviral vector system, we have previously reported improved retroviral vectors (Nakao et al., 1995), which contains the polyomavirus early region and the extended Psi region. We also improved retroviral packaging cell line, and produced ΨMP34 (Yoshimatsu et al., 1998) by stably introducing the polyomavirus early region into the usual packaging cell line Ψ2. The activity of the LTR promoter is stimulated at the transcriptional level by the polyomavirus early region protein(s), and this stimulation might have caused the increase in titer of recombinant retrovirus (Fig. 1).

In this study, supernatant of a retroviral producer cell, ΨMP34 SV/pIP201$^+$ (Yoshimatsu et al., 1998), was concentrated by the low-speed centrifugation (Bowles et al., 1996), resulting in the retroviral titer up to 10^{12}cfu/ml. Using this high-titer retrovirus, the *lacZ* gene was transferred at sufficient quantities not only to neural cells *in vitro* but also to the glioma cells *in vivo*.

MATERIALS AND METHODS

Cell Lines

Murine fibroblast NIH-3T3 was cultured in Dulbecco's modified Eagle's medium (DMEM) containing 10% fetal bovine serum (FBS). Oligodendrocyte-precursor cell line CG4 (Espinosa de los Monteros et al., 1994) provided by Monteros were cultured in GDM medium (1/1 (v/v) mixture of DMEM and Ham's F12 supplemented with 5 mg/l of insulin, 16.1 mg/l of putrescine, 50 mg/l of apo-transferrin, 4.6 mg/l of D-galactose, and 8 µg/l of sodium selenite) containing 2% FBS. Astrocyte-precursor cell line HB108-10 (Baba et al., 1997) was cultured in 1/1 (v/v) mixture of DMEM and Ham's F12 containing 10% FBS supplemented with 100 U/ml of interferon-γ and 500 µg/ml of G-418. Pheochromocytoma cell line PC12h (Hatanaka, 1981) was cultured in DMEM containing 5% FBS and 5% horse serum.

Dissociated Primary Culture of Embryonic Mouse Cerebra

Cerebral cortices from embryonic day 17 (E17) ICR mouse purchased from Japan SLC were finely minced and dissociated by pipetting after treatment with 0.25% trypsin and DNase I. After being separated from the debris by low-speed centrifugation (700 rpm, 1 min, Tomy LC06-SP), the cells were grown in 1/1 (v/v) mixture of DMEM and Ham's F12 containing 10% FBS for mixed glial culture on 4-well culture plates coated with poly-L-lysine.

Retroviral Producer Cell Culture and Preparation of High-Titer Retrovirus

The retroviral stable producer cell, ΨMP34 SV/pIP201$^+$, was established from the packaging cell line, ΨMP34, by introducing SV/pIP201$^+$ vector (Yoshimatsu et al., 1998) by LipofectAMINE™ Reagent (GIBCO BRL). ΨMP34 SV/pIP201$^+$ was maintained in 250 ml of DMEM supplemented with 10% FBS in 500 cm^2 triple flask (Nunc) at 32°C. The supernatant of ΨMP34 SV/pIP201+ at confluent was concentrated at 6,000 × g (8,000 rpm, BECKMAN J2-HS) for 16 hr at 4°C in 250 ml bottles (NALGEN) (Bowles et al., 1996). Viral pellet was resuspended in 1/100 volume of cell culture media or Hank's balanced salt solution (HBSS), resulting in a 100-fold concentration. The viral suspension was filtered through a 0.45-µm filter (MILLIPORE). When higher titer was needed, the same centrifugation was repeated in 1.5 ml Eppendorf tubes. The viral suspension was either used directly or stored at −80°C. High-titer retrovirus was stable at −80°C for more than 6 months (data not shown).

For virus titration, NIH-3T3 was seeded in 4-well plates at 1×10^4 cells/well, with serial dilutions of virus supernatant in the presence of 8 µg/ml of polybrene (Aldrich). For the recombinant retrovirus containing the *lacZ* gene, the cells were stained 48 hr later with X-Gal (Ikenaka et al., 1992), and the titer was estimated according to the following formula: Titer of retrovirus (cfu/ml) = number of X-Gal-positive cell/virus volume (ml) × 4 (replication factor of NIH-3T3) (Cepco, 1992).

Retrovirus Transduction into Cells *in vitro* and Glioma *in vivo*

NIH-3T3, CG4, HB108-10, and PC12h were seeded in 4-well plates at 1×10^4, 3×10^4, 3×10^4, and 5×10^4 cells/well, respectively. Cells were transduced with the viral suspension without polybrene for 48 hr. The culture media were not replaced because there was no cytotoxicity in the absence of polybrene. After infection, the transduced cells were detected by the X-Gal staining.

The RSV-glioma cells were transplanted as described previously (Tamura et al., 1997). 3 weeks later 3 µl of the retrovirus suspension was injected into the glioma.

Tissue Fixation and X-Gal Staining

Mice perfused transcardially with 4% paraformaldehyde, 0.2% glutaraldehyde, 0.02% Nonidet P-40 in phosphate-buffered saline (PBS). The brains were removed and sectioned transversely at 200-µm on a microslicer (DTK-3000; DOSAKA EM CO., LTD). Sections were postfixed with the perfusion solution for 30 min at 4°C and rinsed in PBS containing 1 mM $MgCl_2$ for 3 times every 60 min at 4°C. For X-Gal histochemistry, sections were incubated for 6 hr at 37°C in a solution containing

0.1% 5-bromo-4-chloro-3-indolyl-beta-D-galactopyranoside (X-Gal; Wako), 30 mM potassium ferrocyanide, 30 mM potassium ferricyanide, 0.01% deoxycholic acid sodium salt, 2% Nonidet P-40, and 1 mM $MgCl_2$ in PBS. After X-Gal staining, sections were washed with PBS twice, and soaked in PBS with 1 mM EDTA in order to stop the X-Gal reaction.

RESULTS

In vitro Transduction of Cells with Retrovirus

NIH-3T3, CG4, HB108-10, and PC12h were transduced with 10 µl of retroviral suspension containing usual retroviral titer (1.0×10^6 cfu/ml) of SV/pIP201$^+$ virus. Few X-Gal positive cells were observed in the neural cell lines, CG4, HB108-10, and PC12h in contrast with NIH-3T3 (Fig. 2).

We next used a higher-titer retrovirus from ΨMP34 SV/pIP201$^+$ (10 µl of 1.0×10^{10} cfu/ml) after concentrating once. The *lacZ* gene was transferred to all of the neural cell lines tested at sufficient quantities (Fig. 2).

Next, dissociated primary cultured brain cells were transduced with 10 µl of retrovirus containing 1.0×10^{10} cfu/ml SV/pIP201$^+$. Many cells in dissociated primary culture were positive for the X-Gal staining (Fig. 3A) and the percentage of the transduced cells was almost comparable to that of the bromodeoxyuridine-incorporated cells (Fig. 3B). This suggested that most of the proliferating cells had been transduced with the retrovirus. These proliferating cells were mostly astrocytes as shown by the GFAP staining (Fig. 3C) and a few oligodendrocyte as shown by the O4 staining (data not shown).

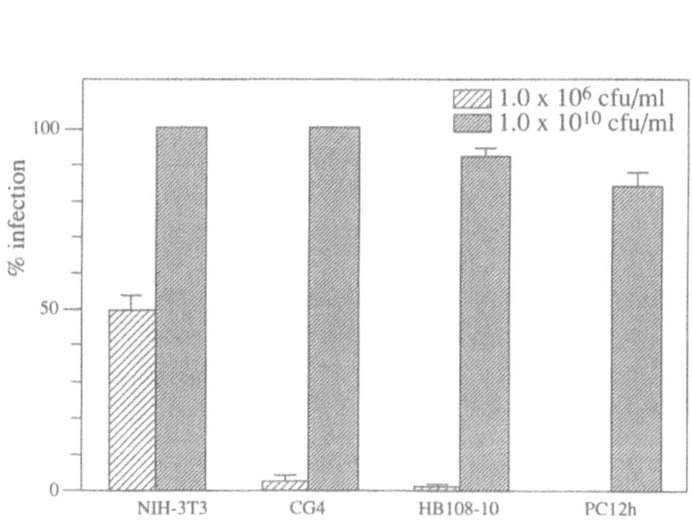

Figure 2. Efficient transduction of neural cell lines by high-titer retrovirus. NIH-3T3, CG4, HB-108-10, and PC12h were transduced with usual (1.0×10^6 cfu/ml) and high-titer (1.0×10^{10} cfu/ml) retrovirus. Using high-titer retrovirus, highly efficient (>80%) gene transfer into all cell lines was achieved.

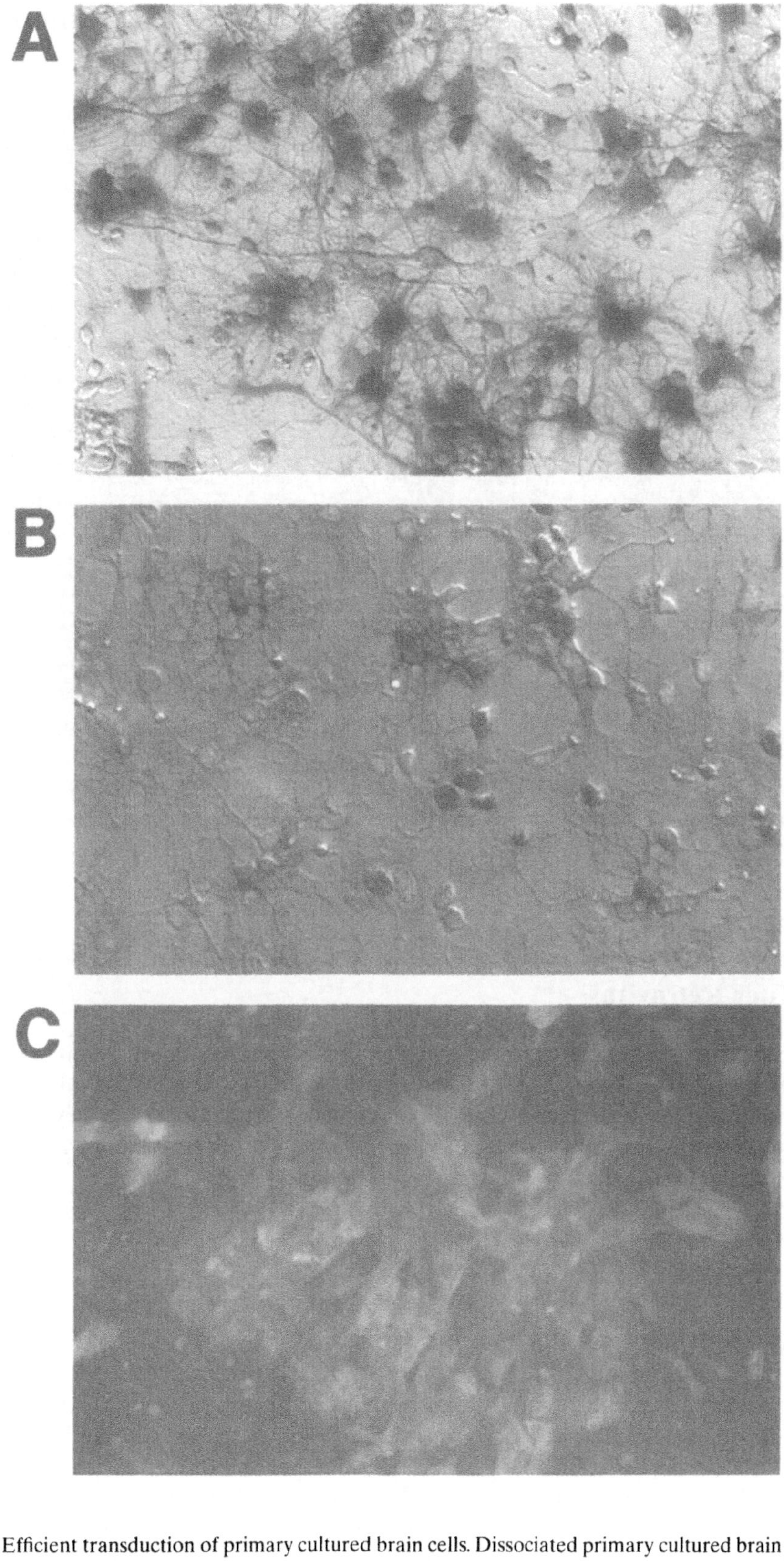

Figure 3. Efficient transduction of primary cultured brain cells. Dissociated primary cultured brain cells were transduced with the *lacZ* gene at sufficient quantities with high-titer retrovirus (A). The percentage of X-Gal positive cells was almost comparable to that of the bromodeoxyuridine-incorporated proliferating cells detected by the DAB staining (brown). Non-proliferating cells were shown by the counter staining with propidium iodide (red) (B). Cells detected by anti-β-Gal antibody (FITC) were mostly astrocytes as shown by the GFAP staining (texas red) (C).

Figure 4. Efficient transduction of glioma cells *in vivo*. *In vivo* gene transfer of RSV-glioma transplanted into mouse brain with high-titer retrovirus (1.0×10^{12} cfu/ml, 3 μl). A huge number of X-Gal positive cells were observed in RSV-glioma.

In vivo Gene Transfer into RSV-Glioma Cells in Mouse Brain with High-Titer Retrovirus

To show efficient gene transfer by the retroviral vector *in vivo*, we transduced the RSV-glioma cells transplanted into mouse brain with 3 μl of high-titer retrovirus (1.0×10^{12} cfu/ml after concentrating twice). Sections made from brains six days after transduction were stained with X-Gal. A huge number of glioma cells were transduced with the *lacZ* gene (Fig. 4).

DISCUSSION

Cells of neural origin tend to be resistant to the retroviral infection as shown in Fig. 2. It was not known whether the reason for their low transduction efficacy is due to the absence of receptor expression or the inefficient processing of transduction in these cells. Our present work clearly showed that it is the latter case and upon increasing the titer we can efficiently transduce neural cells.

We also showed that the glioma cells transplanted into the mouse brain can be transduced with a high efficacy *in vivo*. Because of the low titer of the retrovirus, virus-producing mouse fibroblast cells have been implanted into the glioma for the human gene therapy. This xenogenic transplantation might result in unfavorable side effect. On the other hand direct injection of retrovirus should be much safer. Our present paper shows that it is now possible to treat and possibly cure the brain tumors by direct retroviral injection.

REFERENCES

Baba H, Nakahira K, Morita N, Tanaka F, Akita H, Ikenaka K (1997). GFAP gene expression during development of astrocyte. Dev. Neurosci. 19, 49–57.

Bowles NE, Eisensmith RC, Mohuiddin R, Pyron M, Woo SLC (1996). A simple and efficient method for the concentration and purification of recombinant retrovirus for increased hepatocyte transduction *in vivo*. Hum. Gene Ther. 7, 1735–1742.

Cepco C (1992). Preparation of a specific retrovirus producer cell line. In Current Protocols in Molecular Biology, Supplement 17. FM Ausubel, R Brent, RE Kingston, DD Moore, JG Seidman, JA Smith, and K Struhl, eds. (John Wiley and Sons, New York) pp. 9.11.1–9.11.12.

Espinosa de los Monteros A, de Vellis J (1994). Brain-specific expression of the human transferrin gene: similar elements govern transcription in oligodendrocytes and neuronal cells. J. Biol. Chem. 269, 24504–24510.

Hatanaka H (1981). Nerve growth factor-mediated stimulation of tyrosine hydroxylase activity in a clonal rat pheochromocytoma cell line. Brain Res. 222, 225–233.

Ikenaka K, Nakahira K, Nakajima K, Fujimoto I, Kagawa T, Ogawa M, Mikoshiba K (1992). Detection of brain-specific gene expression in brain cells in primary culture: A novel promoter assay based on the use of a retrovirus vector. New Biol. 4, 53–60.

Nakao J, Yamada M, Kagawa T, Kim SU, Miyao Y, Shimizu K, Mikoshiba K, Ikenaka K (1995). Expression of proteolipid protein gene is directly associated with secretion of a factor influencing oligodendrocyte development. J. Neurochem. 64, 2396–2403.

Tamura M, Shimizu K, Yamada M, Miyao Y, Hayakawa T, Ikenaka K (1997). Targeted killing of migrating glioma cells by injection of HTK-modified glioma cells. Hum. Gene Ther. 8, 381–391.

Yoshimatsu T, Tamura M, Kuriyama S, Ikenaka K (1998). Improvement of retroviral packaging cell lines by introducing the polyomavirus early region. Hum. Gene THER. 9, 161–172.

HEMATOLOGICAL ASPECTS OF COMMON MARMOSET MONKEY TRANSPLANTED WITH AUTOLOGOUS MDR1 GENE TRANSDUCED PERIPHERAL BLOOD STEM CELLS

Kenzaburo Tani,# Hitoshi Hibino,# Hajime Sugiyama,#
Ming-Shiuan Wu,# Kiyoko Izawa,# Tsuyoshi Tanabe,# Hidenori Hase,#
Yukoh Nakazaki,# Hajime Ishii,* Shuzo Suzuki,* Yoshikuni Tanioka,*
Yoshikazu Sugimoto,** and Shigetaka Asano#

Department of Hematology/Oncology
The Institute of Medical Science
The University of Tokyo
Tokyo, Japan
* Central Institute for Experimental Animals
216 Kawasaki, Japan
** Cancer Chemotherapy Center
Japanese Foundation for Cancer Research
Tokyo, Japan

ABSTRACT

We have developed a small primate model system for human gene therapy using the common marmoset to evaluate the safety and efficacy of gene therapeutic approaches. We demonstrated that retroviral transduction of multidrug resistance gene (MDR1) into G-CSF-mobilized peripheral blood progenitor cells (PBPC) was improved by culturing PBPC with the mixed populations of bone marrow stroma cells

Corresponding author: Kenzaburo Tani, MD., PhD., Department of Hematology/Oncology, The Institute of Medical Science, The University of Tokyo, 4-6-1 Shirokanedai Minato-ku, Tokyo 108-8639, Japan, Phone: 81-3-5449-5542, Fax: 81-3-5449-5429, E-Mail: taniken@ims.u-tokyo.ac.jp

Molecular Biology of Hematopoiesis 6, edited by Abraham *et al.*
Kluwer Academic/Plenum Publishers, New York, 1999.

and viral producer cells. We performed autologous PBPC transplantation using MDR1-transduced PBPC in two marmosets. The transduction efficiency to PBPC was 5.9–6.7% and provirus DNA could be detected in peripheral blood granulocytes and lymphocytes by PCR method in all gene transduced marmoset maximally up to 412 days after transplantation. MDR1-transduced autologous PBPC were tranasplanted to irradiated common marmosets and reconstituted their own hematopoiesis. Provirus MDR1 DNA could be detected in peripheral blood granulocytes and lymphocytes by PCR method in all gene transduced marmosets. Replication competent retrovirus was not detected in all marmosets by $S^{+}L^{-}$ assay. Our preclinical stem/progenitor gene therapy system using marmoset is considered to be helpful for the development of human gene therapy.

INTRODUCTION

To investigate the feasibility of gene therapy, it is necessary to study the effectiveness and safety of each vector system. Preclinical studies using mice is considered to be insufficient because of various remarkable differences between two species. Therefore, the introduction of non-human primates is considered to be very critical to obtain more direct data for human in the course of various preclinical studies. Here in Japan, however, it is difficult to obtain and use large primates such as Macaca or Baboon for experiments because of much legal restriction. Based on this background, we have currently focused on common marmoset, a small new world primate (body weight: 200–400 g), because of the easiness to handle with and to propagate, and tried to establish new evaluation system on the effectiveness and safety of gene therapy.

Although there have been a number of studies on the efficacies of various human cytokines in large non-human primates, there have been very few such studies in New World monkeys including common marmoset. Also, little is known about the characteristics of marmoset hematopoietic stem and progenitor cells or about the effects of new reagents on marmoset hematopoiesis. Before the results on newly developed cytokines or gene therapy vectors can be extrapolated to humans, a systematic characterization of marmoset hematopoietic stem and progenitor cells is necessary. The physiological differences between humans and marmosets must also be elucidated.

Recent advances in the field of gene therapy have facilitated studying various intractable disorders including hematological disorders of adenosine deaminase deficiency, thalassemia, and leukemia / lymphoma. A number of successful gene transfers into long term reconstitutive hematopoietic stem cells in murine and canine models have been reported. In addition, several investigators have used monkey models for evaluating the safety and efficacy of retroviral vector systems. However, the transduction efficiency of retroviral vectors into the hematopoietic stem cells of primates, including humans, remains low compared with that of mice. To evaluate the *in vivo* efficacy of newly-developed vectors for human use, it is important to first establish the optimal conditions for gene transduction into hematopoietic stem cells using non-human primates both *in vitro* and *in vivo*.

In this study, we examined the effectiveness of human cytokines on marmoset bone marrow progenitor cells and thereafter examined the transduction efficiency of retroviral vector into marmoset peripheral blood progenitor cells. We also transplanted the MDR(multidrug resistant)1-gene transduced marmoset progenitor cells to marmosets and studied their hematological reconstitution *in vivo*.

MATERIALS AND METHODS

Animals and Cells

Common marmosets were bred in the Central Institute for Experimental Animals, Kawasaki, Japan. All experiments were performed according to the guidelines of The Institute of Medical Science, The University of Tokyo. The mononuclear cells (MNCs) fraction was isolated from femoral bone marrow cells using the Ficoll-Hypaque density gradient centrifuge method. Cells were frozen using a programmed freezer, and stored in liquid nitrogen until use.

G-CSF Mobilized Peripheral Blood Progenitor Cells

Marmosets were treated with human granulocyte colony-stimulating factor (G-CSF) (10μg/kg/day) subcutaneously for 5 days. Peripheral blood (500μl–1ml) was obtained from the femoral vein and the number of leukocytes, neutrophils, red cells, and platelets, and the hematocrit level and hemoglobin concentration were measured with an automatic blood cell analyzer (Sysmex K-1000, Toa-iyoudenshi, Kobe, Japan).

Progenitor Cell Assay

Erythropoietin (Epo) and G-CSF were provided by Chugai Pharmaceutical Co., Tokyo, Japan; granulocyte macrophage colony-stimulating factor (GM-CSF), by Hoechst Japan, Tokyo, Japan; stem cell factor (SCF) and IL-3, by Amgen, Thousand Oaks, CA; thrombopoietin (Tpo), by Kirin Brewery Co., Maebashi, Japan; and IL-6 and IL-11 by Ajinomoto Co., Kawasaki, Japan, and Genetic Institute, Cambridge, MA, respectively.

Bone marrow and peripheral blood MNCs were cultured under standard methylcellulose culture conditions. Briefly, MNCs were plated in 35mm-petri dishes in 1ml of α-minimal essential medium (α-MEM, GIBCO BRL, Grand Island, NY) containing 1.2% methylcellulose (Wako Chemical, Osaka, Japan), 30% fetal bovine serum (FBS, Hyclone, Logan, UT), 1% bovine serum albumin (BSA, Sigma, St. Louis, MO), and 10^{-4} M 2-mercaptoethanol (Wako Chemical) with human SCF (10ng/ml), IL-3 (10 ng/ml), Epo (2u/ml), and G-CSF (10ng/ml). To examine the colony forming effects of each cytokine, cells were incubated at 37°C for 7–14 days in 5% CO_2, in the presence of G-CSF or GM-CSF, and in 5% CO_2, 5% O_2, in the presence of Epo, IL-3, or SCF. Burst-forming unit-erythroid (BFU-E), colony-forming unit-granulocyte (CFU-G), colony-forming unit-macrophage (CFU-M), colony-forming unit-granulocyte / macrophage (CFU-GM), and colony-forming unit-granulocyte / erythroid / megakaryocyte / macrophage (CFU-Mix) were essentially enumerated following the morphological criteria for human colonies. For colony-forming unit-megakaryocyte (CFU-MK) assays, bone marrow MNCs were cultured in a methylcellulose mixture (1.2% methylcellulose, 1% BSA, 30% human plasma, and 10^{-4} M 2-mercaptoethanol). Culture dishes were incubated at 37°C in 5% CO_2 for 7–10 days. To confirm each colony component, several representative colonies were picked up and then stained by the standard May Grünwald Giemsa staining method. Our results demonstrated that the morphological criteria for human colonies were applicable to marmoset colonies.

Gene Transduction into Bone Marrow and Peripheral Blood Progenitor Cells

An amphotropic retroviral vector of HaMDR1, carrying the wild type MDR1 driven by Harvey murine sarcoma virus long terminal repeat, was produced from retroviral producer cells of PA317-HaMDR1. The producer cell lines of PA317-HaMDR1 were maintained in Dulbecco's modified Eagle's medium (Difco) with 10% FBS.

When the producer cells reached 80% confluence, the culture medium was renewed, and the supernatant was harvested after 24 hours. The fresh viral supernatant was then filtered through a 0.45 μm filter unit (Millipore, Bedford, MA) and used for experiments. Cells were cultured with or without allogeneic marmoset bone marrow stromal cells that were prepared in advance, and with the fresh viral supernatant containing 4 μg/ml of protamine sulfate (Sigma) and human cytokines (100 ng/ml SCF, 20 ng/ml IL-3, 80 ng/ml IL-6). The medium was renewed four times every 24 hours. For gene transduction by cocultivation, peripheral blood MNCs were cultured on feeder layers of viral producer cells on a mixed population of viral producer and preliminarily prepared allogeneic marmoset bone marrow stromal cells at a ratio of 1:1, in DMEM supplemented with 10% FBS. The culture medium was supplemented with 4 μg/ml protamine sulfate, with or without human cytokines (SCF, IL-3, IL-6) and was changed every 48 hrs. The culture was performed for 48 to 144 hrs. The feeder layers were preliminarily treated with 5 μg/ml mitomycin C (Sigma) for 2 hr before the cocultivation with bone marrow or peripheral blood MNCs. These cells were cultured in α-MEM with 10% FBS with or without cytokines for 48 hr after the cessation of infection, and then successively plated in methylcellulose culture medium as described above. The colony formations were assayed with or without 10 ng/ml vincristine (VCR, Sigma). The inhibitory effects of VCR on the colony formation of untransduced bone marrow and peripheral blood MNCs were examined using various concentrations of VCR. The results showed that 10 ng/ml of VCR is a high enough concentration to completely inhibit colony formation. To confirm the correspondence between VCR-resistance and the integrated MDR1 gene, some representative VCR-resistant colonies were picked up and provirus was detected by PCR analysis, as described below. The transduction efficiency among various treatment conditions were compared using a Student's t-test.

Transplantation of Peripheral Blood Progenitor Cells

Three two-years-old marmosets (two males and one female), bred in the Central Institute for Experimental Animals, were used. Autologous PBPCs were obtained from G-CSF-treated marmosets at two months and one month before PBPCT. Briefly, 10 μg/kg/day of G-CSF was subcutaneously administered to individual marmosets for 5 days. Peripheral blood was then collected from each marmoset once a day for 5 consecutive days after the final G-CSF administration. MNCs were isolated by Ficoll-Hypaque *centrifugation*, frozen using a programmed freezer, and stored in liquid nitrogen until use. Two-thirds of the peripheral blood MNCs obtained from two marmosets were transduced with the MDR1 gene by the cocultivation method as described above. Briefly, cells were cultured on feeder cells consisting of the same number of retroviral producer and stromal cells in the presence of human cytokines (IL-3, IL-6, SCF) and 4 μg/ml protamine sulfate for 48 hrs from day –2 to day 0. On day 0, all three marmosets received two fractionated doses (2.5 Gy each) of total body irradiation from ^{60}Co sources on days −1 and 0. The marmosets were then infused intravenously with the PBPCs, with or without gene transduction via the femoral vein, within 2 hrs after the second irradiation. One control male marmoset received only untransduced peripheral blood MNC, while the other two marmosets (1 male and 1 female) received the mixed cellular populations of transduced and untransduced peripheral blood MNCs. All marmosets were orally administered 1.5×10^4 u/body polymyxin B (Pfizer, Tokyo, Japan) and 0.5 mg/body fluconazole (Pfizer) daily from day −5, and a daily intramuscular injection of 5 mg/body ampicillin (Meiji-Seika, Tokyo, Japan) from day 0 until the peripheral neutrophil count exceeded 500/μl. Peripheral blood was collected in tubes containing EDTA every two or three days from day 0 to day 30, and thereafter, every one or two weeks. Complete blood cell counts were measured for each blood sample. The S^+L^- assay was performed on peripheral blood samples obtained on day 46 or day 58, to rule out the appearance of replication-competent retrovirus (RCR) using PG4 cells, as described elsewhere.

PCR Analysis of Transduced Gene in Transplanted Animals

The provirus of HaMDR1 was detected by PCR analysis of genomic DNA extracted from peripheral blood cells or bone marrow cells in transplanted marmosets. DNA was extracted from whole peripheral blood cells, granulocytes or lymphocytes which were isolated by Ficoll Hipaque gradient centrifugation using a DNA extraction kit (Microtubogen, Invitrogen, San Diego, CA). A 453-base-pair DNA sequence specific for the HaMDR1 provirus was amplified using the sense-strand primer (F12) TGTTTCAGAATGGCAGAGTCA and the antisense-strand primer (R24) AAACAGAAAGGCGAGCAGAGA. The F12 primer is specific for the MDR1 region and the R24 primer is specific for the viral long terminal repeat region in the Ha vector. PCR amplification was carried out under the following conditions: initial denaturation at 94°C (5 min), followed by 35 cycles of 94°C denaturation (45 sec), 60°C annealing (45 sec), and 72°C extension (1 min). The PCR products were electrophoresed on a 3% NuSieve (FMC, Rockland, ME) agarose gel and transferred to nylon membranes (HybondN, Amersham). The membranes were then cross-linked by UV. Hybridization was performed overnight at 65°C in Church's phosphate buffer containing 1 mM EDTA and 7% SDS using a ^{32}P-labeled MDR1 cDNA as a probe. The amplified products were visualized by autoradiography. Using a Bio-Imaging Analyzer (FUJIx, Tokyo), the copy numbers of these samples were densitometrically compared with those of a standard sample, obtained from serial dilutions of DNA from the packaging cell line (PA317-HaMDR1) containing a single copy per cell of the provirus.

RESULTS

Marmoset Bone Marrow Progenitor Cell Responses to Human Cytokines

Human IL-3 and GM-CSF stimulated CFU-GM formation dose-dependently and IL-3 also stimulated BFU-E formation. The formation of CFU-G and CFU-E was stimulated dose-dependently by G-CSF and Epo, respectively. The marmoset and human BMPCs responded similarly to human cytokines. The number of CFU-MK increased dose-dependently by treatment with Tpo (data not shown). The effects of single

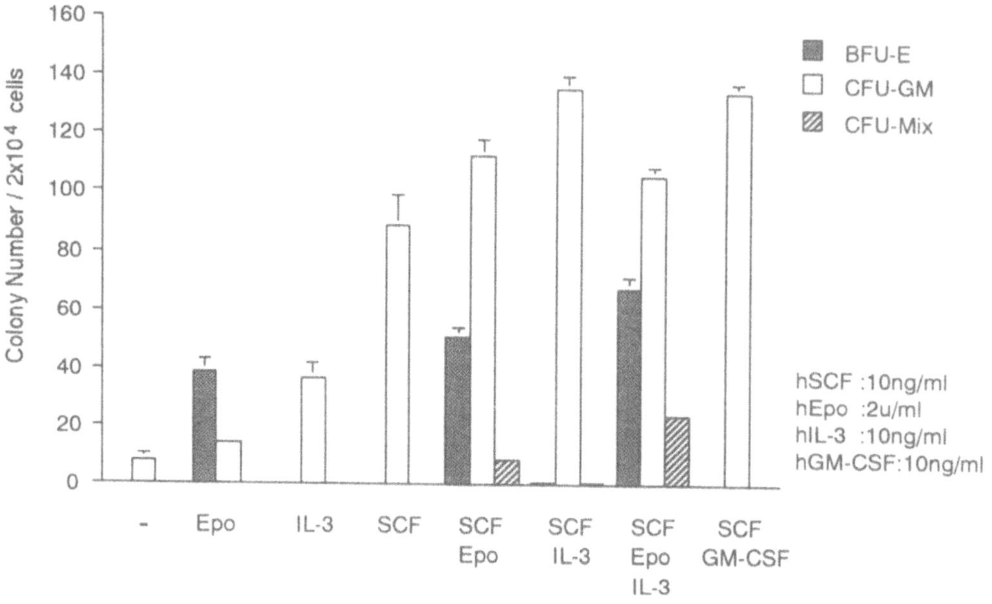

Figure 1. Effects of human cytokines on colony formation of marmoset bone marrow progenitor cells. The effects of EPO, IL-3, SCF, and GM-CSF on the colony formations of BFU-E, CFU-GM, and CFU-Mix were studied. Combination of SCF, EPO, and IL-3 stimulated the CFU-Mix formation best.

cytokines or their combination stimulated the colony formations of marmoset bone marrow BFU-E, CFU-GM, and CFU-Mix as shown in Fig. 1. Particularly, the combination of SCF, EPO, and IL-3 stimulated the CFU-Mix colony formation as seen in human bone marrow progenitor cells. Also, the combination of human IL-6 and IL-3 or IL-3 and IL-11 stimulated the colony formation of CFU-MK (Fig. 2).

G-CSF Mobilized Peripheral Blood Progenitor Cells

10 μg/kg/day of human G-CSF induced the peripheral leukocytosis and mobilized CFU-C (CFU-GM, BFU-E, and CFU-Mix) to the peripheral circulation better than 5 μg/kg/day of human G-CSF. Both the leukocyte and neutrophil count increased immediately after the administration of G-CSF and remained high during treatment and the numbers of peripheral blood CFU-C were highest on days 8–10 (Fig. 3). These results demonstrated that human G-CSF had potent mobilizing activity at 10 μg/kg/day, suggesting the availability of G-CSF-mobilized PBPC for autologous hematopoietic progenitor cell transplantation in marmosets.

Efficiency of Transduction into Bone Marrow and Peripheral Blood Progenitor Cells

The inhibitory effects of VCR on bone marrow colony formation were examined using three different marmosets. Ten ng/ml of VCR completely inhibited the colony formation and we chose this concentration for further experiments. As the provirus was

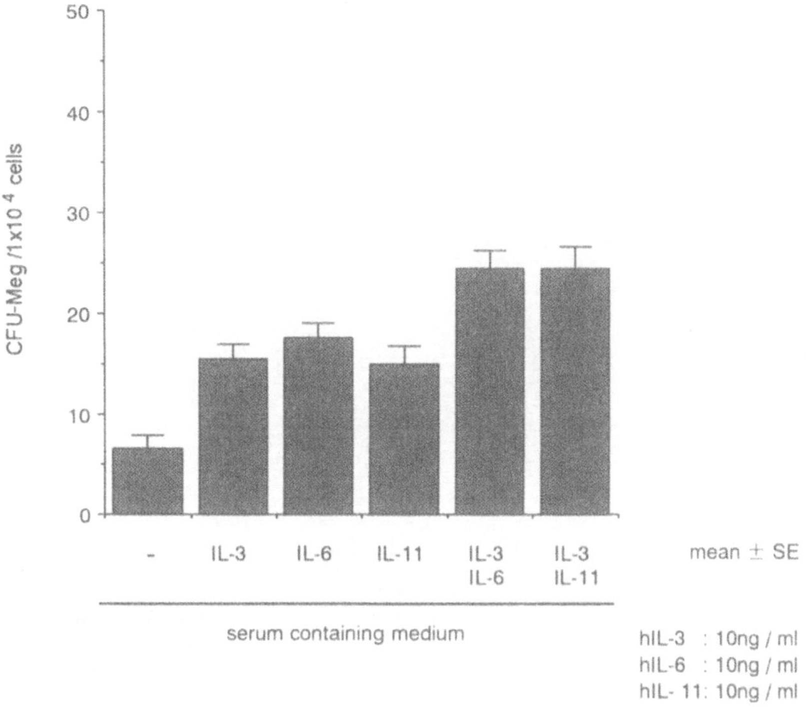

Figure 2. Effects of human cytokines on colony formation of marmoset bone marrow megakaryocyte colony formation (CFU-Meg). The effects of IL-3, IL-6, and IL-11 on the colony formation of CFU-Meg was studied. The combination of IL-3 plus IL-6 or IL-3 plus IL-11 stimulated the CFU-Meg formation best.

demonstrated by PCR analysis in all VCR-resistant colonies randomly analyzed in some of the representative gene transfer experiments, the appearance of VCR-resistant colonies were used for estimating gene transduction efficiency. With the supernatant transduction method used, 1 to 6% of CFU-GM colonies became VCR-resistant. The transduction efficiency increased when the cells were transduced with the viral supernatant in the presence of bone marrow stromal cells, although it was not statistically significant (Fig. 4). The frequency of VCR-resistant colonies in the ordinary coculture systems using viral producer cells was low in spite of an extended culture period (Fig. 4). In an attempt to increase the transduction efficiency, we mixed marmoset bone marrow stromal cells with retroviral producer cells. The frequency of VCR-resistant colonies was significantly higher when target cells were cocultured with viral producer cells and bone marrow stromal cells in each transduction period. Under these conditions, human cytokines significantly enhanced the transduction efficiency when the periods of transduction were 96 hrs and 144 hrs. The coculture of retroviral producer cells and stromal cells was also effective for the transduction of PBPCs. Transduction of the MDR1 gene was performed using this system with human cytokines for 48 and 96 hrs. The frequency of VCR-resistant colonies was about 15% regardless of the transduction period (Fig. 5). These results indicated that cocultivation using mixed populations of retroviral producer cells and stromal cells enhanced the efficiency of transduction of the MDR1 gene into marmoset BMPCs and PBPCs.

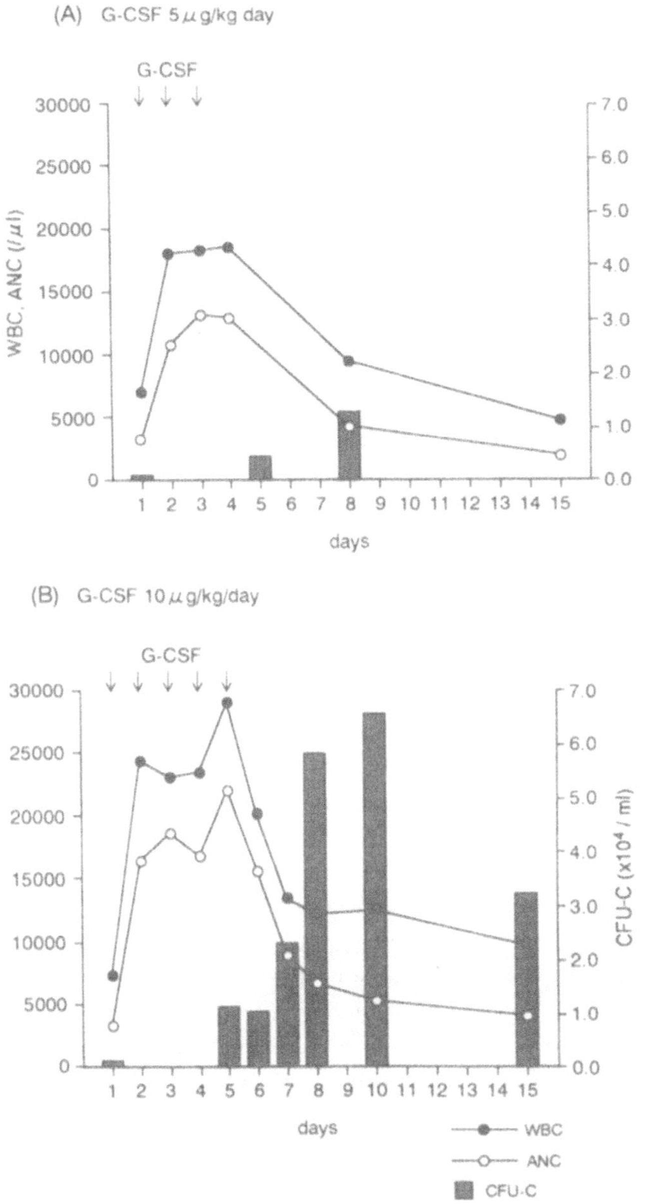

Figure 3. Mobilization of marmoset progenitor cells to peripheral blood with human G-CSF. The effects of human G-CSF to marmoset peripheral blood leukocytes (●) and neutrophils (○) number and progenitor cells were studied. CFU-C numbers (colony numbers of CFU-GM, BFU-E, and CFU-Mix were included) in the peripheral blood were also determined as shown in the text. 5μg/kg/day of human G-CSF for 3 days (A) and 10μg/kg/day of human G-CSF for 5 days (B) was administered. The maximal numbers of CFU-GM, BFU-E, and CFU-Mix in peripheral blood were obtained on day 8, and were approximately 21-, 25-, and 13-fold, respectively, of the value obtained in the unstimulated state. The data shown were obtained from four animals and are expressed as the mean ± SEM.

Transplantation of Gene Transduced or Untransduced Peripheral Blood Progenitor Cells

The total number of autologous peripheral blood MNCs and CFU-GM harvested and frozen ranged from 2.6×10^7 to 4.5×10^7 cells and from 2.0×10^4 to 1.4×10^5 cells, respectively. The actual numbers of peripheral blood MNCs and CFU-GM transplanted to three marmosets are summarized in Table 1. One control animal was transplanted with untransduced CFU-GM. Three marmosets were transplanted with both trans-

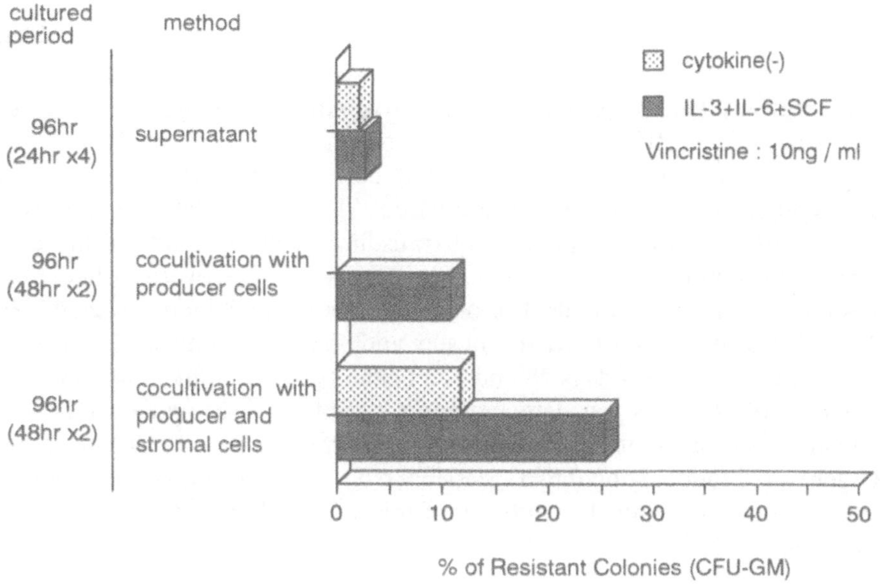

Figure 4. Retroviral transduction of human MDR1 gene into marmoset bone marrow CFU-GMs. The transduction efficiencies were studied with or without IL-3, IL-6, and SCF. Marmoset bone marrow mononuclear cells were transduced by retroviral supernatant alone, cocultivated with retroviral producer cells or cocultivated with both retroviral producer cells and marmoset allogeneic bone marrow stromal cells for 96 hrs. Percentage of vincristine resistant CFU-GM number is shown.

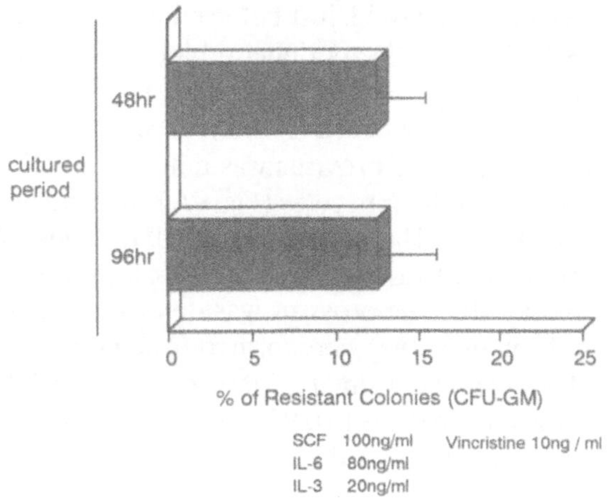

Figure 5. Retroviral transduction of human MDR1 gene into marmoset G-CSF mobilized peripheral blood CFU-GMs. The transduction efficiencies were studied with IL-3, IL-6, and SCF. Marmoset peripheral blood mononuclear cells were cocultivated with both retroviral producer cells and marmoset allogeneic bone marrow stromal cells for 48 or 96 hrs. Percentage of vincristine resistant CFU-GM number is shown.

Table 1. The actual numbers of G-CSF-mobilized MDR1 gene-transduced or -untransduced peripheral blood mononuclear cells (MNC) or CFU-GMs transplanted to three marmosets

	MNC ($\times 10^7$/kg)			CFU-GM ($\times 10^4$/kg)			days of Neu	days of Plt >
No.	transduced	untransduced	total	transduced	untransduced	total	>500/µl	5×10^4/µl
1	—	7.8	7.8	—	5.9	5.9	35	25
2	4.0	3.0	7.0	1.5 (6.7%)*	2.5	4.0 (2.5%)**	25	25
3	5.7	2.8	8.5	2.5 (5.9%)*	3.3	5.8 (2.1%)**	39	39

*Transduction efficiency (percentage of VCR resistant colony number).

**Transduction efficiency in transplanted PBPC that were calculated with transduced and untransduced CFU-GM.

duced and untransduced CFU-GM. *In vitro* progenitor cell assays demonstrated that 6.7% and 5.9% of the PBPC from Nos.2 and 3 marmosets became VCR-resistant after MDR1 gene transduction, respectively. The numbers of CFU-GM were reduced during the culture procedure for retroviral transduction. Also the calculated final transduction efficiency into CFU-GM by a mixture of transduced cells and untransduced cells in these three marmosets was 2.5% and 2.1%, respectively. Recovery of hematopoiesis was observed in all three animals. The peripheral neutrophil count reached 500/μl on days 25 and 39 in gene-transduced marmosets, and on day 35 in the control. The platelet count reached 5×10^4/μl on days 28 and 39 in gene-transduced marmosets and on day 28 in the control. No RCR was detected in the serum obtained on day 46 or day 58 in any animal by S^+L^- assay using PG4 cells. One marmoset (No.3) transduced with the MDR1 gene died suddenly on day 63, possibly from mechanically-induced ileus due to ingestion of vinyl fragments. The autopsy of this animal showed full recovery of bone marrow hematopoiesis.

Detection of Transduced Gene in Transplanted Animals

DNA from peripheral blood cells or bone marrow cells of the transplanted marmosets was extracted and examined for the presence of the transduced MDR1 gene. DNA from peripheral blood cells of marmoset No.3 was examined for provirus-specific sequences by PCR followed by Southern blotting analysis. Peripheral blood cells from day 28 of transplantation showed the appearance of provirus DNA which was also detected on day 53, just before the animal's death (Fig. 6A). The estimated percentage of MDR1 positive peripheral blood cells ranged from 0.2% to 0.4% from the densitometric comparison with standard DNA. The provirus was detected in marmoset No.2 transplanted with MDR1-transduced PBPCs on days 35 and 81 (Fig. 6A). In this animal on day 142, the provirus was detected in DNA isolated from peripheral blood granulocytes and lymphocytes (Fig. 6A), and the provirus was also detected on day 412 (data not shown). The percentage of MDR1 positive granulocytes and lymphocytes ranged from 0.2% to 1.0%. Bone marrow aspiration was performed for this marmoset (No.2) on day 105. The provirus was detected in bone marrow MNCs and in pooled colonies (10 colonies) that were formed from bone marrow MNCs on day 105 in marmoset No.2 (Fig. 6B). The estimated percentage of MDR1 positive granulocytes and lymphocytes ranged from 0.2% to 0.9%. These results indicated that MDR1 gene-transduced PBPCs were engrafted to the bone marrow and reconstituted the marmosets' hematopoiesis *in vivo*.

DISCUSSION

Our current studies have demonstrated that a small primate, the common marmoset, is also useful as a primate model for various hematological research related to the development of new cytokines and new vectors for gene therapy. In Japan, they are available at one third or half of the price of macaques because we have large domestic marmoset colonies in Japan. Marmosets become sexually active at the age of eighteen months, usually bear 2 to 4 offspring per year with a 140–150 day gestation period and they require half or one third the size of cage that macaques do. Also they can be taken care by untrained researchers, which make this model even more useful. As there

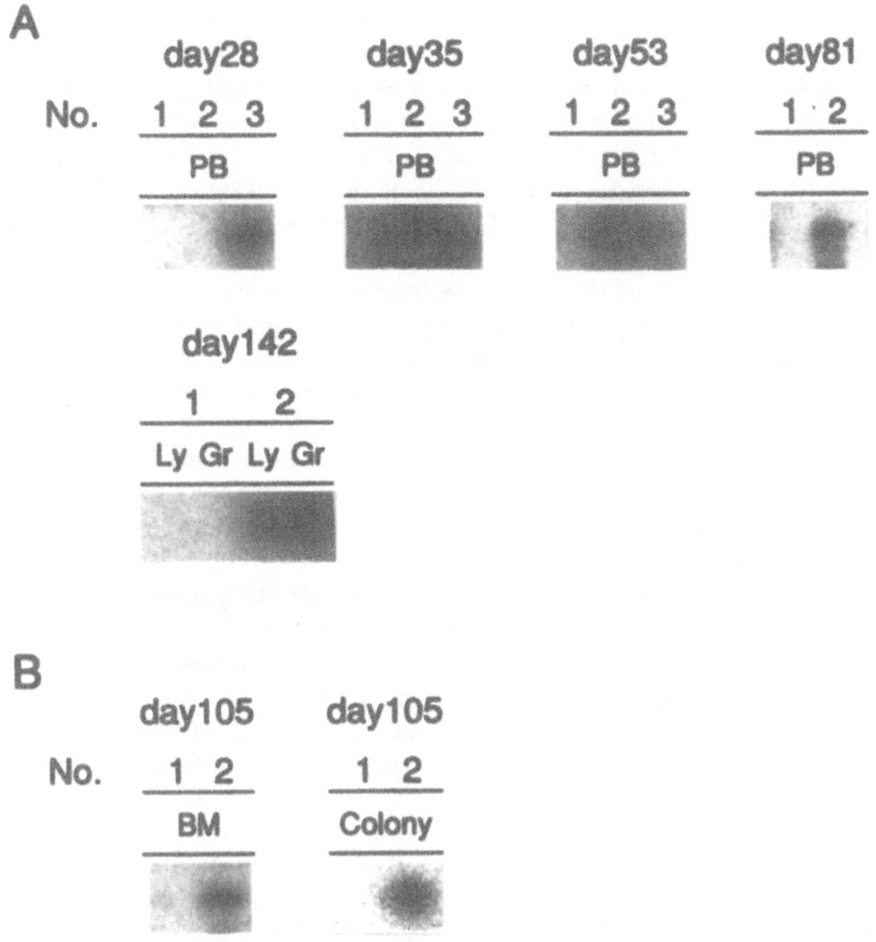

Figure 6. Detection of provirus DNA in peripheral blood cells or bone marrow cells of the transplanted marmosets. On the indicated days, DNA was obtained from each sample and analyzed for the presence of provirus DNA, followed by Southern blot analysis. (A) Peripheral blood cells from days 35, 53, and 81 of transplantation showed the presence of provirus DNA in marmoset No.2. The provirus was detected in marmoset No.3 in peripheral blood cells on days 28, 35, and 53. In No.2 animal, the provirus was detected both in peripheral blood granulocytes and lymphocytes on day 142. (B) A bone marrow aspiration was performed for No.2 marmoset on day 105 ; the provirus was detected in bone marrow MNCs and in pooled colonies (10 colonies). PB: Peripheral whole blood cells, Gr: Granulocytes, Ly: Lymphocytes, BM: Bone marrow MNCs

have been no reported systematic information about the hematological characteristics of marmosets, we analyzed them to clarify the advantages and disadvantages of their use as a primate model for preclinical studies.

The transduction of genes into hematopoietic cells by retroviral vectors was mostly induced by cytokines. Gene transduction efficiency has been reported to be enhanced by the combined treatment of cytokines such as SCF, IL-3, and IL-6 to obtain an efficient proliferation of hematopoietic cells by retroviral vectors. Our results first demonstrated that both the marmoset and human BMPCs responded dose-dependently and in similar fashion to the human IL-3, GM-CSF, G-CSF, and Epo. Both the human IL-3 and GM-CSF are known to react in a species-specific manner. Human SCF, also known to be highly species-specific, stimulated the colony formation of marmoset BMPCs in the presence of other human cytokines. Our present findings strongly support the utility of New World monkeys for hematological investigations, in contrast with the use of large animals such as canine or swine which require the cloning of their own SCF to stimulate hematopoietic cells for gene transduction. In addition, our *in vivo* results demonstrated that the hematopoietic progenitor cells were effectively mobilized to peripheral circulation by G-CSF, and restored hematopoiesis in marmosets as previously reported in humans.

Although hematopoietic stem cells, including PBPCs, are considered to be important targets for gene therapy, the efficiency of transduction into hematopoietic stem cells in primates and humans is still poorer than that in mice. Previous reports demonstrated that the transduction efficiency by retroviral vectors into hematopoietic cells was enhanced by the viral producer cells, bone marrow stromal cells or extracellular matrix molecules in the presence of viral supernatant. Here we designed a novel transduction system to make use of the both advantages of viral producer and stromal cells. Our present results demonstrated that this system was potent for gene transduction to both BMPCs and PBPCs.

We performed PBPCT in marmosets using MDR1 gene-transduced PBPCs. Hematological recovery after PBPCT was observed in all animals. Provirus DNA was detected in DNA from granulocytes and lymphocytes in marmoset No.2 up to day 412 (data not shown). The results suggested that the MDR1 gene was transduced to reconstitutive hematopoietic progenitor cells and proved that our novel transduction method was also effective to enhance the transduction efficiency. However, the hematological recovery in this study seemed to be slower than previous reports using other nonhuman primates even with PBPCT. We presumed that the number of transplanted progenitor cells were few or primitive progenitor cells differentiated during the transduction procedure. Thus, further study is required to clarify the relationship between the number of transplanted PBPCs and hematological recovery in marmosets.

Previous reports suggest that the MDR1 gene can be used as a selectable marker gene *in vivo* because cells transduced with the MDR1 gene can be enriched by treatment with chemotherapeutic agents, including paclitaxel, in mice. Also, the second gene that was coexpressed with the MDR1 gene, using a bicistronic vector, enhanced its expression after the transduced cells were treated with VCR. It has been reported that paclitaxel-induced-leukopenia was completely prevented in MDR1 gene transduced mice. In our preliminary study, docetaxel-induced-neutropenia was not completely prevented, but the percentage of MDR1-positive neutrophils was slightly increased after treatment with docetaxel although the percentage was still very low (data not shown). These levels were not considered to be high enough to prevent the neutropenia induced by docetaxel treatment. Long term follow-up of these marmosets will give us more information on the in vivo kinetics of MDR1 gene transduced marmoset hematopoietic progenitor cells. All of the in vivo studiese were performed without detecting any replication competent retrovirus by S+L− assay.

In summary, we demonstrated that the common marmoset is a useful alternative to large primates in the course of preclinical studies to develop new human cytokine and gene therapy vectors. This animal model can be used for non-human primate studies even in institutes without any special primate centers or any special trainers, and will potentially contribute to the future development of human cytokine and gene therapy study.

REFERENCE

Hibino, H, Tani, K, Ikebuchi, K, Wu, M-H, Sugiyama, H, Nakazaki, Y, Tanabe, T, Takahashi, S, Tojo, A, Suzuki, S, Tanioka, Y, Sugimoto, Y, Nakahata, T, and Asano, S. The Common Marmoset as a Target Preclinical Primate Model for Cytokine and Gene Therapy Studies Blood 1999 93:2839–2848, 1999.

EFFECT OF RETROVIRUS AND ADENOVIRUS MEDIATED INTERFERON GENE TRANSFER ON HEMATOPOIETIC PROGENITOR CELL GROWTH

N. G. Abraham, E. Feldman, and T. Ahmed

Department of Pharmacology
Gene Therapy Program
New York Medical College
Valhalla, New York 10595, USA

ABSTRACT

Gene therapy following gene transfer into hematopoietic cells (CD34+) is now being investigated for several genetic disorders. We have applied a similar approach using interferon gene transfer for treatment of chronic myelogenous leukemia (CML). IFN-α produces a hematologic and cytogenetic response in CML with a survival advantage for cytogenetic responders. Therefore, we examined the effect of transient overexpression of IFN-α using the adenovirus gene transfer approach. The ability of the adenovirus (Adv)-IFN-α gene construct to transfect normal and CML stem cells, CD34+, was examined. The peripheral blood mononuclear fraction from patients with CML treated with G-CSF or GM-CSF/G-CSF was enriched in CD34+ cells. Adv-cytomegalovirus (pCMV) promoter driven IFN-α at multiple doses was assessed to transfect highly purified CD34+ cells in the presence of matrix protein and in co-cultures with the stromal adherent cell layer. The use of cytokines enhances Adv-mediated IFN-α gene transfer into stem cells. Southern blot analysis demonstrated that the Adv-pCMV-IFN-α construct and IFN-α were expressed in cultured CD34+. Transient expression of the IFN-α gene did not suppress proliferation of CD34+ progenitors, CFU-Meg, BFU-E or CFU-CM growth. Reverse transcriptase/polymerase chain reaction (RT/PCR) analysis of RNA from CFU-GM progenitor cells demonstrated transient IFN-α mRNA expression in CD34+ cells. Immunoassay of IFN-α shows that IFN-α suggests that selective expression of IFN-α may be beneficial for CML therapy. We also report on the establishment of novel conditions which permit high efficiency

Molecular Biology of Hematopoiesis 6, edited by Abraham *et al.*
Kluwer Academic/Plenum Publishers, New York, 1999.

of the retrovirus IFN-α gene into CD34+ cells. Stem cells transfected with retrovirus were infused intravenously to irradiated mice and spleen colony forming units were evaluated for IFN-α marked clones by Southern blot analysis. These studies demonstrate that the overexpression of IFN-α gene can be used suppression of K562 cell proliferation and suggest its possible use as gene therapy of CML. For permanent transfection, we constructed two IFNα cDNA-containing retroviral vectors LSN-IFNα was constructed by cloning human IFNα cDNA at the downstream of the 5′ long terminal repeat (LTR) promoter of the retroviral vector LXSN. Retrovirus-containing supernatants from PA317 cell line was used to infect NIH3T3 fibroblasts and K562 cells. The IFNα gene expression in transfected cells was confirmed by Northern Blot, RT-PCR, radioimmunoassay (RIA), and biological assay. Retroviral mediated IFN-α gene transfer infection results showed that cell proliferation and cell viability were significantly suppressed in LSN-IFNαa-infected K562 cells as compared to LXSN-infected K562. Furthermore, we found that retroviral mediated IFN-α causes upregulation of adhesion molecules such as VLA-4.

INTRODUCTION

IFN-α have become important anti-proliferative agents used in cancer therapy. Several clinical trials have demonstrated that administration of IFN-α can be an effective treatment for CML in chronic phases. In those trials, cytogenetic remission can be induced and long-term survival improved compared with other standard therapies.[1,2] A number of *in vitro* effects of IFN-α on CML cells have been reported, such as direct inhibition of CML BM progenitor growth[1,3] and restoration of the adherence to stroma.[4,5] In addition, it has been found that IFN-α regulates the paracrine release of growth factors from the stroma by inhibiting the production of stimulating cytokines such as GM-CSF, G-CSF, IL-1b as well as by increasing the production of inhibiting cytokines.[6,7] It is thought that inhibitory cytokines and cellular immune surveillance play an important role in the control of expansion of the leukemic clone in CML.

Recently, this laboratory has utilized retrovirus and adenovirus vectors as vehicles for gene transfer into hematopoietic stem cells *in vitro* and *in vivo*.[8–12] Furthermore, we have shown that stromal adherence cells permit high efficiency of retrovirus-mediated gene transfer of adenosine deaminase (ADA) into bone marrow hematopoietic progenitor cells.[9–10] Successful transplantation of the genes were further documented by their appearance in hematopoietic clones of cells from animals many months after reconstitution.[10] Over the past six years, investigators have begun to apply retroviral gene transfer technology directed at hematopoietic stem cells in preliminary human clinical trails.[13–14] Recombinant adenoviruses are the second most commonly used viral vector for gene delivery,[15–19] thus minimizing the risk of permanently altering the cellular genotype or of insertional mutagenesis. Adenoviruses are preferentially used when the transient expression of a transgene is sufficient to achieve the therapeutic goals.[14–15]

The current paper investigates the development of gene transfer vectors and assess the effect of retrovirus and adenovirus-mediated IFN-α gene transfer into $CD34^+$ cells on proliferation and differentiation. Successful gene transfer into $CD34^+$ cells monitored by Southern blot analysis and Northern blot analysis of genomic DNA and total RNA from the $CD34^+$ cells. Adherent stromal cell layers will be tested to determine their effect or gene transfer efficiency of retrovirus mediated IFN-α gene transfer. The results will allow us to determine the feasibility of developing a gene therapy protocol.

MATERIAL AND METHODS

The replication-deficient adenovirus vector endoding IFN-a (AdCMV-IFN-α) and retrovirus (LXSN-IFN-α) was constructed in our laboratory, as described previously.[18,20] Bone Marrow Transduction for Retrovirus and Adenovirus was described.[13,20]

Human IFN-α Analysis

The presence of IFN-α in $CD34^+$ cell media was determined using a commercially available quantitative ELISA assay.[18]

Southern and Northern Blot Analysis of Transfected K562 and $CD34^+$ Colony Assay

Genomic DNA was extracted from 5×10^7 $CD34^+$ and K562 cells and digested with BamHI, Hind III, and Xho I. The digested DNA was electrophoresed on a agarose gel and transferred to nirocellular filters as described.[9–10]

RESULTS AND DISCUSSION

A schematic representation of the adenoviral vector containing the IFN-a gene is represented in Fig. 1. We examined the effect of multiple doses of adenovirus containing CMV IFN-α (AdCMV-IFN-α) were used to transfect human $CD34^+$ cells.

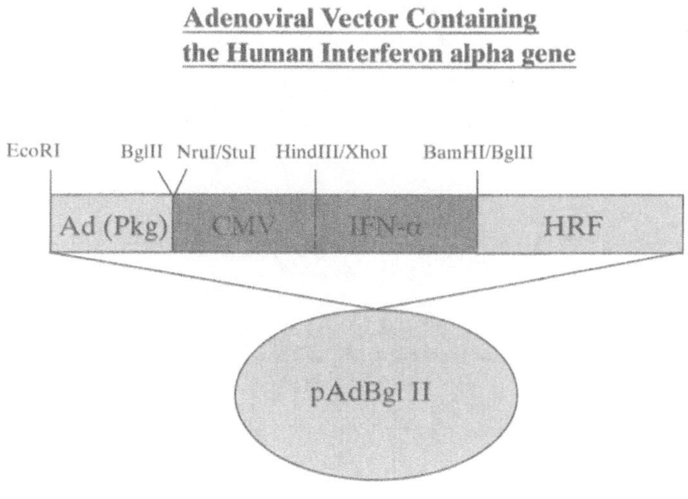

Figure 1. A schematic diagram of the adenoviral vector containing the IFN-α gene.

Table 1. Secretion of IFN-α by Ad CMV-IFN-α transduced human $CD34^+$ cells

Condition	Interferon levels (UNIT/ml)
Control $CD34^+$	0.3 ± 0.1
Transduced $CD34^+$ AdCMV-IFN-α	4.1 ± 0.6
Transduced $CD34^+$ AdCMV-HO-1	0.3 + 0.1

Concentration of released IFN-α in U/10^6 cell/24 hr. Value below assay sensitivity limit were recorded as <0.3 μ/control $CD34^+$ cells/24 hr. Hemeoxygenase gene (HO-1) was used as a control to IFN-α gene

As seen on Table 1, AdCMV-IFN-α caused elevation of IFN-α protein in the condition medium suggesting that the IFN-α gene is expressed into $CD34^+$ cells resulting in the release of IFN-α into the medium (4.1 unit/ml). On the other hand, AdCMV-LACZ on control cells do not express significant levels of IFN-α, the value listed for these two cells represent only about 0.3 unit/ml.

Concentration of released IFN-α in U/10^6 cell/24hr. Value below assay sensitivity limit were recorded as <0.3μ/control $CD34^+$ cells/24hr. Hemeoxygenase gene (HO-1) was used as a control to IFN-α gene

Effect of AdCMV-IFN-α on $CD34^+$ Cells

$CD34^+$ cells were plated in Iscoves Modified Medium (IMDM) at 2×10^6 cells well and incubated with AdCMV-IFN-α 5,10,50,100 pfu/$CD34^+$ cell 5% CO_2 and then washed and used for assessment of mRNA and clonal efficiency. IFN-α mRNA was measured by RT/PCR in $CD34^+$ cells and/or CFU-GM clones. To evaluate the effect of AdCMV-IFN-α on CFU-GM growth we compared the number of CFU-GM clones in infected and non-infected cells. Results shows that there was no significant difference in CFU-GM growth of infected vs control cells for any of the exposure times. Although colony numbers were slightly less for the 24hr cell groups as compared with 4 or 8hr, maximal gene transfer occurred with the 24hr exposure cells (Fig. 2).

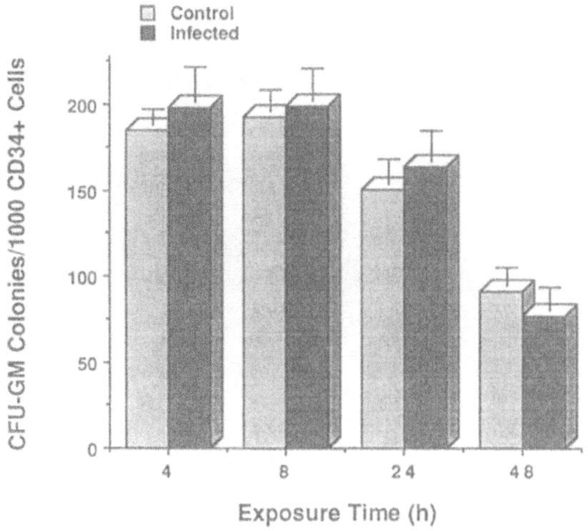

Figure 2. CFU-GM growth by CD34+ cells exposed to AdCMV-IFN-α for different times (CFU-GM) colony assay was performed as described.[18]

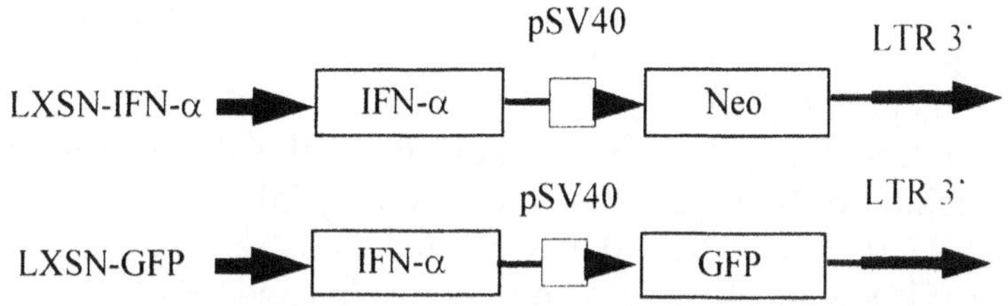

Figure 3. Schematic diagram of retroviral vector LSN-IFN-α, in which (neo r) and GFP used as marker for selection.

Similar results were obtained for BFU-E and CFU-Meg growth. We estimated that the gene transfer efficiency into CD34⁺ cells was 27–30%. Restriction endonuclease analysis of DNA extracted from CFU-GM clones in CD34⁺ transfected cells with AdCMV- IFN-α gene was performed. However, due to the limitaiton of transient expression and immune response, adenoviral vectors seem to be ineligible for clinical CML gene therapy projects.

In this regard, we have been developing a gene transfer strategy in which human IFNα genes were expressed in K562 cells and CD34+ hematopoietic cells via retrovirus-mediated gene transfer techniques (Fig. 3).

Retrovirus Mediated IFN-α Gene-induced K562 Cell Growth Inhibition

To demonstate that the secreted-hIFNα was biologically active and could inhibit K562 cell growth, we assayed the cell proliferation and cell viability after infection with the supernatants of PA317, PA317/LXSN, and PA317/LSN-IFNα. Figure 4 showed that

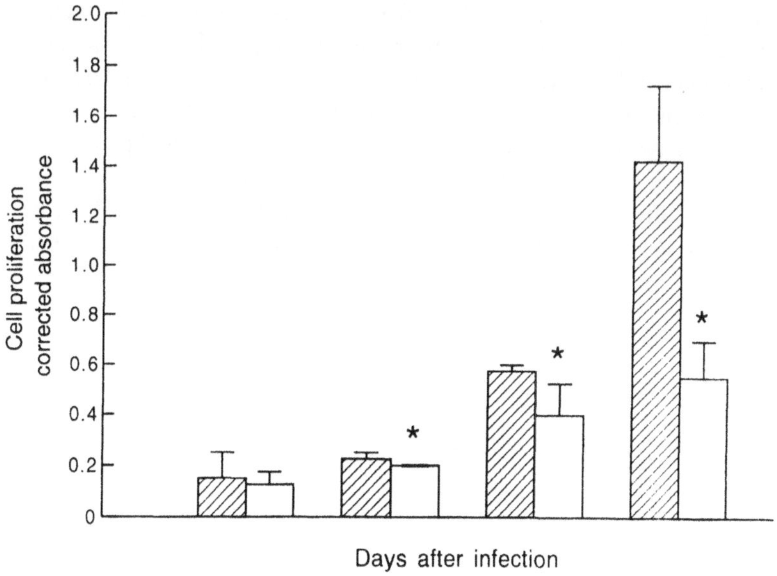

Figure 4. Cell proliferation of K562 cells after infection with the supernatants PA317/LXSN (hatched bar) and PA317/LSN-IFNα (open bar). Experiments were performed as described in materials and methods. At the indicated time points, absorbency values were read at 490 nm after adding the 96 aqueous reagent for 2 hr. Standard deviation was calculated from 6 individual assay values. Corrected absorbance values equal to the samples absorbance minus absorbance of medium.

the K562 cell proliferation was effectively suppressed by the supernatant of PA317/LSN-IFNα, with a 42.6% decrease of absorbance values at day 4 compared with control K562 cells. No significant difference was observed between the non-infected K562 cells and K562 cells infected with the supernatant of PA317/LXSN. Furthermore, we found the partial inhibition of K562 cell growth followed by infection with the supernatant of PA317/LSN-IFNα as compared to non-infected or LXSN infected K562 cells.

To test if IFNα could affect retroviral production and result in promoter attenuation or extinction, we transfected and expressed IFNα cDNA in several cell lines (PA317, PG13, NIH3T3, and K562) with retroviral vectors LSN-IFNα and/or LNC-IFNα in which IFNα cDNA was regulated by 5'LTR promoter and CMV promoter, respectively. Following G418 selection, individual cell clones were isolated, and measured for IFNα expression and retroviral production in the culture medium. In LSN-IFNα-transduced PA317 or PG13, and LNC-IFNα-transduced PA317 cell clones, retroviral titers reached 7.1×10^5 ($4 - 11 \times 10^5$, n = 10), 5.7×10^5 ($2 - 9 \times 10^5$, n = 8), and 6.6×10^5 ($3 - 8 \times 10^5$, n = 10) cfu/ml, respectively. No significant difference was found when compared with viral titers 7.4×10^5 (n = 6), 5.9×10^5(n = 6), and 6.5×10^5 (n = 6) in the control retroviral vector LXSN-transdued PA317 or PG13, LNCX-transduced PA317 cell clones, respectively. There was also no correlation between the IFNα production and viral titers.[21]

Retrovirus Mediated IFN-α-induced Adhesion Molecules

To test if gene transfer of IFNα could affect adhesion molecule expression in IFNα gene-transduced K562 cells, we measured several adhesion molecule production (VLA-4, Mac-1) in bulk-selected K562/LSN-IFNα, clone K562/LSN-IFNα, K562/LXSN, and K562 cells using FACS analysis results showed that IFN-α transfer increases adhesion molecules such as VLA-4, and Mac-1 (Fig. 5).

A previous study found that IFNα gene-transduced K562 cells would lose tumorigenicity in nude mice.[19] Another report showed that transduction of the IFN-α gene could induce MHC-1 expression on a subset of K562 cells (15.5%).[22] As downregulation of adhesion molecul expression is well known to play a critical role in the premature release of malignant Pho-positive progenitors into the circulation.[23–25]

The upregulation of VLA-4 and ICAM-1 can increase adhesion of CML progenitors to stroma through a direct effect on the progenitors.[25–27] Further, inhibition of

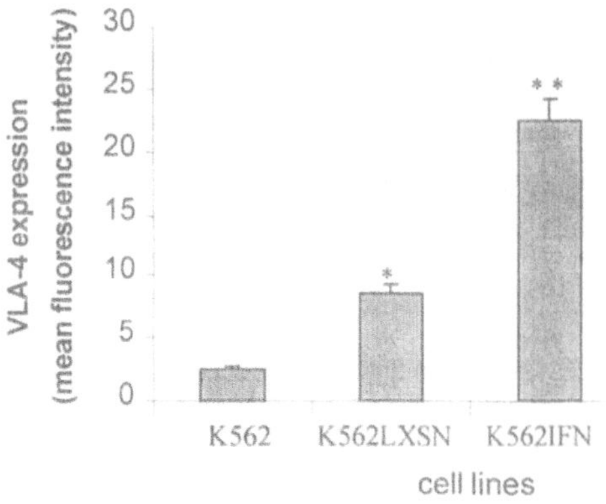

Figure 5. VLA-4 expression in K562 cell lines transduced with retroviral vectors LSN-IFN-α or LXSN. VLA-4 adhesion molecules were measured as described (14).

BCR-ABL expression with either antisense oligodeoxynucleotide or administration of IFN-α can restore β1 integrin-mediated adhesion and proliferation inhibition in CML hematopoietic progenitors.[27] As alterations in normal adhesion behavior are considered to be a pathogenetic factor in CML, the restoration of adhesion in CML progenitors can block the release of premature progenitors into the circulation. Further studies on adhesion molecule interaction of CML progenitors may give us useful insights into the pathogenesis of leukemia. The mechanism by which retrovirus mediated IFN-α upregulate adhesion molecules remains to be investigated.

REFERENCES

1. Italian Cooperative Study Group on Chronic Myeloid Leukemia. Interferon alfa-2α as compared with conventional chemotherapy for the treatment of chronic myeloid leukemia. N Engl J Med 30:820, 1994.
2. Allan NC, Richards SM, Shepard PCA: UK Medical Research Council randomised. Multicentre trial of interferon-α for chronic myeloid leukemia: Improved survival irrespective of cytogenetic response. Lancet 345:1392–7, 1995.
3. Fialkow PJ, Jacobson RJ, Papayannopoulou T: Chronic myelocytic leukemia: Clonal origin in a stem cell common to the granulocyte, erythrocyte, platelet, and monocyte/macrophage. Am J Med 63:125, 1972.
4. Martin PJ, Najfield V, Hansen JA, Penfold JK, Fialkow PJ: Involvement of the B-lymphoid system in chronic myelogenous leukemia. Nature 287:49, 1980.
5. Nitta M, Kato Y, Strife A, Wachter M, Fired J, Perez A, Jhanwar S, Duigou-Ostenfdorf R, Changanti RSK, Clarkson B: Incidence of involvement of the B and T lymphocyte lineages in chronic myelogenous leukemia. Blood 66:1053, 1985.
6. McGlave PB, Bartsch G, Anasetti C, Ash R, Beatty P, Gajewski J, Kernan NA: Unrelated donor bone marrow transplantation therapy for chronic myelogenous leukemia: Initial experience of the National Marrow Donor Program (NMDP). Blood 81:543, 1993.
7. Synder DS, McGlave PB: Treatment of chronic myelogenous leukemia with bone marrow transplantation. Hematol Oncol Clin North Am. 4:5357, 1990.
8. Harrison J, Levere RD, Abraham NG: The hematopoietic stromal microenvironment promotes retrovirus-mediated gene transfer into hematopoietic stem cells. Blood 80:400a, 1992.
9. Chertkov JL, Jiang S, Lutton JD, Harrison J, Preti R, Levere RD, Abraham NG: The hematopoietic stromal microenvironment promotes retrovirus-mediated gene transfer into hematopoietic stem cells. Stem Cells 11:161–175, 1993.
10. Cook PC, Jiang S, Chertkov JL, Fan Y, Levere RD, Abraham NG: Transfer of human AdA gene into murine hematopoietic stem cells: Sequential study of CFU-S from bone marrow of living mice and the requirement of the microenvironment. Acta Haematologica 96:57–63, 1996.
11. Abraham NG, da Silva J-L, Lavrovsky Y, Stoltz RA, Kappas A, Dunn MW, Laniado-Schwartzman M: Adenovirus-mediated heme oxygenase-1 (HO-1) gene transfer into rabbit ocular tissues. Invest. Ophthalmol. Vis. Sci. 36:2202–2210, 1995.
12. Abraham NG, da Silva J-L, Dunn MW, Kigasawa K, Shibahara S: Retinal pigmented epithelial cells (RPE) based gene therapy against hemoglobin toxicity. Int. J. Molec. Med., 1:657–663, 1998.
13. Dunbar CE, Fox M, O'Shaughnessey J, Doren D, Emmons RVB, Soma T, Yu JM, Carter C, Seller S, Hines K, Cowan K, Young NS, Nienhuis AW: Gene transfer to hematopoietic progenitor and stem cells: progress and problems. Hematology Branch, National Heart, Lung, and Blood Institute, National Institute, National Institutes of Health, Bethesda, MD USA.
14. Mulligan RC: The basic science of gene therapy. Science 260:926–932, 1994.
15. Hanania EG, Kavanagh J, Hortogagyi G, Giles RE, Champlin R, Deisseroth AB: Recent advance in the application of gene therapy to human disease. Am J Med 99:537–552, 1995.
16. Christal RG, Mcelvaney NG, Rosenfeld MA, Chu C, Mastrangeli A, Hay JG, Brody SL, Jaffe H, Eissa NT, Danel C: Administration of an adenovirus containing the human CFTR cDNA to the respiratory tract of individuals with cystic fibrosis. Nature Genet 8:42–51, 1994.
17. Ahmed T, Lutton JD, Feldman E, Tani K, Asano S, Abraham NG. Gene transfer of alpha interferon into hematopoietic stem cells. Leukemia Res., 22:119–124, 1998.

18. Feldman E, Ahmed T, Lutton JD, Farley T, Tani K, Freund F, Asano S, Abraham NG: Adenovirus mediated alpha interferon (IFN-α) gene transfer into CD34+ cells and CML mononuclear cells. Stem Cells 15:386–395, 1997.
19. Geng Y, Yu D, Blatt LM, Taylor MW: Tumor suppressor activity of the human concensus type I interferon gene. Cytokines Mol. Ther. 1:289, 1995.
20. Abraham NG: Retrovirus and adenovirus mediated interferon gene transfer into stem cells of normal and CML. Ann. Hematol. 74(I):20, 1997.
21. Yang L, Feldman E, Quan S, Wagener F, Farley T: Abraham NG, Ahmed T: Retroviral-mediated gene transfer of human interferon alpha efficiently inhibits K562 cell growth and modulates adhesion molecule expression. Blood Vol 9, suppl. abstract 1046 (1998).
22. Austruy E, Bagnis C, Carbuccia N, Maroc C, Birg F, Dubreuil P, Mannoni P, Chabannon C: A defective retroviral vector encoding human interferon-α-2 can transduce human leukemic cell lines. Cancer Gene Therapy 5:247, 1998.
23. Dowding CR, Gordon MY, Goldman JM: Primitive progenitor cells in the blood of patients with chronic granulocyte leukemia. Int. J Cell Cloning 4:331, 1986.
24. Eaves AC, Gashman JD, Gaboury LA, Kalousek DK, Eaves CJ: Unregulated proliferation of primitive chroic myelogenous leukemia progenitors in the presence of normal marrow adherent cells. Proc Natl Acad Sci USA. 83:5306, 1986.
25. Verfaillie CM, McCarthy JB, McGlave PB: Mechanisms underlying abnormal trafficking of malignant progenitors in chronic myelogenous leukemia. J Clin Invest 90:1232, 1992.
26. Oostendorp RAJ, Reisbach G, Spitzer E, Thalmeier K, Dineman H, Mergenthaler HG, Dormer P: VLA-4 and VCAM-1 are the principal adhesion molecules involved in the interaction between blast colony-forming cells and bone marrow stromal cells. Br J Haemat 91:275, 1995.
27. Bhatia R, Wayner EA, McGlave PB, Verfaillie CM: Interferon α restors normal adhesion of chronic myelogenous leukemia hematopoietic progenitors to bone marrow sroma by correcting impaired β1 integrin receptor function. J Clin Invest 94:384, 1994.

PROSTAGLANDIN'S SUPPRESSING EFFECT ON BONE MARROW ERYTHROPOIESIS IN RATS

Vania Kalaidjieva

Dept. of Physiology
Faculty of Medicine
Thracian University
Stara Zagora 6000
Bulgaria

ABSTRACT

In order to study the role of prostaglandins on hematopoietic recovery after suppression, we examined erythropoiesis in bone marrow erythroblastic islands in hypertransfused rats, injected with cyclooxygenase inhibitor indomethacin (4 mg/kg/5 d). The study included estimation of plasma erythropoietin level (by RIA), peripheral blood and morphological analysis of the erythroblastic islands (EI) and related to them indices of erythropoiesis during a 10-day's period. We observed a depressed EI development after hypertransfusion which was potentated after indomethacin treatment—an increased number of circulating reticulocytes, of EI/femur, an enhanced reconstruction in involuted EI and a reduced number of maturing EI ($p < 0.01$). Signs of activation of central EI' macrophages were expressed after prostaglandin's inhibition—rate of their repeated participation in EI reconstruction, rate of CFU-E differentiation into erythroblasts in EI were found elevated on days 3–5 ($p < 0.01$). Our study favours the concept about the suppressing effect of prostaglandins on erythropoiesis within bone marrow erythroblastic islands.

INTRODUCTION

Studies in animals have convincingly shown that prostaglandins (PG) are involved both in erythropoietin (EPO) synthesis, and in the generation of red cell mass in bone

Molecular Biology of Hematopoiesis 6, edited by Abraham *et al.*
Kluwer Academic/Plenum Publishers, New York, 1999.

marrow.[8,10,12] Experimental evidence suggests that prostaglandins, particularly those of the E-series, are inhibitors of hematopoietic progenitor cells.[7,13] Administration of drugs, suppressing PG production, was shown to stimulate hematopoiesis in normal mice.[7,9] Pre- or post-irradiation administration of PG inhibitors was found to be effective in enhancing hematopoietic recovery in irradiated mice.[4,7] Although prostaglandins are well known as modulators of erythropoiesis, their precise action is still a subject of controversy.

The aim of the study was to explore PG's involvement in the evolution of erythropoiesis in rat bone marrow EI following hypertransfusion. The erythropoietic response of the organism to changes in PG production may be evaluated by studying the reticulocytes release into circulation and the formation of erythroblastic islands (EI) in bone marrow.[2,11,14,15]

MATERIALS AND METHODS

We used 58 male Wistar rats (110–130g) with free access to food and water throughout the experiment. Induction of absolute erythrocytosis was done by intravenous hypertransfusion with 2.5ml/100g b.w. of a 70% suspension of washed, packed, isogenous red cells on two consecutive days. PG inhibition was performed by injecting animals with indomethacin 4mg/kg s.c./3 days (Sigma, USA) simultaneously with the erythrocyte transfusion. We compared erythropoietic development in only transfused rats (group I, $n = 27$) with erythropoiesis in indomethacin-pretreated animals (group II, $n = 26$). Blood was obtained by caval inferior vein under light ether anesthesia on days 1, 3, 5, 7, and 10 after procedures. We evaluated the parameters listed below.

Peripheral blood analysis: Hematocrits and absolute reticulocytes were determined by standard procedures. Radioimmunoassay was applied for quantitative evaluation of plasma EPO level (INCSTAR, Epo-trac, USA).

Bone marrow examination: We estimated the absolute count of erythroblastic islands/femur as well as the classification of EI by morphological study.[16] In the hematopoietic tissue, there are EI of the I, II, and III classes of maturity, having respectively up to 8, from 9 up to 16 and above 16 nuclei-containing erythroid cells in their composition. In addition, there are involutive EI (class IV)—nuclei containing erythroid cells, which are at late stages of differentiation, not able to divide and reconstructing EI (class V)—non-dividing erythroid cells, and cells at early stages of differentiation. We analyzed the following indices, based on the study of EI: rate of CFU-E differentiation into erythroblasts in EI, i.e. I class + V reconstructive class = A2; rate of repeated participation of macrophages in new EI reconstruction, i.e. EI reconstructive / EI involutive = A5.[17]

Statistical analysis was carried out using one-way analysis of variance (ANOVA). P value differences of less than 0.05 were considered significant. Linear regression analysis was performed for all parameters examined.

RESULTS

In the hypertransfused recipient rats the values of parameters explored were compared to those of the intact animals. The hematocrits increased up to 0.64 ± 0.04 after transfusion of 2.5ml/100g b.w. of 70% red blood cells and were maintained at over 0.55 during the monitored period ($p < 0.01$ vs. untreated rats). Reticulocytes grad-

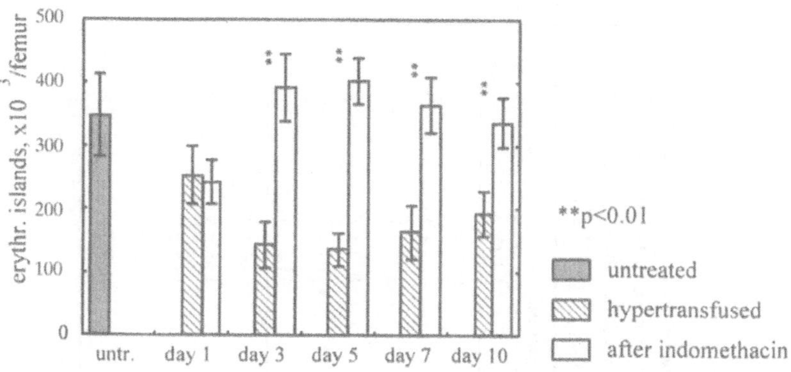

Figure 1. Bone marrow erythroblastic islands (EI) (×10³/femur)—absolute count per femur, in rats after i.v. transfusion with 2.5 ml/100 g b.w. of a 70% suspension of washed, packed, isogenous red cells/2 days and with concomitant indomethacin injection 4 mg/kg/3days s.c.; mean values ± SD.

ually decreased in number and disappeared from the peripheral blood on the seventh day. Plasma EPO level was found reduced to 13.5 ± 3.4 mU/ml vs. 24.6 ± 5.3 mU/ml ($p < 0.01$) on the first day after transfusion, accompanied by a significant attenuation of bone marrow EI formation—253.2 ± 46.3 × 10^3/femur vs. 347.6 ± 65.2 × 10^3/femur during the following week ($p < 0.01$) (Fig. 1; Table 1). The study of EI' classes of maturity showed a persistent reduction of younger islands (classes I and II) and the reconstructive ones (class V) and the related indices of erythropoiesis A2 and A5 (Figs. 2, 3; Table 1).

In indomethacin-pretreated animals EPO values changed insignificantly, the reduction of reticulocyte count showed a tendency to a rapid restoration after the 3rd day, the total number of EI increased on days 3 and 5 (391.5 ± 53.2 and 402.1 ± 36.4 × 10^3/femur vs. 142.6 ± 36.2 and 135.6 ± 25.6 × 10^3/femur in only transfused rats ($p < 0.01$) (Fig. 1; Table 1). The analysis of EI classes of maturity in those animals showed a reduction of class IV (involutive) even on day 1 and an enhanced reconstruction within EI (class V) on days 3–5 ($p < 0.01$) vs. the corresponding islands in only hypertransfused

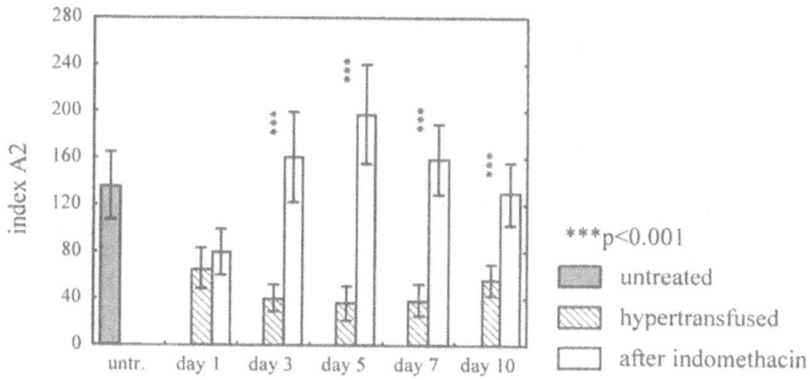

Figure 2. Rate of CFU-E differentiation into erythroblasts in EI, i.e. I class + V reconstructive class = A2 (×10³/femur); in rats after i.v. transfusion with 2.5 ml/100 g b.w. of a 70% suspension of washed, packed, isogenous red cells/2 days and with concomitant indomethacin injection 4 mg/kg/3days s.c.; mean values ± SD.

Table 1. Hematopoietic parameters of peripheral blood: hematocrit (Hct), reticulocytes (Ret), plasma erythropoietin (EPO), and bone marrow erythroblastic islands (EI)—classes of maturity (I-V) ($\times 10^3$/femur) in rats after i.v. transfusion with 2.5 ml/100 g b.w. of a 70% suspension of washed, packed, isogenous red cells/2 days and with concomitant indomethacin injection 4 mg/kg/3days s.c.; mean values ± SD; statistical significance versus untreated rats *p < 0.05, **p < 0.01, ***p < 0.001; statistical significance versus only hypertransfused rats °p < 0.05, °°p < 0.01, °°°p < 0.001

		Hct [l/l]	Ret [$\times 10^9$/l]	EPO [mU/ml]	Bone marrow EI—classes of maturity				
					EI I	EI II	EI III	EI IV	EI V
Untreated		0.42 ± 0.03	272 ± 53	24.6 ± 5.3	37.7 ± 5	28.4 ± 6	97.3 ± 16	85.8 ± 21	98.3 ± 22
Hypertransfused	Day 1	**0.64 ± 0.04	**86 ± 21	**13.5 ± 3.4	***13.5 ± 5	**19.5 ± 4	100.8 ± 23	67.2 ± 13	**52.2 ± 11
	Day 3	**0.62 ± 0.02	***23 ± 6	16.4 ± 3.2	***7.0 ± 3	***8.7 ± 3	**51.3 ± 15	**42.2 ± 8	***33.4 ± 7
	Day 5	**0.58 ± 0.04	***5 ± 2	10.2 ± 4.2	***7.2 ± 3	***7.8 ± 4	**51.8 ± 14	***40.2 ± 8	***28.6 ± 9
	Day 7	**0.55 ± 0.03	***0 ± 0	14.2 ± 5.2	***13.5 ± 4	***11.5 ± 3	*61.7 ± 16	**51.3 ± 14	***25.2 ± 8
	Day 10	*0.47 ± 0.02	***7 ± 3	19.3 ± 4.4	**19.3 ± 5	**10.4 ± 4	*70.6 ± 17	*55.4 ± 12	***36.7 ± 7
Indomethacin	Day 1	°°0.55 ± 0.03	°110 ± 32	°17.3 ± 4.2	15.3 ± 4	17.3 ± 5	92.2 ± 18	53.4 ± 10	64.2 ± 14
	Day 3	°0.52 ± 0.02	°°°175 ± 22	20.5 ± 5.5	°°20.3 ± 5	°°°28.2 ± 6	°°°160.2 ± 28	42.3 ± 9	°°140.5 ± 32
	Day 5	°0.47 ± 0.03	°°°155 ± 33	°19.4 ± 6.2	°°°27.4 ± 6	°°°20.4 ± 6	°°°148.5 ± 23	35.6 ± 9	°°170.2 ± 36
	Day 7	°0.45 ± 0.04	°°°132 ± 41	°21.5 ± 7.3	°°28.5 ± 6	°°22.5 ± 8	°°135.2 ± 26	48.3 ± 9	°°°130.4 ± 22
	Day 10	0.43 ± 0.03	°°°162 ± 28	23.2 ± 5.1	18.8 ± 6	°°°25.2 ± 9	°°142.2 ± 30	40.5 ± 7	°°°110.4 ± 19

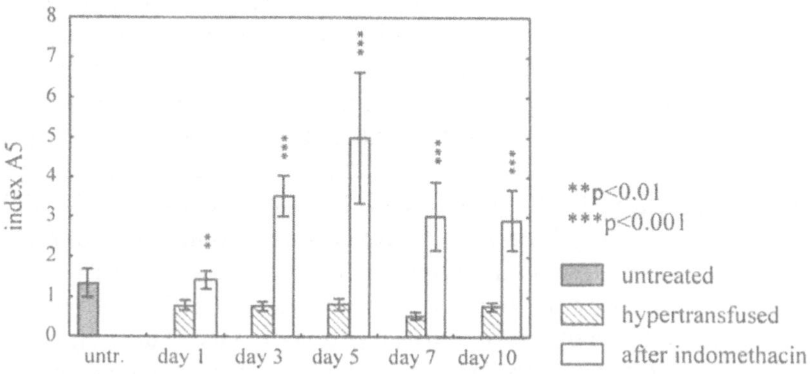

Figure 3. Rate of repeated participation of macrophages in new EI reconstruction, i.e. EI reconstructive / EI involutive = A5 in rats after i.v. transfusion with 2.5 ml/100 g b.w. of a 70% suspension of washed, packed, isogenous red cells/2 days and with concomitant indomethacin injection 4 mg/kg/3days s.c.; mean values ± SD.

rats. The increase of the rate of CFU-E differentiation into erythroblasts in EI (A2), as well as of the rate of repeated participation of macrophages in new EI reconstruction (A5) was remarkably expressed on days 3–5 and persisted during the whole studied period ($p < 0.001$) (Figs. 2, 3; Table 1).

DISCUSSION

Polyglobulia induced by hypertransfusion of packed red cells is known to produce a depression of erythropoiesis due to the increased oxygen supply.[1,3] We found a well-expressed suppression of erythropoiesis after transfusion of packed red cells. This finding was supported by the strong reticulocyte decrease in the circulation and EI reduction in bone marrow. The decrease in both examined indices of erythropoiesis—rate of CFU-E differentiation into erythroblasts in EI (A2), as well as rate of repeated participation of macrophages in new EI reconstruction (A5), supports the concept about the important macrophage participation in erythropoietic processes. We assumed an inhibition of macrophage activity to form EI or reduction of macrophage count in bone marrow leading to suppressed erythropoiesis within EI. Similar results have been documented by Brookoff D. (1982) who describes a sequence of ultrastructural changes in marrow stromal cells after repeated hypertransfusions resulting in a disappearance of medullary macrophages and a shift from erythropoietic to granulopoietic tissue development.[3] Other authors have found a remarkably reduced number of erythroid cells in macrophage-depleted cultures, which suggests that the erythroblast-macrophage contact promotes proliferation and terminal maturation of erythroid cells.[5]

The effect of prostaglandin's inhibition on erythropoiesis is a subject of controversy.[12,13] In our previous study we demonstrated that the cyclooxygenase inhibition by indomethacin did not attenuate the adequate erythropoietin production after hypoxic exposure and enhanced the development of bone marrow erythroblastic islands.[6] The activation of centrally situated macrophage cells within erythroblastic islands was shown to play a significant role in the local regulation of bone marrow erythropoiesis. Our present study confirms the concept of the sustained proliferation in bone marrow after cyclooxygenase inhibition even after increased oxygen supply by hypertransfusion. Such effect has been seen in gamma irradiated rats, pretreated with indomethacin.[4]

The terminal stage of erythropoiesis may be estimated by exploring the interactions between the central macrophage and erythroblasts within bone marrow EI.[2,14,15,17] The enhanced development of EI in indomethacin-treated animals is demonstrated by the increase of absolute EI count and the simultaneous accelerated release of reticulocytes into the circulation. Furthermore, in animals with cyclooxygenase inhibition, an observable stimulation of erythropoiesis was expressed by diminution of the involutive class IV of EI, as well as by an enhanced generation of new EI (EI reconstructive). This finding is supported by the increase of both rate of CFU-E differentiation into erythroblasts (A2) and rate of repeated participation of macrophages in new EI reconstruction (A5). The later is in accordance with the concept of the important role of the centrally situated macrophage via potentiation of its adhesive capacity to erythroblasts after activation.[1,14]

An increased proliferation potential of the red cell line in bone marrow as early as one day after indomethacin administration has been found by other authors as well.[4,13] According to Miller SC. (1992), erythroid cells in bone marrow of adult mice were also increased throughout indomethacin treatment.[7] Considering the negative role of PG in the control of red cells proliferation, we can assume that PG inhibitors can shift the regulatory balance towards the predominance of a positive control. It has been demonstrated that the increase of splenic myelopoiesis induced by interleukin-1 could be greatly amplified by pretreatment of mice with indomethacin. This fact suggests a regulatory relationship between interleukin-1 and PG. A similar relationship has been found to exist between PG and granulocyte-macrophage colony-stimulating activity.[8] The relationship between interleukin-1 and PG and between granulocyte-macrophage colony stimulating activity and PG could be the reason for the intense proliferation in indomethacin-pretreated mice.[9] Inhibition of PGE2 production by indomethacin causes the release of erythroid cells from PGE2-mediated suppression.[7,12] PGE2 appear to mediate their effects by raising cAMP levels in macrophages and therefore reduce macrophage activity.[10]

In conclusion, our study favours the concept about the suppressing effect of prostaglandins on erythropoiesis within bone marrow erythroblastic islands even after hypertransfusion. The activation of centrally situated macrophage cells within erythroblastic islands plays a critical role in the maintenance of erythroid homeostasis.

REFERENCES

1. Albina JE, Henry WL Jr, Mastrofrancesco B, Martin BA, Reichner JS: Macrophage activation by culture in an anoxic environment. J Immunol 155: 4391, 1995.
2. Bernard J: The erythroblastic island: past and future. Blood cells 17: 5, 1991.
3. Brookoff D, Weiss L: Adipocyte development and the loss of erythropoietic capacity in the bone marrow of mice after sustained hypertransfusion. Blood 60: 1337, 1982.
4. Fedorocko P, Mackova O: Radioprotective effects of combination broncho-vaxom, a macrophage activator, and indomethacin, an inhibitor of prostaglandin production: relationship to myelopoiesis. Eur J Haematol 56: 54, 1996.
5. Hanspal M, Hanspal JS: The association of erythroblasts with macrophages promotes erythroid proliferation and maturation: a 30-kD heparin-binding protein is involved in this contact. Blood 84: 3494, 1994.
6. Kalaidjieva V: Modulation of erythropoiesis in rat bone marrow erythroblastic islands by cyclooxygenase inhibition. General Pharmacol 32: 423, 1999.

7. Miller S: Age-related differences in the effect of in vivo administration of indomethacin on hemopoietic cell lineages of the spleen and bone marrow of mice. Experimentia 48: 674, 1992.
8. O'Reilly M, Gamelli RL: Indomethacin augments granulocyte-macrophage colony-stimulating factor-induced hematopoiesis following 5-FU treatment. Exp Hematol 18: 974, 1990.
9. Pospisil M, Hofer M, Pipalova I, Viklicka S, Netikova J, Sandula J: Enhancement of hematopoietic recovery in gamma-irradiated mice by the joint use of diclofenac, an inhibitor of prostaglandin production, and glucan, a macrophage activator. Exp Hematol 20: 891, 1992.
10. Rutherford M, Schook L: Macrophage function in response to PGE2, L-arginine deprivation, and activation by colony-stimulating factors is dependent on hematopoietic stimulus. J Leukoc Biol 52: 228, 1992.
11. Sasaki K, Iwatsuki H: Origin and fate of the central macrophages of erythroblastic islands in the fetal and neonatal mouse liver. Microsc Res Tech 39: 398, 1997.
12. Suk K, Sipes DG, Erickson KL: Enhancement of B-cell translocation gene-1 expression by prostaglandin E2 in macrophages and the relationship to proliferation. Immunology 91: 121, 1997.
13. Taniguchi S, Shibuya T, Harada M, Niho Y: Prostaglandin-mediated suppression of in vitro growth of erythroid progenitor cells. Kidney Int. 36: 712, 1989.
14. Wilson JG: Adhesive interactions in hemopoiesis. Acta Haematol. 97: 6, 1997.
15. Zakharov YM, Medianik BV: Patterns of erythropoietic reconstruction in the erythroblastic islands. Fizio. Zh Im I M Sechenova 80: 76, 1994 (Russian).
16. Zakharov YM, Melnikov IY, Rassokhin AG: Study on erythropoiesis by modification of method for isolation of erythroblastic islets. Hematol Transfusiol 4: 52, 1984.
17. Zakharov YM, Melnikov IY, Rassokhin AG: Classification of erythroblastic islets of the bone marrow and the study of their cellular composition. Arkh Anat Gistol Embriol 98: 38, 1990.

HEMATOLOGICAL AND BIOCHEMICAL PARAMETERS IN HEREDITARY SPHEROCYTOSIS UNDER OXIDATIVE STRESS

Elisa Granjo, Alice Santos-Silva, Irene Rebelo, Ana Nóvoa,
Elísio Costa, José Barbot, Letícia Ribeiro, and Alexandre Quintanilha

Departamento de Hematologia do Hospital S.João do Porto, Porto
the Instituto de Biologia Molecular e Celular, Univ. Porto, Porto
the ICBAS
the Departamento de Bioquímica Faculdade Farmácia
Univ. Porto, Porto
the Serviço de Hematologia
Hospital Maria Pia, Porto
and the Centro Hospitalar de Coimbra, Coimbra, Portugal

A broad spectrum of clinical manifestations in hereditary spherocytosis (HS) has been documented.

The aim of this study was to investigate a correlation between the development of an oxidative stress, during inflammatory or infectious stress, and the development of a hemolytic event in HS patients. We evaluated hematological and biochemical changes i) imposed by infection in 2 HS patients and in 2 HS patients who had undergone splenectomy (more than six months before) ii) in 5 HS patients immediately after splenectomy and iii) in 4 HS patients, more than six months after splenectomy. In both inflammatory and infectious stresses we found a reduction in RBCs concentration and in MCHC, a rise in MCV, and neutrophilic monocytic leukocytosis. The changes in RBCs seemed to correlate with neutrophils and monocytes values. To clarify this, we studied the same parameters and a few others in iiii) a group of 6 HS patients who had undergone splenectomy (more than six months) and compared the results with those presented by a healthy control, including 29 individuals with no HS and no infection.

Address reprint requests to: Alice Santos-Silva, PhD, Faculdade de Farmácia, Departamento de Bioquímica, Rua Aníbal Cunha, 164, 4050-Porto, Portugal. Tel.: + 351 2 2002564; Fax: + 351 2 2003977; e-mail: assilva@ff.up.pt

Molecular Biology of Hematopoiesis 6, edited by Abraham *et al.*
Kluwer Academic/Plenum Publishers, New York, 1999.

Our data suggest that inflammatory and infectious stress by inducing leukocytosis and leukocyte activation, may induce oxidative and proteolytic modifications upon RBCs, leading to an accelerated aging process, to a premature removal or even to a hemolytic event.

In Europe, Hereditary Spherocytosis (HS) is the most common congenital hemolytic anemia. A broad spectrum of clinical manifestations has been documented, which can range from mild to extremely severe. A set of mutations, namely in spectrin, ankirin and, band 3, may underlie HS.[1–3] The mode of transmission is dominant or recessive, with intermediate situations.

During its life span any normal red blood cell (RBC) is exposed to various insults and undergoes physical and chemical changes.[4–7] These modifications become more pronounced with cell age, and with the development of unusual physiologic circumstances, such as increased oxidative and proteolytic stress. In these circumstances, the accelerated and premature development of senescent modifications may trigger the premature removal of RBCs or even a hemolytic event. Cells, such as spherocytes, which develop intracellular defects early during their life span, are also removed from circulation at an earlier stage.

RBC senescence includes several modifications, namely a reduction in its metabolic activity, resulting from a general decreased enzymatic activity. The pathway for the removal of senescent or damaged RBCs involves the development of a senescent cell antigen immunologically related to band 3 protein,[8,9] which marks the RBC for death, by triggering the binding of specific auto-antiband 3 antibodies[10,11] and complement activation.[12] The modified antigenicity of band 3 may result from proteolytic cleavage, clustering or even by exposure of unusual epitopes.

Inflammatory and infectious diseases often present a rise in leukocytes, namely in neutrophils, and a reduction in total RBCs.[13,14] The exposure of circulating RBCs to oxygen metabolites and proteases, which are known to be produced by activated neutrophils, may account for some oxidative and proteolytic RBC damage, leading to an accelerated aging process and premature removal.[7,15–19] In the case of spherocytes presenting a destabilized membrane structure, with loss of membrane components, and with an abnormal reduced life span, the impact of an oxidative and proteolytic stress is probably enhanced, triggering a hemolytic event.

The aim of this study was to evaluate whether a correlation can be drawn between an inflammatory or an infectious stress and the development of a hemolytic event in HS patients. Splenectomy (SPL) was used as the model of inflammatory stress, and bacterial infection as the model of infectious stress. In searching for correlation we studied several hematological and biochemical parameters, that included: concentration of total WBC and of the several WBC types; concentration of RBCs; hematocrit (Ht); hemoglobin (Hb) concentration; hematimetric indexes; direct bilirrubin (DB) and total bilirrubin (TB). The same parameters and a few others were studied in HS patients who had undergone splenectomy several months before (more than 6 months) and compared to a healthy control, to investigate the impact of splenectomy in the spherocyte aging process and to compare it to the normal RBC aging. The other studied parameters included: membrane bound hemoglobin (%MBH); the band 3 profile (high molecular weight aggregates, monomer, and total proteolytic fragments) as a cumulative marker of RBC aging; the RBC glucose-6-phosphate dehydrogenase (G6PD) activity as an index of RBC age; lactoferrin (Lactof) concentration, a product of neutrophil activation, as a marker of leukocyte activation.

MATERIALS AND METHODS

Subjects. We evaluated the referred hematological and biochemical parameters imposed by infection in 2 HS patients and in other 2 HS patients who had undergone splenectomy several months before (more than 6 months). The changes imposed by inflammatory stress induced immediately after surgical spleen removal were evaluated in 5 HS patients. The same parameters were studied in 4 HS patients, to evaluate the modifications imposed by the removal of the spleen (not less than 6 months before). To clarify the results obtained with the previous groups, we studied the same parameters and a few others in 6 HS patients, who had undergone splenectomy more than 6 months before and in 29 healthy individuals (with no HS and no infection).

Assays. Whole blood and plasma (EDTA as anticoagulant) and serum were used for hematological and biochemical procedures.

WBC and RBC count, Ht, Hb concentration, MCV, MCH, MCHC, RDW, and the WBC differential count were evaluated by a Ortho Counter. The morphology of the blood cells was evaluated in a Wright stained blood film.

To evaluate erythrocyte G6PD activity, RBCs were isolated after centrifugation on a double density gradient (Histopaque 1119 and 1077, Sigma), washed with phosphate buffered solution (PBS) pH 7.4 and lysed by thermic shock. The activity was then evaluated using a commercially available kit (Test-Combination G6PDH, Boehringer Mannheim), with modifications.[20] The G6PD activity was expressed as a function of hemolysate hemoglobin content (UI/g Hb), which was evaluated by a colorimetric method (Hb-Boehringer Mannheim).

RBC membrane ghosts were prepared by hypotonic lysis[21] using phenylmethylsulfonyl fluoride as a protease inhibitor (final concentration of 0.1 mM). Protein concentration of the prepared membranes was determined.[22]

%MBH was measured spectrophotometrically,[15] after dissociation of membrane components with Triton X-100 (5% in Dodge buffer) at 415 nm, and compared with membrane protein concentration.

To study band 3 profile RBC membranes were treated with an equal volume of a solubilization buffer (0.125 M Tris HCl pH 6.8, 4% sodium dodecyl sulfate (SDS), 20% glycerol, 10% 2-mercaptoethanol) heat denatured and subjected to SDS PAGE (20 μg of protein/lane) using the discontinuous Laemmli system[23] (9% separating gel; 4,5% stacking gel). Membrane proteins were then electrophoretically transferred from SDS gels to nitrocellulose[24] (22 μm porosity). Additional reactive sites were blocked (3% gelatin and 0.1% Triton-X 100 in PBS pH 7.0/1 h). Monoclonal antibodies anti-human band 3, produced in mouse, recognizing an epitope located in the cytoplasmatic pole of the band 3 molecule (Sigma) were then added (dilution 1 : 3,000) and incubated for 4 h; the washing of the nitrocellulose was followed by the addition and incubation with anti-mouse IgG peroxidase-linked (Amersham) for 1 h (dilution 1 : 15,000). Both incubations were carried out at room temperature in PBS pH 7.0 containing 0.1% detergent and 0.5% gelatin. The washes used the same buffer without gelatin. The immunoblot was developed by the addition of hydrogen peroxide and horseradish peroxidase color developer reagent. The band 3 profile was quantified by densitometry (Cybertech CS1).

Plasma lactoferrin was evaluated by an enzyme immunoassay (Bioxytech lacto f enzyme immunoassay, Oxis International, Inc.).

Statistics. All measurements are expressed as mean value ± standard deviation. The calculations were performed using the StatView software. The significance of the differences between each value presented by two groups was evaluated by the Student t-test and $P < 0.05$ was considered statistically significant. To draw graphs we used the Excel Microsoft.

RESULTS

Figure 1a shows the percentual changes in some hematological and biochemical parameters, imposed by infectious stress in 2 HS-SPL patients and in 2 HS-no SPL patients. In both cases we observed a reduction in RBC concentration, which is accompanied by an increase in bilirrubins. These modifications are followed by an increase in WBC concentration. This rise is specially evident in HS-SPL patients and results mainly from a rise in neutrophils and less in monocytes. In HS-no SPL patients the WBC concentration rise is minor and results mainly from a rise in monocytes and less in neutrophils. Along with these changes the MCV rises and the MCHC reduces. There was an increase in the RDW and this rise was higher in HS-SPL patients than in HS-no SPL patients.

In Fig. 1b we present the percentual changes observed in the same hematological and biochemical parameters, imposed by inflammatory stress induced immediately after SPL in 5 HS patients. RBC modifications were similar to those we found for infectious stress, though with smaller values. As expected the removal of the spleen was followed by a reduction in bilirrubins. WBCs presented a rise similar to that imposed by infectious stress in HS-SPL patients. This rise is mainly due to the increase in monocytes and less in neutrophils.

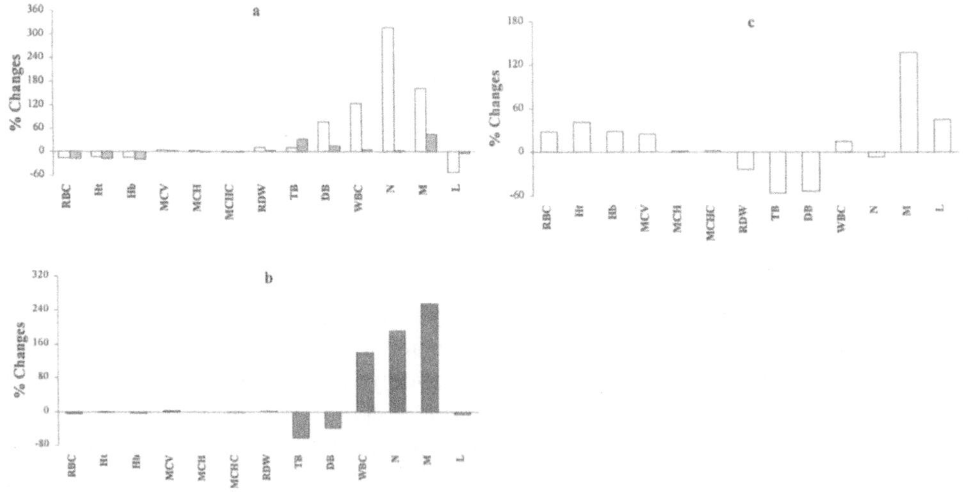

Figure 1. Changes (%) observed in hematological and biochemical parameters;

a) imposed by infectious stress in 2 HS patients who had undergone splenectomy (□) and in 2 HS patients who had not undergone splenectomy (●);
b) imposed by inflammatory stress, induced immediately after splenectomy, in 5 HS patients;
c) imposed by removal of the spleen (not less than 6 months after splenectomy), in 4 HS patients.

RBC ($\times 10^{12}$/l); Hb, MCHC (g/dl); Ht, RDW (%); MCV (fl); MCH (pg); TB, DB (mg/l); WBC, N, M, L ($\times 10^{9}$/l).

Figure 1c shows the impact of SPL (more than six months after splenectomy) in the same hematological and biochemical parameters, observed in 4 HS patients. We found a rise in RBC concentration and in the hematimetric indexes; the RDW and the bilirrubins presented a decrease. In WBCs we found a rise, resulting from an increase in lymphocytes and monocytes. To clarify the above changes, we evaluated in 6 HS-SPL patients the same parameters (Fig. 2a and Fig. 2b) and a few others, namely erythro-

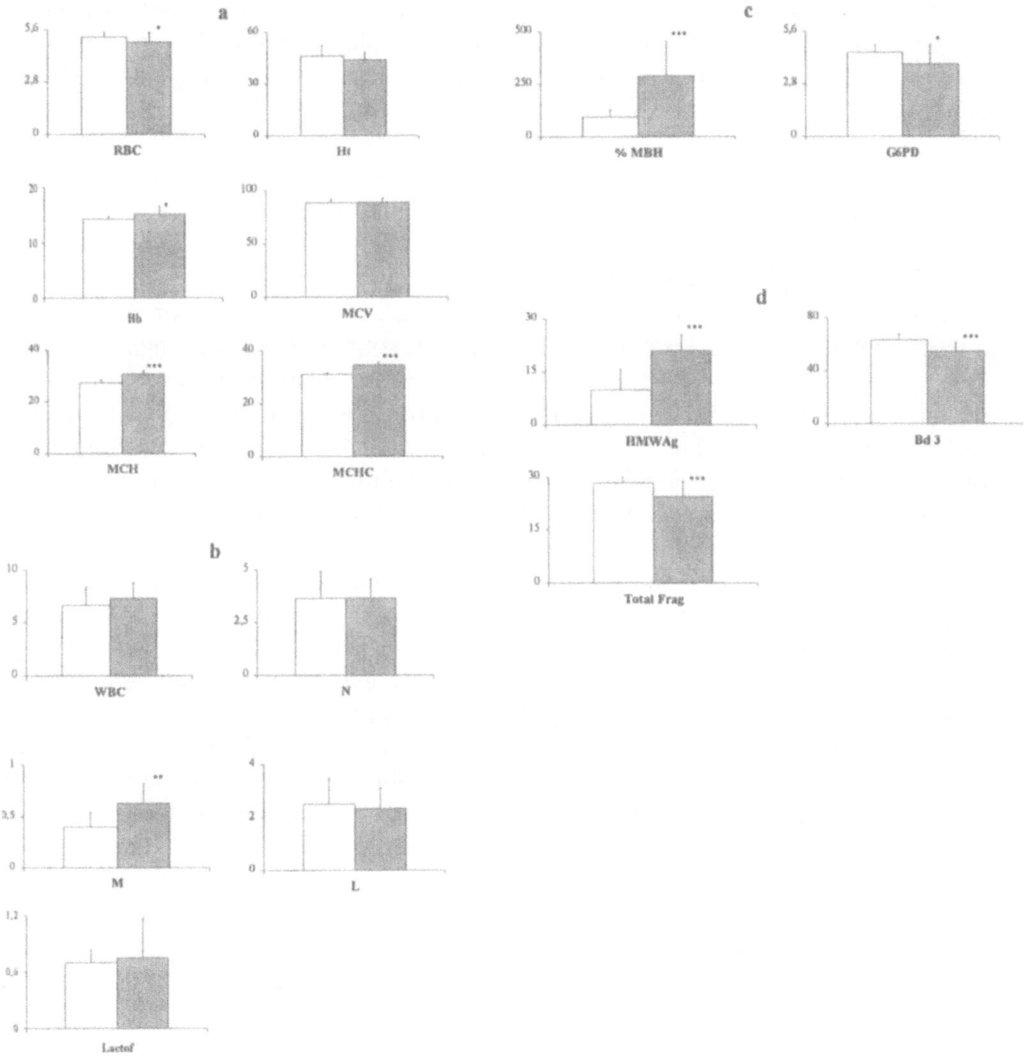

Figure 2. Hematological and biochemical parameters in HS patients who had undergone splenectomy (more than 6 months) and in a healthy control.
□ Controls (n = 29) ◍ Patients HS-SPL (n = 6)
$*P < 0.05$ $**P < 0.01$ $***P < 0.001$

a) RBC ($\times 10^{12}$/l); Ht (%); Hb, MCHC (g/dl); MCV (fl); MCH (pg);
b) WBC, N, M, L ($\times 10^{9}$/l); Lactof (μg/ml);
c) %MBH ($\times 10^{-4}$); G6PD (UI/g Hb);
d) HMWAg, Bd 3, Total Frag (relative %).

cyte G6PD activity and the %MBH (Fig. 2c); band 3 profile (Fig. 2d); and plasma lactoferrin concentration (Fig. 2b). The results were compared with those presented by a control including 29 healthy individuals (with no HS and no infection). We observed an evident improvement in RBC population though some of the characteristics of HS were still observed, namely a high MCHC. A significant reduction in erythrocyte G6PD, a significantly higher %MBH, and a different band 3 profile were also observed. Actually, HS-SPL patients presented a significant increase in high molecular weight aggregates (HMWAg) and a significant decrease in band 3 monomer (Bd 3) and in its proteolytic fragments (Total frag). Concerning leukocytes we found increased values for WBC concentration and a significantly higher value in monocyte concentration. Plasma lactoferrin presented no statistical significance, although presenting a higher value than that presented by the control.

DISCUSSION

Hereditary spherocytes, on account of its membrane modifications, are more sensitive to oxidative and proteolytic stresses. During an inflammatory or an infectious process, activated leukocytes may be an important source of products, namely oxygen metabolites, and proteases. Either of these leukocyte activation products can damage the neighbouring RBCs, imposing oxidative and proteolytic lesions to RBCs, which may lead to an accelerated aging and removal process. Therefore, the impact of inflammatory or infectious stress in spherocytes is probably enhanced and may impose an even more accelerated aging and removal process, leading to a hemolytic event.

We showed (Fig. 1a) that the changes in the hematological and biochemical parameters studied, imposed by infectious stress in both HS-SPL and HS-no SPL patients, were similar. The reduction in RBC concentration was accompanied by a rise in bilirubins, revealing an increased RBC removal. Along with these changes we observed a rise in WBC concentration, which was specially evident in HS-SPL patients and resulted mainly from a rise in neutrophils and less in monocytes. HS-no SPL patients presented a minor rise in WBC concentration and resulted mainly from a rise in monocytes and less in neutrophils. These results suggest the involvement of these leukocytes in accelerated RBC aging, and premature removal. In addition, we showed that the MCV presented a rise and the MCHC a reduction in both HS patients, suggesting that molecular modifications occurred. As expected, there was an increase in the RDW and the rise was higher in HS-SPL patients. This is probably the result of the increase in RBCs presenting the referred molecular changes. The change in the RDW value of HS-no SPL patients was lower since the spleen will filter the RBCs presenting those molecular changes. In fact, the reduction in RBC concentration and the rise in TB was higher in HS-no SPL, denoting an enhanced splenic RBCs removal. It seems that the presence of the spleen, acting as a filter of damaged or senescent RBCs, renders the HS patients more susceptible to develop a hemolytic event during an inflammatory or an infectious stress.

The percentual changes (Fig. 1b) in the same hematological and biochemical parameters, imposed by inflammatory stress induced immediately after splenectomy in HS patients, were similar to those found for infectious stress, though with smaller values. As expected after the removal of the spleen, a reduction in bilirubin was observed denoting reduced RBC destruction.

HS-SPL patients, who had undergone splenectomy (not less than 6 months before) presented the expected changes for RBC concentration and bilirrubins. Actually, the rise in RBC concentration and the reduction in bilirubin was followed by clinical and hematological improvement of the patients. The removal of the spleen, the filter of the damaged or senescent RBCs, seems to underlie these changes. However, we must expect that the absence of that filter may account for the accumulation of damaged RBCs. Contrary to the previous stressfull situations, in this case we observed a rise in WBCs, resulting from an increase in lymphocytes and monocytes. This suggests that the absence of the spleen may trigger a different mechanism to command the granulo-monocytopoiesis or even the immune system.

To clarify the above changes in RBCs, denoting accelerated RBC damage or senescence and the involvement of activated leukocytes in its development, we evaluated in HS-SPL patients and in a healthy control the same parameters and a few others, which could give us some information about the age of RBCs, about the amount of oxidative and proteolytic lesions suffered by RBCs during their life span and about leukocyte activation. Several months after splenectomy the HS patients presented an evident improvement in RBC concentration, showing that the absence of the spleen reduced the removal of damaged or senescent RBCs. We must note, however, that both MCH and MCHC maintained higher values when compared to the control, as it is characteristic of spherocytes. The reduction of RBC removal seems to be followed by the accumulation of older circulating RBCs, as shown by the significant decrease in erythrocyte G6PD activity (Fig. 2c). Strengthening this hypothesis we found also a significantly higher %MBH and a different band 3 profile (Fig. 2d). HS-SPL presented a significant increase in HMWAg and a significant decrease in band 3 monomer and in its proteolytic fragments. When compared this profile with that presented by the control, it appears that HMWAg might result from oxidation and crosslinking of band 3 monomer and its proteolytic fragments or even from a further proteolysis of band 3 followed by oxidation and crosslinking of proteolytic fragments. These results provide further evidence of the premature senescence of RBCs in these patients. We must emphasize that these changes for band 3 profile are similar to those found in patients under chronic stressfull situations, such as cardiovascular disease,[15] and similar to those found for RBC aging *in vitro*, i.e. exposed to activated neutrophils or neutrophilic elastase.[16] Concerning leukocytes, we found higher values for HS-SPL patients for all leukocyte types, though only for monocytes we did observe a significantly higher value when compared to the control (Fig. 2b). We must emphasize that monocytes are the WBC which present a higher rise in both inflammatory and infectious stress in HS-no SPL.

The destabilized membrane structure of spherocytes seems to underlie membrane vesiculation, premature aging and premature removal of the cells. Whenever an inflammatory or an infectious stress develops, a rise in leukocytes, mainly in neutrophils and monocytes occurs. It is known that leukocytes present a reduced deformability and that its activation may produce and release oxygen metabolites and proteases. The rise in leukocytes of reduced deformability may slow the blood flow and favour the interaction of the released leukocyte activation products with the surrounding cells, namely with the spherocytes. It seems, therefore, reasonable to assume that the change in blood flow and the easier interaction of leukocyte activation products with the circulating RBCs, is even more enhanced within the splenic microvasculature, favouring oxidative, and proteolytic damage of spherocytes. The oxidative and proteolytic lesions, by inducing splenic sequestration, may lead to premature removal of a higher number of spherocytes, or even to a hemolytic event.

In conclusion, our data show that inflammatory and infectious stress by inducing leukocytosis and eventually leukocyte activation, may underlie and trigger hemolytic events in HS patients. In addition, %MBH and band 3 profile may provide further evidence of the oxidative and proteolytic modifications imposed upon RBCs by leukocyte activation products.

ACKNOWLEDGMENT

Supported in part by research grants from the "Fundação Gomes Teixeira", Universidade do Porto, and PRAXIS XXI n°2/2.1/SAU/1240/95 (JNICT).

REFERENCES

1. Delaunay J, Alloisio N, Morlé L, Carré G: La sphérocytose héréditaire en 1995: l'apport de la génétique moléculaire. Hématologie 2:115, 1995.
2. Cynober T, Mohandas N, Tchernia G: Red cell abnormalities in hereditary spherocytosis: relevance to diagnosis and understanding of the variable expression of clinical severity. J Lab Clin Med 128:259, 1996.
3. Delaunay J: Hereditary hemolytic anemias due to defects in membrane proteins. Curr Opinion Hematol 59, 1993.
4. Lutz HU: Erythrocyte clearance. In: Harris JR, editor. Blood cell biochemistry. Vol 1 Erythroid cells. New York: Plenum Press, 1981, p 81.
5. Brovelli A, Castellana MA, Minetti G, Piccinini G, Seppi C, Renzis MR, Balduini C: Conformational changes and oxidation of membrane proteins in senescent human erythrocyte. In: Magnani M and De Flora A, editors. Red Blood Cell Aging. New York: Plenum Press, 1991, p 59.
6. Castellana MA, Piccinini G, Minetti G, Seppi C, Balduini C, Brovelli A: Oxidation of membrane proteins and functional activity of band 3 in human red cell senescence. Arch Gerontol Geriatr Suppl 3:101, 1992.
7. Bosch FH, Were JM, Schipper L, Roerdinkholder-Stoelwinder B, Huls T, Willekens FLA, Wichers G, Halie MT: Determinants of red blood cell deformability in relation to cell age. Eur J Haematol 52:35, 1994.
8. Kay MMB, Goodman S, Whitfield C, Wong P, Zaki L, Rudloff V: Senescent cell antigen in immunologically related to band 3. Proc Natl Acad Sci 80:1631, 1984.
9. Kay MMB: Localization of senescent cell antigen on band 3. Proc Natl Acad Sci 81:5753, 1984.
10. Lutz HU: Naturally occurring anti-band 3 antibodies. Transf Med Rev 7:201, 1992.
11. Kay MMB: Role of physiologic autoantibody in the removal of senescent human red cells. J Supra Mol Struct 9:555, 1978.
12. Lutz HU, Bussolino S, Flepp R, Fasler S, Stammler P, Kazatchkine MD, Arese P: Naturally occurring anti-band 3 antibodies and complement together mediate phagocytosis of oxidatively stressed human erythrocytes. Proc Natl Acad Sci USA 84:7368, 1987.
13. Weiss DJ, Klausner JS: Neutrophil-induced erythrocyte injury: a potential cause of erythrocyte destruction in anaemia associated with inflammatory disease. Vet Pathol 25:450, 1988.
14. Rebelo I, Guerra FC, Leite LP, Quintanilha A: Lactoferrin as a sensitive blood marker of neutrophil activation in normal pregnancies. Eur J Obstet Gynecol Reprod Biol 62:189, 1995.
15. Santos-Silva A, Castro EMB, Teixeira NA, Guerra FC, Quintanilha A: Altered erythrocyte membrane band 3 profile as a marker in patients at risk for cardiovascular disease. Atherosclerosis 116:199, 1995.
16. Santos-Silva A, Castro EMB, Teixeira NA, Guerra FC, Quintanilha A: Erythrocyte membrane band 3 profile imposed by cellular aging, by activated neutrophils and by neutrophilic elastase. Clin Chim Acta 275(2):185, 1998.
17. MucCurdy PR, Sherman S: Irreversibly sickled cells and red cell survival in sickle cell anemia. Am J Med 64:253, 1978.
18. Rachmilewitz EA, Shinar E, Shalev O, Galili U, Schrier SL: Erythrocyte membrane alterations in β-thalassemia. Clin Haematol 14:163, 1985.

19. Kaneko JJ: Comparative erythrocyte metabolism. Adv Vet Sci Comp Med 18:117, 1974.
20. Santos-Silva A, Guerra FC, Ribeiro MM, Teixeira N: Deficiência em glucose-6-fosfato desidrogenase (G6PD). Estabelecimento de valores de referência da sua actividade no eritrócito. Estudo de dois hemizigotos. Rev Port Farm 39:84, 1989.
21. Dodge JT, Mitchell C, Hanahan J: The preparation and chemical characteristics of hemoglobin-free ghosts of human erythrocytes. Arch Biochem Biophys 100:119, 1963.
22. Bradford MM: A rapid and sensitive method for the quantitation of microgram quantities of protein utilizing the principle of protein dye binding. Anal Biochem 72: 255, 1976.
23. Laemmli UK: Cleavage of structural proteins during the assembly of the head of bacteriophage T4. Nature 227:680, 1970.
24. Towbin H, Stahelin T, Gordon J: Electrophoretic transfer of proteins from polyacrylamide gels to nitrocellulose sheets: procedure and some applications. Proc Natl Acad Sci 76:4350, 1979.

42

LYMPHOCYTE POPULATIONS IN HEREDITARY SPHEROCYTOSIS PRE AND POST SPLENECTOMY AND UNDER OXIDATIVE STRESS

Elisa Granjo, Ilídia Moreira, Alice Santos-Silva, Irene Rebelo, Ana Nóvoa, Letícia Ribeiro, Manuela Fraga, Margarida Lima, Alexandre Quintanilha and Jorge Candeias

Departamento de Hematologia do Hospital de S. João do Porto, Porto
the Instituto de Biologia Molecular e Celular, Univ. Porto, Porto
the Instituto de Ciências Biomédicas Abel Salazar
the Departamento de Bioquímica da Faculdade de Farmácia
Univ. Porto, Porto
the Hospital de Santo Antínio, Porto
the Royal Free Hospital, London, UK
and the Centro Hospitalar de Coimbra
Coimbra Portugal.

Hereditary Spherocytosis (HS) is the most common congenital hemolytic anemia due to membrane defects. Its main clinical features are hemolytic episodes triggered by infection, pregnancy and other stresses and responsiveness to splenectomy (SPL). In order to clarify immunological disturbances associated to HS, the authors analyzed B, T, and natural killer (NK) lymphocyte populations in the peripheral blood of 18 patients with HS (age 12–68 years; 10 males, 8 females) before and/or at different times after SPL. Our results seem to indicate that immunological abnormalities found in HS may be of multifactorial origin. Some of these immunological abnormalities may be related to the disease itself, or to the continuous stimulation of the immune system by infectious and/or hemolytic episodes. Other abnormalities seem to result from thera-

Address reprints requests to: Alice Santos-Silva, PhD, Faculdade de Farmácia, Departamento de Bioquímica, Rua Aníbal Cunha, 164, 4050-Porto, Portugal. Tel.: +3512 2002564; Fax: + 351 2 2003977; e-mail: assilva@ff.up.pt

Molecular Biology of Hematopoiesis 6, edited by Abraham *et al.*
Kluwer Academic/Plenum Publishers, New York, 1999.

peutic splenectomy. Changes occurring immediately after SPL are probably associated to surgical stress. SPL is followed by a variable period of time during which peripheral blood lymphocyte population change, probably resulting from alteration in lymphocyte circulation and homing. After this period, a new state of equilibrium is achieved in asplenic HS patients. Most of these patients developed persistent lymphocytosis post SPL.

Hereditary Spherocytosis (HS) is the most common congenital hemolytic anaemia due to membrane defects. Its main clinical features are hemolytic episodes triggered by infection, pregnancy and other stresses and responsiveness to splenectomy (SPL).

The spleen selectively retains HS red cells. This custody is detrimental, leading to a loss of membrane surface that fosters further splenic trapping and eventual destruction. It appears that the primary membrane defect involves deficiencies or defects on spectrin, ankyrin, or band 3. Splenic clearance of abnormal red cells induces proliferation of lymphocytes and plasma cells as well as macrophages. True neoplasm-multiple myeloma (MM) has been reported in HS who had not undergone splenectomy. Continuous stimulation of the immune system may underlie this association (HS/MM). Natural killer (NK) cell expansions with normal cell activity has been described post-splenectomy in healthy persons.[1] By the other side, abnormal clones of lymphocytes with NK cell markers were recognized only after splenectomy, but retrospective bone marrow analysis showed that the CD8+, CD57+ lymphocyte proliferation previously existed.[6] It is possible that persistent stimulus of immune cells (T, B, and NK lymphocytes) leads to T, B, and NK lymphocytosis along with the clinical of HS patients predisposed to hemolysis by viral infections and other stresses. Therapeutic splenectomy cures all patients with HS eliminating anemia except for the most severe forms of the disease that also benefit greatly from the procedure. However, the specific surgery origins a new homing and distribution of lymphoid cells which is detected by immunophenotyping of peripheral blood.

In order to clarify immunological disturbances associated to HS, we analyzed B, T, and NK lymphocyte subpopulations in the peripheral blood of patients with HS pre and post splenectomy at different times.

MATERIALS AND METHODS

Subjects. We studied 18 patients with HS, including 10 males and 8 females, 12 to 68 years old. The studies were performed before and/or at different times after SPL.

Sample collection. Peripheral blood samples were collected into K3-EDTA containing tubes.

Leukocytes count. An Ortho Counter was used to evaluate the total white blood cells (WBCs) and the WBC differential count.

Flow cytometric studies. Cells were double stained for T, B, and NK-associated markers by using a whole blood immunofluorescence technique. Briefly, for 15 minutes at room temperature, 1×10^6 cells (100 µl) were incubated in the presence of 10 µl of mouse anti-human CD3, CD4, CD8, CD16, CD19 or CD56, conjugated with phycoerythrin or with fluorescein. Cell fixation and erythrocyte lysis were performed with FALS-lysing solution (Becton-Dickinson), according to the manufacture's instructions. After washing, cells were ressuspended in 500 µl of phosphate buffered solution and acquired in a Facscalibur flow cytometer (Becton-Dickinson). Data were analyzed using the Paint-a-Gate program.

RESULTS

When compared 9 adult HS non-SPL patients to age-matched controls, we found a tendency to a slight decrease of absolute numbers of total lymphocytes, T lymphocytes, B lymphocytes, and NK cells (Fig. 1 a, b, c, and d). On the contrary, 6 adult HS patients studied 1 year or more after SPL showed a tendency to present increased absolute numbers of total lymphocytes, T lymphocytes, B lymphocytes, and NK cells, when compared to both HS non-SPL patients and normal controls (Fig. 1 a, b, c, and d). The evaluation of the T cell/B cell ratio showed that HS non-SPL presented a reduction in its mean value, when compared to the control, and that this decrease was even more accentuated for HS SPL patients (Fig. 1 e). Concerning CD4/CD8 ratio (Fig. 1 f) we found also a reduction in both HS non-SPL and in HS SPL patients, though this change was not so evident. In the case of 2 HS non-SPL children studied, we found no differences when compared to age-matched controls (data not shown).

Four HS patients were analyzed sequentially before and after splenectomy (one day to 3 months after). We found that in the day after splenectomy there was an

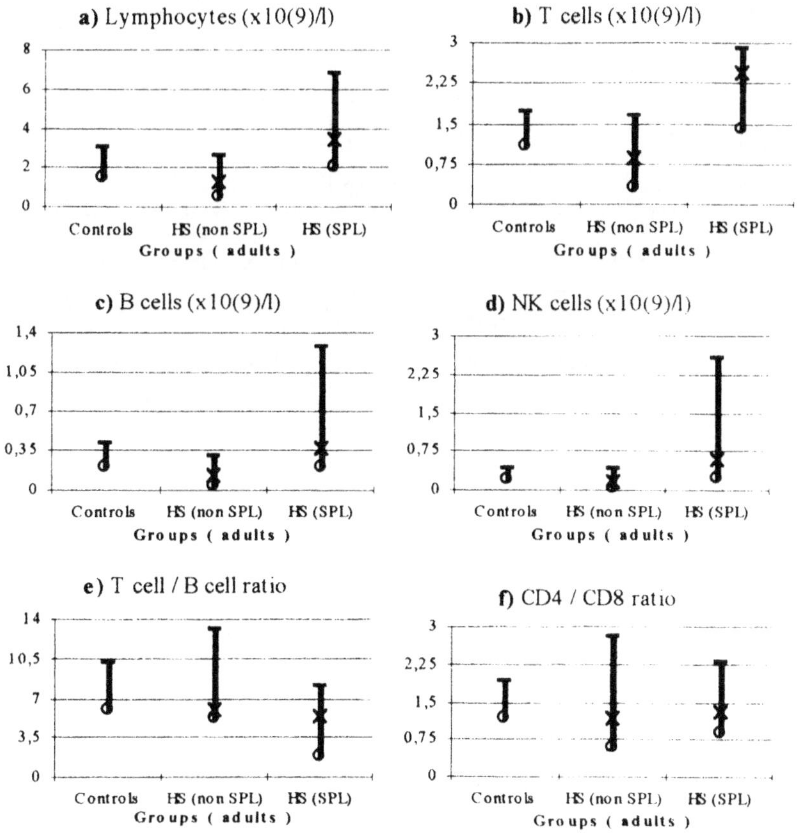

Figure 1. Lymphocyte populations in non-SPL and SPL (1 year or more after SPL) HS patients.

absolute leukocytosis with neutrophilia and a slight lymphopenia, mainly due to a decrease of T and NK cells (Fig. 2 a, b, c, d, and e). One to three months later there was a progressive normalization of the total leukocyte values (Fig. 2 a). The absolute number of lymphocytes presented a tendency to increase above the control values (Fig. 2 b). This rise in lymphocytes seems to result from an increase in T, B, and NK cells. All the studied patients presented higher values of T, B, and NK cells, than that presented before splenectomy. As referred, some HS SPL patients were studied sequentially 1 year or more after splenectomy. In three of these patients we did not find relevant changes in lymphocyte populations during the follow-up period (data not shown). On the contrary, three HS SPL patients developed absolute lymphocytosis ($>=3.5 \times 10^9/l$). In two of them, the rise resulted from an increase in T, NK, and B cells. In the other patient, the rise resulted from an increase in T and NK cells (Table 1). Two of these patients had increased numbers of large granular lymphocytes (LGL), mainly due to NK cell lymphocytosis. In both cases there was a long-standing history of hemolytic disease and recurrent infections. Actually, the highest LGL lymphocytosis was observed in a 23 years old male with recurrent chronic leg ulcers who had transitional neutropenia. These patients showed an expansion of LGL, comprising both Nka(+), CD3(−), TCRαβ(−), TCRγδ(−) and abnormal Nka(+), CD3(+,dim) TCRαβ(−), TCRγδ(−) cells. TCRVβ repertoire and TCR β and δ rearrangement studies showed a germ-line pattern.

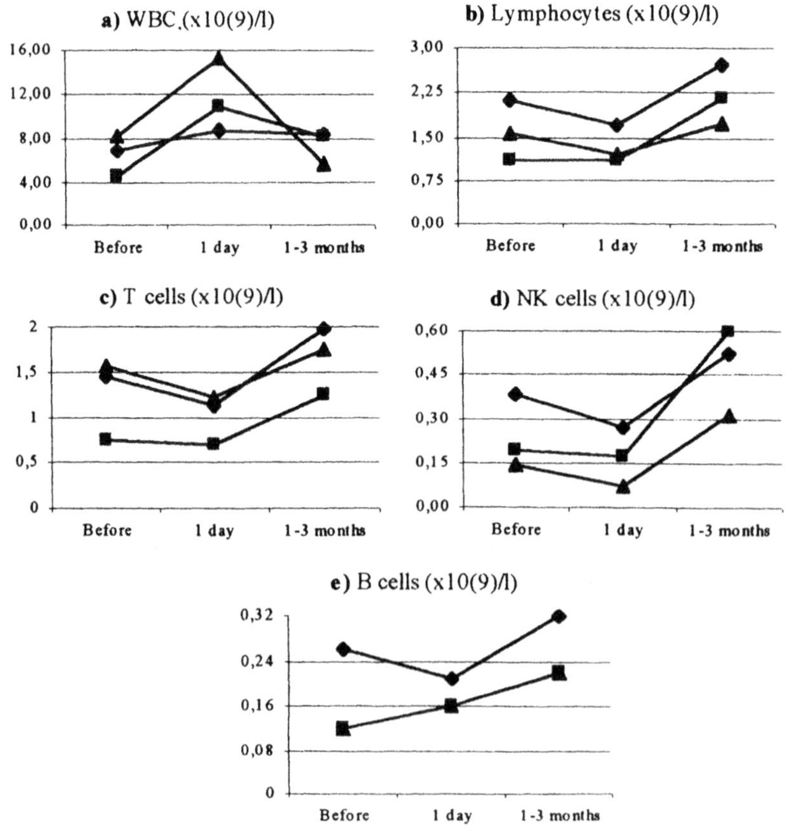

Figure 2. Leukocyte changes induced by SPL. Patients were studied before SPL, 1 day and 1 to 3 months after SPL.

Table 1. Patients with lymphocytosis after SPL

	Patient 1	Patient 2	Patient 3
Lymph × 10(9)/l	3.60↑	6.74↑	3.88↑
T cells × 10(9)/l	2.41↑	2.49↑	2.90↑
B cells × 10(9)/l	0.54↑	1.28↑	0.39
NK cells × 10(9)/l	0.68↑	2.56 ↑	1.32↑
CD4+ cells × 10(9)/l	1.40↑	2.43↑	1.28↑
CD8+ cells × 10(9)/l	1.04↑	1.88↑	1.36↑
T/B cell ratio	4,5	1,9↓	5.4
CD4/CD8 ratio	1,3	1,3	0.9↓

DISCUSSION

These preliminary results seem to indicate that immunological abnormalities found in HS patients may be of multifactorial origin. Some of these abnormalities may be related to the disease itself, or to the continuous stimulation of the immune system by infectious and/or hemolytic episodes. Actually, the patient above described, with chronic recurrent leg ulcers, developed peripheral NK lymphocytosis with bone marrow infiltration of 23% cells with large azurophilic granules-CD3(low), CD5(low), CD8(low), CD2(+), CD7(+), CD16(+), CD56(+), TCRαβ(−), TCRγδ(−). After splenectomy the clinical course of this patient has been characterized by multiple pulmonary infections easily treated in ambulatory by antibiotic *P.O.* It seems reasonable to assume that NK cells may be functionally deficient and/or that a minor clone of abnormal cells emerged after splenectomy.[1,4] T lymphocytes and NK cells may exert both stimulatory and suppressive effects that regulate growth and differentiation of hematopoietic cells. Lymphocytes can regulate hematopoietic cell growth by direct cellular contact or by releasing soluble factors.[2,5] In addition, the neutropenia found post splenectomy in this patient might be explained by the same effects. Other abnormalities seem to result from therapeutic splenectomy. Changes occurring immediately after splenectomy are probably associated to surgical stress. This is followed by a variable period of time characterized by peripheral blood lymphocyte changes that probably result from alterations in lymphocyte circulation and homing.

ACKNOWLEDGMENT

Supported in part by research grants from "the Fundação Gomes Teixeira", Universidade do Porto, and PRAXIS XXI n°2/2.1/SAU/1240/95 (JNICT).

REFERENCES

1. Airo P, Cattaneo R, Franceschini F, Migliosi G, Ferrari CM, Airo R: Lymphocytosis of Large Granular Lymphocytes in Splenectomized Subjects. Boll Is Sieoter Milan. 1998; 67(4): 290–294.
2. Mangan KF. Annu. Rev. Med. 1987; 38:61–70.
3. Mulligan SP, Wills EJ, Young GA: Exercise Induced CD8 Lymphocytosis: a phenomenon associated with large granular lymphocyte leukaemia. Br. J. Haematol. 1990 Jun; 75(2): 175–180.
4. Scott CS, Richards SJ, Sivakumaran M, Short M, Child JA, Hunt KM, McEvoy M, Steed AJ, Balfour

IC, Parapia LA. Transient and persistent expansions of large granular lymphocytes (LGL) and NK-associated (Nka) cells: the Yorkshire Leukaemia group Study. Br. J. Haematol. 1993 Mar; 83(3): 505–515.
5. Trinchieri G, Murphy M, Perussia B. Cril. Rev. Oncol. Hematol. 1987; 7(3): 219–265.
6. Pellegrin JL, Merlio JP, Conymakhoul P, Ferrer AM, Astiergin T, De Mascarel A, Beylot C, Bronstet A, Reiffers J, Leny B. Rev. Med. Interne 1992 Jul; 13(4): 261–267.

43

THE HEREDITARY HEMOCHROMATOSIS GENE AND IRON HOMEOSTASIS

J. N. Feder,*# D. M. Penny,* A. Irrinki,* G. A. Mintier,* J. A. Lebron,‡ C. N. Gross,§ L. Lee,* Z. Tsuchihashi,* C. A. Enns,§ P. J. Bjorkman,‡ and R. C. Schatzman*

* Progenitor, Inc.
Menlo Park,
California USA
‡ California Institute of Technology and
Howard Hughes Medical Institute
Pasadena, California, USA
§ Oregon Health Sciences University
Portland, Oregon USA
Corresponding author

Hereditary hemochromatosis (HH) is an autosomal recessive disorder of iron metabolism and represents one of the most common inherited disorders in individuals of Northern European descent with an estimated carrier frequency between 1 in 8 and 1 in 10.[1-3] In patients with HH, excessive iron deposition in a variety of organs leads to multiorgan dysfunction.[4,5] The symptoms of HH are often similar to other diseases and for this reason, the diagnoses of HH is frequently not made until later in life. HH has commonly been identified with a combination of blood iron measurements and liver biopsy. It has been argued that early detection of the HH and a simple program of therapeutic phlebotomy would be of benefit in preventing the onset of this disease. With the advent of the gene cloning, it is now possible to detect HH with a DNA based test so that diagnosis can be made before the deleterious effects of excess iron are realized.

1. CLONING OF THE HH GENE

The gene responsible for this disorder was identified using a linkage disequilibrium/positional cloning approach,[6] that utilized four basic steps: 1) physical mapping and cloning of genomic DNA covering the largest presumed candidate region; 2) generation of genetic markers and haplotype analysis to identify historic recombination

Molecular Biology of Hematopoiesis 6, edited by Abraham *et al.*
Kluwer Academic/Plenum Publishers, New York, 1999.

events; 3) transcript mapping and 4) mutation analysis to identify the actual disease alterations.

The initial boundaries for the construction of the physical map were D6S275, near HLA-A and the telomeric marker, D6S276. These markers were chosen because they demonstrated lower allelic association with HH than the existing intervening markers. The YAC map of this general region consists of 43 YACs and spans approximately 8 Mb of DNA. The physical distance between D6S265 and D6S276, which are genetically less than 1 cM apart is approximately 6 Mb. Hence, this region appears to have a suppressed rate of recombination. Curiously, a comparison of our map with published MIT/Whitehead and CEPH YAC map revealed a hole in the published maps that included the region around the hemochromatosis gene itself.[7]

The YAC map was further refined by the parallel construction of a large-insert bacterial clone contig. Together the DNA from these clones were used to generate 45 short tandem repeat polymorphic markers (CA-repeat markers) and single-base-pair substitution markers across the entire 6 Mb region. These 45 markers were then used to genotype 101 HH patients and 64 controls. The allele with the highest frequency in the patients compared to the controls was defined as the ancestral allele at that marker. This allowed for the re-construction of the ancestral haplotype on which the HH founder mutation occurred. An inspection of the haplotypes of 46 HH chromosomes that had been separated in somatic-cell hybrid lines, to allow unambiguous marker phasing along the chromosome, defined an area of approximately 400 kb as being the likely location of the HH gene. An analysis of 3 additional chromosomes, which had a minimum of 5 contiguous loci with ancestral alleles, suggested that the probable location for the HH gene was the 250 kb interval between the markers D6S2238 and D6S2241.

A transcript map of this region was constructed by cDNA selection, exon trapping, and sample genomic sequencing. Three novel cDNAs were identified. One with amino acid similarity to the 52 kD Ro/SSA ribonucleoprotein and the RET finger protein, which we named RoRet, one with putative structural similarity to a type 1 sodium phosphate transporter and finally, a cDNA with similarity to MHC class I genes. In addition, this 250 kb region contained 12 previously undiscovered histone genes and 8 ESTs which were not associated with any of the 3 bona fide cDNAs cloned from the region. Analysis of the genomic DNA surrounding these 8 ESTs revealed that they probably arose from oligo d(T) priming of genomic encoded poly (A) stretches and were therefore artifacts.[8]

All 15 genes within this 250 kb interval were screened for mutations by comparing the DNA sequence from 2 HH patients and 2 normal controls. The criteria for the discovery of the founder HH mutation was that the mutation be found on all chromosomes containing the ancestral haplotype and not be found in normal controls at a frequency higher than 5–10%, the HH carrier frequency. Only one cDNA contained a sequence difference that fulfilled these criteria, a G to A transition at nucleotide 845 of the open reading frame in the MHC class I-like cDNA. This base change results in a cysteine to tyrosine substitution at amino acid 282 of the unprocessed protein. A second missense mutation, a C to G change, was found at nucleotide 187 of the open reading frame, which results in a His to Asp substitution at amino acid 63 of the unprocessed protein. The allele frequency of this variant in the general population is approximately 20%. However, it was found that on 8 out of 9 HH patients in our collection were compound heterozygotes for the H63D/C282Y mutations.

(Based on the strong homology of this cDNA to MHC class I-like genes, this gene was initially named *HLA-H*, but was later re-named to *HFE* to avoid confusion with a pseudogene with the same designation).

2. EXPRESSION AND REGULATION OF HFE GENE

Genomic sequencing of the *HFE* gene shows that it is comprised of seven exons, with the basic exon/intron structure of the gene throughout the coding region being identical to other MHC class I genes.[9] However, unlike other class I genes, the basic components of the promoter appear to have significantly diverged or are located in a different arrangement relative to the start of translation.[10] Genomic sequencing of approximately 20 kb 5′ of the end of the existing largest cDNA in our laboratory, cDNA 24, has not revealed any recognizable cis-acting elements known to promote transcription (unpublished data). Attempts to map the promoter using 5′ RACE, Rnase protections and cloning the region upstream of the first exon into various reporter assay vectors, have all been unfruitful (unpublished data). Therefore, it is certainly possible that a large intron and additional 5′ untranslated exon exists separating what is now considered the first exon from the actual start of transcription. Further work will be needed to resolve these issues.

The *HFE* gene is expressed at relatively low levels in most tissues as a 4.0 kb major transcript (Fig. 1).

Two other minor transcripts are also observed at 5.7 and 2.4 kb. It is not known if these transcripts arise from alternative splicing or from alternative 3′ end processing. Rnase protection assays have shown that the liver, mammary gland, small intestine, and spleen express higher levels than bone marrow, kidney, testis, and thymus. Expression of the *HFE* gene is notably absent from the fetal brain (Fig. 2).

The presumed role of the HFE gene product in the regulation of iron homeostasis suggests that the gene itself may be regulated by varying iron levels. To investigate this, we have carried out experiments in undifferentiated human intestinal epithelial Caco-2 cells, whereby the amount of iron available to the cell is modulated first by the chelation of intracellular iron with desferrioxamine and then by the repletion of iron with ferric ammonium citrate. Total RNA was isolated, and the steady-state level of the *HFE* RNA was measured by Rnase protection. No appreciable change in *HFE* RNA levels

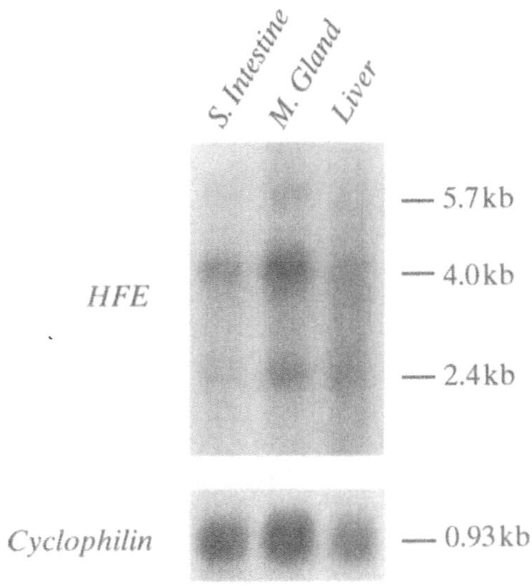

Figure 1. Northern blot analysis of *HFE* mRNA. Poly (A)+ RNA was isolated from the small intestine (lane 1), mammary gland (lane 2) and liver (lane 3) and 2 μg were electrophoretically separated and transferred to a nylon membrane. The membrane was hybridized with radiolabeled probes for *HFE* and cyclophilin. The *HFE* probe reveals three different size transcripts, a major one at 4.0 kb and two minor ones at 5.7 and 2.4 kb (overnight exposure). The cyclophilin probe shows the relative amount of RNA loaded in each lane.

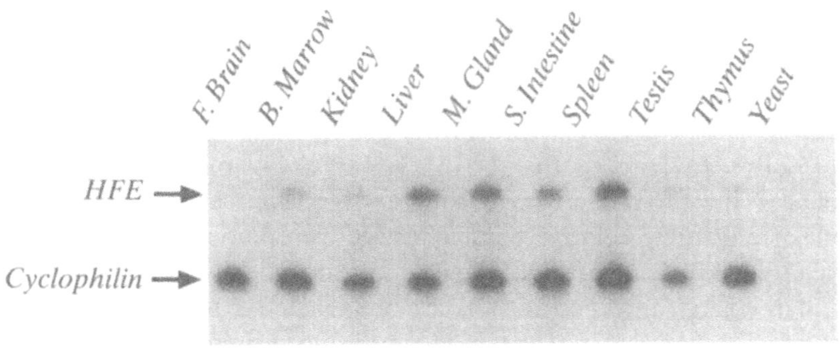

Figure 2. Rnase protection analysis of *HFE* RNA in various tissues. Total RNA from a variety of tissue sources were hybridized with a radiolabeled *in vitro* transcribed probes representing the second exon of the HFE gene and portion of the cyclophilin gene. After digestion with Rnases, the protected products were separated on a denaturing polyacrylamide gel and autoradiograhy performed. *HFE* is expressed at higher levels in the liver, mammary gland, small intestine and spleen, than in the other tissues. The cyclophilin protected product shows the relative amount of RNA in each lane.

was detected, although a slight increase, unrelated to iron status, could be detected in cells that have grown to confluency (Fig. 3).

As a positive control for the iron treatments, the steady-state levels of the transferrin receptor (TfR) was also measured by Rnase protection and found to display characteristic behavior (i.e., increased levels in the absence of iron and dramatic decrease during iron repletion for review see).[11] No change in the level of the human homolog to the rat iron transporter, DCT1[12] was observed in these experiments. We concluded that at least in these cells, the steady-state levels of *HFE* RNA is not regulated by intracellular iron levels in a manner analogous to TfR.

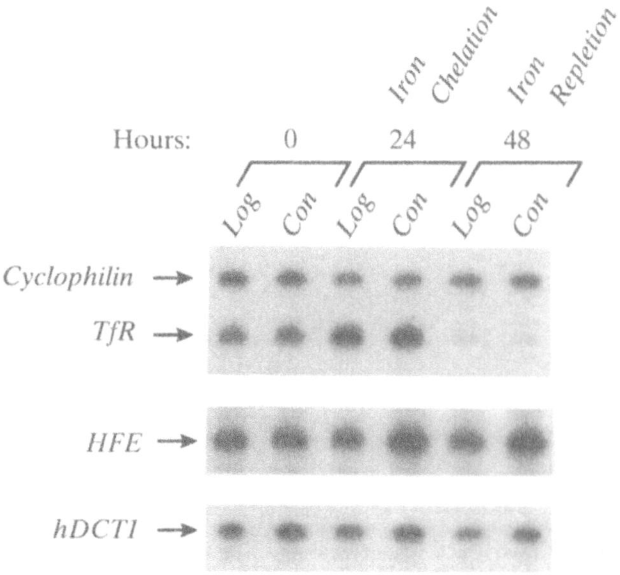

Figure 3. Rnase protection analysis of *HFE* RNA in Caco2 cells. Undifferentiated Caco2 cells were treated with desferrioxamine for 24 hrs. This was followed by the addition of ferric ammonium citrate for an additional 24 hrs. Total RNA was then assayed by Rnase protection. No change in the steady-state levels of HFE or DCT1 were observed. The TfR exhibits characteristic behavior during periods of low and high cellular iron content. The cyclophilin protected product shows the relative amount of RNA in each lane.

3. THE HFE PROTEIN

The analysis of the HFE protein has focused on three main areas: determining the structure of the protein; analyzing the consequences of over-expressing the protein in tissue culture systems and; surveying tissues that express HFE by immunohistochemistry.

The HFE crystal structure has proven, as suggested by primary amino acid sequence, that HFE is indeed a MHC class I-like molecule.[13] Like the classical MHC class I protein HLA-A2, $\alpha1$ and $\alpha2$ domains of HFE from a platform composed of eight antiparallel β strands topped by two antiparallel α helices positioned on top of two immunoglobulin constant-like domains: $\alpha3$ and light chain, $\beta2$ m. The residue mutated in the majority of HH patients, Cys-282 (Cys260 in the numbering system of the mature protein), disulfide bonds with Cys-203. Alterations at this position would have a dramatic effect on overall protein structure. The second residue mutated in some HH patients, His-63, (His-41 in the numbering system of the mature protein) forms a salt bridge with Asp-73. Substituting another residue of similar charge at this position could lead to local rearrangements in a loop in the $\alpha1$ domain. HFE does not bind endogenous peptides and the crystal structure reveals the reason for this. Although HFE has an $\alpha1$–$\alpha2$ platform, the groove between the helices is closer together than classical antigen presenting class I molecules making it impossible for those peptide-interacting residues conserved between HFE and antigen presenting class I to be in a functioning orientation. Hence, HFE represents another example, like FcRn and CD1, of a protein with overall structural similarity to the MHC class I molecule, whose function appears to be non-immune related.

Over-expression of the *HFE* gene in human embryonic 293 cells has shown that the protein is found on the cell surface, whereas over-expressed Cys282Tyr HFE is found exclusively intracellularly.[14] The His63Asp mutation was shown to have no effect on the cell surface presentation of HFE. These studies demonstrate the functional significance of the major HH causing mutation Cys282Tyr, with regard to the trafficking of HFE. Immunohistochemical studies on normal tissue samples have shown that HFE is found on the epithelial cells along the alimentary canal from the esophagus to the rectum. Similar to that observed in cultured cells, HFE can be found on the entire plasma membrane of nonpolarized epithelial cells and on the basolateral surfaces of the polarized epithelial cells of the stomach and colon.[15] However, in the area of the deep crypts of the intestine, the protein appears to be localized intracellularly, with pronounced perinuclear staining. It is the cells of the crypts that are thought to regulate the amount of iron absorbed by the small intestine and transported into the serum. Hence, the observation that HFE is not normally on the cell surface in this region of the gastrointestinal tract suggests that HFE participation in iron homeostasis is one that operates intracellularly.

4. HFE INTERACTING PROTEINS

The question as to how a protein with the characteristics of a MHC class I-like molecule, could influence body iron homeostasis is perplexing. One possibility is that HFE binds metals directly from the serum on the basolateral surface of the cells in the small intestine and signals the iron absorbing machinery on the lumen side of the intestine to either increase or decrease iron uptake. Although this model has not been formally disproven, the structure of the HFE does not suggest that it is capable of binding free metals and furthermore, as discussed above, HFE does not appear to be located on the basolateral surfaces of the cells in the crypts. A more likely model is that HFE interacts with other components of the iron homeostasis pathway and influences their ability to sense and transport iron across the cells of the deep crypts into the serum.

As a first approximation to the question of what other proteins HFE may interact with, cell-surface proteins were labeled with biotin and HFE was immunoprecipitated with anti-HFE antibodies.[16] If associating proteins bind to HFE tightly, they may be detected as unique bands on western blots stained with streptavidn-HRP. In 293 cells which over-express wildtype HFE on the cell surface, a band of approximately 96kD was found to co-immunoprecipitate with the expected HFE bands of 49kD (heavy chain) and 12kD (the β_2m light chain). This 96kD band was absent when similar experiments were performed on cells over-expressing the Cys282Tyr mutant, which is not expressed in a functional form. The 96kD band was also absent in the immunoprecipitate when the experiments were carried out using parental 293 cells, which do not express detectable levels of HFE. This 96kD protein is similar in mass to that of the monomeric subunit of the TfR. Therefore, the immunoprecipitation experiments were repeated using anti-TfR antibodies. Only in the cells over-expressing wildtype HFE was the TfR band of 96kD found to co-immunoprecipitated the 49kD and 12kD components of HFE. It was concluded that the HFE protein forms a stable complex with TfR.

5. BIOLOGICAL SIGNIFICANCE OF THE HFE/TFR INTERACTION

To determine the effect of HFE on TfR, transferrin binding studies were carried out on the cells over-expressing the Cys282Tyr mutant protein, the wild type protein, and the His63Asp mutant protein.[16] Scatchard analysis determined that in cells over-expressing the Cys282Tyr mutant protein, TfR bound transferrin with a K_D of between 12 and 14nM. However, in cells expressing the wildtype HFE protein the TfR bound transferrin with a K_D of between 40–180nM. The presence of the HFE protein had a pronounced effect on the affinity of the transferrin receptor for transferrin. Cells expressing the His63Asp mutant protein had an intermediate effect on the affinity of the TfR for transferrin, with the K_D measured at approximately 20nM. Collectively, these data suggest that cells expressing wild type HFE would be expected to bring less iron bound transferrin into the cell at a steady state.

As a second means of determining the effect of HFE on TfR, various concentrations of soluble HFE/β_2m heterodimers (the same reagent used in the determination of the crystal structure) were added to HeLa cells incubated at 4°C. Scatchard analysis revealed that the affinity of TfR for transferrin decreased in a dose-dependent manner in the presence of soluble HFE/β_2m heterodimers, such that at 1000nM soluble HFE/β_2m heterodimers, essentially no transferrin was bound to the receptor. These data demonstrated that like HFE expressed endogenously, the soluble version of the protein can complex with the TfR and function in decreasing receptor affinity of transferrin. These data also underscore the possible therapeutic applications of using soluble HFE/β_2m heterodimers in the modulation of cellular iron uptake.

A third methodology was employed to study the effect of the HFE/TfR interaction. HeLa cell lines were constructed whereby the *HFE* coding sequence was put under control of the Tet promoter.[17] The expression of *HFE* can thus be regulated by the presence (no HFE) or absence of Tet (HFE). Figure 4 shows a Rnase protection assay carried out on total RNA isolated from HeLa cells in which HFE expression is either repressed (+Tet) or induced (−Tet).

How the other major proteins involved in cellular iron homeostasis, the TfR and ferritin, are affected by the presence of HFE in these cell lines was also addressed.[17] Western blot analysis showed a slight increase in the level of TfR. Scatchard analysis of the transferrin binding data from these cell line corroborated these results by showing an increase in transferrin binding sites in the presence of HFE. This result would imply that the cell is sensing low amounts of available iron and is responding by increasing the steady-state levels of the TfR to bring more transferrin into the cell. The fact that the cell is sensing low iron is demonstrated even more dramatically by the amount of ferritin found in cells over-expressing HFE, which is decreased 10-fold compared to cells grown in the absence of HFE. Again, in the presence of excess HFE, the cell behaves as if the amount of available iron is diminished, and has no need to store excess iron in the form of ferritin.

A similar effect of HFE on ferritin levels can be observed in the HFE over-expressing 293 cell lines (Fig. 5).

Compared to parental 293 cells grown to mid log phase, the amount of ferritin found in cells over expressing the wildtype HFE protein is greatly reduced. Cells over expressing both mutant forms of the protein have higher steady-state levels of ferritin, suggesting that the mutant proteins are less efficient in decreasing intracellular iron levels. These data provide additional support for the argument that over-expressed HFE decreases transferrin uptake and hence intracellular iron.

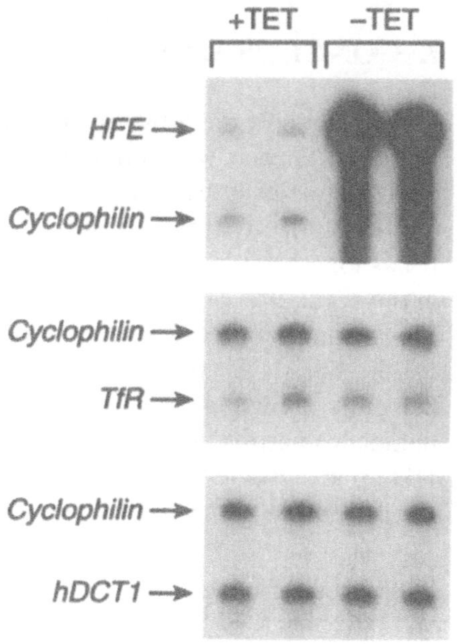

Figure 4. Inducible expression of HFE in fwTHFE/TAHeLa cells. Cells were cultured in the presence (HFE off) or absence of Tet (HFE on). Total RNA was isolated and analyzed by Rnase protection. In the absence of Tet, HFE expression is induced approximately 50-fold. No change is observed in the level of DCT1. The cyclophilin protected product shows the relative amount of RNA in each lane.

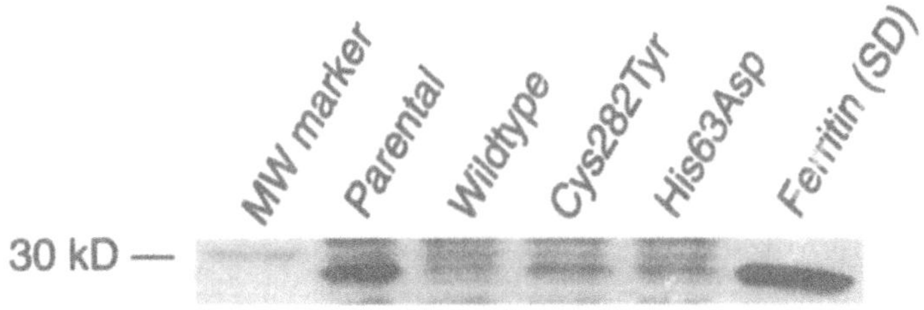

Figure 5. Western blot analysis of ferritin levels in HFE over-expressing 293 cell lines. Equal amounts of protein (120 μg) were blotted and incubated with an anti-ferritin antibody. Parental 293 cells show the highest relative amount of ferritin (lane 2). Cells expressing the wildtype HFE protein have diminished levels of ferritin (lane 3). The mutant proteins both have higher levels (lane 4 and 5). Lane 6 contains purified ferritin protein as an additional marker.

6. THE HIS63ASP MUTANT PROTEIN

The second mutation associated with HH, the His63Asp mutation, is found throughout the general population at a fairly high frequency (20%), yet when found in combination with the Cys282Tyr mutation, hemochromatosis can result. We initially hypothesized that the His63Asp mutant would have some subtle effect on how HFE interacted with a ligand. The crystal structure supports this theory. The results obtained by over-expression of the protein in 293 cells demonstrate that the mutation has an effect on the ability of HFE to modulate the affinity of TfR for transferrin (see above). To extend these observations we have determined the K_i (the concentration of protein needed to achieve half maximal inhibition of transferrin binding measured at the normal K_D of the TfR for transferrin) for a soluble version of the His63Asp protein to 105 nM. This is in contrast to the K_i of the wildtype protein which we have previously measured to 39 nM. Hence, these data further demonstrate the effect of the His63Asp substitution on HFE function.

7. CONCLUSIONS

In spite of the advances made in the field of hemochromatosis since the cloning of the HFE gene, we do not completely understand how the HFE gene product precisely regulates intestinal iron absorption. The observation that HFE complexes with the transferrin receptor in some ways only adds to enigma of this interesting protein. It has long been held that the transferrin/transferrin receptor pathway is not involved in the regulation of intestinal iron absorption, nor in the etiology of hemochromatosis. This assumption must now be reconsidered. A particular challenge is to put the observation that HFE can decrease TfR affinity for transferrin, into the context of what is currently known about iron absorption kinetics in normal individuals and those with hemochromatosis. As shown in one study by Mclaren, the defect in iron absorption in HH appears to be the continued transfer of iron out of the intestinal mucosa and into the serum, even in the presence of excess body iron.[18] No difference in the uptake of iron from the intestinal lumen was observed between normal individuals and those with hemochromatosis. Although the protein responsible for the uptake of iron from the intestinal lumen, DCT1/Nramp2 has recently been cloned,[12] molecules responsible for iron transfer into the serum have yet to be identified. One candidate is the iron binding protein transferrin. Immunohistochemical studies have found transferrin to be present throughout the absorptive cells of the villus including the apical regions, whereas the TfR is generally thought to cycle only from the basolateral surfaces. Within these same cells HFE appears to remain intracellular.[15] It has been suggested that transferrin can act as the shuttle protein for iron between the lumen and basolateral surfaces.[19,20] If specialized pH conditions exist within these cells, it would be possible for the TfR to participate in the transfer of iron-bound tranferrin from the cell into the serum. The role of HFE would be to modulate this transfer event by altering the affinity of the receptor for intracellular transferrin. Recent data of Gross et al.,[17] demonstrates that HFE and TfR form a complex shortly after the biosynthesis of each protein, suggesting that intracellular modulation of TfR function by HFE is a possibility. In the absence of HFE, transferrin would bind at higher affinity to the intracellular receptor and the transfer of iron out of the intestinal mucosa would proceed in an un-regulated manner.

The cloning and preliminary analysis of the HFE gene has stimulated renewed interest in the field of the regulation of iron homeostasis. We anticipate that within a short period of time, many of the major questions stimulated by the cloning of the HFE gene will be answered.

REFERENCES

1. Dadone, MM, Kushner, JP, Edwards, CQ, Bishop, DT, & Skolnick, MH. Hereditary Hemochromatosis analysis of laboratory expression of the disease by genotype in 18 pedigrees. *Am. J. Clin. Pathol.* **78**, 196–207 (1982).
2. Edwards, CQ, et al. Prevalence of hemochromatosis among 11,065 presumably healthy donors. *N. Engl. J. Med.* **318**, 1355–1362 (1988).
3. McLaren, CE, et al. Prevalence of heterozygotes for hemochromatosis in the white population of the United States. *Blood* **86**, 2021–2027 (1995).
4. Bothwell, TH, Charlton, RW, & Motulsky, AG in *The metabolic and molecular basis of inherited disease*. (eds. Scriver, CR, Beaudet, AL, Sly, WS & Valle, D.) 2237–2269 (McGraw-Hill, New York, 1995).
5. Bacon, BR & Tavill, AS in *Hepatology. A textbook of liver disease* (eds. Zakim, D & Boyer, TD) 1439–1472 (WB Saunders, Philadelphia, 1996).
6. Feder, JN, et al. A novel MHC class 1-like gene is mutated in patients with hereditary haemochromatosis. *Nature Genetics* **13**, 399–408 (1996).
7. Lauer, P, et al. Clone-Contig and STS maps of the hereditary hemochromatosis region on human chromosome 6p21.3–p22 *Genome Research* **7**, 457–470 (1997).
8. Ruddy, DA, et al. A. 1.1-Mb transcript map of the hereditary hemochromatosis locus *Genome Research* **7**, 441–456 (1997).
9. Malissen, M, Malissen, B, & Jordan, BR. Exon/intron organization and complete nucleotide sequence of an HLA gene *Proc. Natl. Acad. Sci. USA* **79**, 893–897 (1982).
10. Ting, JP & Baldwin, AS Regulaton of MHC gene expression *Current Opinion in Immunology* **5**, 8–16 (1993).
11. Richardson, DR & Ponka, P. The molecular mechanisms of the metabolism and transport of iron in normal and neoplastic cells *Biochim. Biophys. Acta* **1331**, 1–40 (1997).
12. Gunshin, H, et al. Cloning and characterization of a mammalian proton-coupled metal-iron transporter **388**, 482–488 (1997).
13. Lebron, JA, et al. Crystal structure of the hemochromatosis protein HFE and characterization of its interaction with transferrin receptor *Cell* **93**, 111–123 (1998).
14. Feder, JN, et al. The hemochromatosis founder mutation in HLA-H disrupts β_2-microglobulin interaction and cell surface expression. *J. Biol. Chem.* **272**, 14025–14028 (1997).
15. Parkkila, S, et al. Immunohistochemistry of HLA-H, the protein defective in patients with hereditary hemochromatosis, reveals unique pattern of expression in gastrointestinal tract. *Proc. Natl. Acad. Sci. USA* **94**, 2534–2539 (1997).
16. Feder, JN, et al. The hemochromatosis gene product complexes with the transferrin receptor and lowers its affinity for ligand binding *Proc. Natl. Acad. Sci. USA* **95**, 1472–1477 (1998).

17. Gross, CN, Irrinke, A, Feder, JN, & Enns, CA Co-trafficking of HFE with transferrin receptor implies role in intracellular iron regulation *J. Biol. Chem.* **In press** (1998).
18. McLaren, GD, Nathanson, MH, Jacobs, A, Trevett, D, & Thomson, W. Regulation of intestinal iron absroption and mucosal iron kinetics in hereditary hemochromatosis. *J. lab. Clin. Med.* **117**, 390–401 (1991).
19. Banerjee, DB, Flanagan, PR, Cluett, J, & Valberg, LS. Transferrin receptors in the human gastrointestinal tract. *Gastroenterology* **91**, 861–869 (1986).
20. Huebers, HA, Huebers, E, Csiba, E, Rummel, W, & Finch, CA. The significance of transferrin for intestinal iron absorption. *Blood* **61**, 283–290. (1983).

44

IRON AND ERYTHROPOIESIS

Lessons from Anemic Mice

Nancy C. Andrews and Mark D. Fleming

Howard Hughes Medical Institute and
Division of Hematology/Oncology
Children's Hospital
Departments of Pathology and Pediatrics
Harvard Medical School, Boston MA
USA

1. DISORDERS OF IRON METABOLISM

Iron deficiency and iron overload disorders are among the most common diseases afflicting modern man. Iron deficiency affects up to one billion individuals worldwide. Women and children are most commonly iron deficient, as a result of increased iron needs attributable to regular blood loss and rapid growth, respectively. Untreated iron deficiency in infants and toddlers can lead to irreversible cognitive deficits. Internationally, most iron deficiency can be traced to dietary insufficiency or subacute gastrointestinal hemorrhage associated with chronic parasitic infections. Rare patients have inherited defects in iron absorption or iron utilization. Iron deficiency is usually treatable with oral iron salts. While this treatment is inexpensive, it may be difficult to achieve compliance, because iron repletion takes several months to complete.

Iron overload disorders are also prevalent. When untreated, they lead to tissue iron deposition and consequent fibrosis, most commonly affecting the liver, heart, and endocrine organs. The best understood is hereditary hemochromatosis (HHC), which is prevalent in individuals from northwestern Europe. HHC is the most common genetic disease in Caucasians. It is due to a mutation in HFE, an atypical MHC class I-like molecule.[1] Nearly all HHC patients are homozygous for the same mutation (C282Y), and are descended from a common ancestor in the remote past. A second HFE mutation (H63D) is more widely distributed around the world, but its significance is less clear.[2] The role of HFE in iron metabolism remains to be defined.

There are other inherited iron loading disorders, that do not involve mutations at

Molecular Biology of Hematopoiesis 6, edited by Abraham *et al.*
Kluwer Academic/Plenum Publishers, New York, 1999.

the HFE locus. African iron overload is prevalent in sub-Saharan Africa. It was first thought to be due to environmental factors, because it was initially described among people who drank beer brewed in iron pots. It is now clear that there is an inherited predisposition to iron overload, though the genetics remains poorly understood.[3] Recently, it has become clear that African Americans also have a high frequency of iron loading.[4] It is not yet known whether this is related to African iron overload.

As described elsewhere in this volume, several Italian families have been described who suffer from particularly severe iron overload, termed Juvenile Hemochromatosis.[5] This disorder is also distinct from HHC, though its features suggest that patients carry a mutation in a protein acting in the same iron regulatory pathway as HFE.

2. MODEL SYSTEMS TO STUDY IRON TRANSPORT

In order to develop novel treatments for iron deficiency and iron overload disorders, it is first necessary to work out the details of mammalian iron transport and iron metabolism. Until recently, little was known of the steps involved in intestinal iron absorption or cellular iron uptake. We chose to focus on two aspects of iron transport: luminal transfer of iron into the enterocytes of the small intestine, and endosomal delivery of iron to developing erythroid precursors. We were particularly interested in four questions: (1) What intestinal transporter brings dietary iron into the body? (2) How is that transporter regulated? (3) How is its regulation perturbed in hereditary hemochromatosis? (4) How, precisely, does iron enter red blood cells?

To begin to answer these questions, we chose to use rodents as animal models. Rodents have several clear advantages for this work. First, their iron transport and metabolism appear to be quite similar to that of humans. Second, there are several well-characterized strains that have genetic defects in iron transport pathways. The genetics and physiology of these strains have been studied in detail, and clearly the affected genes are involved in key steps in iron metabolism. Only one of these strains, hypotransferrinemia (*hpx*), has a defect in a protein that was previously known to play a role in iron utilization. The *hpx* mice have a splicing defect that prevents synthesis of normal transferrin protein.[6] The other mutant strains must define novel components of iron pathways. Dense genetic marker maps have been developed for mice (and to some extent for rats), facilitating positional cloning efforts. Finally, mice have the advantage that their genome can be manipulated. By producing transgenic mice and knockout mice the roles of known genes can be studied in new detail.

2.1. Microcytic Anemia Mice

Microcytic anemia (*mk*) mice were first described in 1964 at the Jackson Laboratory.[7] They were discovered as pale newborns with severe skin lesions, and poor viability. The anemia is present at least as early as 15 days gestation, and it is characterized by hypochromic, microcytic red blood cells with increased total red cell numbers.[8] Blood smears show marked anisocytosis, poikylocytosis, and reticulocytosis, consistent with severe iron deficiency anemia. Careful work in a number of laboratories demonstrated that the mice had defects in both intestinal iron absorption and erythroid iron utilization.[8–12] The *mk* locus was mapped to the telomeric portion of mouse chromosome 15 using classical mouse genetics.[13] Three independent alleles have arisen, designated mk^{1J}, mk^{2J}, and mk^{van}.[14-16] Of these, only mk^{1J} has been maintained; the other *mk* strains are now extinct, though DNA has been saved from mk^{2J} animals.

This mutant was particularly interesting because its intestinal defect localized to the apical surface of the enterocyte, suggesting that the mutation had occurred in a gene encoding an intestinal iron transporter.[10] Very little iron entered the intestinal epithelium from the gut lumen, but that iron that did enter was efficiently transferred across the basolateral membrane.

To find the *mk* gene, we set up two large mouse backcross experiments. Homozygous *mk/mk* male mice were outcrossed to either BALB/c mice or *M. musculus castaneus* wild mice. Both BALB/c and *castaneus* are highly polymorphic with the *MK/ReJ* background strain carrying the *mk* mutation. The offspring from these crosses were backcrossed to their *mk/mk* fathers. As a result of this breeding strategy, all N2 backcross progeny carried one *mk* chromosome and a second chromosome that could be *mk*, wild type (BALB/c or *castaneus*) or a mixture of the two, for any chromosome examined. This allowed analysis of one meiotic event in each backcross animal.

By comparison of phenotypes (affected *mk/mk* versus unaffected *mk*/+) and genotypes for a large panel of genetic markers, it was possible to generate a genetic map surrounding the *mk* locus (Fig. 1). The *mk* gene was positioned relative to several anonymous markers (D15Mit 42, 44, 14, 77, 97, and 173) and several known genes (*Scn8a*, *Glycam1*, and *Nramp2*).[16] Strikingly, in 1000 backcross animals, *mk* was invariably non-recombinant with a marker derived from the 3′ untranslated region of the *Nramp2* gene. This prompted us to consider *Nramp2* a candidate for *mk*.

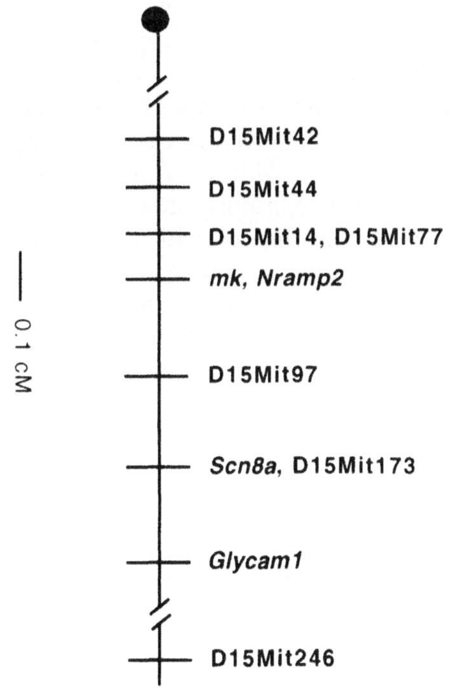

Figure 1. Genetic map of the *mk* locus. A genetic map of distal mouse chromosome 15 was produced from backcross data as described in the text. Positions of anonymous DNA markers and known genes are shown.

Nramp2 was originally isolated as a cross-hybridizing homolog of *Nramp1*.[17] *Nramp1* was one of the first murine genes to be identified by positional cloning.[18] Philippe Gros and co-workers were interested in host resistance to infection. They focused on animals that were particularly susceptible to infections with *Mycobacteria*, *Salmonella*, and *Leishmania*. They determined that a single genetic locus, *bcg*, was important for resistance to these diverse microorganisms. They mapped *bcg*, and showed that the phenotype was attributable to a single basepair substitution in *Nramp1*, that converted a glycine residue to aspartic acid (G169D). When first discovered, *Nramp1* was unrelated to any known proteins. It has twelve predicted transmembrane domains, similar to many transporters, but it could not be shown to transport any molecules tested. Nramp1 protein is exclusively expressed in macrophages, and derives its name from the description Natural Resistance Associated Macrophage Protein 1.

Even less was known about the function of Nramp2, but it had been shown to be expressed in all tissues, making it a plausible candidate gene for *mk*. A transport activity had been shown for a different Nramp-related protein, the yeast protein SMF1. SMF1 was isolated in a screen for mitochondrial defects, and it was shown to function as a manganese transporter.[19] Manganese is adjacent to iron in the periodic table of the elements, and it has long been known that iron and manganese share a common intestinal transport system.[20] This information was a valuable clue in support of the idea that the *mk* locus was *Nramp2*.

We determined the entire coding sequence of the *Nramp2* gene in *mk* mice, and found a single basepair substitution that converted a glycine residue to an arginine (G185R).[16] This mutation was found in DNA from both mk^{1J} and mk^{2J} mice. However, it was not found in eighteen other wild type DNA samples, including samples from the strains upon which the *mk* mutations had arisen. This data, in combination with functional data described below, provided strong evidence that *Nramp2* is the site of the *mk* mutation, and Nramp2 protein functions as a transmembrane iron transporter.

In parallel with our genetic experiments, Hiromi Gunshin and colleagues were attempting to isolate iron transporters by expression cloning using a *Xenopus* oocyte assay system. They found that a single cDNA encoding rat Nramp2 (which they designated DCT1) stimulated iron transport into oocytes.[21] They went on to show that Nramp2/DCT1 functioned as a proton-coupled transporter for a variety of divalent metals, including manganese, cobalt, zinc, copper, and lead as well as iron.

Our genetic studies complement Gunshin's biochemical studies very nicely. Taken together, the data strongly support the idea that Nramp2 is the major transporter of inorganic iron that functions at the apical surface of the absorptive enterocytes in the small intestine.

2.2. Belgrade Rats

The primary site of iron utilization in mammals is the erythroid bone marrow. Almost three quarters of the normal body iron endowment is found in erythroid precursors and mature red blood cells. In the terminal stages of erythroid differentiation, as proerythroblasts mature to form hemoglobinized precursors and finally anucleate red cells, there is an enormous surge in erythroid iron uptake. This is paralleled by a transient increase in expression of cell surface transferrin receptors. Indeed, the transferrin cycle is essential for normal iron transport into erythroid precursors. In normal individuals, nearly all iron in the plasma circulates as a chelate with the abundant protein transferrin. Each transferrin molecule binds two iron atoms. Diferric transferrin interacts with specific cell surface transferrin receptors, to trigger receptor-mediated endocytosis. The transferrin-transferrin receptor complex is internalized in specialized endosomes, that become acidified through the action of a proton pump. Acidification facilitates iron release, and strengthens the protein-protein interaction between transferrin and transferrin receptor. The fate of the protein components is well established; they are recycled back to the plasma membrane for transferrin release and reutilization. In contrast, much less was known of the fate of the iron. Because the inside of the endosome is topologically equivalent to the exterior of the cell, it is necessary for iron to cross the endosomal membrane to enter the cytoplasm. The transporter accomplishing this transmembrane iron transport step was not known.

To determine the identity of the endosomal iron transporter, we took advantage of a second animal model. Belgrade (*b*) rats also have an autosomal recessive, hypochromic, microcytic anemia that is due to a primary defect in iron metabolism.[22-26] Reticulocyte iron uptake had been studied in detail by two groups, and shown to be defective.[27,28] Although the animals can endocytose diferric transferrin, iron does not leave endosomes normally, and much of the internalized iron is inappropriately recycled back to the cell surface (Fig. 2). This phenotype suggested that the *b* defect affects the transporter that exports iron from transferrin cycle endosomes.

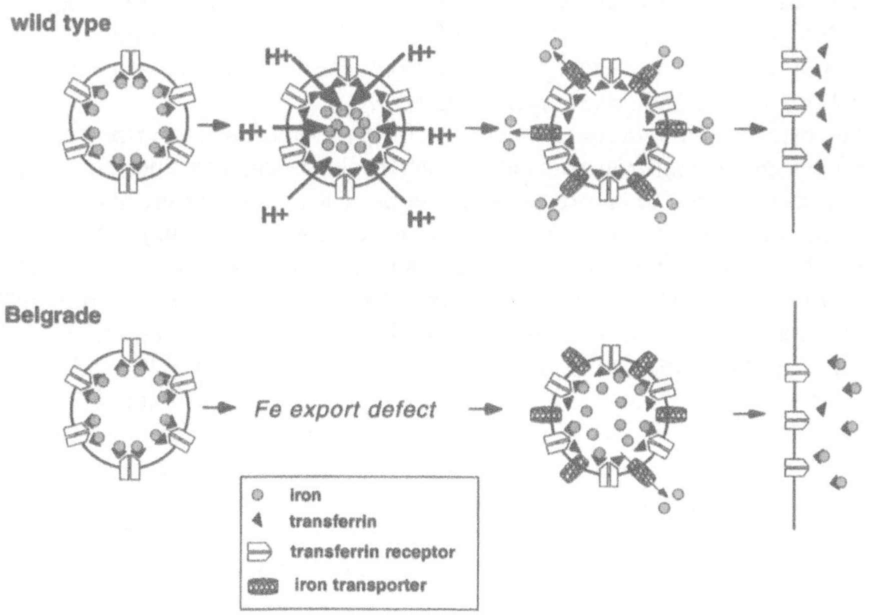

Figure 2. Transferrin cycle in *b* rats. Endosomal steps of the reticulocyte transferrin cycle are depicted. Iron, bound to transferrin, has been internalized with the transferrin-transferrin receptor complex. Wild type endosomes are acidified, and iron is exported through a transmembrane iron transporter. Very little iron exits the endosomes of Belgrade reticulocytes. In all cases, transferrin and transferrin receptor are recycled to the plasma membrane, where transferrin is released. Belgrade animals have a futile cycle, in that iron is also returned to the membrane, rather than retained within the cell. This figure summarizes work from several groups.[27,31]

In order to map the *b* gene, we set up a backcross experiment similar to that used to map *mk*. Breeding was carried out by our collaborators, Michael Garrick, Laura Garrick, and Michelle Romano, at the State University of New York in Buffalo. We found that the *b* phenotype was nonrecombinant with markers near the centromere of rat chromosome 7.[29] This region is syntenic with the telomeric region of mouse chromosome 15, and it contains the rat *Nramp2* gene. After determining that *b* and *Nramp2* were nonrecombinant in approximately 100 backcross animals, we examined the DNA sequence of *b Nramp2*. Surprisingly, we found exactly the same G185R mutation that was found in mk^{1J} and mk^{2J} alleles of *mk* mice.[29] Thus, *Nramp2* is also the gene mutated in *b* rats. Considering the earlier studies of iron uptake by *b* reticulocytes, we conclude that Nramp2 is also important for transmembrane transport of iron out of transferrin cycle endosomes.

3. FUNCTION AND LOCALIZATION OF NRAMP2

To create an assay for Nramp2 iron transport in living cells, we transfected expression constructs carrying *Nramp2* cDNA into human embryonic kidney (HEK293T) cells. Two days post transfection, cells were incubated for twenty minutes with ^{55}Fe-nitrilotriacetic acid at pH 6.0, in the presence of ascorbic acid.[30] Under these conditions, approximately 70% of the iron is in the ferrous state. Iron uptake was stimulated approximately 75-fold in cells transfected with a wild type *Nramp2* construct, as compared to cells transfected with an antisense control construct.[30]

To study the effect of the G185R mutation, we tested an *Nramp2* construct carrying cDNA derived from the *mk* allele of *Nramp2*. We found that the G185R mutation severely affected iron transport. Mutations made elsewhere in transmembrane domain 4 of Nramp2 had less impact on activity.[30]

We postulate that Nramp2 is important both for apical iron transport in the intestine, and for endosomal iron transport in erythroid precursors. Therefore, we would expect Nramp2 protein to be present both on the plasma membrane and within transferrin cycle endosomes. At present, there is no good antiserum directed against Nramp2. However, a FLAG epitope tag, inserted at the C-terminus of Nramp2, has no deleterious effect on iron transport capability, and allows detection of transfected Nramp2. Intact transfected cells were treated with biotin to label surface proteins, immunoprecipitated with anti-FLAG antiserum, fractionated and blotted onto membranes, and probed with streptavidin. Both wild type and G185R Nramp2 proteins were easily detected, indicating that they are present on the cell surface.[30]

Cells transfected with FLAG-tagged Nramp2 were also examined by immunofluorescence. First, cells were incubated with Texas Red transferrin for uptake. They were then permeabilized, and incubated with anti-FLAG antiserum, to stain Nramp2. Both wild type and G185R Nramp2 co-localize with Texas Red transferrin in transferrin cycle endosomes.[30]

4. CONCLUSIONS

Identification of the gene defective in *mk* mice and *b* rats has led to new insights into mammalian iron transport. Nramp2 appears to be the major transmembrane iron transporter. It is responsible for key steps in both intestinal iron absorption and extraction of iron from transferrin cycle endosomes. Nramp2 mRNA is found in all tissues examined, suggesting that Nramp2 may have a role in iron transport in other cell types as well. Future studies must be directed towards understanding the regulation of Nramp2 protein in normal individuals, and in patients with iron deficiency and iron overload.

REFERENCES

1. Feder JN, Gnirke A, Thomas W, Tsuchihashi Z, Ruddy DA, Basava A. Dormishian F, Domingo R, Ellis MC, Fullan A, Hinton LM, Jones NL, Kimmel BE, Kronmal GS, Lauer P, Lee VK, Loeb DB, Mapa FA, McClelland E, Meyer NC, Mintier GA, Moeller N, Moore T, Morikang E, Prass CE, Quintana L, Starnes SM, Schatzman RC, Brunke KJ, Drayna DT, Risch NJ, Bacon BR, Wolff RK: A novel MHC class I-like gene is mutated in patients with hereditary haemochromatosis. Nature Genetics 13:399, 1996.
2. Adams PC, Chakrabarti S: Genotypic/phenotypic correlations in genetic hemochromatosis: evolution of diagnostic criteria. Gastroenterology 114:319, 1998.
3. Gordeuk V, Mukiibi J, Hasstedt SJ, Samowitz W, Edwards CQ, West G, Ndambire S, Emmanual J, Nkanza N, Chapanduka Z, al. e: Iron overload in Africa. Interaction between a gene and dietary iron content. N Engl J Med 326:126, 1992.
4. Gordeuk V, McLaren C, Looker A, Hasselblad V, Brittenham G: Distribution of transferrin saturations in the African-American population. Blood 91:2175, 1998.
5. Camaschella C, Roetto A, Cicilano M, Pasquero P, Bosio S, Gubetta L. Di Vito F. Girelli D, Totaro A, Carella M, Grifa A, Gasparini P: Juvenile and adult hemochromatosis are distinct genetic disorders. Eur J Hum Genet 5:371, 1997.
6. Huggenvik JI, Craven CM, Idzerda RL, Bernstein S, Kaplan J, McKnight GS: A splicing defect in the mouse transferrin gene leads to congenital atransferrinemia. Blood 74:482, 1989.
7. Nash DJ, Kent E, Dickie MM, Russell ES: The inheritance of "mick," a new anemia in the house mouse [abstract]. American Zoologist 4:404, 1964.
8. Russell ES, McFarland EC, Kent EL: Low viability, skin lesions, and reduced fertility associated with microcytic anemia in the mouse. Transpl Proc 2:144, 1970.
9. Bannerman RM, Edwards JA, Kreimer-Birnbaum M, McFarland E, Russell ES: Hereditary microcytic anaemia in the mouse: studies in iron distribution and metabolism. British Journal of Haematology 23:235, 1972.
10. Edwards JA, Hoke JE: Defect of intestinal mucosal iron uptake in mice with hereditary microcytic anemia. Proc Soc Exp Biol Med 141:81, 1972.
11. Harrison DE: Marrow Transplantation and Iron Therapy in Mouse Hereditary Microcytic Anemia. Blood 40:893, 1972.
12. Edwards JA, Hoke JE: Red cell iron uptake in hereditary microcytic anemia. Blood 46:381, 1975.
13. McFarland EC, Russell ES: Microcytic anemia (mk) has been located close to Ca on chromosome 15. Mouse News Letter 53:35, 1975.
14. Russell ES, Nash DJ, Bernstein SE, Kent EL, McFarland EC, Matthews SM, Norwood MS: Characterization and genetic studies of microcytic anemia in house mouse. Blood 35:838, 1970.
15. Handa S, Ferguson JM, Wallace ME, Bulfield G: Characterization and mapping of a viable anaemic mutant in the mouse: a new allele, mk^{van}, at the microcytic anaemia locus. Genetical Research 51:41, 1987.
16. Fleming MD, Trenor CC, Su MA, Foernzler D, Beier DR, Dietrich WF, Andrews NC: Microcytic anemia mice have a mutation in Nramp2, a candidate iron transporter. Nature Genetics 16:383, 1997.
17. Grunheid S, Cellier M, Vidal S, Gros P: Identification and characterization of a second mouse Nramp gene. Genomics 25:514, 1995.
18. Vidal SM, Malo D, Vogan K, Skamene E, Gros P: Natural resistance to infection with intracellular parasites: isolation of a candidate for Bcg. Cell 73:469, 1993.
19. Supek F, Supekova L, Nelson H, Nelson N: A yeast manganese transporter related to the macrophage protein involved in conferring resistance to mycobacteria. Proc Natl Acad Sci USA 93:5105, 1996.
20. Thomson ABR, Olatunbosun D, Valberg LS: Interrelation of intestinal transport system for manganese and iron. J Lab Clin Med 78:642, 1971.
21. Gunshin H, Mackenzie B, Berger UV, Gunshin Y, Romero MF, Boron WF, Nussberger S, Gollan JL,

Hediger MA: Cloning and characterization of a mammalian proton-coupled metal-ion transporter. Nature 388:482, 1997.

22. Sladic-Simic D, Martinovich PN, Zivkovic N, Pavic D, Martinovic J, Kahn M, Ranney HM: A thalassemia-like disorder in Belgrade laboratory rats. Annals of the New York Academy of Sciences 165:93, 1969.
23. Edwards JA, Garrick LM, Hoke JE: Defective iron uptake and globin synthesis by erythroid cells in the anemia of the Belgrade laboratory rat. Blood 51:347, 1978.
24. Garrick LM, Edwards JA, Hoke JE: The effect of hemin on globin synthesis and iron uptake by reticulocytes of the Belgrade rat. FEBS Letters 93:109, 1978.
25. Edwards JA, Sullivan AL, Hoke JE: Defective delivery of iron to the developing red cell of the Belgrade laboratory rat. Blood 55:645, 1980.
26. Bowen BJ, Morgan EH: Anemia of the Belgrade rat: evidence for defective membrane transport of iron. Blood 70:38, 1987.
27. Farcich EA, Morgan EH: Uptake of transferrin-bound and nontransferrin-bound iron by reticulocytes from the Belgrade laboratory rat: comparison with Wistar rat transferrin and reticulocytes. American Journal of Hematology 39:9, 1992.
28. Garrick MD, Gniecko K, Liu Y, Cohan DS, Garrick LM: Transferrin and the transferrin cycle in Belgrade rat reticulocytes. Journal of Biological Chemistry 20:14867, 1993.
29. Fleming MD, Romano MA, Garrick LM, Garrick MD, Andrews NC: Anemia of the Belgrade rat is caused by a mutation in iron transporter protein Nramp2. Proc Natl Acad Sci USA 95:1148, 1998.
30. Su MA, Trenor CC, Fleming JC, Fleming MD, Andrews NC: The G185R mutation disrupts function of iron transporter Nramp2. Blood in press, 1998.
31. Garrick MD, Gniecko K, Liu L, Cohan DS, Garrick LM: Transferrin and the transferrin cycle in Belgrade rat reticulocytes. Journal of Biological Chemistry 268:14867, 1993.

45

JUVENILE HEMOCHROMATOSIS

Search for Candidate Genes

C. Camaschella,° A. Roetto, M. Cicilano, S. Bosio, D. Girelli, A. Totaro,* M. Carella,* A. Grifa,* and P. Gasparini*

Dipartimento di Scienze Cliniche e Biologiche
Università di Torino Azienda Ospedaliera S. Luigi, Orbassano
Torino, Italy
' Clinica Medica, Università di Verona, Italy
* IRCCS CSS San Giovanni Rotondo, Foggia, Italy
° Corresponding author

Juvenile Hemochromatosis (JH) is a rare recessive disorder of iron metabolism which leads to early-onset, severe iron overload. As compared to HFE-associated hemochromatosis (HH),[1,2] JH presents specific features: it affects equally both sexes and it is characterized by a prevalence of hypogonadism and heart failure. Also, the clinical picture is extremely severe and death can occur for cardiac complications if patients are not intensively treated by phlebotomies (for review see reference 3). Genetic evidence indicates that JH is a disorder distinct from the more widespread HH. JH patients do not have mutations in the HFE gene (4, Cazzola M. and Cox T., personal communications), even after screening of all the coding region of the gene.[4] These patients have unusual HLA-A haplotypes.[4–5] The possibility of a second gene on chromosome 6p has been ruled out, since linkage to the interval of chromosome 6p, where the HFE gene resides,[6] has been excluded in Italian families.[4] The disorder seems more frequent in Italy (4,5 and unpublished data), where HFE related hemochromatosis is less frequent than in Northern Europe.[7]

SEARCH FOR CANDIDATE GENES

Theoretically all the genes involved in iron metabolism might be candidates for JH. We have attempted to identify the JH gene using a candidate-positional approach. First, a genome wide search was performed using 375 microsatellites at approximately

Molecular Biology of Hematopoiesis 6, edited by Abraham *et al.*
Kluwer Academic/Plenum Publishers, New York, 1999.

10 cM distance along the entire genome. To assess linkage, homozygosity mapping was used in families with consanguineous parents and identity by descent in families with two affected siblings.[8] This analysis carried out in a small number of patients provided several potentially candidate regions. A Gene Bank search indicated that two of them contained "iron genes". One was 18q21.3, a region where the ferrochelatase (FECH) gene was previously mapped and the second, a region on12q13, was close to the Nramp2 metal transporter.[9,10]

EXCLUSION OF FECH GENE CANDIDACY

FECH gene codifies for the last enzyme involved in the biosynthesis of heme.[11] Although FECH deficiency causes erythropoietic protoporphyria, a dominant disorder characterized by light-sensitive dermatitis and liver abnormalities,[12] its involvement in iron metabolism, the possibility of pleiotropy, and the presence of a 2Fe-2S cluster in FECH[13] prompted us to analyze it as candidate for JH. To this aim we used the following approaches:

a) FECH cDNA from one patient was fully sequenced
b) two FECH intragenic polymorphisms were evaluated: the first was a microsatellite in intron 2[10] and the second a G→T change at position 798 of the cDNA, identified by sequencing and investigated with the use of Mae III enzyme
c) FECH gene was mapped on a YAC contig of the 18q21.3 region developed on the basis of Gene Bank and personal data, using techniques previously described.[14]

Sequencing all the coding region of FECH[11] in one patient did not reveal any causal mutation, but only polymorphic changes (not shown). One out of five patients expected to be homozygotes at the disease locus was heterozygote for both intragenic polymorphisms. FECH gene was mapped outside the "homozygosity region" in a YAC contig of 18q21.3. Taken together these data exclude FECH as candidate gene for JH.

EXCLUSION OF NRAMP2 CANDIDACY

A second potentially candidate region was on 12q13, where the iron transporter gene Nramp2 is mapped. Nramp2 was recently demonstrated to be the gene responsible of mk mouse[9] and Belgrade rat.[15] Although when the gene function is disrupted by inactivating mutations the resulting phenotype in rodents is that of iron deficient microcytic anemia, a mutation in control or regulatory sequences might theoretically produce iron overload. The IRE element at the 3′UTR of the gene was sequenced and found normal in two JH patients. Also, two out of 5 patients studied were heterozygous for the alleles at an intragenic microsatellite in intron 3.[16] Finally, the Nramp2 gene was mapped on a YAC contig outside the homozygosity region. Our results exclude both FECH and Nramp2 as candidate genes for JH and are consistent with the exclusion of all the other iron genes. Thus is likely that the JH gene represents a new iron-related protein.

The distinction between HFE and JH disorders raises the possibility that the different clinical presentation of JH is not only age-related, but likely depends on a different biochemical defect. Identification of the JH gene, beside allowing early diagnosis of the disease, that may avoid cardiac complications, will provide further insights in iron metabolism.

ACKNOWLEDGMENTS

We are indebted to C. Sala and D. Toniolo—DIBIT—YAC Screening Center—Milan, for providing YAC clones. This work was partially supported by MURST ex 40% 1997, E.U.BIOMED Contract BMH4-CT96-0994, CNR contract 97.01054.PF49. M. C. is a recipient of a fellowship from Regione Piemonte.

REFERENCES

1. Powell LP, Jazwinska E, Halliday JW: Primary iron overload. In: Iron Metabolism in health and disease (Brock H, Halliday JW, Pippard MJ, Powell LP eds) London: Saunders 1994: 227–270.
2. Camaschella C, Piperno A: Hereditary Hemochromatosis: recent advances in molecular genetics and clinical management. Haematologica 82: 77–84, 1997.
3. Camaschella C: Juvenile Hemochromatosis. Baillière's Clinical Gastroenterology. 12: 227–235, 1998.
4. Camaschella C, Roetto A, Cicilano M, Pasquero P, Bosio S, Gubetta L, Di Vito F, Girelli D, Totaro A, Carella M, Grifa A, Gasparini P: Juvenile and adult hemochromatosis are distinct genetic disorders. European Journal of Human Genetics 5: 371–375, 1997.
5. Cazzola M, Ascari E, Barosi G et al.: Juvenile idiopathic haemochromatosis: a life-threatening disorder presenting as hypogonadotropic hypogonadism. Human Genetics 65:149–154, 1983.
6. Feder JN, Gnirke A, Thomas W et al.: A novel MHC class I-like gene is mutated in patients with hereditary haemochromatosis. Nature Genetics 13: 399–408, 1996.
7. Carella M, D'Ambrosio L, Totaro A et al.: Mutation analysis of HLA-H gene in Italian hemochromatosis patients. American Journal of Human Genetics 60: 828–832, 1997.
8. Lander ES, Botstein ••: Homozygosity mapping: a way to map human recessive traits with the DNA of inbred children. Science 236:1567–1570, 1987.
9. Fleming MD, Trenor CC, Se MA et al.: Microcytic anaemia mice have a mutation in Nramp2, a candidate iron transporter gene. Nature Genetics 16: 383–386, 1997.
10. Gunshin H, Mackenzie B, Berger UV et al.: Cloning and characterization of a mammalian proton-coupled metal-ion transporter. Nature 388: 482–487, 1997.
11. Taketani S, Inazawa J, Nakahashi Y, Abe T, Tokunaga R: Structure of human ferrochelatase gene. Eur J Biochem 205: 217–222, 1992.
12. Bloomer JR, Bonkowsky HL, Ebert PS, Mahoney MJ: Inheritance in protoporphyria. Lancet 2: 226–228, 1976.
13. Ferreira GC, Franco R, Lloyd SG, Pereira AS, Moura I, Moura JJG, Huynh BH: Mammalian Ferrochelatase, a new addition to the metalloenzyme family. J Biol Chem 296, 10: 7062–7065, 1994.
14. Malaspina P, Roetto A, Trettel F, Jodice C, Blasi P, Frontali M, Carella M, Franco B, Camaschella C, Novelletto A: Construction of a YAC contig covering human chromosome 6p22. Genomics 36: 399–407, 1996.
15. Fleming MA, Romano LM, Garrick LM, Garrick MD, Andrews NC: The iron transporter Nramp2 is mutated in the anemic belgrade (b) rat: evidence of a role for nramp2 in endosomal iron transport. Proc Natl Acad Sci USA 95: 1148–1153, 1998.
16. Lee PL, Gelbart T, West C, Halloran C, Beuler E: The human Nramp2 gene: characterization of the gene structure, alternative splicing, promoter region, and polymorphisms. Blood cells, Molecules and Diseases 15: 199–215, 1998.

46

NEW INSIGHTS ON CONGENITAL DYSERITHROPOIETIC ANEMIA TYPE II

A. Iolascon,[a] E. Miraglia del Giudice,[b] S. Perrotta,[b] R. Carbone,[a] J. Delaunay,[c] M. Carella,[d] and P. Gasparini[d]

[a] Dipartimento di Biomedicina dell' Età Evolutiva
Università degli Studi di Bari
(Email: a.iolascon@bioetaer.uniba.it)
Bari, Italy
[b] Dipartimento di Pediatria
Seconda Università degli Studi di Napoli
Napoli, Italy
[c] Service d'Hématologie
d'Immunologie et de Cytogénétique -INSERM U 473
Le Kremlin-Bicêtre, France
[d] Servizio di Genetica Medica
IRCCS-Casa Sollievo Sofferenza
San Giovanni Rotondo, Italy

Epidemiology and clinical findings. Congenital dyserythropoietic anemia type II (CDA II or CDAN2) is an autosomal recessive disease,[1] which represents the most frequent form of CDA. Like other forms it is characterized by a mild to moderate life-long anemia, ineffective erythropoiesis, and morphologic abnormalities of mature red blood cells and their precursors,[1-3] the reticulocyte count can be normal or slightly elevated; furthermore anisopoichilocytosis, anisochromasia, and spherocytes are present. Splenomegaly and jaundice are the most evident clinical signs.[3,4] Differential diagnosis between CDA II and not dominant Hereditary Spherocytosis (HS) seems to be, in our experience, an intriguing and till now little known problem. Alteration of the normally utilized indices of hemolysis, i.e. indirect bilirubin, hemopessin and haptoglobin, associated with the presence of spherocytes on the blood smear and with the observation of a frequently increased osmotic fragility could erroneously direct some CDA II patients toward HS diagnosis. This is possible particularly in the CDA II subjects showing reticulocytes numbers ranging from 120 to 180 $\times$ 10^9/L. Careful evaluation of the ratio between reticulocyte count and levels of hemoglobin and often the

Molecular Biology of Hematopoiesis 6, edited by Abraham *et al.*
Kluwer Academic/Plenum Publishers, New York, 1999.

electrophoretic studies of the red cell membrane are of help for the correct diagnostic setting.[3,4]

Serology. The erythrocytes from patients with CDA-II showed an increased agglutinability and lysis to anti-i and anti-I sera.[2] Ii are carbohydrate antigens present on the erythrocyte surface and they are determined by branched (I) and linear (i) poly-N-acetyllactosamines.[5] Moreover they are lysed when incubated in acidified serum by an alloantibody present in normal sera (Ham-test).[2,6] Consequently, CDA-II is also called HEMPAS (hereditary erythroblastic multinuclearity associated with a positive acidified serum test).

In only one other syndrome have the red blood cells been shown to be lysed in acidified serum: paroxysmal nocturnal hemoglobinuria (PNH). This syndrome is so clearly different from CDA II that no difficulty in distinguishing them is apparent. Furthermore the serum of patients with CDA II always lacks the antibody and thus, unlike the cases with PNH, lysis does not occur in autologous serum. Compelling evidence indicates that hemolysis of CDA II cells in acidified serum is dependent on the presence of a naturally occurring IgM antibody. This antibody, present in the plasma of approximately 30% of the population, presumably recognizes a particular antigen that arises as a consequence of the abnormal glycosilation of erythrocyte membrane constituents. Recently some experimental data showed that also an aberrant regulation of complement contributes to the susceptibility of CDA II erythrocytes to acidified serum lysis.

Bone marrow observation. This is for long time the principal observation for the CDA-II diagnosis. Bone marrow examination demonstrates five to ten times more erythroblasts than normal (erythroid hyperplasia); early erythroblasts are relatively normal, but about 10–40% of more mature erythroblasts (late polycromatophilic and oxyphilic erythroblasts) are bi- or multinucleated.[2,4]

Extensive morphological anomalies have been observed in electron microscopy on CDA II erythroblastic cells; the most peculiar is the presence of the so-called double membranes, i.e. peripheral cisternae running parallel to and beneath the plasma membrane. The same alteration can also be observed in CDA II erythrocytes. Studies by immunogold electron microscopy showed that the second membrane of the red cell double membrane lacks plasma membrane components. These results suggest that the second membrane can be the extension of the endoplasmic reticulum rather than a duplication of plasma membranes.[7,8]

Biochemical findings. Acrylamide gel electrophoresis in the presence of sodium dodecylsulphate (SDS-PAGE) reveals numerous biochemical changes in circulating CDA II erythrocytes. Band 3 was found with a narrower aspect and a faster migration. As a matter of fact the slightly decreased molecular weight of band 3 in CDA II erythrocytes is consistent with the decreased glycosylation of this protein. The defect in glycosylation is not an exclusive finding of band 3 but concerns the major membrane glycoproteins of CDA II erythrocytes. On the contrary glycolipids appear overglycosylated.[9,10]

Recently, in red cell membrane of CDA II patients, the presence of three minor proteins with apparent molecular weights of 74, 59, and 58 Kd respectively was shown. These bands were lacking in normal controls, in CDA II obligate carriers, in patients presenting with CDA I, HS (with or without band 3 deficiency), and various cases of HE.[11] Amino acid sequencing revealed that the 74 Kd. protein was the glucose-regulated protein referred to as GRP 78, the 59 Kd protein was the protein disulfide isomerase (PDI) and the 58 Kd protein was calreticulin, a high affinity Ca^{2+} binding protein.[12,13] All these proteins are major and constitutive proteins of the endoplasmic reticulum lumen. Their presence strongly supports the hypothesis that the second membrane of the well known double membrane system present in CDA II erythrocytes must arise from the endoplasmic reticulum.

Antibodies against these three proteins are available; detection of GRP 78, calreticulin and PDI in CDAII red blood cells by immunofluorescence microscopy technique has been suggested as a novel, non invasive and straightforward diagnostic tool.

Pathogenesis of hemolysis. More recently we analyzed the effects of reduced glycosylation on the functionality of band 3. Analysis of the anion transport (inhibition of sulphate flux by H2-DIDS) demonstrated that in the CDA-II erythrocytes there was a decrease in the activity of the anion transport for band 3 molecule.[14]

Furthermore we demonstrated that the latter cells contain higher amounts of aggregated band 3 than control eryhtrocytes.

As aggregated band 3 was reported to bind naturally occurring antibodies which are able to mediate the phagocytic removal of red blood cells; we evidenced that both the phagocytic index (RBCs/macrophage) and the amount of membrane bound IgG was increased in CDA-II eryhtrocytes.

Our results suggested that the mild hemolysis showed by CDA-II patients may be ascribed to clusterization of band 3 which leads to IgG binding and phagocytosis and not to a secondary modification of the RBC cytoskeletal structure.

Molecular analysis. At present, it is not clear whether CDA-II is genetically heterogeneous or homogenous, but it appears as a heterogeneous enzymatic defect of glycosylation enzymes.[14–15] Cloning and mapping of the putative genes makes possible molecular analysis; however no linkage studies with this disorder have previously been performed. The knowledge of the localization of candidate genes allowed for the study of linkage in families with multiple affected children using highly polymorphic markers (microsatellites).[14] In most of cases the affected members of each of the families shared different alleles demonstrating the absence of co-segregation within each family. Microsatellite markers flanking the GnT-II, alpha-Man II, and alpha-Man IIx genes gave always negative lod scores at recombination frequencies of 0.0, 0.01, and 0.05. Only marker D5S421 (linked to alpha-Man II) showed a positive but not significant lod score of 0.04 at a recombination frequency of 0.1. Anyway, in this case, the two other markers defining the same chromosomal region gave negative lod scores at the same recombination frequency. Negative results obtained indicated that most likely none of the investigated regions contain the gene involved in CDA-II.[15]

After the exclusion linkage with the three candidate genes, we decided to perform linkage analysis to localize the gene for CDA II. To facilitate the genome search, the screening was initially limited to 12 Italian and 1 French unrelated kindreds. Six of them were characterized by at least two affected individuals, while the remaining six by one affected and one or more normal sibs. Finally a consaguineous pedigree was also selected and included. After analyzing data from 145 markers, which excluded about 50% of the genome, evidence for linkage was obtained with markers D20S195 and D20S107. An additional six markers from the region were then typed in the six families. Pairwise lod scores for the eight markers are shown in Table 1. The highest lod score obtained was with D20S863 ($Zmax = 5.41$ at $q = 0.00$, but no recombinants were detected also with D20S195 ($Zmax = 4.5$ at $q = 0.00$). Positive results were also obtained with the remaining six markers, at recombination frequencies higher than 0.[16]

Several recombinations were detected that help in defining the limits for the CDA II gene location. The centromeric limit is clearly defined by marker D20S890 which show a recombinational event in one family, while additional recombinants have been detected with markers located further centromeric such as D20S200, and D20S912. A recombinant in family CDACE indicates that the CDAII gene is located proximal to D20S908. The region spanning from D20S890 to D20S908 is approximately 5 cM. Since these two markers are located in the long arm of chromosome 20 (20q11.2), we propose this region as the location for the CDA II gene. Refinement of the locus could be obtained by the preparation of additional markers or more easily by new family cases. New families means new recombinants hence the possibility to narrow the area of mapping.

Genetic heterogeneity. Very Recently we described two unrelated families in which CDA-II disease was not linked with the CDAN2 locus demonstrating for the first time the presence of genetic heterogeneity.

The first family comes from a little town on the Ionian see (southern Italy). The parents could be related since 3 of the grandparents had the same family name. Clinical findings are those of severe anemia.

The second family comes from Lecce (southern Italy) and it consisted of two affected and one unaffected sibs. The mother and the sons had beta-thalassemic trait. In probands anemia was documented since the newborn period when jaundice required exchange-transfusion.[17]

Linkage analysis by means of microsatellite markers localized at long arm of chromosome 20 performed in these families demonstrated that the CDA-II locus was not linked to chromosome 20. At the same time we excluded the involvment of Mana-II and GnT-II genes.

This constitutes the first demonstration that CDA-II is genetically heterogenous.

ACKNOWLEDGMENTS

We thank the families and the collagues who collaborated in the CDAN2 Project. This work is supported in part by Ministero Italiano della Sanita' (to PG), by Telethon (to AI) project E-645.

REFERENCES

1. Heimpel H, Wendt F. Congenital dyseryhtopoietic anemia with karyorrehexis and multinuclearity of erythroblasts. Helv.Med.Acta 34: 103, 1968.
2. Crookston JH, Crookston MC, Burnie KL, Fracombe WH, Dacie JV, Davis JA, Lewis SM. Hereditary erythroblastic multinuclearity associated with a positive acidified-serum test: a typical congenital dyserythropoietic anemia. Br.J.Haemat. 17: 14, 1969.
3. Marks PW, Mitus AJ. Congenital dyserythropoietic anemias. Am.J.Hemat. 51: 55, 1996.
4. Iolascon A, D'Agostaro G, Perrotta S, Izzo P, Tavano R, Miraglia del Giudice E. Congenital dyserythropoietic anemia type II: molecular basis and clinical aspects. Haematologica 81: 542, 1996.
5. Hakamori S. Bood group ABH and Ii antigens of human erythrocytes: chemistry, polymorhism and their developmental change. Semin.Hemat. 18: 39, 1981.
6. Rosse WF. Dr. Ham's test revisited. Blood 78, 547, 1991.
7. Wong KY, Hug G, Lampkin BC. Congenital dyserythropoietic anemia type II. Ultrastructural and radioautographic studies of blood and bone marrow. Blood 39: 23, 1972.
8. Fukuda MN, Klier G, Scartezzini P. Congenital dyserythropoietic anemia type II (HEMPAS): characterization of aberrant intracellular organelles by immunogold electron microscopy. Br.J.Haemat. 67: 95, 1987.
9. Alloisio N, P Texier, L Denoroy, C Berger, E Miraglia del Giudice, S Perrotta, A Iolascon, F Gilsanz,

G Berger, J Guichard, JM Massé, N Debili, J Breton-Gorlus, J Delaunay. The cisternae decorating the red blood cell membrane in congenital dyserythropoietic anemia (Type II) originate from the endoplasmic reticulum. Blood 87: 4433, 1996.
10. Anselstetter V, Horstmann HJ, Heimpel H. Congenital dyserythropoietic anemia type I and II: aberrant pattern of erythrocyte membrane proteins in CDA II, as revealed by two-dimensional polyacrylammide gel electrophoresis. Brit.J.Haemat. 35: 209, 1977.
11. Majeed HA, Kalaawi M, Mohanty D, Teebi AS, Al-Gazzar AH. Congenital dyserythropoietic anemia and chronic recurrent multifocal osteomyelitis in three related children and the association with Sweet syndrome in two siblings. J.Pediat. 115: 730, 1989.
12. Fukuda MN, Dell A, Scartezzini P. Primary defect congenital dyserythropoietic anemia type II: failure in glycosylation of eryhtrocyte lactosaminoglycan proteins caused by lowered N-acteylglucosaminyl-transferase II. J.Biol.Chem. 262: 7195, 1987.
13. Fukuda MN, Masri KA, Dell A, Thonar EJM, Klier G, Lowenthal RM. Defective glycosylation of erythrocyte membrane glycoconjugates in a variant of congenital dyserythropoietic anemia type II: association of low level of membrane-bound form of galactosyltransferase. Blood 13: 1331, 1989.
14. De Franceschi L, Turrini F, Miraglia del Giudice E, Perrotta S, Olivieri O, Corrocher R, Mannu F, Iolascon A. Decreased band 3 anion transport activity and band 3 clusterization in congenital dyserythropoietic anemia type II. Experimental Hematol. 26: 869, 1998.
15. Iolascon A, Miraglia del Giudice E, Perrotta S, Granatiero M, Zelante L, Gasparini P. Exclusion of three candidate genes as determinants of congenital dyserithropoietic anemia type II (CDA II). Blood 10: 4197, 1997.
16. Gasparini P, Miraglia del Giudice E, Delaunay J, Totaro A, Granatiero M, Melchionda S, Zelante L, Iolascon A. Localization of the congenital dyserythropoietic anemia II locus to chromosome 20q11.2 by genomewide search. Am.J.Hum.Genet. 61: 1112, 1997.
17. Iolascon A, Perrotta S, Carella M, Lambertenghi Deliliers G, De Mattia D, Gasparini P.—Genetic heterogeneity of Congenital Dyserythropoietic anemia II (CDA II). Blood 1998 (in press).

47

IRON CHELATION IN CANCER THERAPY

Suzy V. Torti,[1,6] R.P. Planalp,[2] M.W. Brechbiel,[3] G. Park,[2] and F.M. Torti[4,5,6]

Department of Biochemistry,[1] Cancer Biology[4]
and Internal Medicine[5]
Wake Forest University School of Medicine
and the Comprehensive Cancer Center of Wake Forest University
Winston-Salem, North Carolina 27157-1082
Department of Chemistry,[2] University of New Hampshire
Durham, NH 03824
Radiation Oncology Branch[3]
National Institutes of Health
Bethesda, MD 20892
Supported in part by Grant No. DK 42412
from the National Institutes of Health[6]

INTRODUCTION

There is a large body of experimental and clinical literature that attests to the importance of iron in tumor cell growth. High levels of dietary iron have been linked epidemiologically to the increased development of tumors in man;[1,2] dietary iron increases the rate of tumor formation in experimental animals.[3-5] Although these observations may in part reflect the role of iron in tumor initiation through the generation of oxygen free radicals, there is abundant evidence that an iron-rich environment contributes to enhanced tumor cell growth. For example, administration of iron-dextran to mice inoculated with leukemia cells dramatically enhanced leukemic cell proliferation; at the higher levels of supplemental iron, the leukemic animals died 25% faster than controls.[6] In a study of subcutaneously implanted tumors in three strains of mice maintained on a low or normal iron supplemented diet, the low iron diet resulted in less rapid tumor growth.[7] Similar observations have been made in spontaneous mammary tumors in rats.[8]

Pathways of iron uptake and regulation in cells are depicted in Fig. 1. Circulating iron is predominantly found associated with the serum transport protein, transferrin.

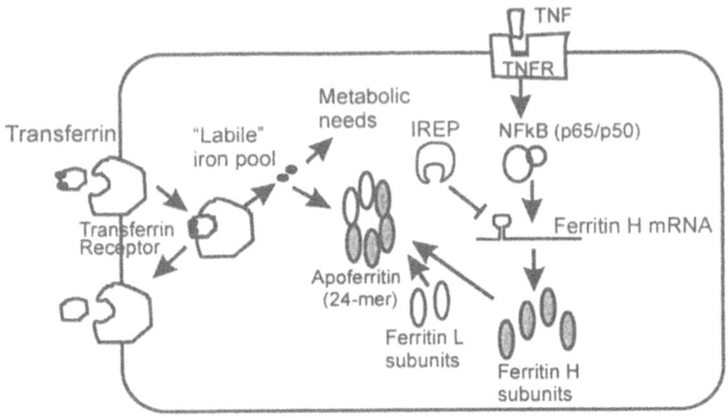

Figure 1. Proteins involved in intracellular iron regulation.

Diferric transferrin binds to transferrin receptors present on cell surfaces, and is endocytosed. Once in the acidic environment of the endosome, iron is released from the transferrin/transferrin-receptor complex to enter a poorly characterized "regulatory" or "labile" iron pool.[9] Iron is delivered from this pool to sites of utilization, including proteins of the respiratory chain, ribonucleotide reductase, and numerous other enzymes that require iron as a cofactor. Iron that is not required for metabolic needs is stored in the protein ferritin. Mammalian ferritin is a 24 subunit protein composed of two subunit types, termed H and L, that assemble in a tissue-specific way to form a protein capable of storing as many as 4500 atoms of iron in a ferrihydride mineral core.[10,11] Intracellular iron levels are sensed and communicated to transferrin receptor and ferritin at least in part through the action of iron regulatory proteins or IRPs. These intracellular proteins act at a translational level to repress ferritin synthesis and augment transferrin receptor synthesis when iron levels are low, and conversely, to permit ferritin synthesis and reduce transferrin receptor synthesis when the cell is iron replete.[12–14] In addition, ferritin levels are subject to transcriptional control. In particular, transcription of the H subunit of ferritin is augmented by TNF[15] via a responsive element that binds NFκB; ferritin H transcription is also repressed by the oncogene E1a.[16] By selectively modulating the H subunit of ferritin, these environmental and oncogenic stimuli alter the subunit composition of newly synthesized ferritin molecules, an effect that in turn modulates the intracellular availability of iron.[17,18] Recently, additional proteins involved in iron uptake have been identified, including the product of the hemochromatosis gene, HFE, which may act to modulate the activity of the transferrin receptor,[19] and Nramp2, a metal transporter with a role in iron transport in both intestinal and erythroid cells.[20,21] Some of the relationships among these proteins are depicted in Fig. 1.

The exuberant demand of some tumors for iron has led to the suggestion that iron deprivation may be of utility in cancer therapy.[22,23] Indeed it has been argued that the hypoferremia and hyperferritinemia that frequently characterize patients with malignant disease represents a host strategy designed to limit tumor growth by reducing available iron.[24] Therapeutic strategies which utilize iron depletion to selectively target tumor cells in humans are currently being explored. Some of these strategies are depicted in Fig. 2. These include the use of the iron chelator deferoxamine in the treatment of neuroblastoma,[25,26] neonatal acute leukemia[27] and prostate cancer, and I^{131} and

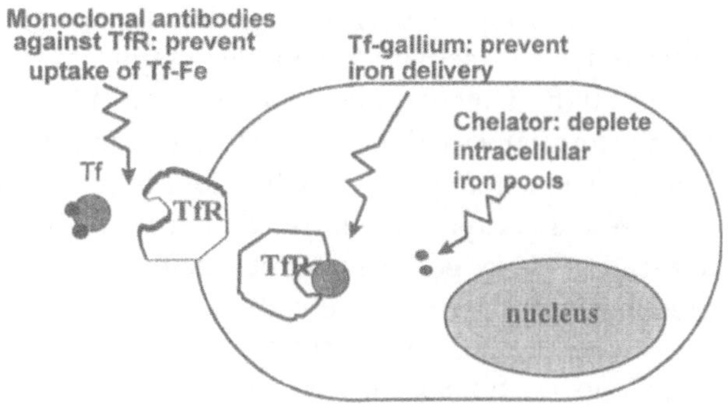

Figure 2. Strategies for iron depletion in tumor cells.

Y^{90} antiferritin antibodies in Hodgkin's disease.[28,29] In addition, gallium, which replaces iron in transferrin and leads to iron deprivation, is in clinical use in the treatment of bladder cancer and other malignancies;[30] and anti-transferrin receptor antibodies, which block the entry of iron into cells, have been investigated in clinical trials (Fig. 2).

However, although iron deprivation strategies to inhibit tumor growth have a sound theoretical underpinning and have shown encouraging results in some clinical trials, the design of iron chelators for use in tumor treatment remains a relatively unexplored field. In fact, iron chelator design has largely focused on treatment of iron overload as a biological endpoint. As illustrated in Table I, the desirable characteristics of a therapy for iron overload may differ considerably from those required for successful anti-tumor therapy. In particular, such criteria as oral availability, lack of toxicity when delivered chronically over a period of many years, and low cost—criteria that have rendered the search for new iron chelators exceedingly challenging—are all of lesser importance in the design of an anti-cancer agent. Anti-tumor drugs are expected to be delivered over a much shorter time interval, and mild toxicities will be correspondingly more tolerable.

Some studies of anti-proliferative effects of iron chelators have been performed. Many of these studies have utilized desferioxamine, a bacterial siderophore and potent iron chelator used in conditions of iron overload.[7,23,31] This has been an attractive drug to consider, since its widespread use in the treatment of iron overload means that its clinical effects are well understood. However, although it possesses some activity, as a single agent in the treatment of human cancer desferioxamine has met with limited success.[25] Ponka, Richardson, and collegues have therefore begun to examine chelators of the pyridoxyl isonicotonyl hydrazide (PIH) family[32] for their potential as anti-proliferative agents. These studies have been promising, and have shown this group of compounds to be effective in inhibiting the growth of cultured cells. The systematic

Table I. Chelator design depends on chelator use: treating cancer is different from treating iron overload

Property	Treatment of iron overload	Treatment of cancer
Oral availability	Important	Not important
Ability to be delivered continuously (daily for years)	Important	Not important
Fe(III) selectivity	Important	Less important
Toxicity	Many toxicities unacceptable	More acceptable

study of a group of PIH analogs has begun to identify key features important for antiproliferative activity. Lipophilicity appears to be one such feature. Interestingly, a recent study demonstrated that for some PIH derivatives, cytotoxicity was not uniformly correlated with iron chelation efficiency, suggesting that the ability to mobilize iron from essential pools may be more important than overall iron affinity.[33]

We have explored a novel chelator, which we term tachpyr, for its potential as an anti-cancer drug. For these experiments, we used cultured bladder cancer cells as a model system. Use of these cells as a model was prompted by previous experiments suggesting that this tumor might be sensitive to iron depletion therapies: gallium nitrate is toxic to J82 human bladder cancer cells cultured *in vitro*.[30] In addition, the expression of transferrin receptors on bladder cancer cells, and the absence of detectable transferrin receptors on normal bladder urothelium[34,35] suggest that agents which target iron metabolism might be both effective and selective agents in bladder cancer chemotherapy.

METHODS

Preparation of tachpyr and its derivatives: Tachpyr, $(N\text{-}Me)_3$tachpyr, and $(N\text{-}Et)_3$tachpyr were synthesized from *cis*-1,3,5-triaminocyclohexane[36,37] and their nitrate salts prepared.[38] Metal complexes of tachpyr were prepared from reaction of the appropriate metal salt with the chelator in methanol solution. Metal salts were recrystallized and purity determined by combustion analyses[39] (G Park, N Ye, N Tsoupas, RD Rogers, MW Brechbiel, and RP Planalp, in preparation). Compounds were dissolved in phosphate buffered saline, pH 7.4, before use in cytotoxicity assays.

Cell lines and cytotoxicity assays. MBT2 and T24 are mouse and human bladder cancer cell lines, respectively. MBT2 cells were a generous gift of Dr. N. Lattime, and T24 cells were obtained from the American Type Culture Collection. Cells were grown in RPMI medium (MBT2) or DME medium (T24) containing 10% fetal bovine serum in a humidified chamber containing 5% CO_2. For cytotoxicity assays, 5×10^3 cells were plated in 96 well tissue culture dishes and allowed to attach overnight before test compounds were added. Six replicate cultures were used for each point. After 72 hours, viability was assessed using an MTT assay.[40] In some experiments, clonegenic[38] or sulforhodamine B[41] assays were used to confirm the results of the MTT assays. IC_{50} was calculated using the Multiple Drug-Effect Analysis Program (Biosoft).

RESULTS AND DISCUSSION

The structure of tachpyr and several other representative iron chelators are shown in Fig. 3. Tachpyr is a novel metal chelator based on the framework triamine cis-1,3,5-triaminocyclohexane.[36] Tachpyr has three aminopyridine arms attached to a cyclohexyl ring (Fig. 1). It undergoes a cyclohexyl ring flip (Fig. 3B) to give a chelator with hexadentate octahedral geometry. Tachpyr accepts metals of radius from 0.76 Å to 0.94 Å with minimal distortion.[42]

Tachpyr has a unique chemical structure relative to other metal ligands that have been explored as anti-tumor therapeutics. It differs from the PIH chelator family[33] in containing exclusively nitrogen donor ligands (Fig. 3). In this regard tachpyr also differs from natural siderophores such as desferioxamine, as well as most chemical chelators (e.g. L1, CP94, DTPA, DBED, EDTA), which generally contain oxygen donor atoms[41]

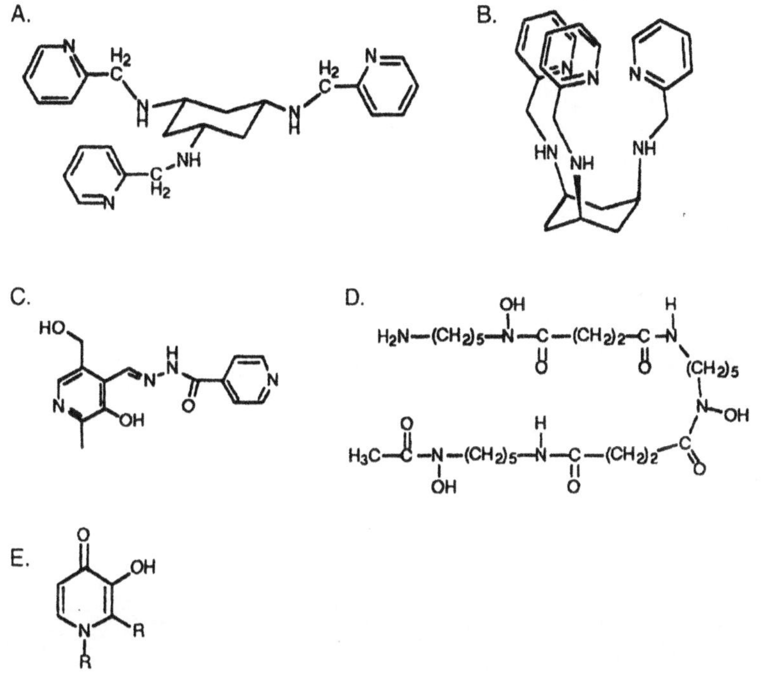

Figure 3. Structure of representative iron chelators. A. Tachpyr, open conformation. B. Tachpyr, closed conformation. C. PIH. D. Desferioxamine. E. hydroxypyridone L1 (deferiprone, R = CH_3).

(Fig. 3). Additionally, tachpyr differs from PIH in denticity, being hexadentate rather than tridentate. As discussed by Singh,[44] due to entropy considerations, hexadentate chelators may afford increased kinetic stability over bidentate chelators. Hexadentate geometry also has the advantage of reducing the concentration of ligand required for effective metal chelation, since it results in a 1:1 rather than 3:1 chelator:metal stoichiometry. This may represent an advantage of tachpyr over other experimental chelators such as L1 (bidentate) and PIH (tridentate). An additional advantage of tachpyr is conferred by its change in conformation following metal binding: as shown by Hou,[45] this property of predisposition—whereby a chelator easily accesses a metal in the open conformation but then closes completely around the metal—may benefit its rate of metal complexation and ability to remove coordinated metals.

Our first experiments testing the potential of tachpyr as an anti-cancer drug measured the ability of tachpyr to kill cultured bladder cancer cells. Stock solutions of tachpyr were dissolved in phosphate buffered saline, added to cells in growth medium at various concentrations, and viability assessed after 72 hours using an MTT assay. As shown in Fig. 4, tachpyr exerted a cytotoxic effect on both a murine and human bladder cancer cell line. These results have been confirmed using other measures of cell killing, including a clonegenic assay and sulforhodamine B assay (not shown). In four independent experiments, the mean IC_{50} of tachpyr for MBT2 cells was 4.6μM.

As expected, we have been able to prepare tachpyr complexes with a number of biologically important metal ions, including calcium, magnesium, manganese, copper, and zinc. However, based on hard and soft acid and base (HSAB) considerations plus metal inherent acidity (Irving-Williams order) considerations, it is anticipated that tachpyr will bind Ca(II), Mn(II), and Mg(II) less tightly than Fe(II), Zn(II), or Cu(II).[46] Tachpyr is therefore more likely to exert its cytotoxic effect by depleting cells of either iron, copper or zinc than by depleting them of calcium, manganese or magnesium. To

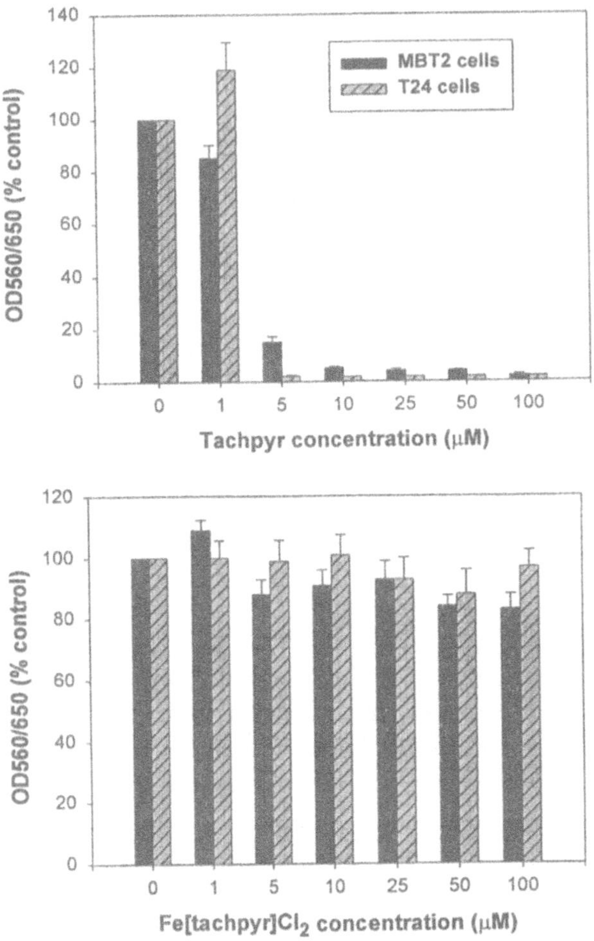

Figure 4. MBT2 and T24 cells were treated with the indicated concentrations of tachpyr or Fe[tachpyr] for 72 hours in growth medium and an MTT assay was performed. The OD560/650 of untreated cultures was defined as 100%.

test this prediction, we prepared complexes of tachpyr with all of these metals and tested the effects of the metal complexes on the viability of MBT2 cells. As seen in Fig. 4, the iron complexes of tachpyr were virtually without cytotoxic effect on either MBT2 cells or T24 cells. Table II shows the results for the remaining metal complexes as tested on MBT2 cells. As predicted, the magnesium, calcium, and manganese complexes of

Table II. Effects of metal complexes of tachpyr on viability

Drug (all at 100 μM)	Percent viability
None	100
Tachpyr	11.3 ± 0.8
[Mg]tachpyr	7.6 ± 0.7
[Ca]tachpyr	5.5 ± 0.5
[Mn]tachpyr	8.5 ± 1.1
[Cu]tachpyr	107 ± 7.0
[Zn]tachpyr	114 ± 4.2
[Fe]tachpyr	82.3 ± 5.0

MBT2 cells were treated with the indicated compounds for 72 hours and viability assessed using an MTT assay. The average of six replicate cultures and standard errors are shown.

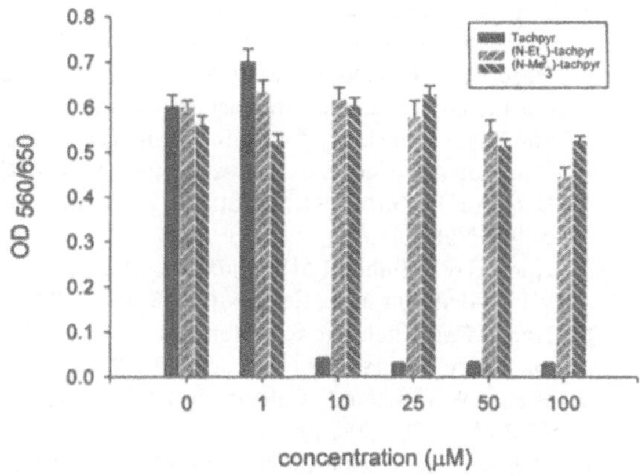

Figure 5. MBT2 cells were treated with the indicated concentrations of tachpyr or N-alkylated derivatives of tachpyr for 72 hours and viability assessed using an MTT assay.

tachpyr were as cytotoxic as tachpyr itself. In contrast, the copper, zinc, and iron complexes of tachpyr showed virtually no cytotoxicity, even at the highest dose tested. Since cytotoxicity can be prevented by formation of the copper, zinc or iron complexes of tachpyr, these results suggest that depletion of one or more of these metals is an important component of the cytotoxic mechanism of tachpyr.

In order to assess further the role of iron chelation in the cytotoxic mechanism of tachpyr, we prepared $(N\text{-}Me)_3$-tachpyr and $(N\text{-}Et)_3$-tachpyr, alkylated derivatives of tachpyr containing either methyl or ethyl groups at the three secondary amines of tachpyr. X-ray crystal structure of these derivatives and their nickel complexes demonstrate that N-alkylation enlarges the metal binding pocket of tachpyr, lengthening the bond distance between the metal and the amine nitrogens from 2.102(4) to 2.168(4) Å (G. Park, N. Ye, N. Tsoupas, R.D. Rogers, M.W. Brechbiel, and R.P. Planalp, in preparation). This steric effect weakens the association of $(N\text{-}Me)_3$-tachpyr with metal ions, and in particular affects the reaction of tachpyr with iron. Thus it is possible to prepare the complexes $Cu[(N\text{-}Me)_3tachpyr][ClO_4]_2$, $Zn[(N\text{-}Me)_3tachpyr][ClO_4]_2$ and $Zn[(N\text{-}Et)_3tachpyr][ClO_4]_2$, while it has not been possible to observe a strong association between Fe(II) or Fe(III) and $(N\text{-}Me)_3$tachpyr (G Park and R Planalp, unpublished observations). We then compared the cytotoxicity of $(N\text{-}Me)_3$-tachpyr and $(N\text{-}Et)_3$-tachpyr to tachpyr itself. As seen in Fig. 5, the N-alkylated derivatives of tachpyr were virtually non-toxic. These results strongly suggest that the ability to interact with iron plays an important role in the cytotoxicity of tachpyr. In support of these results, we have observed that tachpyr selectively represses the synthesis of both the H and L subunits of ferritin, a distinctive property of iron chelators (not shown).

Our studies of tachpyr are still in their early stages. However, they suggest that the further evaluation of tachpyr and/or tachpyr derivatives as anti-tumor agents may be fruitful. More broadly, we believe that our results and those of others herald the emergence of iron chelators as a new class of antitumor therapeutics.

REFERENCES

1. Stevens RG, Jones DY, Micozzi MS: Body iron stores and the risk of cancer. N. Engl. J. Med. 319: 1047, 1988.
2. Stevens RG, Graubard RI, Micozzi MS, Neriishi K, Blumberg BS: Moderate elevation of body iron level and increased risk of cancer occurrence and death. Int. J. Cancer 56: 364, 1994.

3. Siegers CP, Bumann D, Trepkau HD, Schadwinkel B, Baretton G: Influence of dietary iron overload on cell proliferation and intestinal tumorigenesis in mice. Cancer Lett. 65: 245, 1992.
4. Siegers CP, Bumann D, Baretton G, Younes M: Dietary iron enhances the tumor rate in dimethylhydrazine-induced colon carcinogenesis in mice. Cancer Lett. 41: 251, 1988.
5. Smith AG, Francis JE, Carthew P: Iron as a synergist for hepatocellular carcinoma induced by polychlorinated biphenyls in *Ah*-responsive C57BL/10ScSn mice. Carcinogenesis 11: 437, 1990.
6. Bergeron RJ, Streiff RR, Elliott GT: Influence of iron on *in vivo* proliferation and lethality of L1210 cells. J. Nutr. 115: 369, 1985.
7. Hann HWL, Stahlhut MW, Blumberg BS: Iron nutrition and tumor growth: decreased tumor growth in iron-deficient mice. Cancer Res. 48: 4168, 1988.
8. Hann HWL, Stahlhut BA, Menduke H: Iron enhances tumor growth: observation on spontaneous mammary tumors in mice. Cancer 68: 2407, 1991.
9. Breuer W, Epsztejn S, Cabantchik ZI: Dynamics of the cytosolic chelatable iron pool of K562 cells. FEBS Lett. 382: 304, 1996.
10. Harrison PM, Arosio P: The ferritins: molecular properties, iron storage function, and cellular regulation. Biochim. Biophys. Acta 1275: 161, 1996.
11. Theil EC: Ferritin: structure, gene regulation, and cellular function in animals, plants, and microorganisms. Annu. Rev. Biochem. 56: 289, 1987.
12. Phillips JD, Guo B, You Y, Brown FM, Leibold EA: Expression and biochemical characterizatin of iron regulatory proteins 1 and 2 in Saccharomyces cerevisiae. Biochemistry 35: 15704, 199613.
13. Theil EC: The iron responsive element (IRE) family of mRNA regulators. Regulation of iron transport and uptake compared in animals, plants, and microorganisms. Metal Ions in Biol. Syst. 35: 403, 1998.
14. Rouault TA, Klausner RD: Iron-sulfur clusters as biosensors of oxidants and iron. TIBS 21: 174, 1996.
15. Torti SV, Kwak EL, Miller SC, Miller LL, Ringold GM, Myambo KB, Young AP, Torti FM: The molecular cloning and characterization of murine ferritin heavy chain, a tumor necrosis factor-inducible gene. J. Biol. Chem. 263: 12638, 1988.
16. Tsuji Y, Kwak E, Saika T, Torti SV, Torti FM: Preferential repression of the H subunit of ferritin by adenovirus E1A in NIH-3T3 mouse fibroblasts. J. Biol. Chem. 268: 7270, 1993.
17. Miller LL, Miller SC, Torti SV, Tsuji Y, Torti FM: Iron-independent induction of ferritin H chain by tumor necrosis factor. Proc. Natl. Acad. Sci. USA. 88: 4946, 1991.
18. Tsuji Y, Miller LL, Miller SC, Torti SV, Torti FM: Tumor necrosis factor-α and interleukin 1-α regulate transferrin receptor in human diploid fibroblasts: relationship to the induction of ferritin heavy chain. J. Biol. Chem. 266: 7257–7261, 1991.
19. Feder JN, Penny DM, Irrinki A, Lee VK, Lebron JA, Watson N, Tsuchihashi Z, Sigal E, Bjorkman PJ, Schatzman RC: The hemochromatosis gene product complexes with the transferrin receptor and lowers its affinity for ligand binding. Proc. Natl. Acad. Sci. USA 95: 1472, 1998.
20. Fleming MD, Trenor CC, Su MA, Foernzler D, Beier DR, Dietrich WF, Andrews NC: Microcytic anaemia mice have a mutation in Nramp2, a candidate iron transporter gene. Nat. Genet. 16: 383, 1997.
21. Gunshin H, Mackenzie B, Berger UV, Gunshin Y, Romero MF, Boron WF, Nussberger S, Gollan JL, Hediger MA: Cloning and characterization of a mammalian proton-coupled metal-ion transporter. Nature 388: 482, 1997.
22. Kemp J, Cardillo T, Stewart BC, Kehrberg E, Weiner G, Hedlund B, Naumann PW: Inhibition of lymphoma growth *in vivo* by combined treatment with hydroxyethyl starch deferoxamine conjugate and IgG monoclonal antibodies against the transferrin receptor. Cancer Res. 55: 3817, 1995.
23. Hsieh-Ma S, Shi T, Reeder J, Ring DB: *In vitro* tumor growth inhibition by bispecific antibodies to human transferrin receptor and tumor-associated antigens is augmented by the iron chelator deferoxamine. Clin. Immunol. Immunopathol. 80: 185, 1996.
24. Weinberg ED: The role of iron in cancer. Eur. J. Cancer Prev. 5: 19, 1996.
25. Donfrancesco A, Deb G, Dominici C, Pileggi D, Castello M, Helson L: Effects of a single course of deferoxamine in neuroblastoma patients. Cancer Res. 50: 4929, 1990.
26. Frantz CN, Bernstein M, Cstelberry R, Vietti T: Dose escalation of desferrioxamine (DFO) in children with refractory neuroblastoma: a pediatric Oncology Group study. Proc. Am. Soc. Clin. Oncol. 416, 1994 (abstr).
27. Estrov Z, Tawa A, Wang XH, Dube ID, Sulh H, Cohen A, Gelf EW, Freedman MH: *In vitro* and *in vivo* effects of deferoxamine in neonatal acute leukemia. Blood 69: 757, 1987.
28. Order SE, Klein JL, Leichner PK, Frincke J, Lollo C, Carlo DJ: ^{90}Yttrium antiferritin-a new therapeutic radiolabeled antibody. Int. J. Radiat. Oncol. Biol. Phys. 12: 277, 1986.

29. Order SE, Stillwagon GB, Klein JL, Leichner PK, Siegelman SS, Fishman EK, Ettinger DS: Iodine 131 antiferritin, a new treatment modality in hepatoma: a radiation therapy oncology group study. J. Clin. Oncol. 3: 1573, 1985.
30. Dreicer R, Propert KJ, Roth BJ, Einhorn LH, Loehrer PJ: Vinblastine, ifosfamide, and gallium nitrate—an active new regimen in patients with advanced carcinoma of the urothelium. Cancer 79: 110, 1997.
31. Seligman PA, Schleicher RB, Siriwardana G, Domenico J, Gelfand EW: Effects of agents that inhibit cellular iron incorporation on bladder cancer cell proliferation. Blood, 82: 1608, 1993.
32. Ponka P, Borova J, Neuwirt J, Fuchs O: Mobilization of iron from reticulocytes. Febs. Lett. 97: 317, 1979.
33. Richardson DR, Tran EH, Ponka P: The potential of iron chelators of the pyridoxal isonicotinoyl hydrazone class as effective antiproliferative agents. Blood 86: 4295, 1995.
34. Seymour GJ, Walsh MD, Lavin MF, Struitton G, Gardiner RA: Transferrin receptor expression by human bladder transitional cell carcinomas. Urol. Res. 15: 341, 1987.
35. Fujiwara M: Expression of transferrin receptor on transitional cell cancer. Acta. Urol. Jpn. 37: 11, 1991.
36. Bowen T, Planalp RP, Brechbiel MW: An improved synthesis of cis,cis-1,3,5-triaminocyclohexane. Synthesis of novel hexadentate ligand derivatives for the preparation of gallium radiopharmaceuticals. Bioorg Med Chem Lett 6: 807, 1996.
37. Park G, Lu FH, Ye N, Brechbiel MW, Torti SV, Torti FM, Planalp RP: Novel iron complexes and chelators based on cis,cis-1,3,5-triaminocyclohexane: iron-mediated ligand oxidation and biochemical properties. JBIC 3: 449, 1998.
38. Torti SV, Torti FM, Whitman SP, Brechbiel MW, Park G, Planalp RP: Tumor cell cytotoxicity of a novel metal chelator. Blood 92: 1384, 1998.
39. Ye N, Rogers RD, Brechbiel MW, Planalp RP: Synthesis and X-ray crystal structure of N,N′,N″-tris(2-thienylmethyl)-cis-1,3,5-triaminocyclohexane copper(II) dichloride. Polyhedron 17: 603, 1998.
40. Mosmann T: Rapid colorimetric assay for cellular growth and survival: application to proliferation and cytotoxicity assays. J. Immunol. Methods 65: 55, 1983.
41. Papazisis KT, Geromichalos GD, Dimitriadis KA, Kortsaris AH. Optimization of the sulforhodamine B colorimetric assay. J. Immunol. Meth. 208: 151, 1997.
42. Hilfiker KA, Rogers RD, Brechbiel MW, Planalp RP: Tricationic Metal Complexes ([ML][NO3]3, M = Ga, In) of N, N′, N″-Tris(2-pyridylmethyl)-cis-1,3,5-triaminocyclohexane: Preparation and Structure. Inorg Chem 36: 4600, 1997.
43. Galey JB: Potential use of iron chelators against oxidative damage. Adv. Pharmacol. 38: 167, 1997.
44. Singh S, Khodr H, Taylor MI, Hider RC: Therapeutic iron chelators and their potential side effects. Biochem. Soc. Symp. 61: 127, 1995.
45. Hou Z, Stack TDP, Sunderland CJ, Raymond KN: Enhanced iron(III) chelation through ligand predisposition: syntheses, structures, and stability of tris-catecholate enterobactin analogs. Inorg. Chim. Acta. 263: 341, 1997.
46. Huheey JE, Keiter EA, Keiter RL: *In* Inorganic Chemistry: Principle of Structure and Reactivity, New York, Harper, p 347, 1993.

TRANSCRIPTIONAL REGULATION OF THE FERRITIN H-CHAIN AND TRANSFERRIN RECEPTOR IN HEMATOPOIETIC CELLS

Giovanna Marziali, Edvige Perrotti, Ramona Ilari,
Eliana M. Coccia, Ugo Testa, and Angela Battistini

Departments of Virology
Hematology and Oncology, Immunology
Istituto Superiore di Sanità, Rome

ABSTRACT

The mechanisms that regulate the ferritin and transferrin receptor (TfR) expression in response to variation in the cellular iron supply have been extensively studied and represent a model of coordinate post-transcriptional regulation. However, additional mechanisms are operative in hematopoietic cells where high levels of iron are required to sustain specific cell requirements. We investigated the transcriptional regulation of the two main molecules involved in iron metabolism i.e. TfR and ferritin, which are both upregulated during erythroid differentiation. Friend leukemia cells (FLCs) treated with heme or induced to terminally differentiate with dimethylsulfoxide (DMSO), were used as experimental model. By transfection experiments, we identified a minimum region of 77 nucleotides (nt) upstream the TATA box of the ferritin H-gene promoter able to confer transcriptional regulation to a reporter gene in heme-treated FLCs. The CCAAT element present in reverse orientation in this promoter region is necessary for gene activity and binds a protein complex, greatly enhanced in heme-treated FLCs, and containing the heteromeric and ubiquitous transcription factor NF-Y. Similarly we identified in the TfR gene promoter, a region of 500 base pairs (bp) starting from the transcription start site able to confer transcriptional inducibility in FLCs induced to differentiate by DMSO treatment. We conclude that

---

Address reprint requests to: Angela Battistini, PhD, Department of Virology, Istituto Superiore di Sanità, Viale Regina Elena, 299, 00161, Rome, Italy. Phone: (3906) 49903266, FAX (3906) 49902082.

Molecular Biology of Hematopoiesis 6, edited by Abraham *et al.*
Kluwer Academic/Plenum Publishers, New York, 1999.

transcriptional induction of both ferritin and TfR plays a relevant role in the regulation of the synthesis of these two molecules in erythroid cells and is not operative in other cell types (e.g. fibroblastic cell lines).

INTRODUCTION

Iron is required by all cell types as essential for DNA synthesis, electron-transfer reactions, nitrogen fixation. In addition iron serves as metal cofactor for many enzymes either non heme- or heme-proteins. In contrast iron can be a potent harmful agent and therefore the regulation of its metabolism and homeostasis is an essential feature of living organisms both at the systemic and cellular level.

Iron homeostasis is maintained by specialized mechanisms and molecules, which coordinately regulate cellular iron uptake via the transferrin receptor (TfR) and the detoxification of iron excess through sequestration by ferritin. This regulation is mainly exerted through a feed-back control mechanism mediated at the post-transcriptional level by specific high affinity mRNA-protein interactions between the iron regulatory proteins (IRP-1 and IRP-2) and iron responsive elements (IREs) contained in the 5′ and in the 3′ untranslated regions (UTR) of ferritin mRNA and TfR, respectively. In iron deprived cells this binding inhibits ferritin translation by preventing the stable association of the small ribosomal subunit with these messages and stabilizes the TfR transcript against targeted endonucleotidic degradation. On the contrary, when iron levels in the medium rise, the IRE binding is prevented, the TfR mRNA is degraded and ferritin mRNA is actively translated.[1]

This mechanism represents the best characterized level of regulation in the majority of cell types studied, however increasing evidences indicate that in erythroid cells to satisfy their needs for iron and to specifically utilize this metal, additional and/or distinct control mechanisms are operative. Hemoglobin synthesis in red cells is in fact the major iron utilization pathway in the human body and accounts for more 80% of systemic iron turnover.

Despite the extensive analysis and identification of mechanisms and factors involved in the post-transcriptional regulation, the mechanisms and the trans-activating factor(s) by which TfR and ferritin transcription is regulated, in terms of either basal expression, induction, or tissue specific expression are poorly understood. Friend leukemia cells (FLCs) provide a very useful model for studies of erythroid cell differentiation at both cellular and molecular level. They are erythroid precursors blocked in their differentiation pathway at the proerythroblastic stage and treatments with several compounds leads to a sequence of differentiative events that mimic normal differentiation.[2] We previously showed that in FLCs treated with heme or induced to terminally differentiate with dimethylsulfoxide (DMSO) a relevant transcriptional control is responsible for the up-modulation of both ferritin and TfR observed in these cells.[3–6] Here we report the characterization of the minimal proximal region of the ferritin H-chain promoter able to induce transcriptional inducibility to a reporter gene in erythroid cells treated with heme. We also identified the trans-activating factor responsible for this inducibility. Moreover, an analysis of cis-elements responsible for TfR transcriptional control in erythroid cells induced to terminally differentiate with DMSO is reported.

MATERIALS AND METHODS

Cell cultures and treatments. FLC cells were grown in RPMI 1640 supplemented with 5% fetal calf serum (FCS) and antibiotics. F4-12B2 is a thymidine Kinase$^-$ (TK$^-$) FLC clone (kindly provided by W. Ostertag)[7] which was grown in monolayers in minimal essential medium containing 10% FCS and antibiotics. NIH-3T3 cells were grown in Dulbecco modified Eagle medium supplemented with 10% FCS and antibiotics. Cells were treated for the indicated times with 2% DMSO or with 100 µM hemin, prepared as previously described.[8]

Transfection experiments. F4-12B2 cells (1×10^7) were grown in 150 mm culture dishes for 18 hours. Culture medium was then replaced with 30 ml of fresh medium 4 hours prior to the addition of 60 µg of calcium phosphate-precipitated[9] ferritin promoter CAT constructs.[8,10,11] To generate the TfR constructs, a plasmid containing a 3 Kb DNA from the human TfR promoter region (a generous gift of L. Kuhn) was digested to obtain fragments of 500-bp and 300-bp. These fragments were cloned upstream the CAT reporter gene. At 12 hours post-transfection cells were split, to standardize transfection efficiency, the growth medium was then replaced and 6 hours later, cells were treated with 100 µM hemin or 2% DMSO for 24 hours and processed for RNase protection assay.

RNase protection experiments. Total RNA was isolated from transfected cells by the guanidium cesium chloride method[12] followed by a DNase treatment. Total RNA (20 µg) was hybridized for 18 hours to the pCAT[3] and pTRI-GAPDH (AMBION, Texas) riboprobes and processed as previously reported.[8]

Preparation of whole cell extracts. Cells (3×10^7) were washed twice in cold PBS and then collected by centrifugation. The pellet was resuspended in lysis buffer as previously reported[8] incubated for 30 minutes on ice and centrifuged at 10,000 rpm for 10 minutes in an Eppendorf centrifuge. The supernatants were aliquoted and stored at −80°C.

DNA electrophoretic mobility shift assay (EMSA). Oligonucleotides were prepared on an Applied Biosystems DNA synthesizer and labeled with [γ-^{32}P]ATP by T4 polynucleotide Kinase. Total cell lysates (10 µg) were added to the binding reaction mixture[8] and incubated for 20 minutes at room temperature. Samples were then electrophoresed in 6% polyacrylamide gel in 0.5 X TBE buffer for 2 hours at 200 V at 18 °C. Gels were then dried and subjected to autoradiography. Competition studies were performed by adding unlabeled double stranded oligonucleotides at 100-fold molar excess over the labeled probe. Supershift analysis was performed as previously described.[8]

RESULTS

Functional analysis of the proximal region of the ferritin H-chain promoter. We previously showed that treatment of FLCs with heme induces an accumulation of ferritin mRNA between 10 and 15 folds.[3] This accumulation is not due to a stabilization of the transcript but reflects an increase in the rate of transcription of the ferritin H-chain gene dependent upon a 160-bp segment upstream from the transcription start site. To identify the minimal promoter sequence required for transcriptional induction by heme, different fragments of the human ferritin promoter have been cloned upstream the CAT reporter gene and transiently transfected in F4-12B2, a TK^- FLC clone able to grow in monolayer. After transfection, cells were split to standardize the transfection efficiency and treated for 24 hours with control medium or 100 μM heme. RNA was then extracted and analyzed by RNase protection with a specific riboprobe for CAT mRNA. The glyceraldehyde-3-phosphate dehydrogenase (GAPDH) riboprobe was used as an internal control.

As shown in Fig. 1 in FLCs (white and grey bars), the expression of the 77-bp fragment construct was stimulated by the heme treatment to the same extent as was the expression of the entire 160-bp fragment construct, indicating that the 77-bp fragment is sufficient for the transcriptional activation by heme of the ferritin-H gene promoter. When the same constructs were transfected in the fibroblastic cell line NIH-3T3 the 77-bp (black bars) as well as the 160-bp (not shown) construct expression was not enhanced by the heme treatment, indicating that the transcriptional stimulation by heme of the ferritin H-chain gene promoter is specific for erythroid cells.

Analysis of trans-activating factor(s) binding to the functional ferritin H gene promoter region. To identify the regulatory factor(s) able to bind to the 77-bp region of the ferritin H-chain gene promoter, EMSA was performed. Whole cell extracts were incubated with a labeled double stranded oligonucleotide corresponding to a minimal portion of the 77-bp fragment spanning from −77 nt to −43 nt and the complex(es) formed resolved by PAGE. As shown in Fig. 2, a discrete complex formed with this

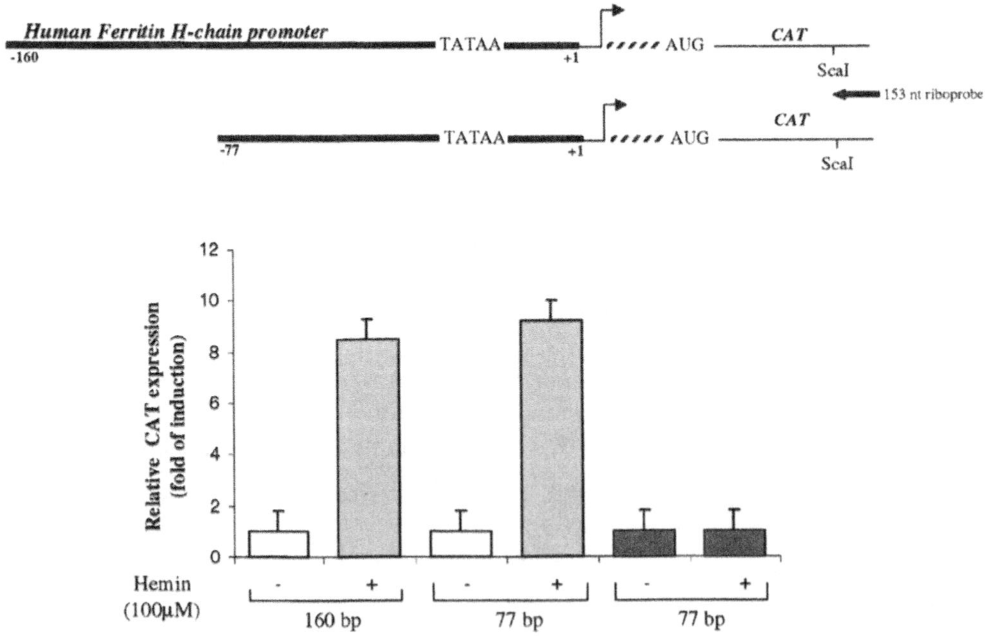

Figure 1. *In vivo* stimulation of the ferritin H-gene proximal promoter by heme. F4-12B2 (white and grey bars) and NIH3T3 cells (dark grey bars) were transiently transfected with constructs containing the 160-bp or the 77-bp ferritin H promoter fragment linked to the CAT reporter gene. After transfection, cells were split and then treated for 24 hours with medium alone or 100 μM hemin. RNA was then extracted and analyzed by RNase protection using specific riboprobes for CAT (see Materials and Methods). The autoradiograms were scanned by an LKB 2202 Ultoscan XL laser microdensitometer and the values, expressed in arbitrary units, are reported in the graph after normalization with GaPDH used as internal control.

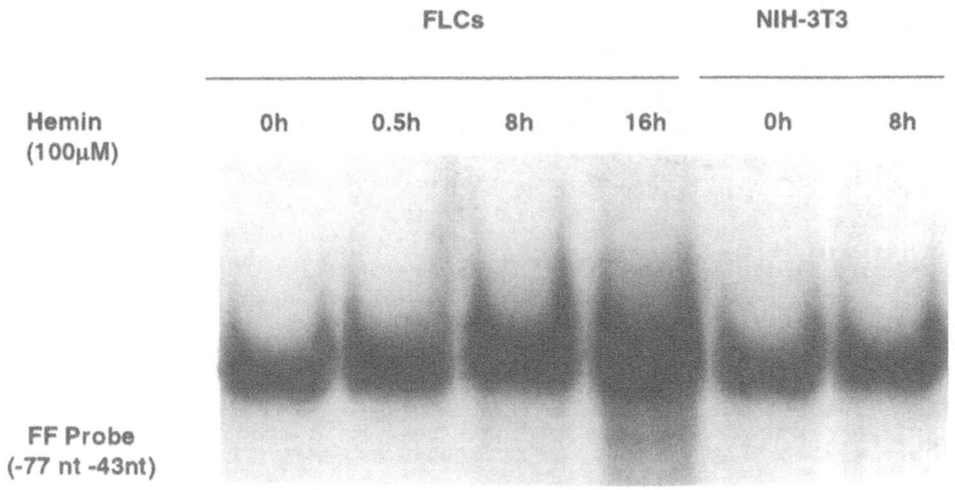

Figure 2. Kinetic of activation of transcription factor(s) able to bind to a minimal fragment (FF) of the 77-bp proximal region of the ferritin H-gene promoter in erythroid (FLCs) and fibroblastic (NIH-3T3) cell lines. Total cell extracts (10μg) from control cells and cells treated with 100μM hemin for the indicated time were incubated with FF radiolabeled probe and analyzed by EMSA (see Materials and Methods).

region, starts to increase after 30 minutes of 100μM heme treatment and lasts several hours. Conversely a complex with the same electrophoretic mobility is present in cell extracts from NIH-3T3 but the heme treatment has no influence on its amount, irrespective of the treatment time.

As shown in the upper panel of Fig. 3, the sequence of the ferritin H proximal promoter region which binds the complex shown in Fig. 2, presents a CCAAT box in reverse orientation surrounded by two Sp1 consensus-binding sites. To determine the relative significance of these two elements in the formation of the complex, specific competitions and supershift assays were performed. As shown in Fig. 3 the binding of the complex was not specifically competed by the consensus sequence for Sp1, nor were specific anti Sp1 antibodies able to supershift the complex. Similarly an oligonucleotide mutated in both the Sp1 sites (4T) competed for the binding. However, when an oligonucleotide, FFmut, containing a mutation of the central CCAAT residue (CC*G*AT) known to be critical for CCAAT factor binding was used, no more competition for the binding was achieved. The importance of this CCAAT box was functionally demonstrated in transfection experiments where, using the 77-bp promoter construct bearing the CC*G*AT point mutation, the expression of CAT mRNA was reduced to background levels.[8]

Sequence specificity of the CCAAT-box-binding protein stimulated by the heme treatment in FLC. To date a multiplicity of CCAAT box binding factors have been identified, some of which are tissue specific whereas others are expressed ubiquitously.[13] To test whether the CCAAT box of the ferritin H-chain proximal promoter can bind proteins that have already been described to interact with the CCAAT motif, specific oligonucleotides containing consensus binding site for cEBP/β, NFI, NF-Y, and CP2 were used for competition experiments. The sequences of these oligonucleotides are indicated on the top of Fig. 4. These experiments demonstrated that only the consensus sequence specific for NF-Y was able to compete for the binding, whereas sequences comprising the binding site for CP2, cEBP/β or NFI, were not.

Sequences of oligonucleotides used as competitors in EMSA

FF	GGCTCGGGGCGGGCGGCGCTG**ATTGG**CCGGGGCGGG
wt Sp1	ATTCAGTCGGGGCGGGGCGAGC
4T	GGCTCGG**TT**CGGGCGGCGCTGATTGGCC**TT**GGCGGG
FF mut	GGCTCGGGGCGGGCGGCGCTGAT**G**GGCCGGGGCGGG

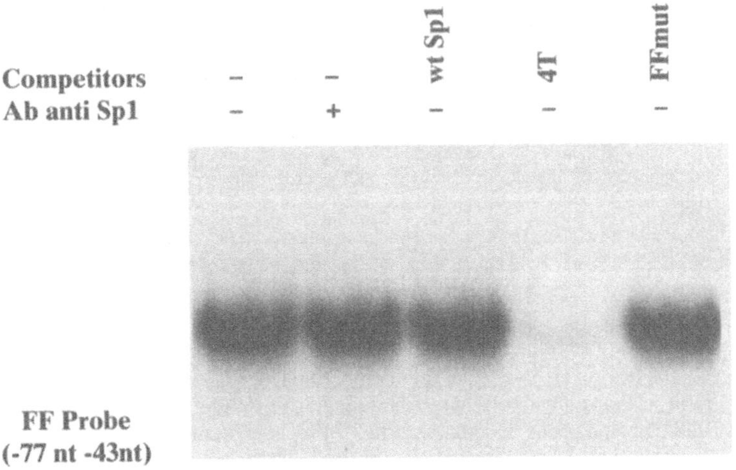

Figure 3. Analysis of the FF sequences necessary for the binding of the heme-stimulated complex in FLC cells. Whole cell extracts from FLCs treated with 100 μM hemin were incubated with the FF labeled probe and analyzed by EMSA. Sequences of oligonucleotides used as probes or competitors are shown at the top of the Figure. A 100 fold molar excess of unlabeled oligonucleotides wt Sp1, 4T and FFmut and polyclonal anti Sp1 antibody were added as indicated.

Sequences of oligonucleotides used as competitors in EMSA

c EBP/β	TGCAGATTGCACAATCTGCA
NFI	TTTGGCTTGAAGCCAATATGAG
CP2	GCCCTAACAAGTTTTACTGGGTAGAGCAAGCACAAACCAGCCAATGAGTA
NF-Y	GTCTGAAACATTTTTCTGATTGGTTAAAAGTTGAGTGCT

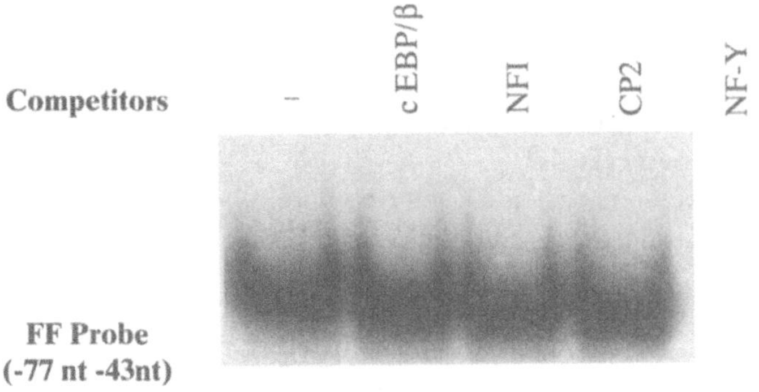

Figure 4. The heme-stimulated complex in FLCs contains the NF-Y transcription factors. Whole cell extracts from hemin treated FLCs were incubated with the FF radiolabeled probe. Competition assays were performed using 100 fold molar excess of unlabeled oligonucleotides containing the consensus binding sites for cEBP/β, NFI, CP2, and NF-Y, whose sequences are indicated on the top of the Figure.

Specificity of a ubiquitous transcription factor. NF-Y is the major CCAAT-box recognizing transcription factor composed by at least three subunits NF-YA, NF-YB, NF-YC, all of which are necessary for DNA binding.[14–16] Its unique structure and evolutionarily conservation suggest a crucial role in transcription of eukaryotic genes. However, although ubiquitous, it also participates in the regulation of some promoters by controlling gene expression in a lineage- and activation-specific manner.[17,18] Increasing evidences from us and other laboratories suggest a modulation of this factor in particular conditions such as, serum starvation, calcium depletion, differentiation, and cell proliferation.[8,19–21] The specificity could be provided in different tissues: **i)** by the combined action with other cell specific transcription factors; **ii)** by the type of NF-YA isoform expressed in different cell types; **iii)** by modulating its activity through the availability of one of its subunits. To investigate the mechanism responsible for NF-Y modulation in heme-treated FLCs Western blot analysis was performed with specific antibodies recognizing the A and B subunits of NF-Y. As shown in Fig. 5 the NF-YB subunit was equally expressed in untreated and 100 μM hemin-treated FLCs, whereas

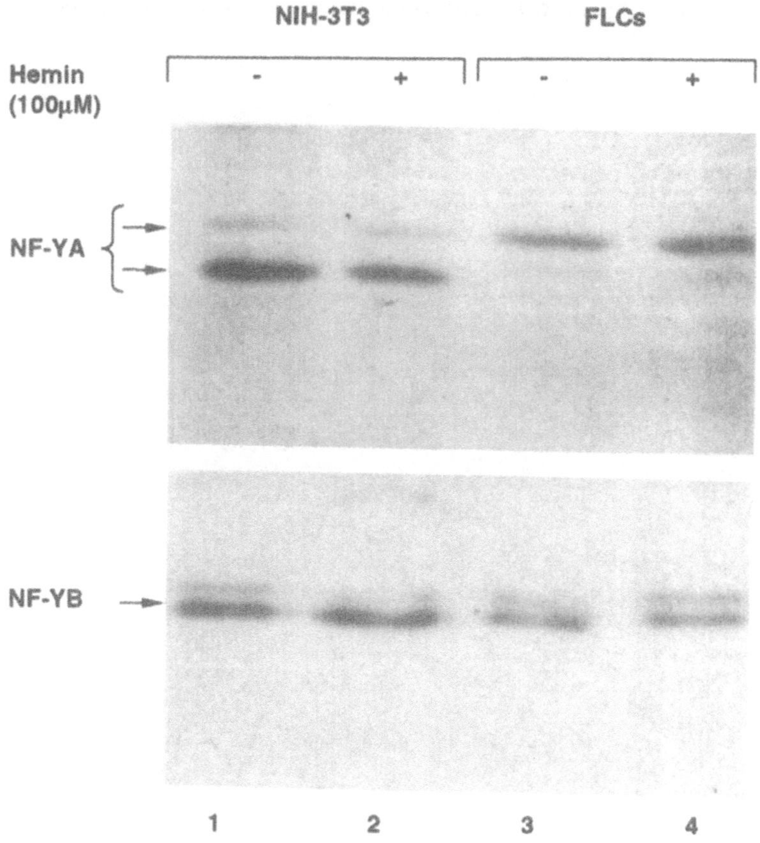

Figure 5. Western blot analysis of NF-Y and its subunits expression in fibroblastic and hemopoietic cell lines. Western blot analysis of whole cell extracts from NIH-3T3 and FLCs treated with control medium or 100 μM hemin, was performed as indicated in the Materials and Methods. Samples (60 μg) were probed with an affinity purified antibody against NF-YB or against NF-YA. The doublet, which presumably corresponds to the long and short form of NF-YA, is indicated.

the NF-YA subunit was present in low amounts in untreated FLCs and increased after the heme treatment.

When cell extracts from the fibroblastic cell line NIH3T3 were examined no modulation for both subunits was observed.

Functional analysis of the proximal region of the TfR promoter. TfR is upregulated in differentiating erythroid cells where the negative feedback able to maintain iron homeostasis seems not operative.[22-25] In differentiating erythroid cells in fact, which must incorporate high levels of iron to sustain hemoglobin synthesis, a several-fold increase in TfRs number is observed even in the presence of high levels of iron.[24,25] This increase precedes the production of hemoglobin and to date only indirect evidences suggest that a transcriptional regulation mechanism may play a relevant role.[22] To assess if the increase in TfR number observed during erythroid differentiation may be due to transcriptional regulation, progressive deletions of the TfR proximal promoter were performed. Fragment of 500-bp and 300-bp of the proximal TfR promoter have been cloned upstream the CAT reporter gene and transiently transfected in F4-12B2 cells. Cells were then treated for 24 hours with 2% DMSO and the CAT mRNA analyzed by RNase protection. As shown in Fig. 6 these two constructs were equally stimulated by the DMSO treatment (white and grey bars) indicating that the 300-bp fragment starting from the transcription start site is sufficient for the transcriptional

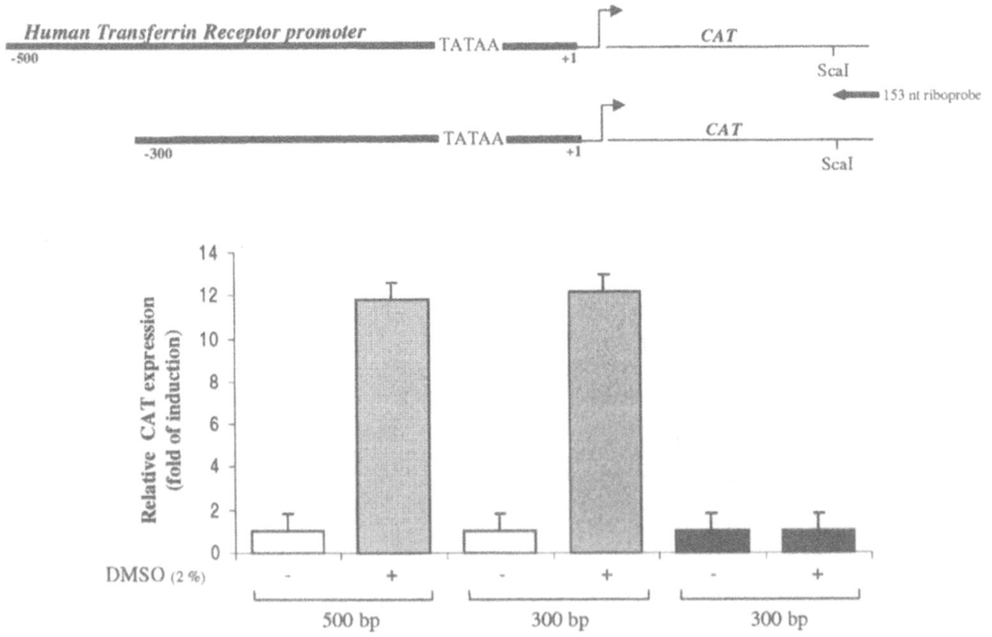

Figure 6. *In vivo* stimulation of the TfR gene proximal promoter by DMSO. F4-12B2 (white and grey bars) and NIH3T3 (black bars) cells were transiently transfected with constructs containing the 500-bp or the 300-bp TfR gene proximal promoter fragment linked to the CAT reporter gene. After transfection, cells were split and then treated for 24 hours with medium alone or 2% DMSO. RNA was then extracted and analyzed by RNase protection using specific riboprobes for CAT (see Materials and Methods). The autoradiograms were scanned by an LKB 2202 Ultoscan XL laser microdensitometer and the values, expressed in arbitrary units, are reported in the graph after normalization with GaPDH used as internal control.

activation of the TfR gene. This transcriptional stimulation seems to occur only in erythroid cells induced to differentiate, indeed when the same construct (300-bp) was transfected in NIH-3T3 cells (black bars), CAT mRNA expression was not enhanced by the DMSO treatment.

Experiments are in progress to identify both the minimal region of the promoter necessary for transcriptional regulation and the transcriptional factor(s) able to bind to this region and drive transcription during erythroid differentiation.

DISCUSSION

The present studies focus on the transcriptional regulation of the ferritin H-chain and TfR genes during erythroid differentiation.

Despite the extensive analysis and identification of the mechanisms and factors responsible for the coordinate post-transcriptional regulation of these two molecules, increasing evidences indicate that in some specialized cells such as macrophages, which are physiologically involved in iron storage,[23] and FLCs undergoing erythroid differentiation,[2,24,25] the regulation of the expression of these genes is highly dependent on increased transcription rate. We previously identified[3] a 160-bp segment upstream from the transcription start site of the human ferritin H-chain gene that confers transcriptional regulation by heme to a reporter gene. We extended the analysis of this segment of the ferritin proximal promoter and identified a 77-bp minimal region sufficient to drive transcription in erythroid cells in response to heme. Additional regulatory elements located further upstream from the transcription start site have been shown to be necessary for full activation for the mouse gene. In MEL induced to differentiate or in hemin-treated K562 cells transcriptional activation of the mouse H-ferritin gene is mediated by an enhancer element located 4.5 kb upstream of the transcription start site.[26] This region includes a perfect match to the consensus NF-E2 binding site, however the exact role of this hematopoietic transcription factor in the erythroid-specific activation of the H-ferritin gene is not yet clear. We instead identified a transcription factor, which specifically binds to the proximal promoter region of the human ferritin H gene and increases expression of a reporter gene after the heme treatment. This factor, called NF-Y, is one of the best characterized CCAAT box-binding protein. It is a ubiquitous heterotrimeric protein composed of three subunit (A, B, C) all necessary for the DNA binding and activates a high number of eukaryotic genes.[14–16] NF-Y acts participating in the correct positioning of other transcription factors at the transcription start site as well as increasing the activity of neighbouring enhancer motifs. It also participates in the regulation of some promoters by controlling gene expression in a lineage- and activation-specific manner.[15,16] Even if the molecular basis of this specificity have to be clearly defined, there are increasing evidences which account for a modulation of the NF-Y binding activity in different systems. In particular by Western blot analysis we showed that the expression of the A subunit of NF-Y is modulated during the heme treatment. Even if the combined action with other cell specific transcription factors can not be definitively excluded, these results suggest a mechanism of specificity due to the limited availability of one subunit of NF-Y. Similarly we showed that in another differentiating system, i.e. monocytes maturing to macrophages, the increased accumulation of ferritin mRNA observed, is largely dependent on a progressive stimulation of NF-YA expression.[8] We thus conclude that in few specialized cell types in which the transcriptional control plays a pivotal role in the regulation of ferritin expression, this control is largely exerted by the availability of the A subunit of the NF-Y transcription factor.[present report and 27] Structural studies demonstrate that NF-YB and NF-YC show sequence similarity with histone fold motifs[28] and that the YB/YC dimer can be found in high molecular weight complexes even in the absence of NF-YA.[29] This observation allow us to speculate that NF-YB and NF-YC are devoluted to play some fundamentally basic roles in gene activation, whereas NF-YA may represent the regulatory subunit of the entire complex.

In the present paper we also analyzed the transcriptional regulation of TfR during erythroid differentiation. As demonstrated for ferritin also TfR is upregulated in erythroid cells and it has been shown by us[6,25] and by other authors,[30–33] that the negative feedback mechanism able to maintain iron homeostasis in a variety of cell types does not seem to be operative in erythroid cells. It has been suggested that the high rate of transcription operative in erythroid cells can override the IRE-dependent control.[33,34] The transcriptional regulation of TfR has been investigated especially in response to mitogens. In many cell types in fact TfR expression correlates directly with their rate of growth, being high in proliferating cells, and declining during growth arrest. However, if TfR gene expression is related to proliferation in non-erythroid cells it seems also directly related to erythroid differentiation. Until now transcriptional activation of the TfR gene has been demonstrated by run-on assays in phytohemagglutinin-stimulated T lymphocytes,[35] in DMSO-treated MEL[33] and in chick embryo erythroblasts[34] but no indications of the cis and trans elements involved in this control have been, to date, identified. Here we show that a minimal portion of 500-bp of the TfR proximal promoter is able to drive the transcription of a reporter gene in response to the DMSO in FLCs, but not in fibroblastic cell lines. Several consensus binding sites for known transcription factors have been identified in this TfR promoter region. Beside them, sequences able to potentially bind factors of the Ets[36] and AP1 family and of the ubiquitous transcription factor Sp1 have been reported.[37] Interestingly, a similar region seems involved in the mitogenic induction. Moreover, in the case of the mitogenic stimulus the major complex induced involves Sp1.[37] Our preliminary results indicate that also during differentiation Sp1 sequences play an essential role probably mediating or favouring the binding of specific transcription factors. We could thus speculate that depending on cell types and stages of cell growth and differentiation the same elements can mediate different activities.

In conclusion, our results provide evidence that transcriptional stimulation is a relevant mechanism of control of ferritin and TfR gene expression in a few specialized cell types such as erythroid cells. In these cells these genes are hyperexpressed and the negative feedback mechanism exerted in other cell types by heme is not operative. Moreover, the identification of the ubiquitous transcription factor NF-Y as able to modulate the activity of the ferritin promoter through the availability of one of its subunit shed light into a novel mechanism of specificity in gene regulation. Similarly the identification of the transcription factors able to drive TfR transcription during differentiation versus mitogenic stimulation could unravel mechanisms of tissue-specific gene regulation.

ACKNOWLEDGMENTS

We gratefully acknowledge L. Kuhn for providing the transferrin receptor gene promoter. We thank S. Tocchio for editorial assistance and R. Gilardi for graphics.

Supported by grants from Italy-USA program on "Therapy of Tumors" and from the Istituto Superiore di Sanità "special program on AIDS" to A. Battistini.

REFERENCES

1. Klausner RD, Rouault TA, Harford JB: Regulating the fate of mRNA: The control of cellular iron metabolism. Cell 72: 15, 1993.
2. Reuben RC, Rifkind RA, Marks PA. Chemically induced murine erythroleukemic differentiation. Biochim Biophys Acta 605: 325, 1980.
3. Coccia EM, Profita V, Fiorucci G, Romeo G, Affabris E, Testa U, Hentze MW, Battistini A: Modulation of ferritin H-chain expression in Friend erythroleukemia cells: transcriptional and translational regulation by hemin. Mol Cell Biol 12: 3015, 1992.
4. Coccia EM, Stellacci E, Perrotti E, Marziali G, Battistini. A Differential regulation of ferritin expression in Friend leukemia cells by iron. J Biol Reg Hom Agents 8: 81, 1994.
5. Coccia EM, Stellacci E, Orsatti R, Testa U, Battistini A: Regulation of ferritin H-chain expression in differentiating Friend leukemia cells. Blood 86: 1570, 1995.
6. Coccia EM, Stellacci E, Marziali G, Orsatti R, Perrotti E, Del Russo N, Testa U, Battistini A: Iron regulation of transferrin receptor and ferritin expression in differentiating Friend leukemia cells. Proceedings of the 9th Simposium "Molecular Biology of Hematopoiesis". Edited by Abraham N.G., Asano S., Brittinger G., and Shadduck R. Plenum Press, New York, 1996.
7. Greiser-Wilke I, Ostertag W, Goldfarb P, Lang A, Furusawa M, Conscience JF: Inducibility of spleen focus forming virus by BrdUrd is controlled by the differentiated state of the cell. Proc Natl Acad Sci USA 75: 2995, 1981.
8. Marziali G, Perrotti E, Ilari R, Testa U, Coccia EM, Battistini A: Transcriptional regulation of ferritin H-gene: the activity of the CCAAT binding factor NF-Y is modulated in heme-treated FLCs and during monocyte to macrophage differentiation. Mol Cell Biol 17: 1387, 1997.
9. Wigler M, Sweet R, Sim GK, Wold B, Pollicer A, Lacy E, Maniatis T, Silverstain S, Axel R: Transformation of mammalian cells with genes from procaryotes and eucaryotes. Proc Natl Acad Sci USA 77: 3570, 1979.
10. Hentze MW, Keim S, Papadopoulos P, Brien SO', Modi W, Drysdale J, Leonard WJ, Harford JB, Klausner RD: Cloning, characterization, expression, and chromosomal localization of a human ferritin heavy chain gene. Proc Natl Acad Sci USA 83: 7226, 1986.
11. Hentze MW, Caughman SW, Rouault TA, Barriocanal J, Dancis A, Harford JB, Klausner RD: Identification of the iron-responsive element for the translational regulation of human ferritin mRNA. Science 238: 1570, 1987.
12. Glisin V, Crkvenjakov R, Byus C: Ribonucleic acid isolated by cesium chloride centrifugation. Biochemistry 13: 2633, 1974.
13. Mantovani R: A survey of 178 NF-Y binding CCAAT boxes. Nucleic Acids Res 26: 1135, 1998.
14. Coustry F, Maity SN, de Crombrugghe B: Studies on transcription activation by multimeric CCAAT-binding factor CBF. J Biol Chem 270: 468, 1995.
15. Kim In-S, Sinha S, de Crombrugghe B, Maity SN: Determination of functional domains in the C subunit of the CCAAT-binding factor (CBF) necessary for formation of a CBF-DNA complex: CBF-B interacts simultaneously with both the CBF-A and CBF-C subunits to form a heterotrimeric CBF molecule. Mol Cell Biol 16: 4003, 1996.
16. Sinha S, Maity SN, Lu J, de Crombrugghe B: Recombinant rat CBF-C, the third subunit of CBF/NFY, allows formation of a protein-DNA complex with CBF-A and CBF-B and with yeast HAP 2 and HAP 3. Proc Natl Acad Sci USA 92: 1624, 1995.
17. Benoist C, Mathis D: Regulation of major histocompatibility complex class-II genes: X, Y, and other letters of the alphabet. Annu Rev Immunol 8: 681, 1990.
18. Mantovani R, Pessara U, Tronche F, Li XY, Knapp AM, Pasquali JL, Benoist C, Mathis D: Monoclonal antibodies to NF-Y define its function in MHC class II and albumin gene transcription. EMBO J 11: 3315, 1992.
19. Roy B, Lee AS: Transduction of calcium through interaction of the human transcription factor CBF with the proximal CCAAT regulatory element of the grp/BiP promoter. Mol Cell Biol 15: 2263, 1995.
20. Chang ZF, Liu CJ: Human thymidine kinase CCAAT-binding protein is NF-Y, whose A subunit expression is serum-dependent in human IMR-90 diploid fibroblasts. J Biol Chem 269: 17893, 1994.

21. Katula KS, Wright KL, Paul H, Surman DR, Nuckolls FJ, Smith JW, Ting JPY, Yates J, Cogswell JP: Cyclin-dependent kinase activation and S-phase induction of the cyclin B1 gene are linked through the CCAAT elements. Cell Growth Differ 8: 811, 1997.
22. Ponka P: Tissue-specific regulation of iron metabolism and heme-synthesis: distinct control mechanisms in erythroid cells. Blood 89: 1, 1997
23. Testa U, Petrini M, Quaranta MT, Pelosi-Testa E, Mastroberardino G, Camagna A, Boccoli G, Sargiacomo M, Isacchi G, Cozzi A, Arosio P, Peschle C: Iron upmodulates the expression of TfRs during monocyte-macrophage maturation. J Biol Chem 264: 13181, 1989.
24. Battistini A, Marziali G, Albertini R, Habetswallner D, Bulgarini D, Coccia EM, Fiorucci G, Romeo G, Orsatti R, Testa U, Affabris E, Peschle C, Rossi GB: Positive modulation of hemoglobin, heme, and transferrin receptor synthesis by murine interferon-α and -β in differentiating Friend cells: pivotal role of heme synthesis. J Biol Chem 266: 528, 1991.
25. Battistini A, Coccia EM, Bulgarini D, Scalzo S, Fiorucci F, Romeo G, Affabris E, Testa U, Rossi GB, Peschle C: Intracellular heme coordinately modulates globin chains synthesis, transferrin receptor number and ferritin content in differentiating Friend erythroleukemia cells. Blood 78: 2098, 1991.
26. Beaumont C, Seyhan A, Yachou AK, Grandchamp B, Jones R: Mouse ferritin H subunit gene. Functional analysis of the promoter and identification of an upstream regulatory element active in erythroid cells. J Biol Chem 269: 20281, 1994.
27. Marziali G, Perrotti E, Ilari R, Coccia EM, Mantovani R, Testa U, Battistini A: The activity of the CCAAT-box binding factor NF-Y is modulated through the regulated expression of its a subunit during monocyte to macrophage differentiation: regulation of tissue-specific genes through a ubiquitous transcription factor. Blood 93: 519, 1999.
28. Sinha S, Kim IS, Sohn KY, de Crombrugghe B, Maity SN: Three classes of mutations in the A subunit of the CCAAT-binding factor CBF delineate functional domains involved in the three-step assembly of the CBF-DNA complex. Mol Cell Biol 16: 328, 1996.
29. Bellorini M, Lee DK, Dantonel JK, Zemzoumi K, Roeder RG, Tora L, Mantovani R: CCAAT-binding NF-Y-TBP interactions: NF-YB and NF-YC require short domains adjacent to their histone fold motifs for association with TBP basic residues. Nucleic Acids Res 25: 2174, 1997.
30. Busfield SJ, Tilbrook PA, Callus BA, Spadaccini A, Kuhn L, Klinken SP: Complex regulation of transferrin receptors during erythropoietin-induced differentiation of J2E erythroid cells. Elevated transcription and mRNA stabilisation produce only a modest rise in protein content. Eur J Biochem 249: 77, 1997.
31. Rao K, Harford JB, Rouault T, McClelland A, Ruddle FH, Klausner DD: Transcriptional regulation by iron of the gene for the transferrin receptor. Mol Cell Biol 6: 236, 1986.
32. Chan LL, Gerhardt EM: Transferrin receptor gene is hyperexpressed and transcriptionally regulated in differentiating erythroid cells. J Biol Chem 267: 8254, 1992.
33. Chan RYY, Seiser C, Schulman HM, Kuhn LC, Ponka P: Regulation of transferrin-receptor mRNA expression: distinct regulatory feature in erythroid cells. Eur J Biochem 220: 683, 1994.
34. Chan L-NL, Gerhardt EM: Transferrin receptor gene is hyperexpressed and transcriptionally regulated in differentiating erythroid cells. J Biol Chem 267: 8254, 1992.
35. Kronke M, Leonard WJ, Depper JM, Greene WC: Sequential expression of genes involved in human T lymphocyte growth and differentiation. J Exp Med 161: 1593, 1985.
36. Sieweke MH, Tekotte H, Frampton J, Graf T: MafB is an interaction partner and repressor of Ets-1 that inhibits erythroid differentiation. Cell 85: 49, 1996.
37. Ouyang Q, Bommakanti M, Miskimins WK: A mitogen-responsive promoter region that is synergistically activated through multiple signalling pathways. Mol Cell Biol 13: 1796, 1993.

INDEX

GPSR Compliance

The European Union's (EU) General Product Safety Regulation (GPSR) is a set of rules that requires consumer products to be safe and our obligations to ensure this.

If you have any concerns about our products, you can contact us on ProductSafety@springernature.com

In case Publisher is established outside the EU, the EU authorized representative is:

Springer Nature Customer Service Center GmbH
Europaplatz 3
69115 Heidelberg, Germany

Batch number: 10755805

Printed by Printforce, the Netherlands